Kurt Gödel
COLLECTED WORKS
Volume I

Dᵣ Kurt Gödel

Kurt Gödel
COLLECTED WORKS

Volume I
Publications 1929–1936

EDITED BY
Solomon Feferman
(Editor-in-chief)
John W. Dawson, Jr.
Stephen C. Kleene
Gregory H. Moore
Robert M. Solovay
Jean van Heijenoort

Prepared under the auspices of the
Association for Symbolic Logic

OXFORD UNIVERSITY PRESS • NEW YORK
CLARENDON PRESS • OXFORD
1986

Oxford University Press

Oxford New York Toronto
Delhi Bombay Calcutta Madras Karachi
Petaling Jaya Singapore Hong Kong Tokyo
Nairobi Dar es Salaam Cape Town
Melbourne Auckland

and associated companies in
Beirut Berlin Ibadan Nicosia

Library of Congress Cataloging-in-Publication Data
Gödel, Kurt.
Collected works.
Bibliography: v. 1, p.   Includes index.
Contents: v. 1. Publications 1929-1936.
1. Logic, Symbolic and mathematical—Collected works.
I. Feferman, Solomon. II. Title.
QA9.G5313 1986 511.3 86-15501
ISBN 0-19-503964-5

British Library Cataloging-in-Publication Data
Gödel, Kurt
Collected works.
Vol. 1: Publications 1929-1936
1. Logic, symbolic and mathematical
I. Title II. Feferman, Solomon
511.3 BC135
ISBN 0-19-503964-5

Printing (last digit): 9 8 7 6 5 4 3 2
Printed in the United States of America

Preface

Here is the first volume of a comprehensive edition of the works of Kurt Gödel, the most outstanding logician of the twentieth century. This volume includes all of Gödel's published writings in the period 1929–1936, and begins with his dissertation of 1929, available previously only through the University of Vienna. Volume II will contain the remainder of his published works, which go through 1974. Succeeding volumes are to contain Gödel's unpublished manuscripts, lectures, lecture notes, and correspondence, as well as extracts from his scientific notebooks.

While this edition as a whole will stand as a testimonial to the great significance, scope and depth of Gödel's thought and achievements, honoring Gödel has not been the editors' primary aim. Rather, our purpose has been to make the full body of his work accessible and useful to as wide an audience as possible, without sacrificing the requirements of historical and scientific accuracy in any way. This has led to a plan for our edition which is in many respects unique, one that takes the following particular form in the first two volumes.

First of all, each article or closely related group of articles is preceded by an introductory note which elucidates it and places it in historical context. These notes have been written by the members of the editorial board as well as a number of outside experts. Secondly, each article originally written in German is reproduced here with an English translation on facing pages. (Gödel wrote only in German and English, and in his notebooks he frequently employed Gabelsberger script, a now obsolete German shorthand.) Furthermore, the original articles—whose printing inevitably involved a variety of typefaces and layouts and whose appearances were of variable quality—have been typeset anew in a uniform, attractive and more readable form. This has been achieved by means of the computerized TEX typesetting system, and the same system has been utilized throughout for all the new material. Incidentally, almost all textual annotations have been relegated either to the introductory notes or to textual notes at the end of the volumes, so as not to interrupt the normal flow of reading. Finally, for this first of our volumes, I have written—as an introductory chapter—a biographical essay on Gödel's life and work. Adding to the personal aspect, a number of photographs of Gödel, his family, friends and colleagues are included in both this and the next volume, some of which appear publicly for the first time.

We expect these volumes to be of interest and value to professionals and students in the areas of logic, mathematics, philosophy, history of science, computer science, and even physics, as well as to many non-specialist readers with a broad scientific background. Obviously, not all the mate-

rial can be made equally accessible to such a variety of readers. In order to maintain the appropriate scientific and historical level, some technical sophistication had to be presumed in various instances. The introductory chapter on Gödel's life and work is the most general and can stand as an initial gauge for those who would venture further. Up to a point, many of the introductory notes can also be read without special training—enough at least to gain some appreciation of what Gödel accomplished in each instance.

Naturally, the more sophisticated readers who are familiar with the main body of Gödel's contributions will be most interested in the unpublished material to appear in future volumes. But even in the first two volumes they may find new or less familiar items. More importantly, by compiling for critical examination the entire corpus of Gödel's publications, we offer such readers the opportunity to put his work in historical and mathematical perspective. Furthermore, the published material (which in a number of cases is not easily available, even for scholars at major universities) serves as the reference base for the unpublished material to appear later, facilitating future study of the relationships between both bodies of work.

The reader will find below more detailed information concerning the format and particular features of these volumes, which should be of help in their use. In the remainder of this Preface I want to explain the genesis and development of our project for this edition and, in the process, to express my appreciation and that of my co-editors for all that has been done to assist with and support its preparation.

The idea of publishing the collected works of Kurt Gödel was discussed by the Council of the Association for Symbolic Logic in 1979, and initial steps were taken in that direction, though with inconclusive results. At the beginning of 1982, while seeking to renew the previous efforts during my term as President of the Association, I was urged by many colleagues to lead the project myself. In particular, the encouragement of Gregory H. Moore and Jean van Heijenoort and their willingness to assist in the basic planning and formation of the editorial board was of pivotal importance both at that stage and thereafter. At first our project was funded by the Association for Symbolic Logic through its Endowment Fund. The latter was augmented by sizable contributions from two donors (both of whom preferred to remain anonymous) who saw value in supporting just such efforts as this. After a while, however, it became clear that more substantial outside funding would be needed, and several foundations were then approached, with successful results as described below. Because of the important initial role of the Association, its support during the first two years, and the significance of this project generally for the community of logicians, it was agreed to identify the publication as having been prepared under the auspices of the Association for Symbolic Logic.

At the end of 1983 we obtained the first of two grants from the Sloan

Foundation, a short-term one to carry us over a crucial period, and then a more substantial long-term one beginning in the summer of 1984. The latter came as a significant supplement to a major grant (also beginning in 1984) from the National Science Foundation, under its program for History and Philosophy of Science. We are enormously indebted to both these foundations for freeing us from financial worries and allowing us to concentrate on our work and in particular, at this juncture, to see the first part of it brought to completion. The assistance of Steven Maurer and James D. Koerner at Sloan and of Ronald J. Overmann, director of the history program at the National Science Foundation, has been of particular importance to us. Both of the indicated grants have been funded through the Association for Symbolic Logic, and in this period the continued encouragement of its President, Ruth Barcan Marcus, and the practical help with the grants by its Secretary-Treasurer, C. Ward Henson, have been much appreciated.

Though perhaps not so evident in Volumes I and II, a critical factor in the long-term success of our project has been the co-operation of the Institute for Advanced Study in Princeton, its Director, Harry Woolf, and the committee of professors there responsible for Gödel's *Nachlass* (literary estate): Enrico Bombieri, Armand Borel and John Milnor. We are grateful that the Institute has given us general permission to draw on the *Nachlass* for our future work. In addition, it was distinctly advantageous to have John Dawson as a Visiting Member of the Institute in the period 1982–1984 to catalogue the *Nachlass*. As a result we learned early on of correspondence whose contents enriched the introductory chapter below. More importantly, we were able to begin one very substantial aspect of our project necessary for future volumes, namely the transcription of Gödel's notebooks, written in Gabelsberger shorthand script, into standard German. This has been carried out since 1982 by Cheryl Dawson in collaboration with Hermann Landshoff[a]; they were joined in this work by Tadashi Nagayama in 1985. Finally, access to the *Nachlass* made possible our relatively early planning of the structure of future volumes.

The staffs of the libraries at the Institute for Advanced Study and at Princeton University have both been very helpful; in particular, at the latter Jean Preston, Curator of Manuscripts at Firestone Library, assisted with the *Nachlass* material deposited there (some details of which will be described in the information section below).

From the beginning we have been able to draw on the valuable general advice of two among those who were closest to Gödel in his later years,

---

[a]Cheryl Dawson also gave greatly valued assistance to John Dawson in his work of cataloguing the *Nachlass* and with his work as editor throughout the project. In addition, Mr. Landshoff, a photographer by profession, helped John Dawson with the photographs for the volumes.

Georg Kreisel and Hao Wang. Use of their fund of specific knowledge about him will be evident in a number of places in these volumes, particularly in the introductory chapter. Rudolf Gödel, Kurt's brother, was also informative on biographical matters. Many more such sources are credited at the end of the first chapter below. Mention should also be made of the cooperation provided by a group in Vienna who are preparing a German-language volume on Gödel's life and work: Eckehart Köhler, Werner Schimanovich and Peter Weibel. Their volume is in many respects intended to be complementary to our own edition.

The publisher of this edition, Oxford University Press, has made every effort to accommodate our overall conception and specific wishes. From the outset we have worked personally with Donald Degenhardt, the Science Editor, who has been extremely responsive with needed advice and forbearing as to our progress, all the while gently prodding us to bring this first part of our work to completion in a reasonable period of time.

As mentioned above, each of Gödel's articles (or closely related group of articles) is preceded by an introductory note; the same is provided for some of Gödel's reviews of special interest and even for brief published remarks. In addition to the six members of the editorial board, the authors are: Burton Dreben, Jens-Erik Fenstad, Warren D. Goldfarb, S. W. Hawking, Rohit Parikh, Charles Parsons, W. V. Quine, Howard Stein, A. S. Troelstra, Robert L. Vaught and Judson Webb. Details of the editorial procedure for the preparation of these notes are provided in the information section below. The general plan took for its initial model the volume *From Frege to Gödel: a source book in mathematical logic, 1879–1931*, edited by Jean van Heijenoort (*1967*), and the present enterprise was conceived as a natural and direct continuation of that work. However, the present notes are somewhat more variable in length than the ones found there: some are just a paragraph or even only a few lines long, while others are extensive essays, not infrequently longer than the items to which they apply. (Such expansiveness is justified both by the concision of Gödel's published writings and the far-reaching influence and implications of his work and thought.) We are extremely grateful to the authors who joined us in the preparation of these introductory notes, which are essential to the success of these volumes. Their willingness to engage in the editorial process has been uniformly high, even when protracted through several rounds. They, like us, did not fully realize what they were getting into at the outset of the project; nevertheless, they accepted without reservation the responsibilities that emerged.

Another of our basic intentions was to provide a first-rate English translation for each German original. Wherever possible, existing translations have been used, but in some cases new translations had to be supplied in order to achieve the desired quality. A few items, in particular Gödel's 1929 dissertation and some of the shorter notes, as well as all the reviews,

are here translated for the first time. Responsibility for all the translations lay primarily with Jean van Heijenoort and secondarily with John Dawson; in this they received major collaboration from Stefan Bauer-Mengelberg. Credit for the translations varies from item to item, as is explained in more detail in the information section below.

Bauer-Mengelberg also provided considerable assistance on all legal matters, particularly those having to do with publication agreements and copyrights. John Dawson, the editor responsible for securing the permissions for original works and photographs (detailed in a section below) worked closely with him on these matters as well.

Turning now to those directly responsible for the physical production of these volumes, we must first thank Yasuko Kitajima, who did all the TEX typesetting work. This task involved enormous time and effort, often under great pressure; we soon took for granted the care and perfection of her work under these conditions. Both she and Richard Weyhrauch were extremely helpful with technical details of the TEX system during the initial period when we were experimenting with various typefaces and layouts, and further on when special symbols had to be created. In the final stages of typesetting, Carolyn Talcott joined us to give crucial assistance. The computer facilities used for these purposes were those of the Computer Science Department and the Center for the Study of Language and Information at Stanford University. We are grateful to the Center for donating computer time in the latter part of our project.

The Department of Mathematics at Stanford provided work space and staff time. I wish particularly to thank my secretaries Priscilla Feigen and Isolde Field, who have been involved at every stage of the project with such things as preliminary drafts, grant proposals, endless correspondence, and assistance to the members of the editorial board during the various periods of the editors' work at Stanford. Their interest, personal engagement and responsiveness to our needs were all beyond the call of duty and have been constantly appreciated.

My wife, Anita, at first an amused and bemused bystander, was drawn more and more into my part of the work. I want to thank her for all her help, particularly as my rescuer of style, and in too many other ways to explain.

Finally, I wish to express my deep gratitude to my co-editors for their exceptional engagement and dedication at every level and to every aspect of this project, from the largest plans and concerns to the smallest details. Their varied knowledge, experience, talents and thoroughness contributed in every way to bringing these first two volumes to fruition. One of the most taxing parts of this work has been that of the copy-editing, and I must single out for special thanks Gregory H. Moore, whose responsibility that was from the beginning to the very end.

Solomon Feferman

Information for the reader

Introductory notes. The purpose of the notes described in the Preface above is (i) to provide a historical context for the items introduced, (ii) to explain their contents to a greater or lesser extent, (iii) to discuss further developments which resulted from them, and (iv) in some cases to give a critical analysis. Each note was read in draft form by the entire editorial board, and then modified by the respective authors in response to criticisms and suggestions, the procedure being repeated as often as necessary in the case of very substantial notes. No attempt, however, has been made to impose uniformity of style, point of view, or even length; the reason for allowing expansiveness in regard to the last has already been discussed in the Preface. While the editorial board actively engaged in a critical and advisory capacity in the preparation of each note and made the final decision as to its acceptability, primary credit and responsibility for the notes rest with the individual authors.

Introductory notes are distinguished typographically by a running vertical line along the left or right-hand margin and are boxed off at their end.[a] The authorship of each note is given in the Contents and at the end of the note itself.

References. Each volume contains a comprehensive References section which comprises the following three categories of items: (i) a complete bibliography of Gödel's own published work, (ii) all items referred to by Gödel in his publications, and (iii) all items referred to in the chapter on Gödel's life and work or in the individual introductory notes.

In the list of references each item is assigned a date with or without a letter suffix, e.g. "1930", "1930a", "1930b", etc.[b] The date is that of publication, where there is a published copy, or of presentation, for unpublished items such as a speech. A suffix is used when there is more than one publication in that year. (The ordering of suffixes does not necessarily correspond to order of publication within any given year.) Date of composition has *not* been used for references, since that is frequently unavailable or only loosely determined.

Within the text of our volumes, all references are supplied by citing author(s) and date in italics, e.g. *Gödel 1930* or *Hilbert and Bernays 1934*.

---

[a] A special situation occurs when the note ends in mid-page before facing German and English text. Then the note extends across the top half of the facing pages and is boxed off accordingly.

[b] "198?" is used for articles whose date of publication is to be in 1986 or later, or is not yet known.

Where no name is specified or determined by the context, the reference is to Gödel's bibliography, as e.g. in "Introductory note to *1929*, *1930* and *1930a*". Examples of the use of a name to set the context for a reference are: "Frege's formal system presented in *1879*", "Skolem proved in *1920* that ...", and "Skolem (*1920*) proved that ...".

There are two works by Gödel whose dating required special consideration: *1929* and *1972*; they appear in Volumes I and II, respectively. The first of these is Gödel's dissertation at the University of Vienna; here the date is that of the year in which the dissertation was submitted (as distinguished from the date of its acceptance, which was 1930). The second, a translation and revision of Gödel's paper *1958*, was intended for publication in the journal *Dialectica* but was never actually published. It reached the stage of page proofs and was found in that form in Gödel's *Nachlass*. Correspondence surrounding this projected publication shows that Gödel worked on the revision sporadically over a number of years, beginning in 1965; the last date for which we have evidence of his making specific changes is 1972, and that date has therefore been assigned to it in our References. (For more information concerning this work, see the introductory note to *1958* and *1972* in Volume II, to be found preceding the article *1972*.) Appended to the page proofs of *1972* were three short notes on the incompleteness results; they have been assigned the date *1972a* in the References.

To make the References as useful as possible for historical purposes, authors' names are there supplied with first and/or middle names as well as initials, except when the information could not be determined. Russian names are given both in transliterated form and in their original Cyrillic spelling. In some cases, common variant transliterations of the same author's name, attached to different publications, are also noted.

Editorial annotations and textual notes. Editorial annotations within any of the original texts or their translations or within items quoted from other authors are signaled by double square brackets: ⟦ ⟧. Single square brackets [ ] are used to incorporate corrections supplied by Gödel. In some articles, editorial footnotes are inserted in double square brackets for a further level of annotation. Each volume has, in addition, a separate list of textual notes in which other corrections are supplied. Finally, the following kinds of changes are made uniformly in the original texts: (i) footnote numbers are raised above the line as simple numerals, e.g. [2] instead of [2)]; (ii) spacing used for emphasis in the original German is here replaced by italics, e.g., e r f ü l l b a r is replaced by *erfüllbar*; (iii) references are replaced by author(s) and date, as explained above; (iv) initial sub-quotes in German are raised, e.g. „engeren" becomes "engeren".

Translations. The overall aim for the translations, as well as the variety of work required and general responsibility for them has been described in the Preface above. Credit for the individual translations varies from item

to item. For most of the longer pieces, credit is indicated explicitly with
the translation itself. In all cases where that is not shown, the translation
is the work of John Dawson, revised in consultation with Stefan Bauer-
Mengelberg and Jean van Heijenoort.

The reader should take special note of the points made regarding translations at the beginning of the textual notes to this volume.

Spelling. There are in Gödel's texts a number of variations ('Variablen' and 'Variabeln', 'Aussageaxiome' and 'Aussagenaxiome', 'so daß'
and 'sodaß', perhaps a few more) that we have not eliminated; we have
preserved the original text.

Logical symbols. The logical symbols used in Gödel's original articles
are here presented intact, even though these symbols may vary from one article to another. Authors of introductory notes have in some cases followed
the notation of the article(s) discussed and in other cases have preferred
to make use of other, more current, notation. Finally, logical symbols are
sometimes used to abbreviate informal expressions as well as formal operations. No attempt has been made to impose uniformity in this respect. As
an aid to the reader, we provide the following glossary of the symbols that
are used in one way or another in these volumes, where 'A', 'B' are letters
for propositions or formulas and '$A(x)$' is a propositional function of x or
a formula with free variable 'x'.

Conjunction ("A and B"): $A \cdot B$, $A \wedge B$, $A \,\&\, B$

Disjunction ("A or B"): $A \vee B$

Negation ("not A"): $\overline{A}$, $\sim A$, $\neg A$

Conditional, or *Implication* ("if A then B"): $A \supset B$, $A \to B$

Biconditional ("A if and only if B"): $A \supset\subset B$, $A \equiv B$, $A \sim B$, $A \leftrightarrow B$

Universal quantification ("for all x, $A(x)$"): $(x)A(x)$, $\Pi x A(x)$, $x\Pi(A(x))$,
$(\forall x)A(x)$

Existential quantification ("there exists an x such that $A(x)$"): $(Ex)A(x)$,
$\Sigma x A(x)$, $(\exists x)A(x)$

Unicity quantification ("there exists a unique x such that $A(x)$"):
$(E!x)A(x)$, $\Sigma!x A(x)$, $(\exists!x)A(x)$

Necessity operator ("A is necessary"): $\Box A$, $N A$

Minimum operator ("the least x such that $A(x)$"): $\epsilon x(A(x))$, $\mu x(A(x))$

Provability relation ("A is provable in the system S"): $S \vdash A$

Note: (i) The "horseshoe" symbol is also used for set-inclusion, i.e., for
sets X, Y one writes $X \subset Y$ (or $Y \supset X$) to express that X is a subset
of Y. (ii) Dots are sometimes used in lieu of parentheses, e.g. $A \supset . B \supset A$
is written for $A \supset (B \supset A)$.

Typesetting. These volumes have been prepared by the TEX computerized mathematical typesetting system (devised by Donald E. Knuth of
Stanford University), as described briefly in the Preface above. The resulting camera-ready copy was delivered to the publisher for printing. The
computerized system was employed because: (i) much material, including

the introductory notes and translations, needed to undergo several revisions; (ii) proof-reading was carried on as the project proceeded; (iii) the papers could be prepared in a uniform, very readable form, instead of being photographed from the original articles. Choices of the various typesetting parameters were made by the editors in consultation with the publisher. Primary responsibility for preparing copy for the typesetting system lay with Gregory H. Moore, and the typesetting itself was carried out by Yasuko Kitajima.

For all previously published articles, original pagination is indicated herein by numbers in the margins, with vertical bars in the body of the text used to show the exact page breaks. No page bar or number is used to indicate the initial page of an article.

Footnotes. We use a combination of numbering and lettering, as follows. All footnotes for Gödel's texts and their translations are numbered, with only rare exceptions, as in the original. There is, however, one special case, that of *1972*, in which Gödel provided a second series of footnotes, essentially to preserve the original series from *1958* without change of numbering. The new series is here distinguished by boldface lower-case Roman letters. Footnotes in the introductory chapter on Gödel's life and work are numbered, except that footnotes giving credits and sources are lettered and located at the end of the chapter. In the latter case and for all the other material of Volumes I and II, footnotes are indicated by lightface lower-case Roman letters (as in the present Information section).

Gödel's Nachlass. The scientific *Nachlass* of Kurt Gödel was donated to the Institute for Advanced Study in Princeton, N.J., by his widow Adele shortly after his death. The *Nachlass* consists of unpublished manuscripts, lecture notes, course notes, notebooks, memoranda, correspondence, and books from Gödel's library. It was catalogued at the Institute for Advanced Study during the years 1982–1984 by John Dawson. Early in 1985 the *Nachlass* with its catalogue was placed on indefinite loan to the Manuscripts Division (located in the Rare Book Room) of the Firestone Library at Princeton University, where the material is available for scholarly examination. All rights for use still reside, however, with the Institute for Advanced Study. Though the *Nachlass* is referred to only here and there in Volumes I and II, it will be the source of almost all the material in subsequent volumes. For further information concerning its general character, see below, pages 26–28.

Photographs. Primary responsibility for securing these lay with John Dawson. Their various individual sources are credited in the Permissions section, which follows directly.

Copyright permissions

The late Professor Alfred Tarski, of the University of California, for the portrait of him and Gödel in Vienna.

Dr. Werner Schimanovich, of the University of Vienna, who kindly supplied the photograph of the Gödel family.

Just before going to press we learned of the death of Professor Julia B. Robinson (8 December 1919–30 July 1985). An outstanding logician whose achievements brought her some of the highest honors the mathematical community has to bestow, she maintained throughout a rare modesty and appreciation for the work of others.

The loss of Julia Robinson is of special significance for the editors of this volume, since she was one of the two anonymous donors mentioned on page ii above. Not only did we appreciate her financial support, which helped our project through a difficult period in its early stages, but we also valued her continuing interest and encouragement. We are grateful that her family has agreed we can now publicly acknowledge her generosity.

Contents

Volume I

Kurt Gödel
COLLECTED WORKS
Volume I

Gödel's life and work

Kurt Gödel's striking fundamental results in the decade 1929–1939 transformed mathematical logic and established him as the most important logician of the 20th century. His work influenced practically all subsequent developments in the subject as well as all further thought about the foundations of mathematics.

The results that made Gödel famous are the completeness of first-order logic, the incompleteness of axiomatic systems containing number theory and, finally, the consistency of the axiom of choice and the continuum hypothesis with the other axioms of set theory. During the same decade he made other less dramatic but still significant contributions to logic, including work on the decision problem, intuitionism and notions of computability.

In 1940 Gödel emigrated from Austria to the United States, where he became established at the Institute for Advanced Study in Princeton. In the years following, he continued to grapple with difficult problems in set theory and at the same time began to think and write in depth about the philosophy of mathematics. Later in the 1940s he arrived at his unusual but less well-known contributions to relativistic cosmology, in which he produced solutions of Einstein's equations permitting "time travel" into the past. While Gödel's philosophical interests dominated his attention from 1950 to his death in 1978, enormous advances were made in the subject of mathematical logic during the same period. The Institute for Advanced Study became a focal point for much of this activity, largely because of his presence.

Gödel published comparatively little, but almost always to maximum effect; his papers are models of precision and incisive presentation. In his *Nachlass* at the Institute there are masses of detailed notes that he had made on a remarkable variety of topics in logic, mathematics, physics, philosophy, theology and history. These have begun to reveal further the extraordinary scope and depth of his thought.

The present account of Gödel's life and work is divided into two parts. The first of these, which follows directly, is devoted to his life and career. Then the second part presents a survey of Gödel's work and thought as revealed in his publications and, to the extent presently accessible, unpublished material from his *Nachlass*.[1,a]

---

[1] All documentation of sources is given by lettered notes at the end of the chapter, pp. 35–36. In particular, note 'a' details my main sources of material and the variety of assistance I have received in preparing this chapter. Other footnotes accompany the text and are numbered.

I. Life and career

"Kurtele, if I compare your lecture with
the others, there is no comparison."

—Adele Gödel[b]

In the end we search out the beginnings. Established, beyond comparison, as the most important logician of our times by his remarkable results of the 1930s, Kurt Gödel was also most unusual in the ways of his life and mind. Deeply private and reserved, he had a superb all-embracing rationality, which could descend to a maddening attention to detail in matters of everyday life. Physically, Gödel was slight of build and almost frail-looking. Cautious about food and fearful of illness, he had a constant preoccupation with his health to the point of hypochondria, yet mistrusted the advice of doctors when it was most needed. It was a familiar sight to see Gödel walking home from the Institute for Advanced Study, bundled up in a heavy black overcoat, even on warm days.

Genius will out, but how and why, and what serves to nurture it? What consonance is there with the personality, what determines the particular channels taken by the intellect and the distinctive character of what is achieved? As with any extraordinary thinker, the questions we would really like to see answered in tracing Gödel's life and career are the ones which prove to be the most elusive. What we arrive at instead is a mosaic of particularities from which some patterns clearly emerge, while the deeper ones must be left as matter for speculation, at least for the time being.

Kurt Friedrich Gödel was born 28 April 1906, the second son of Rudolf and Marianne (Handschuh) Gödel. His birthplace was Brünn, in the Austro–Hungarian province of Moravia. This region had a mixed population which was predominantly Czech but with a substantial German-speaking minority, to which Gödel's parents belonged. His father Rudolf, an energetic self-made man, had come from Vienna to work in Brünn's thriving textile industry. He was eventually to become managing director and part owner of one of the main textile firms there. The family of Kurt's mother, which came from the Rhineland region, had also been drawn to Brünn for the work in textiles. Marianne Handschuh received a better than ordinary education, partly in a French school in Brünn, in the course of which she developed life-long cultural interests.

Much of our present information concerning the family history comes from Dr. Rudolf Gödel, Kurt's brother and his elder by four years.[c] Rudolf tells us that Kurt's childhood was generally a happy one, though he was timid and could be easily upset. When he was six or seven, Kurt contracted rheumatic fever and, despite eventual full recovery, he came to believe that he had suffered permanent heart damage as a result. Here are the early signs of Gödel's later preoccupation with his health. His special intellectual

talents also emerged early. In the family, Kurt was called *Herr Warum* (Mr. Why), because of his constant inquisitiveness.

Following the religion of his mother rather than that of his father (who was "Old" Catholic), the Gödels had Kurt baptized in the Lutheran Church. In 1912, at the age of six, he was enrolled in the Evangelische Volksschule, a Lutheran school in Brünn.[2] From 1916 to 1924, Kurt carried on his school studies at the Deutsches Staats-Realgymnasium, where he showed himself to be an outstanding student, receiving the highest marks in all his subjects; he excelled particularly in mathematics, languages and religion. (Though the latter was not given much emphasis in the family, Kurt took to it more seriously.[d]) Some of Gödel's notebooks from his young student days are preserved in the *Nachlass*, and among these the precision of the work in geometry is especially striking.

Though World War I took place during Gödel's school years, it had little direct effect on him and his family. The region of Brünn was far from the main fronts and was untouched by the devastation wrought elsewhere by the war in Europe. But the collapse of the Austro–Hungarian empire at war's end and the absorption of Moravia together with Bohemia into the new nation of Czechoslovakia was eventually to affect the German-speaking minority in adverse ways. One of the most immediate signs of the shift in national identity was the displacement of the German name 'Brünn' in favor of the Czech name 'Brno'. For the Gödels, though, life in the years after the war continued much as before, with the family comfortably settled in a villa by that time.

Following his graduation from the Gymnasium in Brno in 1924, Gödel went to Vienna to begin his studies at the University. Vienna was to be his home for the next fifteen years, and in 1929 he was also to become an Austrian citizen. The newly created republic of Austria had entered on a difficult course following the collapse of the Austro–Hungarian empire in 1918. The political, social and economic upheavals which followed the disappearance of monarchy and empire affected all spheres of activity, what with the economic base enormously shrunk and the *raison d'être* for the swollen bureaucracy gone. The extraordinary cultural and intellectual center that had been Vienna before the war was transformed by the changed conditions, but the Viennese spirit and ambience lived on, now sharing in the general revolutionary ferment and excitement of the 1920s. Before long, Gödel was brought into contact with the Vienna Circle, a hot-bed of new thought that proved to be very significant for his work and interests.

At the university, Gödel was at first undecided between the study of mathematics and physics, though he apparently leaned toward the latter.

---

[2]For specific dates of significance in Gödel's life, see the Chronology, prepared by John Dawson, on p. 37.

It is said that Gödel's decision to concentrate on mathematics was due to his taste for precision and to the great impression that one of his professors, the number-theorist Philipp Furtwängler, made on him.[e] A description of the mathematical scene at the University of Vienna in those days is given by Olga Taussky(-Todd) in her reminiscences of Gödel (*198?*), from which the following information is drawn. Besides Furtwängler, the professors were Hans Hahn and Wilhelm Wirtinger. Karl Menger, one of Hahn's favorite students, was an *ausserordentlicher* (associate) professor and among the *Privatdozenten* (unsalaried lecturers) were Eduard Helly, Walter Mayer and Leopold Vietoris. Taussky came to know Gödel as a fellow student in 1925, their first real contact coming in a seminar conducted by the philosopher Moritz Schlick on Bertrand Russell's book *Introduction to mathematical philosophy* (*1919*). Gödel hardly ever spoke, but was very quick to see problems and to point the way through to solutions. Though he was very quiet and reserved, it became evident that he was exceptionally talented. Gödel's help was much in demand and he offered such whenever needed. One could talk to him about any problem; he was always very clear about what was at issue and explained matters slowly and calmly.

Hans Hahn became Gödel's principal teacher. He was a mathematician of the new generation, had returned to Austria from a position in Bonn, and was interested in modern analysis and set-theoretic topology, as well as logic, the foundations of mathematics and the philosophy of science. It was Hahn who introduced Gödel to the group of philosophers around Moritz Schlick, holder of the chair in the Philosophy of the Inductive Sciences (which in earlier years had been held successively by the renowned physicists Ernst Mach and Ludwig Boltzmann). Schlick's group was later baptized the "Vienna Circle" (*Wiener Kreis*) and became identified with the philosophical doctrine called logical positivism or logical empiricism.[3] The aim of this school was to analyze knowledge in logical and empirical terms; it sought to make philosophy itself scientific and rejected metaphysical speculation. Gödel attended meetings of the Circle quite regularly in the period 1926–1928, but in the following years he gradually moved away from it, though he maintained regular contact with some of its members, particularly Rudolf Carnap.

One main reason for Gödel's disengagement from the Circle was that he had developed strong philosophical views of his own which were, in large part, almost diametrically opposed to the views of the logical positivists.[f]

---

[3]For general information on Schlick's Circle and its later developments, see the articles on Moritz Schlick and logical positivism in *Edwards 1967*. *Feigl 1969* gives a lively picture from a more personal point of view and traces the movement of the Circle's members and their ideas. For a detailed discussion of Gödel's relations with the Vienna Circle, see *Köhler 198?*. Hahn's role in the Circle is described by Menger in his introduction to the philosophical papers *Hahn 1980*.

Nevertheless, the sphere of concerns that engaged the Circle surely influenced the direction of Gödel's own interests and work. The logical empiricists had combined ideas from several sources, principally Ernst Mach's empiricist-positivist philosophy of science and Bertrand Russell's logicist program for the foundations of mathematics, both filtered through the *Tractatus logico-philosophicus* of Ludwig Wittgenstein. The logicist ideas had been developed in great detail by Russell and Alfred North Whitehead in their famous magnum opus, *Principia mathematica*, over a decade earlier. Hahn, who was at least as important as Schlick in the formation of the Vienna Circle, gave a seminar on this work in 1924–1925, but Gödel does not seem to have participated, since he reports first studying the *Principia* several years later.ᵍ

Hahn's own mathematical interest in the modern theory of functions of real numbers must also have influenced Gödel, as this involved, to a significant extent, set-theoretical considerations deriving from Georg Cantor and passing through the French school of real analysis. However, it seems that the most direct influences on Gödel in his choice of direction for creative work were Carnap's lectures on mathematical logic and the publication in 1928 of *Grundzüge der theoretischen Logik* by David Hilbert and Wilhelm Ackermann. In complete contrast to the massive tomes of Whitehead and Russell, the *Grundzüge* was a slim, unlabored and mathematically direct volume, no doubt of greater appeal to Gödel, with his taste for succinct exposition. Posed as an open problem therein was the question whether a certain system of axioms for the first-order predicate calculus is complete. In other words, does it suffice for the derivation of every statement that is logically valid (in the sense of being correct under every possible interpretation of its basic terms and predicates)? Gödel arrived at a positive solution to the completeness problem and with that notable achievement commenced his research career. The work, which was to become his doctoral dissertation at the University of Vienna, was finished in the summer of 1929, when he was 23. The degree itself was granted in February 1930, and a revised version of the dissertation was published as *Gödel 1930*.[4] Although recognition of the fundamental significance of this work would be a gradual matter, at the time the results were already sufficiently distinctive to establish a reputation for Gödel as a rising star.

The ten years 1929–1939 were a period of intense work which resulted in Gödel's major achievements in mathematical logic. In 1930 he began to pursue Hilbert's program for establishing the consistency of formal axiom

---

[4]Hahn was nominally Gödel's thesis advisor, but later in life Gödel made it known that he had completed the work before showing it to Hahn and that he made use of Hahn's (essentially editorial) suggestions only when revising the thesis for publication; see *Wang 1981*, pp. 653–654.

systems for mathematics by finitary means. The systems that had already been singled out for particular attention dealt with the general subjects of "higher" arithmetic, analysis and set theory. Gödel started by working on the consistency problem for analysis, which he sought to reduce to that for arithmetic, but his plan led him to an obstacle related to the well-known paradoxes of truth and definability in ordinary language.[h] While Gödel saw that these paradoxes did not apply to the precisely specified languages of the formal systems he was considering, he realized that analogous non-paradoxical arguments could be carried out by substituting the notion of provability for that of truth. Pursuing this realization, he was led to the following unexpected conclusions. Any formal system S in which a certain amount of theoretical arithmetic can be developed and which satisfies some minimal consistency conditions is *incomplete*: one can construct an elementary arithmetical statement A such that neither A nor its negation is provable in S. In fact, the statement so constructed is true, since it expresses its own unprovability in S via a representation of the syntax of S in arithmetic.[5] Furthermore, one can construct a statement C which expresses the consistency of S in arithmetic, and C is not provable in S if S is consistent. It follows that, if the body of finitary combinatorial reasoning that Hilbert required for execution of his consistency program could all be formally developed in a single consistent system S, then the program could not be carried out for S or any stronger (consistent) system. The incompleteness results were published in *Gödel 1931*; the stunning conclusions and the novel features of his argument quickly drew wide attention and brought Gödel recognition as a leading thinker in the field.

One of the first to recognize the potential significance of Gödel's incompleteness results and to encourage their full development was John von Neumann.[i] Only three years older than Gödel, the Hungarian-born von Neumann was already well known in mathematical circles for his brilliant and extremely diverse work in set theory, proof theory, analysis and mathematical physics. Others interested in mathematical logic were slower to absorb Gödel's new work. For example, Paul Bernays, who was Hilbert's assistant and collaborator, although quickly accepting Gödel's results, had difficulties with his proofs that were cleared up only after repeated correspondence.[j] Gödel's work even drew criticism from various quarters, which was invariably due to confusions about the necessary distinctions involved, such as that between the notions of truth and proof. In fact, the famous set-theorist Ernst Zermelo interpreted these concepts in such a way as to arrive at a flat contradiction with Gödel's results. In correspondence during 1931 Gödel took pains to explain his work to Zermelo, apparently without success.[k] In general, however, the incompleteness theorems were absorbed

---

[5] The technical device used for this construction is now called 'Gödel numbering'.

before long by those working in the mainstream of mathematical logic; indeed, one can fairly say that Gödel's methods and results came to infuse all aspects of that mainstream.[6]

Gödel's incompleteness work became his *Habilitationsschrift* (a kind of higher dissertation) at the University of Vienna in 1932. In his report on it, Hahn lauded Gödel's work as epochal, constituting an achievement of the first order.[l] The *Habilitation* conferred the title of *Privatdozent* and provided the *venia legendi*, which gave Gödel the right to deliver lectures at the university, but without pay except for fees he might collect from students. As it turned out, he was to lecture only intermittently in Vienna during the following years.

Meanwhile, significant changes had also been taking place in Gödel's personal life. At the age of 21 he met his wife to be, Adele Nimbursky (née Porkert), but the difference in their situations led to objections to their developing relationship from his parents, especially his father. Adele was a dancer, had been briefly married before and was six years older than Kurt. Though his father died not long after, Kurt and Adele were not to be married for another ten years.

The death of Kurt's father in 1929, at the age of 54, was unexpected; fortunately he left his family in comfortable financial circumstances. While retaining the villa in Brno, Gödel's mother took an apartment in Vienna with her two sons. By then Kurt's brother Rudolf had become successfully established as a radiologist. Rudolf never married, and during their period together in Vienna the three of them frequently went out, especially to the theater. According to his brother, at home Kurt went out of his way to "hide his light under a bushel", despite his growing international fame.[m]

In the early 1930s Gödel steadily advanced his knowledge in many areas of logic and mathematics. He took a regular part in Karl Menger's colloquium in Vienna, which had begun meeting in 1929, and he also assisted in the editing of its reports, *Ergebnisse eines mathematischen Kolloquiums*. In the period 1932–1936 he published thirteen short but noteworthy papers in that journal on a variety of topics, including intuitionistic logic, the decision problem for the predicate calculus, geometry, and length of proofs. Some of the results in logic were to be of lasting interest, though not of the same order as his previous work on completeness and incompleteness. During the same period he was an active reviewer for *Zentralblatt für Mathematik und ihre Grenzgebiete* and, less frequently, for *Monatshefte für Mathematik und Physik*.[7]

---

[6]For the influence of Gödel's work on logicians in the 1930s, see *Kleene 1981* and *198?*.

[7]After 1936, Gödel never reviewed again for these or any other journals.

Menger occasionally invited foreign visitors of interest to speak in his colloquium. Among them was the Polish logician Alfred Tarski, who was shortly to become famous for his work on the notion of truth in formal languages and increasingly, later, for his leadership in the development of model theory. In early 1930 Tarski spent a few weeks in Vienna and was introduced to Gödel at that time; Gödel used the occasion to discuss the results of his 1929 dissertation. Tarski returned for a more extensive visit as a guest of Menger's colloquium during the first half of 1935.[n]

Initially, in his unsalaried position as *Privatdozent*, Gödel had to depend on the resources of his family for his livelihood. However, these means were supplemented before long by income from visiting positions in the United States of America. Gödel's first visit was to the Institute for Advanced Study in Princeton during the academic year 1933–1934. The Institute had been formally established in 1930, with Albert Einstein and Oswald Veblen appointed its first professors two years later by its original director, Abraham Flexner. Veblen, who was a leader in the development of higher mathematics in America and had played a principal role in building up an outstanding mathematics department at Princeton University, was largely responsible for selecting the further "matchless" mathematics faculty at the Institute: James Alexander, Marston Morse, John von Neumann and Hermann Weyl.[o] In addition, he helped arrange postdoctoral visits for rising young mathematicians, including Gödel; no doubt Veblen had heard about Gödel from von Neumann, who regarded him as "the greatest logician since Aristotle".[8,p]

Gödel's visit in 1933–1934 was the first of three that he was to make to the Institute before taking up permanent residence there in 1940. He lectured on the incompleteness results in Princeton in the spring of 1934. Apparently he had already begun to work with some intensity on problems in set theory; at the same time, he felt rather lonely and depressed during this period in Princeton. Following his return to Europe, he had a nervous breakdown and entered a sanatorium for a time. In the following years there were to be recurrent bouts of mental depression and exhaustion. A scheduled return visit to Princeton had to be postponed to the fall of 1935 and then was unexpectedly cut short after two months, again on account of mental illness. More time was spent in a sanatorium in 1936, and Gödel was unable to carry on at the University of Vienna until the spring of 1937.[q] When he was finally able to resume teaching, he lectured on some of his major new results in axiomatic set theory, the development of which we now trace.

---

[8] Apropos of this, Kreisel remarks: "... if Gödel's work is to be compared to that of one of the ancients, Archimedes is a better choice than Aristotle (who invented logic, but proved little about it). Archimedes did not invent mechanics, as Gödel did not invent logic. But both of them changed their subjects profoundly..." (*Kreisel 1980*, p. 219).

Two problems that had preoccupied workers in the field of set theory since its creation by Cantor beginning in the 1870s concerned the well-ordering principle and the cardinality of the continuum. Zermelo had examined the first of these, both informally (*1904*) and then within the framework of his newly introduced system of axioms for set theory (*1908*, *1908a*), and had shown that the well-ordering principle is equivalent to the axiom of choice (*AC*). There was much intense dispute among mathematicians about the evidence for or against this new "axiom". Under its assumption, every infinite set would have a determinate cardinal number in an ordered list of transfinite cardinals. After Cantor proved that the continuum (i.e., the measurement line) is uncountable, he conjectured that its cardinal would be the least among all uncountable cardinal numbers. This conjecture became known as the continuum hypothesis (*CH*).[9]

It was to these problems in set theory that Gödel began to devote himself as his main area of concentration after obtaining the incompleteness results.[r] He considered the statements of *AC* and *CH* in the framework of axiomatic set theory (by then enlarged and made more precise through the work of Fraenkel, Skolem, von Neumann and Bernays), to see whether they could be settled on the basis of the remaining axioms. His major result, finally achieved by the summer of 1937, was that both the axiom of choice and the continuum hypothesis (even in a natural generalized form, *GCH*) are consistent with the Zermelo–Fraenkel axioms (ZF) without the axiom of choice, and hence cannot be disproved from them if the axioms of ZF are consistent. This result at least provided a minimal guarantee of safety in the use of the seemingly problematic statements *AC* and *GCH*.

Underlying Gödel's proof was his definition within ZF of a general notion of constructibility for sets. His plan, which emerged quite early, was to show that the constructible sets form a model for all the axioms of ZF and, in addition, for the axiom of choice and the generalized continuum hypothesis. In 1935 he was able to tell von Neumann that he had succeeded in verifying all the ZF axioms together with *AC* in this model; but, as noted above, it took him two more years to push his work to completion by verifying that *GCH* holds in the constructible sets as well. With the modest techniques then available, the details that Gödel needed to establish were formidable, and this deep and complicated work caused him much effort, especially in its final part. Perhaps that was one reason for the mental stress he suffered throughout much of the period 1934–1937.[s]

The years 1937–1939 brought further significant changes in both Gödel's personal life and career. His mother returned to her home in Brno in 1937, though his brother remained in Vienna to continue his medical practice.

---

[9] A full history of the emergence of the axiom of choice as a fundamental principle of set theory and of the controversies that surrounded it is provided by *Moore 1982*.

That move may have eased the way for Kurt Gödel and Adele Nimbursky to be married finally, in September 1938. The marriage of Kurt and Adele proved to be a warm and enduring one, and for Kurt a source of constant support in the difficult times ahead.[t]

During the 1930s there were many shifts in the lives of Gödel's friends and colleagues in Vienna. Marcel Natkin and Herbert Feigl, early friends from the Circle, had already left Vienna at the turn of the decade, the first for Paris and the second for America. Rudolf Carnap left to teach in Prague in 1931; he was eventually to go to America, too. Gödel's teacher Hans Hahn died in 1934, of natural causes. Then in 1936, Moritz Schlick, the central figure of the Vienna Circle, was murdered by a deranged former student on the way to a lecture; naturally, the case created a sensation. Upset by this turn of events and the general situation in Austria, Karl Menger left the following year to take up a position at Notre Dame. Gustav Bergmann and Abraham Wald, two other contemporaries of Gödel's, also left for America in 1938.

These and related changes were taking place in the context of the difficult economic conditions that had been gripping European nations since the severe depression of 1929 and of the political situation created by the advent to power in 1933 of Adolf Hitler and the Nazis in Germany. In 1934 Austria itself fell under the rule of a semi-fascist regime, led by Engelbert Dollfuss until his assassination by Austrian Nazis later that same year. Dollfuss' murder was a premature attempt by the Nazis to gain power in Austria and to carry out the *Anschluss* (political and economic union) of Austria with Germany, which had been forbidden by the 1919 Treaty of Saint-Germain. There was much sentiment for *Anschluss* among certain groups of Austrians, but the main pressure came from Hitler. That mounted steadily until Hitler's threat of invasion brought down the succeeding Schuschnigg regime in the spring of 1938. Austria thenceforth became a province (*Ostmark*) of a wider Nazi Germany. The year 1938 saw the beginning of a general transformation of Austrian cultural and intellectual life, comparable to that in Germany five years previously. This led to an exodus of intellectuals, particularly those of Jewish background, for whom the move was a matter of survival, while for others emigration was a reaction to the super-nationalistic and racist politics characteristic of the Nazi regime. An incidental result of all this was the final disintegration of the Vienna Circle.[u] As for Gödel, his stance was basically apolitical and non-committal; while he was by no means unaware of what was taking place, he ignored the increasingly evident implications of the transformations around him.

At the urging of Menger, Gödel visited America once more in 1938–1939. He spent the fall term at the Institute for Advanced Study, where he lectured on his new results concerning the consistency of the axiom of choice and the generalized continuum hypothesis. For the spring term he joined

Menger at Notre Dame, where he lectured again on his set-theoretical work and conducted an elementary course on logic with Menger. Gödel then returned to Vienna to rejoin his wife, whom he had left the previous fall only two weeks after their marriage.

Gödel planned to return to the Institute in the fall of 1939, but political events intervened; his life was now directly affected by the Nazi regime in two very different ways. He was called up for a military physical examination and much to his surprise (in view of his generally poor health and his conviction that he had a weak heart) found "fit for garrison duty". Then there was the question of his situation at the University of Vienna. The unpaid position of *Privatdozent* had been abolished by the Nazis, who had replaced it by a new paid position called *Dozent neuer Ordnung*. The latter, however, required a fresh application that could be rejected on political or racial grounds. Although Gödel applied for the new position in September 1939, approval was slow in forthcoming. Questions were raised about his associations with Jewish professors (Hahn in particular), and while it was recognized that he was apolitical, his lack of open support for the Nazis counted against him. In this insecure situation and with the likely possibility that he would be drafted (war having begun in September), Gödel wrote Veblen in desperation in November 1939, seeking assistance to leave. Somehow U.S. non-quota immigrant visas and German exit permits were arranged, and Kurt and Adele managed to leave Vienna in January 1940. As it was too dangerous at that point to cross the Atlantic by boat, they made their way instead by train through Eastern Europe, then via the Trans-Siberian Railway across Russia and Manchuria, and thence to Yokohama. From there they traveled by ship to San Francisco, and in March 1940 they finally proceeded by train to Princeton.[v]

Gödel was never to return to Europe. Ironically, his application for *Dozent neuer Ordnung* was belatedly approved in June 1940.[w] Long afterward he remained bitter about his predicament in Austria in the year 1939–1940, apparently blaming it more on Austrian "sloppiness" (*Schlamperei*) than on the outrageous Nazi conditions. In particular, on the occasion of his 60th birthday in 1966, he turned down an honorary membership in the Austrian Academy of Sciences. However, he couched the refusal in pseudo-legalistic terms which suggested that his U.S. citizenship might be jeopardized if he were to accept membership in the academy of the country of his former citizenship.[x]

In 1940 Gödel was made an Ordinary Member of the Institute for Advanced Study, and he and his wife settled in Princeton, where they established a quiet social life. Among Gödel's closest friends there were Albert Einstein and Oskar Morgenstern. The latter was another ex-Viennese, who had emigrated in 1938 and taken a position at Princeton University. Already established as a mathematical economist, Morgenstern was later to become well known to a wide public through his important and influen-

tial work with von Neumann, *The theory of games and economic behavior* (*1944*). (Von Neumann himself would have been less accessible to Gödel during the early 1940s, since he was frequently away from the Institute in his capacity as consultant for innumerable government war projects.[y])

Morgenstern had many stories to tell about Gödel. One concerned the occasion when, in April 1948, Gödel became a U.S. citizen, with Einstein and Morgenstern as witnesses.[z] Gödel was to take the routine citizenship examination, and he prepared for it very seriously, studying the United States Constitution assiduously. On the day before he was to appear, Gödel came to Morgenstern in a very excited state, saying: "I have discovered a logical-legal possibility by which the U.S.A. could be transformed into a dictatorship." Morgenstern realized that, whatever the logical merits of Gödel's argument, the possibility was extremely hypothetical in character, and he urged Gödel to keep quiet about his discovery at the examination. The next morning, Morgenstern drove Gödel and Einstein from Princeton to Trenton, where the citizenship proceedings were to take place. Along the way Einstein kept telling one amusing anecdote after another in order to distract Gödel, apparently with great success. At the office in Trenton, the official was properly impressed by Einstein and Morgenstern, and invited them to attend the examination, normally held in private. He began by addressing Gödel: "Up to now you have held German citizenship." Gödel corrected him, explaining that he was Austrian. "Anyhow", continued the official, "it was under an evil dictatorship... but fortunately, that's not possible in America." "On the contrary," Gödel cried out, "I know how that can happen!!" All three had great trouble restraining Gödel from elaborating his discovery, so that the proceedings could be brought to their expected conclusion.

Einstein and Gödel could frequently be seen walking home together from the Institute, engaged in rather intense conversations. A number of stories concerning the two have been recounted by the mathematician Ernst Straus, who was Einstein's assistant during the years 1944–1948. He summarized appreciatively their unusual relationship in the following passssage, taken from his reminiscences (*Straus 1982*, page 422).

The one man who was, during the last years, certainly by far Einstein's best friend, and in some ways strangely resembled him most, was Kurt Gödel, the great logician. They were very different in almost every personal way—Einstein gregarious, happy, full of laughter and common sense, and Gödel extremely solemn, very serious, quite solitary, and distrustful of common sense as a means of arriving at the truth. But they shared a fundamental quality: both went directly and wholeheartedly to the questions at the very center of things.

At the Institute Gödel had no formal duties and was free to pursue his research and studies as he pleased. During the first years there he continued his work in mathematical logic, along various lines. In particular, he made strenuous efforts to prove the independence of the axiom of choice and the continuum hypothesis, but only with partial success, and that just on the former problem. His efforts in this direction were never published; they remain to be deciphered (if possible) from notebooks in his *Nachlass*.[10] Another achievement early in this period (though not published until 1958) was a new constructive interpretation of arithmetic that proved its consistency, but via methods going beyond evidently finitary means in Hilbert's sense.

From 1943 on, Gödel devoted himself almost entirely to philosophy, first to the philosophy of mathematics and then to general philosophy and metaphysics. The year 1944 marks the publication of his paper on Bertrand Russell's mathematical logic, which was extremely important both for its searching analysis of Russell's work and for its open statement of Gödel's own "platonist" views of the reality of abstract mathematical objects.[11]

An expository paper on Cantor's continuum problem in 1947 brought out these views quite markedly in the context of set theory. One other writing of a partly philosophical character from this period did not appear until somewhat later, namely, the address in 1946 to the Princeton Bicentennial Conference on Problems of Mathematics. As for general philosophy, Gödel continued his long-pursued reading and study of Kant and Leibniz, turning also to the phenomenology of Edmund Husserl in the late 1950s.[aa] In Gödel's *Nachlass* are many notes on the writings of these philosophers.

An apparent exception to these directions of thought was Gödel's surprising work on the general theory of relativity during the period 1947–1951, in which he produced new and unusual cosmological models that, in theory, permit "time travel" into the past. According to Gödel, this work did not come out of his discussions with Einstein but rather was motivated

---

[10]The full independence results were eventually obtained by Paul Cohen (*1963*).

[11]An amusing aside in this respect has its source in a statement by Bertrand Russell to be found in the second volume of his *Autobiography* (*1968*, pp. 355–356): "I used to go to [Einstein's] house once a week to discuss with him and Gödel and Pauli. These discussions were in some ways disappointing, for, although all three of them were Jews and exiles and, in intention, cosmopolitans, I found that they all had a German bias toward metaphysics [and that] Gödel turned out to be an unadulterated Platonist...".

Gödel's attention was drawn to this in 1971 and he drafted a reply that is preserved in the *Nachlass*, though it was never actually sent: "As far as the passage about me [by Russell] is concerned, I have to say *first* (for the sake of truth) that I am not a Jew (even though I don't think this question is of any importance), 2) that the passage gives the wrong impression that I had many discussions with Russell, which was by no means the case (I remember only one). 3) Concerning my 'unadulterated' Platonism, it is no more 'unadulterated' than Russell's own in 1921...". Fuller quotations are given in *Dawson 1984a*, pp. 13 and 15.

by his own interests in Kant's philosophy of space and time.[bb] Einstein himself was preoccupied, as he had been for a long time, with constructing a unified field theory, a project about which Gödel was skeptical.[cc] In this work Gödel brought to bear mathematical techniques and physical intuitions that one who was familiar only with his papers in logic would not have expected. The mathematics, however, harks back to his brief contributions to differential geometry in the 1930s, as well as to his studies of analysis with Hahn and in Menger's colloquium.

In addition to reflecting Gödel's primary interests in logic, philosophy and, to a lesser extent, mathematics and physics, the notebooks in his *Nachlass* are unexpectedly wide-ranging, revealing for example, sustained interests in history and theology. The latter even included a long-standing fascination with demonology.[dd]

Gödel was made a Permanent Member of the Institute for Advanced Study in 1946. His subsequent promotion to Professor in 1953 required him to take part in some aspects of Institute business.[12] He devoted a good deal of time to the details of these affairs and, in particular, took very seriously the increasingly frequent applications by logicians for visiting memberships.[13] When logic started to flourish in that period, the Institute became a Mecca for younger logicians—many of them rising stars—and drew visits as well from older colleagues of the pre-war generation, such as Paul Bernays. Gödel limited his contacts with most younger visitors, though he would give serious consideration to their work and interests and would volunteer suggestions. A few of the more advanced logicians were able to establish deeper scientific and personal relations with him and were privy to his thoughts and speculations in extensive conversations; most prominent among these were William Boone, Georg Kreisel, Gaisi Takeuti, Dana Scott and Hao Wang. Others whose work impressed him and with whom he had some significant (though less extensive) contact were Clifford Spector and Abraham Robinson. But Gödel never had students or disciples in the usual sense of the word.

Beginning in 1951, Gödel received many honors. Particularly noteworthy are his sharing of the first Einstein Award (with Julian Schwinger) in

---

[12]Questions have been raised about the reasons for the relative lateness of this promotion in *Ulam 1976* (p. 80) and *Dyson 1983* (p. 48). One explanation has it that promotion was held back for Gödel's sake, so as not to burden him with the administrative responsibilities accompanying faculty status. Another has it that there were fears Gödel's exceptional attention to detail and his legalistic turn of mind would hinder the conduct of Institute business if he were to assume those responsibilities. (See also *Dawson 1984a*, p. 15.)

[13]Concerning the latter, Gödel's colleague Hassler Whitney commented as follows: "Gödel was keenly interested in the affairs of the Institute. It was ... hard to appoint a new member in logic since Gödel could not 'prove to himself that a number of candidates shouldn't be members, with the evidence at hand'." (Quotation from *The mathematical intelligencer*, *1* (1978), p. 182.) For a complementary view, see *Kreisel 1980*, p. 159.

1951, his choice as Gibbs Lecturer for 1951 by the American Mathematical Society, and his elections to membership in the National Academy of Sciences (1955), to the American Academy of Arts and Sciences (1957) and to the Royal Society (1968). In 1975 he was awarded the National Medal of Science by President Ford, but because of ill health he could not attend the ceremony. A complete list of awards and honors is given in the Chronology at the end of this chapter.

In the last fifteen years of his life, Gödel was busy with visitors, Institute business and his own philosophical studies; during this time he returned to logic only rarely. Some papers were revised and a few notes were added to new translations. In particular, he expended a good deal of effort over a period of years on a translation and revision of his paper *1958*, which gave a constructive interpretation of arithmetic. The revised work never reached published form, though it was found in galley proof in his *Nachlass*.[14] In the early 1970s there was a flurry of interest and excitement among logicians about notes by Gödel in which he proposed new axioms for set theory that were supposed to imply the falsity of the continuum hypothesis; but essential problems were found in the arguments and the notes were withdrawn. Gödel blamed his having overlooked the difficulties on the drugs he was then taking for illness.

In fact, Gödel's health was poor from the late 1960s on. Among other things he had a prostate condition for which surgery had been recommended, but he would never agree to have the operation done. Along with his hypochondriacal tendencies he also had an abiding distrust of doctors' advice. (Back in the 1940s, for example, he delayed treatment of a bleeding ulcer so long that he would have died, had it not been for emergency blood transfusions.) In addition to prostate trouble, he was still convinced that his heart was weak, although there was no medical substantiation. During the last few years of his life, his wife Adele was unable to help him to the same extent as before, since she herself was partially incapacited by a stroke and was, for a time, moved to a nursing home. Gödel's depressions returned, accompanied and aggravated by paranoia; he developed fears about being poisoned and would not eat. He died in Princeton Hospital on 14 January 1978 of "malnutrition and inanition caused by personality disturbance".[ee] Adele survived him by three years, dying on 4 February 1981. Kurt and Adele had no children, leaving Kurt's brother Rudolf as the sole surviving member of the Gödel family.

---

[14]It is reproduced in Volume II in this edition of his works as *Gödel 1972*.

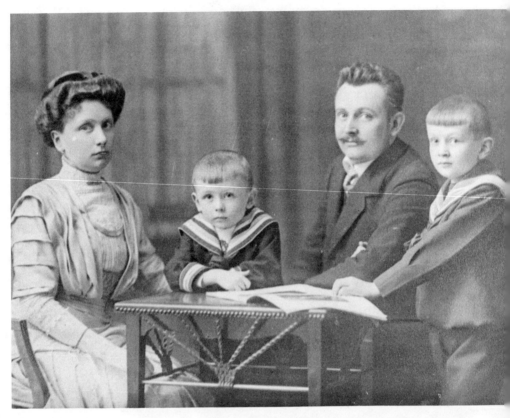

The Gödel family: Mother Marianne, son Kurt, father Rudolf, elder son
Rudolf

School portrait of Kurt Gödel

Kurt Gödel as a young scholar

Maria Lutman-Kokoszyńska.

Alfred Tarski and Kurt Gödel in Vienna, 1935

Adele and Kurt Gödel: wedding portrait, 20 September 1938

II. Works, thought and influence

This part is divided into four Sections. The first Section provides a survey of the works contained in the first two volumes of this collection. Section 2 gives a preliminary picture of the works and notes found in Gödel's *Nachlass*, to appear in a future volume. The third Section outlines Gödel's views on the philosophy of logic and mathematics. Finally, we conclude in Section 4 with a general assessment in terms of the character, impact and influence of Gödel's work. The introductory notes to Gödel's papers in this and the succeeding volume go into the material of Sections 1 and 3 in a much deeper and more detailed way. At the same time, some of the following repeats material from Part I, but more systematically.

1. **Works in the first two volumes.** This Section provides an overview of the entire corpus of Gödel's publications, consisting of his papers, notes, remarks and reviews, together with the closely related but previously unpublished items which are to be found in this and the next volume. Instead of considering these items in order of appearance, I have grouped the material into thirteen categories, as follows:

(a) completeness,
(b) incompleteness,
(c) speed-up theorems,
(d) cases of the decision problem,
(e) intuitionistic logic and arithmetic,
(f) propositional calculus,
(g) geometry,
(h) set theory,
(i) philosophy of mathematics,
(j) functional interpretation,
(k) relativistic cosmology and the philosophy of time,
(l) miscellaneous, and
(m) reviews.

As it happens, this classification broadly follows the chronological order of publication (except for the reviews), although there is much interleaving of items. All classification schemes have some degree of arbitrariness, and mine is no exception. For the most part, however, the scheme adopted here was dictated by the nature of the material itself. Those cases where that was less certain had mainly to do with articles on the borderline between technical work and philosophy.

No attempt is made here to explain the contents of each category in any depth; that is the task of the introductory notes accompanying individual items throughout the first two volumes. Extensive references to related work are also given there.

(a) *Completeness* (*1929, 1930, 1930a*). Gödel's solution of the *completeness problem* posed in *Hilbert and Ackermann 1928* constituted his first major result, and became his doctoral dissertation (*1929*) at the University of Vienna.[15] The question was whether validity in the first-order predicate calculus (or the restricted functional calculus, as it was then called) is equivalent to provability in a specific system of axioms and rules of inference. Gödel's affirmative solution actually established more, implying one version of the "downward" Löwenheim–Skolem theorem: if A is a closed first-order formula then either A is refutable (that is, its negation $\sim A$ is provable in the predicate calculus) or A has a denumerable model (that is, A is satisfiable in a finite or countably infinite domain). Gödel also extended this result to denumerable sets Σ of formulas which, if consistent with the system, have a denumerable model. The paper *1930* largely follows the dissertation *1929*, with two significant exceptions, one being a deletion and the other an addition. First of all, the very interesting informal section with which *1929* began was omitted in *1930*. In that section Gödel had situated his work relative to the ideas of Hilbert and Brouwer, arguing against both in certain respects. Even more noteworthy is that he had already raised the possibility of incompleteness of mathematical axiom systems in the deleted introduction. Secondly, Gödel added a new result closely related to the extended completeness theorem (above) for denumerable sets Σ. Now known as the *compactness theorem*, this stated that if every finite subset of Σ is satisfiable then all of Σ is simultaneously satisfiable in some (denumerable) domain. This result (generalized to uncountable Σ in *Maltsev 1936* and *Henkin 1949*) proved to be fundamental for the subject of model theory some years later.

(b) *Incompleteness* (*1930b, 1931, 1931a, 1932b, 1934*). The paper *1931*, in which Gödel established the incompleteness of sufficiently strong axiomatic theories, is perhaps the most famous one in all of mathematical logic.[16] Its results were formulated for primitive recursive extensions S of a system P obtained from that of *Principia mathematica* by adding Peano's axioms at the ground type. By introducing the device of effective numbering of formal expressions, Gödel arithmetized the syntax of S, thereby showing that its proof relation is equivalent to a primitive recursive relation between numbers. A crucial notion introduced by Gödel is that of a relation being decidable (*entscheidungsdefinit*) in S; in particular, it is shown that this holds of every primitive recursive relation. It follows that one can verify within S itself positive assertions of provability in S. Next, Gödel constructed a sentence G that can be interpreted as expressing of itself (as given by its number) that G is not provable in S. The *first incompleteness*

---

[15] The note *1930a* is an abstract of the dissertation.

[16] An abstract for *1931* is given in *1930b*.

theorem tells us that (i) if S is consistent then indeed G is not provable in S, and (ii) if S is ω-consistent then also the negation $\sim G$ is not provable in S. By formalizing the proof of (i), Gödel arrived at the provability within S of $W \supset G$, where W expresses that S is consistent (*widerspruchsfrei*). Hence if S is consistent then S does not prove W (the *second incompleteness theorem*). This latter result subsequently forced major reconsideration of Hilbert's program aimed at establishing the consistency of formal systems of mathematics by finitary means. (But already the first incompleteness theorem refuted Hilbert's idea that his program would provide an ultimate solution to foundational problems.)   Another noteworthy result of *1931* is that the undecidable proposition constructed there can be expressed in purely arithmetical form (in fact, as a Diophantine equation preceded by mixed universal and existential quantifiers). The proof of the second incompleteness theorem was only sketched in *1931*; Gödel planned to give full details in a sequel to the paper but never did so (leaving to *Hilbert and Bernays 1939* the first detailed exposition).

The remarks *1931a*, made at a meeting in Königsberg in 1930, provided the first public, though informal, announcement of the first incompleteness theorem. The full significance of Gödel's results was apparently not appreciated on the occasion itself.

In the note *1932b*, the incompleteness theorems were formulated more generally for extensions of a form of Peano's arithmetic in the first-order predicate calculus.

In the spring of 1934 Gödel gave lectures on his incompleteness results at the Institute for Advanced Study, and the publication *1934* consists of the resulting lecture notes (taken by Kleene and Rosser and edited by Gödel). These notes cover much the same ground as *1931*, though now starting with a form of second-order arithmetic. One of the main points of interest in *1934* is Gödel's introduction of the notion of *general recursiveness*, following a suggestion of Herbrand. This and (what was later shown to be) an equivalent notion of λ-*definability* were proposed by Church in his paper *1936* as precise explanations of the informal notion of effective calculability. Gödel did not accept "Church's thesis" at the time, only doing so after Turing in his *1937* gave an analysis in terms of abstract computing machines; computability by "Turing machines" was also shown to be equivalent to λ-definability and general recursiveness. These notions became fundamental for the enormously important subject of recursion theory. They also provided the key theoretical concepts for modern general-purpose digital computers, as realized by Turing and von Neumann in the 1940s.

(c) *Speed-up theorems (1936a)*. This note contains the first statement (without proof) of what is now called a "speed-up" theorem. Gödel had pointed out in his earlier work (*1931*, footnote 48a, and *1932b*) that passage to higher types allows one to prove statements undecidable in systems of lower levels, for example the consistency of those systems. Thus, by

successively adding higher types one arrives at a sequence of ever stronger systems S_n. The main result here is that there are formulas which can be proved both in S_n and S_{n+1} but whose shortest proof in S_{n+1} is much shorter than that in S_n (as measured in advance by any given computable function).

Gödel also noted in *1936a* the absoluteness of his notion of *Entscheidungsdefinitheit* (from *1931*) for the systems S_n, that is, the same relations are numeralwise expressible in S_{n+1} as in S_n. In fact this is a much more general phenomenon, and the notion of being *entscheidungsdefinit* in a (recursively axiomatized) formal system turned out to provide still another equivalent to the precise definitions of effective computability mentioned in (b).

(d) *Cases of the decision problem (1932a, 1933i)*. Much attention had been given in the 1920s and early 1930s to the decision problem for satisfiability for various classes of formulas in the first-order predicate calculus (or restricted functional calculus), mainly those grouped according to the nature of their prefix in prenex normal form. An informal notion of decidability had been used in the case of positive results, while some classes of formulas were shown to be *reduction classes* for the decision problem for all formulas. (The exact situation was not clarified until Church (*1936a*) used the thesis referred to above to establish the undecidability of full quantificational logic, from which it follows that every reduction class is also undecidable. Incidentally, Theorem X of *Gödel 1931* implies this undecidability result, once Church's thesis is accepted; see page 136 below.)

Ackermann had shown the ∃...∃∀∃...∃ formulas to be decidable, while Skolem had shown the ∀...∀∃...∃ formulas to constitute a reduction class. In the paper *1932a* Gödel bettered these results by showing the ∃...∃∀∀∃...∃ formulas to be a decidable class and the ∀∀∀∃...∃ formulas to constitute a reduction class, thereby establishing a sharp boundary between the two types of classes. The paper *1933i* strengthened the decidability result by showing that every satisfiable formula has a finite model. Some of these results of Gödel were obtained independently by Kalmár and Schütte.

At the end of *1933i* Gödel claimed that his decidability argument for the ∀∀∃...∃ class carries over to the predicate calculus with identity. That there could be a problem with this assertion was not realized until the mid-1960s; in fact it was eventually shown to be false in *Goldfarb 1984a*.

(e) *Intuitionistic logic and arithmetic (1932, 1933e, 1933f)*. Intuitionistic logic had been set up in *Heyting 1930* as a formalization of the basic reasoning admitted in Brouwer's intuitionistic reconstruction of mathematics. The key difference from classical logic was omission of the law of excluded middle. Beginning with *1932*, Gödel recurrently paid attention to systems based on intuitionistic logic, though he did not subscribe to Brouwer's tenets. (His position on intuitionism and constructivity is philosophically

interesting, as will be discussed in Section 3 below.) One of the results of *1932* is that, if only finite truth tables are used, there is no completeness theorem for intuitionistic propositional calculus analogous to that for the classical calculus.

The paper *1933e*, though short, is important for its interpretation of classical systems within corresponding intuitionistic ones, in particular, systems of arithmetic. This uses the so-called "negative" translation of each formula A into a formula A', for which if A is classically provable then A' is intuitionistically provable. For a wide class of As one has A' the same as A, so for such formulas the systems agree as to their provability. The same results (with slightly different translation functions) were obtained independently by Bernays and Gentzen; Gödel was also anticipated in part of his work by Kolmogorov. Philosophically, Gödel's interpretation is significant in that it shows classical arithmetic to be as "secure" as intuitionistic arithmetic, though the informal grounds for accepting either one of these systems in preference to the other would differ.

In *1933f*, Gödel in a sense reversed the direction of interest by interpreting the intuitionistic propositional calculus within a classical calculus enriched by a unary operator B where "B" was taken to suggest "beweisbar" (or provable). The B-system is actually a form of the system S4 of modal logic. Gödel pointed out that, by his second incompleteness theorem, B cannot in turn be interpreted as representing provability in a formal system. In recent years there has been much work on the logics that *are* verified by such interpretations of B.

(f) *Propositional calculus* (*1932c*, *1933*, *1933a*). The first of these notes on the classical propositional calculus is the most significant of the three, but even it has largely been overlooked. This established a generalization of (what has later come to be called) Lindenbaum's extension theorem to *arbitrary* (possibly uncountable) sets T of sentences in the propositional calculus; that is, if any such T is consistent then it has a complete, consistent extension. It is of interest that transfinite methods were required for the proof. Gödel *could* have inferred from this, with no additional work, the compactness theorem for arbitrary sets of sentences in the propositional calculus, but there is no indication that he did so. There is also no indication that Gödel was aware of Lindenbaum's prior result for countable T.

The remark *1933* suggests a possible definition of "p analytically implies q" as "q is derivable from p... and q contains no notions other than those in p", and asks whether this implication relation has a complete axiomatization by a system which had been proposed by Parry. The subject of "relevance logic" is here foreshadowed. (The question raised at the end of *1933* was answered by Gödel himself in what appeared as *1932*.)

In *1933a*, Gödel was again concerned with possible uses of finite truth tables. Here he showed that not every independence result in the propositional calculus can be established by such means.

(g) *Geometry (1933b, c, d, g, h)*. With the exception of *1933g*, these notes of Gödel on geometry had been missed in all bibliographies of Gödel's work prior to *Dawson 1983*. Though of lesser importance, they reveal another side of Gödel's interests and abilities. Apparently they were stimulated by his participation in Menger's colloquium, as they were among Gödel's thirteen short papers published in *Ergebnisse eines mathematischen Kolloquiums* during the period 1932–1936. The paper *1933h* is a joint one (his only such) with A. Wald that came out of a discussion with Menger. Altogether these notes deal with a variety of technical and axiomatic questions in differential and projective geometry.

(h) *Set theory (1938, 1939, 1939a, 1940)*. Gödel's next major results, and technically his most difficult, were those that established the consistency of the axiom of choice (AC) and the generalized continuum hypothesis (GCH) relative to various systems of set theory. These results were achieved by use of a certain "inner model" called the *constructible sets*. This model was described informally in *1938* (Gödel's first announcement of his results, in the *Proceedings of the National Academy of Sciences, U.S.A.*) as an extension of Russell's ramified theory of types into the transfinite, using arbitrary ordinals. A precise definition of the hierarchy M_α (subsequently denoted L_α) of constructible sets was given in the abstract *1939*. The systems for which Gödel says (in *1938* and *1939*) that his method works are those of von Neumann, *Principia mathematica*, Zermelo (Z) and Zermelo–Fraenkel (ZF). He also indicates that the method is unaffected by the addition of certain large cardinal axioms such as the existence of inaccessible cardinals.

As mentioned in Part I above, Gödel had arrived at the idea of using the model of constructible sets to obtain these results relatively early in the 1930s. He had obtained the consistency of AC by 1935, as he reported to von Neumann, but it took him another two years to prove that GCH holds in this model. The main idea for this is outlined in his note *1938*, where consistency is established relative to Z by using the L_α up to $\alpha = \omega_\omega$ and relative to ZF by using the L_α for all accessible ordinals α (ω_α denotes the αth initial ordinal, that is, the least ordinal of cardinality $\aleph_\alpha$). There is an obvious definable well-ordering of the constructible sets, but the key point is to show that this definition is *absolute*, that is, has the same meaning within the model as when regarded from the outside. It is technically more convenient to use absoluteness to show that the statement A asserting that all sets are constructible is true in the model and that A implies AC. For the proof with respect to the GCH, the crucial lemma is that if x is a constructible subset of L_{ω_α} then $x \in L_{\omega_{\alpha+1}}$ (since the cardinality of L_{ω_α} is $\aleph_\alpha$ within the constructible universe). Here the main tools are a form of downward Löwenheim–Skolem argument and what is called "collapse" to a transitive set.

The monograph *1940* is based on notes of Gödel's lectures that he gave

at the Institute for Advanced Study in the fall of 1938. Here the consistency of AC and GCH is established in complete detail relative to the theory BG of sets and classes introduced by Bernays and modified by Gödel (BG is a descendant of von Neumann's system). Then the statement that all sets are constructible may be expressed by an equation, $V = L$. In BG the metamathematical notions which had to be formalized in order to define constructibility in ZF can be avoided by the use of finitely many class functions that behave like the logical operations. Unfortunately, the resulting treatment is much less intuitive than that outlined in *1938*. The plan of the latter note was eventually elaborated by other logicians, and it is generally considered to be the preferred approach to the subject. As described in the introductory note to *1938–1940* in the second of these volumes, constructible sets have been the object of intensive study by logicians, especially since the 1960s, and have also formed an integral part of other technical methods in set theory, such as the ramified forcing of *Cohen 1963*. Thus not only Gödel's results but also his notions and methods have been of lasting value in the subject. Scott (*1961*) discovered a bound to the large cardinal axioms with which $V = L$ can be shown consistent; namely, the existence of measurable cardinals implies $V \neq L$.

(i) *Philosophy of mathematics (1944, 1946, 1947, 1964)*. Gödel's publications through *1940* hardly gave a hint of his own philosophical position; it emerged for the first time in the article *1944*, written for the Bertrand Russell volume in the Library of living philosophers (*Schilpp 1944*). The piece consists of an involved series of critical remarks on Russell's underlying viewpoint in *Principia mathematica* and a statement of Gödel's own views, which takes that as a point of departure. Against Russell's "no class" ramified theory he poses there his own realistic philosophy, according to which classes have an objective, independent existence (this will be elaborated in Section 3 below). Gödel mentions that a transfinite extension of the ramified theory has technical value, since that is what he used to define the constructible sets in *1938–1940*. The entire article is rich with Gödel's thoughts about the philosophy of logic and mathematics together with a number of interesting suggestions.

The much briefer *1946* was prepared for a conference on the problems of mathematics held at Princeton University on the occasion of its bicentennial. However, the article itself did not appear in print until the collection *Davis 1965*. Here Gödel deals with ideas of absolute definability and demonstrability, independent of any formal languages and theories. The latter notion is only speculative, but for the former Gödel suggests a definition of set-theoretic definability within set theory itself; this permits arbitrary ordinals as primitive terms, and so has since come to be called "ordinal definability". Later it was independently rediscovered by others, in particular *Myhill and Scott 1971*; the notion eventually turned out to be technically very useful. In *1946*, Gödel also made some interesting remarks

about the nature and possible use of strong axioms of infinity (or large cardinal axioms, as they are now called). He thought it possible that every set-theoretical proposition is decidable from the present axioms together with some axiom of infinity. Thus far there have been no precise positive results in this direction.

The piece *1947* is Gödel's only expository work. It explains Cantor's continuum problem and the failure of mathematicians to settle it. As early as 1922 Skolem began to suspect that the continuum hypothesis *CH* might be independent of axiomatic set theory. This was finally established for the Zermelo–Fraenkel axioms on one side (namely that ZF with *AC* does not prove the negation of *CH*) by Gödel in *1938–1940*. The other side (that ZF with *AC* does not prove *CH*) was settled only later by *Cohen 1963*, though Gödel had worked intensively on this problem in the early 1940s. The article *1964* is a revision of *1947* that describes Cohen's achievement, but it also contains a number of other changes from the original. Philosophically, the most significant part of *1947* lies in Gödel's assertion that *CH* has a definite truth value (according to his realistic view of set theory) and that it is likely false. He proposes establishing this by the use of new large cardinal axioms but gives no specific indication what those might be.

(j) *Functional interpretation (1958, 1972)*. Though the publication *1958* is actually Gödel's last, the work itself was done by the early 1940s—he gave a lecture on it at Yale University in 1941. It was also his only publication in German after 1936 and was written in honor of Paul Bernays' 70th birthday. The paper *1972* is a translation of *1958* into English with some revisions and with extensive amplifications, particularly concerning its foundational aspects. It was found in galley proofs in Gödel's *Nachlass*, but, despite repeated urging by Bernays, never appeared in print. Gödel's original purpose for this work was technical, for example to establish the existential realization property for intuitionistic arithmetic. But by the time he wrote *1958* his interests had shifted to more philosophical aims.

Gödel begins *1958* by saying it is virtually certain that finitary mathematics in Hilbert's sense of "concrete intuition" does not go beyond arithmetic; thus, to establish the consistency of arithmetic, some abstract notions are needed. The abstract notion of computable functional of finite type introduced here is argued to be constructive; it is evidently closed under the means of definition of primitive recursive functionals, expressed in a quantifier-free theory T. Gödel then shows how Heyting's arithmetic (and hence also Peano's arithmetic, by *Gödel 1933e*) can be interpreted in T. This interesting functional interpretation was later extended to analysis by *Spector 1962*, although the constructivity of the latter has been disputed. Gödel himself considered Spector's work to be quite significant, as he stated in a Postscript (*Gödel 1962*) to Spector's paper. The method of functional interpretation has found a number of other technical uses.

(k) *Relativistic cosmology and the philosophy of time (1949, 1949a, 1952)*. Gödel's contributions in this area were quite unexpected, both as to the nature of the results and in relation to the major body of his work. It is true that he had had an early interest in physics and had thought to pursue physics as his major area of study while at the university, but we have no indication that Gödel continued to follow the subject. It is possible that his contact with Einstein at the Institute for Advanced Study in the 1940s reawakened his interest in the subject, and more specifically in relativity theory, though by his own account his motivations were rather different from Einstein's.^{ff} On the technical side, Gödel's attention to differential geometry in the 1930s (see (g) above) could have equipped him with the kind of mathematical tools needed for this work.

In his paper *1949* Gödel showed for the first time that there are rotating models of Einstein's field equations. This violates Mach's principle which, in some sense, denies the existence of absolute space. Einstein had hoped Mach's principle would be a consequence of his equations, but more recently physicists seem to be inclined against the principle. Gödel's model in *1949* is non-expanding, thus failing to correspond to the observed red-shift of distant galaxies. It also has the very unusual property that there are closed time-like lines, theoretically permitting time travel into one's past.

The paper *1952* is the text of Gödel's invited address to the 1950 International Congress of Mathematicians. There he considers rotating models more generally, including some that are more physically plausible in that they are expanding and time travel into the past is excluded. Observational evidence for rotation could be possible, but it is now known that the rate of rotation, if such exists at all, must be very low.

Gödel's discoveries in this area stimulated discussion on the relationship between general relativity and causality; they are to be situated within modern work on relativistic cosmology that explores an enormous variety of possibilities of solutions of Einstein's equations, tested by both experimental evidence and theoretical considerations.

In his article *1949a*, written for the Einstein volume in the Library of living philosophers (*Schilpp 1949*), Gödel connects his work with philosophical discussions concerning the nature of time. He argues that his *1949* model supports an idealistic view of time and change as illusory, in the sense that they are a contribution of our own minds rather than an objective aspect of the physical world. The philosophical influences here are Leibniz and Kant. An expanded version of *1949a*, but one showing some differences of viewpoint, has been found in Gödel's *Nachlass*. Apparently Gödel's long-standing interest in philosophy and metaphysics were important to him as motivations for his engagement in this area of work.

(l) *Miscellaneous (1936, 1972a, 1974)*. The entry *1936* in the present volume is simply a two-sentence remark in which Gödel suggests a more realistic generalization of Wald's equations for production in economics.

Wald was a colleague of Gödel's in Vienna and also participated in Menger's colloquium. As will be seen from the introductory note to *1936*, Gödel's remark has not been without interest for economists.

Appended to the galley proofs (*1972*) of Gödel's revised translation of *1958* (discussed in (j) above) are three brief notes on the incompleteness results; they are reproduced in Volume II of this edition as *1972a*. The first of the notes concerns a technical generalization that Gödel had previously formulated in print but felt had been overlooked. The other two notes have to do with the philosophical significance of the incompleteness results, one part concerning the search for new axioms and the other expressing opposition to Turing's mechanistic view of mind.

The remark *1974* is a quoted statement that appears in the preface to the second edition of Robinson's book on non-standard analysis, in which Gödel gives very high praise to the development of that subject; he regards it as a fundamental and natural extension of ordinary analysis through its systematic use of infinitesimals, achieved by means of a non-standard extension of the real number system. (The basic fact that guarantees the existence of such non-standard models is the compactness theorem for the first-order predicate calculus in an uncountable language or, alternatively, the ultrapower theorem.)

(m) *Reviews (1931b–g, 1932d–o, 1933j–m, 1934a–f, 1935, 1935a–c, 1936b)*. As far as has been determined, Gödel wrote thirty-three reviews, all during the period 1930–1936 and those only for *Zentralblatt für Mathematik und ihre Grenzgebiete* and for *Monatshefte für Mathematik und Physik (Literaturberichte)*. The reviews are of interest because they show the kind of attention Gödel paid to parts of logic outside his own areas of research, as well as how he dealt with work related to his own. However, they should by no means be regarded as giving a picture of the state of logic in the 1930s (for example, there are no reviews of the work of Herbrand or Gentzen, of the work in the Tarski school, nor of any work on set theory). The style in these reviews is generally "objective", with an occasional telling point or brief mention being made of a relevant result obtained by Gödel. Among the topics covered are foundational schemes and the philosophy of mathematics, axiomatization of mathematical concepts, algebra of logic, alternative logics, the decision problem in the predicate calculus, the postulates of the λ-calculus, and non-standard models.

From the historical point of view, one of the most interesting of these reviews is that (*1934c*) of *Skolem 1933a*, which Gödel misprized. In his paper, Skolem had provided a mathematical construction of a non-standard extension N^* of the natural number system N that satisfies all true sentences of N. Gödel states in his review that the existence of such a model N^* is a consequence of his *incompleteness* theorem; but the latter (if taken in combination with the *completeness* theorem) would only give a model N' in which some false sentence of arithmetic holds. Gödel should have

128,357

said that the existence of such an N^* is a consequence of his *compactness* theorem.[17] In any case Skolem's particular method, which Gödel did not appreciate at all, turned out to be a forerunner of the powerful and versatile method of model construction by ultrapowers.

An earlier review particularly worth noting is that (*1931c*) of an article by Hilbert (*1931*) where, in order to overcome the incompleteness of arithmetic, Hilbert introduced a new kind of rule which allows the inference of $(x)A(x)$ whenever it has been shown that each instance $A(z)$ (for z a numeral) is correct. This is one of the first forms of what is now called the ω-rule.

One should also take note of the laudatory review (*1933m*) of Hahn's book on real analysis, in which Gödel took the opportunity to pay homage to his former teacher.

2. **Works and notes in the *Nachlass*.** Kurt Gödel's scientific *Nachlass* includes unpublished manuscripts, notes for lecture courses and for individual lectures, notebooks, memoranda and correspondence. All these were donated to the Institute for Advanced Study by Gödel's widow after his death; the literary rights in his estate were subsequently vested in the Institute under the terms of Mrs. Gödel's will. The *Nachlass* was catalogued by John Dawson during the academic years 1982–1984.

The following survey of material from the *Nachlass* is intended only to give a general idea of the character and extent of its contents. A more detailed assessment is reserved for a succeeding volume of Gödel's *Collected works*, in which the most interesting and important portions of the *Nachlass* are to be published. The main sources for this section are *Kreisel 1980* and *Dawson 1983*, together with information communicated personally by Dawson.

(a) *Unpublished manuscripts.* These items are preserved as hand drafts or typescripts.

1° "Simplified proof of a theorem of Steinitz" (prepared for Menger's colloquium in Vienna).

2° "Is mathematics syntax of language?" (intended for *The philosophy of Rudolf Carnap* in the Library of living philosophers).

3° "Some observations about the relationship between theory of relativity and Kantian philosophy" (expanded version of *1949a*).

4° "Some considerations leading to the probable conclusion that the true power of the continuum is $\aleph_2$".

5° "A proof of Cantor's continuum hypothesis from a highly plausible axiom about orders of growth".

---

[17]While it *is* the case that this is an immediate consequence of his compactness theorem, the kind of argument using it was apparently not made until the 1950s.

Item 2° exists in six versions. Item 4° was circulated rather widely in the early 1970s in preprint form. It was intended for publication in the *Proceedings of the National Academy of Sciences* but was withdrawn after errors were found. Item 5° is a later attempt to prove the opposite conclusion, by related considerations; it was apparently not circulated.

(b) *Individual lecture texts.* The following include all those found as clear texts in the *Nachlass*. Lectures preserved only as fragmentary notes are not listed.

1° "The present situation in the foundations of mathematics" (address to the American Mathematical Society, Cambridge, Mass., 30 December 1933).

2° "The existence of undecidable propositions in any formal system containing arithmetic" (Address to the New York Philosophical Society, 18 April 1934).

3° "The consistency of the generalized continuum hypothesis" (address to the American Mathematical Society, Williamsburg, Va., 28 December 1938).

4° "In what sense is intuitionistic logic constructive?" (Lecture on primitive recursive functionals, Yale University, 15 April 1941).

5° "Lecture on rotating universes", Princeton, May 1949.

6° Address to International Congress of Mathematicians, Cambridge, Mass., 31 August 1950.

7° "Some basic theorems on the foundations of mathematics and their philosophical implications" (25th Gibbs Lecture, American Mathematical Society, Providence, R.I., 26 December 1951).

8° Undelivered lecture on undecidable sentences and polynomials (undated draft).

(c) *Lecture notes.* The lectures that Gödel gave at the Institute for Advanced Study in 1934 and 1938 have been published as *1934* and *1940*, respectively. The notes for *1940* are to be found in the *Nachlass*. There are only fragmentary notes for the three courses that Gödel gave as a *Privatdozent* at the University of Vienna. Detailed notes do exist, though, for the two courses he gave at the University of Notre Dame in 1939, with the following titles.

1° A course in elementary logic.

2° Seminar on constructible sets.

Both of these are rather polished, and 2° differs in some interesting respects from the *1940* version.

(d) *Notebooks.* There are over 100 of these altogether, among which are various subseries distinguished by Gödel: 16 *Arbeitshefte* (mathematical workbooks), as well as 14 philosophical and 2 theological notebooks. The *Arbeitshefte* and philosophical notebooks are almost entirely in Gabels-

berger shorthand,[18] a script requiring special transcription. For the *Arbeitshefte* Gödel prepared indexes in longhand that give some indication of the contents. Among the listings are the following headings.

1° Independence of the axiom of choice.

2° Independence of the continuum hypothesis.

3° Corrections to Herbrand.

4° Consistency of analysis.

5° Ordinals of analysis.

6° Attempts to solve the paradoxes.

7° Absolute definability.

8° Axioms of infinity.

9° Consistency of $\neg(p)(\neg p \lor p)$.

10° Interpretations of intuitionistic logic.

There are 55 listings in all, a number of them concerning topics in mathematics and physics. The philosophical notebooks apparently contain both original reflections and much material on philosophers such as Leibniz. The items 1°–3° date from the early 1940s. By Gödel's own later account,[gg] he only succeeded in obtaining the independence of the axiom of choice from the theory of finite types, and by a method different from that of forcing in *Cohen 1963* (which eventually succeeded for both *AC* and *CH* relative to ZF). Concerning 3°, Gödel had detected the substantive errors in Herbrand's work that were not discovered by other logicians until much later (*Dreben, Andrews and Aanderaa 1963*). It should be emphasized that complete transcription will be necessary to see how far each topic was carried in the various notebooks; this work is in progress.

3. **Gödel's philosophy of mathematics**. Gödel is noted for his vigorous and unwavering espousal of a form of mathematical realism (or "platonism"). In this general direction he joins the company of such noted mathematicians and logicians as Cantor, Frege, Zermelo, Church, and (in certain respects) Bernays. These views of mathematics also accord with the implicit working conceptions of most practicing mathematicians (the "silent majority"). However, the preponderance of developed thought on the philosophy of mathematics since the late 19th century has been critical of realist positions and has led to a number of alternative (and opposing) standpoints, going under such names as constructivism, formalism, finitism, nominalism, predicativism, definitionism, positivism and conventionalism. Leading figures identified with one or another of these positions are Kronecker, Brouwer, Poincaré, Borel, Hilbert, Weyl, Skolem, Heyting,

---

[18]This was one of two competing German shorthand systems in widespread use during the early part of this century. It has since been superseded by a unified script. Among famous users of Gabelsberger were Bernays, Husserl, Schrödinger and Zermelo.

Herbrand, Gentzen and Curry, as well as Carnap. Russell veered from a distinctly realist position in his earlier work, *The principles of mathematics*, to a more equivocal predicativist approach in *Principia mathematica*.

Critics of the realist position have raised both ontological and epistemological issues. With respect to the former, the ideas of mathematical objects as independently existing abstract entities, and in particular of infinite classes as "completed" totalities, are considered to be problematic. For the latter, questions have been raised about the admissibility of such principles as that of the excluded middle and the axiom of choice, each in its way leading to non-constructive existence proofs. Especially in the earlier part of this century, the paradoxes of classes found by Cantor, Burali-Forti and Russell were felt in addition to require radical reconsideration of the entire set-theoretic, philosophically platonist approach to the foundations of mathematics. This last receded in importance when it was recognized how Zermelo had rescued set theory from the obvious contradictions by means of his axiomatization and its underlying interpretation in the iterative conception of sets.

Hilbert and Brouwer were perhaps the most influential figures proposing alternative foundational schemes during the period in which Gödel was beginning his work in logic. Hilbert had elaborated a program to "secure" mathematics—including, as he hoped, Cantor's set theory—by means of finitary consistency proofs for formal axiom systems. Brouwer rejected non-constructive existence proofs and Cantorian conceptions of "actual" infinities, seeking to rebuild mathematics according to his own intuitionistic version of constructivism. In Vienna, special attention was naturally also given to the program of the logical empiricists developed by Hahn, Schlick, Carnap and others. Their aim was to place mathematics in a conventionalist role as the "syntax of language", thus separating it from physical science, which itself was to rest finally on empirical observation. According to his own account much later,[hh] Gödel had arrived at a general platonist viewpoint by 1925, around the time he came to Vienna. When he began to take part in the meetings of the Vienna Circle, he did so primarily as an observer, not openly disputing the approach taken, though disagreeing with it. Gödel did remark critically on the positions of Hilbert and Brouwer in the introduction to his dissertation *1929*, but mainly in connection with the completeness problem. In particular, he made some trenchant remarks there concerning the idea of consistency as the criterion for existence. That view could be identified with Hilbert, though it was not a necessary part of Hilbert's program. However, this discussion was omitted from the published version (*1930*) of the dissertation. Gödel openly criticized the related idea, again deriving from Hilbert, of consistency sufficing for correctness when one extends a system of meaningful statements by a system of "ideal" statements and axioms. These remarks were made at the important symposium on the foundations of mathematics held at

Königsberg in 1930 (see *1931a*). Gödel made no further published statements on the nature of his position until the appearance of his substantial article *1944* on Russell's mathematical logic. In retrospect, however, one can recognize some brief remarks or footnotes in earlier papers as providing indications of the directions of his thought.

The main published sources for Gödel's views on the philosophy of mathematics are the papers *1944, 1946, 1964* (a revised and expanded version of *1947*) and *1958*, as well as his personal and written communications to Hao Wang reproduced in *Wang 1974*. Further sources that appear in the present volume but were not previously printed are the introduction to *1929*, the revised and expanded version *1972* of *1958*, and finally the brief notes *1972a*. All this is amplified but not modified in any significant way by unpublished manuscripts and correspondence found in Gödel's *Nachlass*. In particular, the article mentioned in Section 2 above, "Is mathematics syntax of language?", would have been Gödel's first systematic published attack on the program of the logical positivists, had it appeared in the Carnap volume as intended.[19]

The main features of Gödel's philosophy of mathematics that emerge from these sources are as follows. Mathematical objects have an independent existence and reality analogous to that of physical objects. Mathematical statements refer to such a reality, and the question of their truth is determined by objective facts which are independent of our own thoughts and constructions. We may have no direct perception of underlying mathematical objects, just as with underlying physical objects, but—again by analogy—the existence of such is necessary to deduce immediate sense perceptions. The assumption of mathematical objects and axioms is necessary to obtain a satisfactory system of mathematics, just as the assumption of physical objects and basic physical laws is necessary for a satisfactory account of the world of appearance. An example of mathematical "sense data" requiring this kind of explanation is provided by instances of arithmetical propositions whose universal generalizations demand assumptions transcending arithmetic; this is a consequence of Gödel's incompleteness theorem.[20] While mathematical objects and their properties may not be

---

[19]In the penultimate section of his introductory note to *1944* in this collection, Charles Parsons suggests that in one respect, at least, Gödel is more closely engaged with the ideas of the Vienna Circle than is ordinarily viewed. The relationship has to do with the thesis that mathematics is analytic. In *1944* Gödel considers two senses of the notion of analyticity of a statement, respectively (roughly speaking) that of its being true in virtue of the definitions of the concepts involved in it and that of its being true in virtue of the meaning of those concepts. In *1944* Gödel rejects the thesis that mathematics is analytic in its first sense but accepts it in its second sense (at least for the theory of types and axiomatic set theory).

[20]Gödel frequently refers to such propositions as arithmetical problems of Goldbach type (the conjecture that every even positive integer is a sum of two odd primes).

immediately accessible to us, mathematical intuition can be a source of genuine mathematical knowledge. This intuition can be cultivated through deep study of a subject, and one can thus be led to accept new basic statements as axioms. Another justification for mathematical axioms may be their fruitfulness and the abundance of their consequences; however, that is less certain than what is guaranteed by intuition.

Gödel discussed these ideas most explicitly in connection with set theory and Cantor's continuum problem, particularly in *1947* and *1964* (see especially the supplement to *1964*). There he argues that Cantor's notion of (infinite) cardinal number is definite and unique, and hence that the continuum hypothesis *CH* has a determinate truth value, even though efforts to settle it thus far have failed.[21] One can begin by examining the question of its demonstrability with reference to presently accepted axioms for set theory. These axioms (for example, the Zermelo–Fraenkel system ZF) are evidently true for the iterative structure of sets in the cumulative hierarchy. This is a perfectly self-consistent conception that is untouched by the paradoxes. In Gödel's view, the axiom of choice is just as evident for this notion as are the other axioms, and hence Cantor's cardinal arithmetic is adequately represented in the axiom system.

In *1940* Gödel had shown that *AC* and *CH* are consistent with ZF, by use of his model *L* of constructible sets. But he conjectured in *1947* that *CH* is false, hence underivable from the true axioms ZF + *AC*. After Cohen (*1963*) proved the independence of *CH* from ZF + *AC*, Gödel could expand on the anticipated undecidability of *CH* by presently accepted axioms. This confirmed what he had long expected, namely that new axioms would be needed to settle *CH*. In particular he mentioned the possibility of using strong axioms of infinity (or large cardinal axioms) for these purposes, pointing out once more that in view of the incompleteness theorem, such axioms are productive of arithmetical consequences. But he also thought that axioms based on new ideas may be called for.[22] He argued again that such axioms need not be immediately evident, but may be arrived at only after long study and development of the subject.

---

[21]There is one earlier statement by Gödel that apparently presents a different view concerning the question of definiteness of set-theoretical concepts. Namely, at the end of *1938* he says: "The proposition *A* [*V* = *L*] added as a new axiom seems to give a natural completion of the axioms of set theory, insofar as it determines the vague notion of an arbitrary infinite set in a definite way." In a personal communication, Martin Davis has argued: "This is not at all in the spirit of the point of view of *1947*, and ... it suggests that Gödel's 'platonism' regarding sets may have evolved more gradually than his later statements would suggest." There is currently no further evidence available which would help clarify Gödel's intentions in his 1938 remark and its relationship to his later views.

[22]Indeed, he proposed certain new axioms himself in the unpublished manuscripts 4° and 5° mentioned in Section 2(a) above.

There are briefer discussions or indications by Gödel in other of his publications concerning his belief in the objectivity of mathematical notions outside set theory. These are: abstract concepts (in *1944*), absolute demonstrability and definability (in *1946*), and constructive functions and proofs (in *1958* and *1972*). One should also mention the steady interest he showed in intuitionism through his publications described in Sections 2(e) and 2(j) above. Thus his mathematical realism is not necessarily confined to set theory, though that is where it is most thoroughly elaborated.

In the correspondence reproduced in *Wang 1974* (pages 8–11), Gödel credits a large part of his main successes, where others had failed, to his realist views; they are said to have freed him from the philosophical prejudices of the times which had shackled others. In this respect he mentions Skolem's failure to arrive at the completeness of predicate logic and Hilbert's failure to "prove" *CH* in contrast to his own results in *1930* and *1940*; he also mentions his belief in the objectivity of mathematical truth as having led to the incompleteness theorems of *1931*. Whatever their final merits, the efficacy of Gödel's views seems in this respect to be indisputable.

A deeper examination of Gödel's ideas on the philosophy of mathematics will be found in the introductory notes to *1944*, *1946*, *1947* and *1964*, *1958* and *1972* and *1972a*, in the second volume of this collection.

4. Character, impact and influence of the work.

Gödel's main published papers from 1930 to 1940 were among the most outstanding contributions to logic in this century, decisively settling fundamental problems and introducing novel and powerful methods that were exploited extensively in much subsequent work. Each of these papers is marked by a sense of clear and strong purpose, careful organization, great precision—both formal and informal—and by steady and efficient progress from start to finish, with no wasted energy. Each solves a clear problem, simply formulated in terms well-understood at the time (though not always previously formulated as such). Their significance then was in one sense prima facie evident, though their significance more generally for the foundations of mathematics would prove to be the subject of unending discussion. As he has told us, Gödel was strongly motivated by his realist philosophy of mathematics, and he credited it with much of the reason for his success in being led to the "right" results and methods.[ii] Nevertheless, philosophical questions are given bare notice in these papers. In addition, Gödel made special efforts where possible to extract results of potential mathematical (as opposed to logical or foundational) interest—for example, the compactness theorem for the first-order predicate calculus (*1930*), the incompleteness of axiomatic arithmetic with respect to (quantified) Diophantine problems (*1931*), and the existence of non-measurable *PCA* sets of reals in the universe of con-

structible sets (*1938*).[23]

Concerning Gödel's methods, one may say that many of the constructions and arguments were technically difficult for their time, or at any rate too novel or unexpected to be readily absorbed (though the arguments for completeness were largely anticipated by Skolem, and those for incompleteness and completeness both seem much simpler now). But technical ingenuity is never indulged in or displayed for its own sake in Gödel's papers; rather it is always there as a means to an end. We have considerable evidence that Gödel worked and reworked his papers many times, partly to arrive at the most efficient means of presentation.

Gödel's contributions bordered on the two fundamental technical concepts of modern logic: *truth for formal languages* and *effective computability*. With respect to the first he stated in his *1934* lectures at Princeton (and elaborated in some correspondence) that he was led to the incompleteness of arithmetic via his recognition of the undefinability of arithmetic truth in its own language, though he took care to credit Tarski for elucidating the exact concept of truth and establishing its undefinability. In the same lectures he offered a notion of general recursiveness in connection with the idea of effective computability; this was based on a modification of a definition proposed by Herbrand. In the meantime, Church was propounding his thesis, which identified the effectively computable functions with the λ-definable functions. But Gödel was unconvinced by Church's thesis, since it did not rest on a direct conceptual analysis of the notion of finite algorithmic procedure. For the same reason he resisted identifying the latter with the general recursive functions in the Herbrand–Gödel sense. Indeed, in his Princeton lectures Gödel said that the notion of effectively computable function could serve just as a heuristic guide. It was only when Turing, in 1937, offered the definition in terms of his "machines" that Gödel was ready to accept such an identification, and thereafter he referred to Turing's work as having provided the "precise and unquestionably adequate definition of formal system" by his "analysis of the concept of 'mechanical procedure'" needed to give a general formulation of the incompleteness results.[jj] It is perhaps ironic that the various classes of functions (λ-definable, general recursive, Turing computable) were proved in short order to be identical, but Gödel's initial reservations were justified on philosophical grounds.

In general, Gödel shied away from new concepts as objects of study, as opposed to new concepts as tools for obtaining results. The constructible hierarchy may be offered as a case in point, concerning which Gödel says

---

[23]This must be moderated in two respects. It is certainly the case that Gödel himself never made any use of compactness and that we have the benefit of hindsight in assessing its mathematical value. Moreover, according to *Kreisel 1980*, p. 197, the existence of non-measurable *PCA* sets in *L* was suggested to Gödel by Stanislaw Ulam.

that he is only using Russell's idea of the ramified hierarchy, but with an essentially impredicative element added, namely the use of arbitrary ordinals.[kk] Only the concept of effective functional of finite type, which he had arrived at by 1941, comes close to being a new fundamental concept (see *1958* and *1972*).

There is a shift in the 1940s that corresponds to Gödel's changed circumstances and interests. Prior to that time, Gödel was understandably cautious about making public his platonist ideas, contrary as they were to the "dominant philosophical prejudices" of the time.[ll] With his reputation solidly established and with the security provided by the Institute, Gödel felt freer to pursue and publicly elaborate his philosophical vision.

Gödel did the major part of his logical work in isolation, though he had a certain amount of stimulating contact with Menger, von Neumann and Bernays in the pre-war period. As described in the biographical sketch above, in the 1950s the Institute increasingly attracted younger logicians, many of them in the forefront of research, as well as older colleagues of the pre-war generation. Some of them sought Gödel out and established lengthy scientific relations with him that were also personally comfortable and friendly. Yet Gödel never had any students, never established a school, and never collaborated with others to advance his favorite program, namely, the discovery of essentially new axioms for set theory. Nonetheless that program was taken up by many others in the wave of work in set theory from the 1960s on. Gödel's main results proved to be absolutely basic—the sine qua non for all that followed in almost all parts of logic—and it is through the work itself that he has had his major impact and influence.

As much as anything, Gödel's achievement lay in arriving at a very clear understanding of which problems of logic could be treated in a definite mathematical way. Along with others of his generation, but always leading the way, he succeeded in establishing the subject of mathematical logic as one that could be pursued with results as decisive and significant as those in the more traditional branches of mathematics. It is for this double heritage of the content and character of his work that we are indebted to him.

Solomon Feferman

Source notes

[a]In preparing the following I have drawn heavily on a number of sources, of which the main published ones are: *Christian 1980, Dawson 1983, 1984a, Kleene 1976, 1985, Kreisel 1980*, and *Wang 1978, 1981*. Some use has also been made of unpublished material from Gödel's *Nachlass*.

For the biographical material I have relied primarily on *Kreisel 1980* (pp. 151–160), *Wang 1981* and *Dawson 1984a*. Further personal material of value has come from *Quine 1979, Zemanek 1978, Köhler 198?* and *Taussky-Todd 198?*. I have also made use of personal impressions communicated to me by A. Raubitschek (whose father was one of Hahn's best friends and who himself knew Gödel in Princeton), and of my own impressions (from contacts with Gödel during my visit to the Institute for Advanced Study in 1959–1960).

Finally, I am indebted to my co-editors as well as to the following of my colleagues for their many useful comments which have helped appreciably to improve the presentation: J. Barwise, S. Bauer-Mengelberg, M. Beeson, G.W. Brown, M. Davis, A.B. Feferman, H. Feigl, J.E. Fenstad, R. Haller, E. Köhler, G. Kreisel, R.B. Marcus, K. Menger, G. Müller, C. Parsons, W.V. Quine, A. Raubitschek, C. Reid, J. Robinson, P.A. Schilpp, W. Sieg, L. Straus, A.S. Troelstra and H. Wang.

[b]Following Gödel's delivery of the Gibbs lecture in 1951. The story is told by Olga Taussky-Todd in her reminiscences (*198?*).

[c](Dr.) Rudolf Gödel wrote up a family history in 1967, with a supplement in 1978. This was made available to Georg Kreisel along with some correspondence between Kurt Gödel and his mother; see *Kreisel 1980*, page 151. Gödel also communicated information about his family to Hao Wang for the article *Wang 1981*.

Another useful source on this and Gödel's intellectual development is a questionnaire which had been put to Gödel in 1975 by Burke D. Grandjean, then an instructor in sociology at the University of Texas. Its purpose was to gather information on Gödel's background in connection with research that Grandjean was doing on the social and intellectual situation in Central Europe during the first third of the 20th century. The questionnaire was found in Gödel's *Nachlass*, fully filled out, along with a covering letter to Grandjean dated 19 August 1975; however, neither was apparently ever sent. I shall refer to this several times as a source in the following, calling it 'the Grandjean interview'.

[d]*Kreisel 1980*, p. 152.

[e]*Kreisel 1980*, p. 153.

[f]According to Gödel in the Grandjean interview, he had already formed such views before coming to Vienna. See also *Wang 1978* (p. 183) and *Köhler 198?*.

[g]The Grandjean interview.

[h]*Wang 1981*, p. 654.

[i]*Wang 1981*, pp. 654–655. For information on von Neumann's life and work, see *Ulam 1958, Goldstine 1972* and *Heims 1980*.

[j]See *Dawson 1985*.

[k]See *Grattan-Guinness 1979, Moore 1980* and *Dawson 1985, 1985a*. An account of Gödel's first (and perhaps only) personal meeting with Zermelo is given in *Taussky-Todd 198?*.

[l]Hahn's report is quoted in *Christian 1980*, p. 263.

[m]*Kreisel 1980*, p. 154.

[n]Information communicated by E. Köhler.

[o]See *Montgomery 1963* and *Goldstine 1972*, pp. 77–79.

[p]*Goldstine 1972*, p. 174.

[q]The available published information about Gödel's recurrent illness during this period is slim; in this connection see *Kreisel 1980* (p. 154), *Wang 1981* (pp. 655–656) and *Dawson 1984a* (p. 13), as well as *Taussky-Todd 198?*.

[r]*Wang 1981*, p. 656.

^sFor more on how Gödel achieved his results on *AC* and *GCH*, see *Wang 1978* (p. 184), *Kreisel 1980* (pp. 194–198), and *Dawson 1984a* (p. 13). The dates given in these sources do not always square with each other. It is hoped that study of the correspondence and notes in Gödel's *Nachlass* will be of assistance in clearing this up. Already discovered is a shorthand annotation preceding Gödel's notes on the *GCH* in his *Arbeitsheft* 1, which has been transcribed (by C. Dawson) as "Kont. Hyp. im wesentlichen gefunden in der Nacht zum 14 und 15 Juni 1937", in other words that Gödel had "essentially found [the proof for the consistency of] *GCH* during the night of 14–15 June 1937".

^t *Kreisel 1980*, pp. 154–155.

^u *Feigl 1969, Köhler 198?*.

^v For accounts of these events see *Kreisel 1980* (pp. 155–156) and *Dawson 1984a* (pp. 13, 15).

^wIt is a further point of irony that Gödel was listed in the catalogues for the University of Vienna between 1941 and 1945 as "Dozent für Grundlagen der Mathematik und Logik" and under course offerings "wird nicht lesen". (Information communicated by E. Köhler.)

^xGödel's response to the Austrian Academy of Sciences is quoted in *Christian 1980*, p. 266; Kreisel (*1980*, p. 155) says that Gödel refused various honors from Austria after the war, "sometimes for mindboggling reasons".

^y *Ulam 1958* (pp. 3–4) and *Goldstine 1972* (pp. 177–182).

^zThis is recounted (in German) in *Zemanek 1978*, p. 210. Zemanek locates the hearing in Washington but it was more likely Trenton. Since the story is third-hand and translated, quotations are not exact.

^{aa}The Grandjean interview, *Wang 1978* (p. 183) and *Wang 1981* (pp. 658–659).

^{bb} *Wang 1981*, p. 658. Also, *Straus 1982* (pp. 420–421) says that "Gödel ... was really totally solitary and would never talk with anybody while working ...".

^{cc}Unpublished letter from Gödel to Carl Seelig, dated 7 September 1955.

^{dd} *Kreisel 1980*, p. 218.

^{ee}See *Kreisel 1980* (pp. 159–160), *Dawson 1984a* (p. 16). Information in this paragraph on Gödel's last years also comes from an interview that Dawson had with Gödel's friend and colleague, Deane Montgomery. The death of his old friend Oskar Morgenstern in mid-1977 was apparently a shock to Gödel. The quotation giving cause of death is from his death certificate, on file in the Mercer County courthouse, Trenton, New Jersey.

^{ff}See notes bb and cc.

^{gg}See *Wang 1981*, p. 657, and a letter from Gödel to W. Rautenberg published in *Mathematik in der Schule* 6, p. 20.

^{hh}The Grandjean interview.

ⁱⁱSee Gödel's letters to Hao Wang, dated 7 December 1967 and 7 March 1968, quoted in *Wang 1974*, pp. 8–11. The significance of Gödel's convictions for his work is discussed further in *Feferman 1984a*.

^{jj}See the note, added 28 August 1963, to *1931* and the Postscriptum to *1934*.

^{kk}See *1944*.

^{ll}See the letters mentioned in note ii for Gödel's characterization of these "prejudices", and *Feferman 1984a* for a discussion of Gödel's caution in this respect.

A Gödel chronology

Starred items below are based on sources from Gödel's *Nachlass*, as follows: Gödel's replies to a questionnaire from Burke D. Grandjean by *, and notes from Gödel's *Arbeitshefte* by ** (*Heft* 1) or *** (*Heft* 15). Information based on *Aufzeichnungen* from Carnap's *Nachlass* is labelled †.

1906	April 28	Birth of Kurt Friedrich Gödel, Brünn, Moravia
1912	September 16	Enrollment in *Evangelische Privat-Volks- und Bürgerschule*, Brünn
1914	(ca.)	Bout with rheumatic fever
1916	July 5	Graduation from the *Evangelische Schule*
	Fall	Entrance to *Staats-Realgymnasium mit deutscher Unterrichtssprache*, Brünn
1919–21		Formal study of Gabelsberger shorthand
1920–21	(ca.)	Awakening of interest in mathematics*
1922	(ca.)	First study of Kant's work*
1924	June 19	Graduation from *Realgymnasium*
	Fall	Entrance to University of Vienna, with intention to seek degree in physics
1925	(ca.)	Adoption of realist position in philosophy of mathematics*
1926		Change of degree focus to mathematics
1926	(ca.)	Beginning of attendance at meetings of Moritz Schlick's circle (the Vienna Circle)
1927	(ca.)	First acquaintance with future wife, Adele Nimbursky (née Porkert)
1928		Statement of the completeness problem for the first-order predicate calculus in *Hilbert and Ackermann 1928*; later settled by Gödel in his dissertation (*1929*)
1929		First study of *Principia mathematica**
	February 23	Premature death of Rudolf Gödel (father), born on 28 February 1874
	February 26	Granting of release from Czechoslovakian citizenship
	June 6	Granting of Austrian citizenship
	July 6	Approval of dissertation (*1929*) by Hans Hahn and Philipp Furtwängler

	October 15	Submission of dissertation to dean of the philosophical faculty
	October 22	Submission of revised dissertation (*1930*) to *Monatshefte für Mathematik und Physik*
	October 24	First regular meeting of Karl Menger's mathematics colloquium, and Gödel's first attendance there (at Menger's invitation); thirteen contributions by Gödel published in the colloquium proceedings during 1932–1936; volumes 1–5, 7 and 8 edited with the assistance of Gödel (and others).
1930	February 6	Granting of Dr. Phil.
	May 14	Presentation of dissertation results to Menger's colloquium
	August 26	Private announcement to Carnap, Feigl and Waismann of the (first) incompleteness result, during discussion at the Café Reichsrat.[†]
	September 6	Presentation of completeness results to the second *Tagung für Erkenntnislehre der exakten Wissenschaften*, Königsberg (abstract: *1930a*)
	September 7	First public announcement of the existence of formally undecidable propositions in number theory, during discussion session at Königsberg (*1931a*)
	October 23	Communication of abstract (*1930b*) of the incompleteness results to the Vienna Academy of Sciences
	November 17	Receipt of the incompleteness paper (*1931*) by *Monatshefte für Mathematik und Physik*
	November 20	Letter to Gödel from von Neumann, announcing his own independent discovery of the unprovability of consistency
	Late fall	Formal establishment of the Institute for Advanced Study (I.A.S.), Princeton
1931	January	Publication of *1931*
	Spring (?)	Letter from Herbrand (not long before his death on 27 July) to Gödel suggesting a definition of recursive functions

	September 15	Presentation of the incompleteness results to the annual meeting of the *Deutsche Mathematiker-Vereinigung*, Bad Elster; meeting and discussion with Zermelo
1932	June 25	Submission of *1931* to the University of Vienna as Gödel's *Habilitationsschrift*
	December 1	Granting of *Habilitation*
1933	March 11	Granting of *Dozentur*; Gödel's first course (foundations of arithmetic) taught shortly thereafter
	September 30 –October 6	Voyage to America aboard S. S. *Aquitania*
	October 2	Beginning of first year of operation of the Institute for Advanced Study, with Gödel as a visiting member through May 1934
	December 30	Address to the annual meeting of the American Mathematical Society: "The present situation in the foundations of mathematics"
1934	February to May	I.A.S. lectures on the incompleteness results (*1934*)
	April 18	Lecture to New York Philosophical Society: "The existence of undecidable propositions in any formal system containing arithmetic"
	April 20	Lecture to Washington (D.C.) Academy of Sciences: "Can mathematics be proved consistent?"
	May 26– June 3	Return voyage to Europe aboard S. S. *Rex*
	Mid-summer	Election to membership in the American Mathematical Society, by mail ballot of the A.M.S. council
	July 24	Death of Hans Hahn
	Fall	Admission to sanatorium in Purkersdorf bei Wien for treatment of nervous depression
1935	May 4	Start of Vienna lecture course on selected topics in mathematical logic
	June 19	Presentation of length-of-proofs paper (*1936a*), last contribution by Gödel to Menger's colloquium
	September 20–28	Voyage to America aboard Cunard liner *Georgic*

	October	Communication to von Neumann of relative consistency proof for the axiom of choice
	November 17	Sudden resignation from the I.A.S., on account of depression
	November 30–December 7	Voyage to Le Havre aboard steamer *Champlain*; subsequent return to Vienna for rest and recuperation
1936	Winter–spring	More time in sanatoria
	June 22	Schlick murdered by a deranged ex-student
1937	May–June	Lecture course on axiomatic set theory at the University of Vienna
	night of June 14–15**	Discovery of crucial step in proof of relative consistency of the generalized continuum hypothesis
1938	March 13	Hitler's annexation of Austria (*Anschluss*)
	September 20	Marriage in Vienna to Adele Nimbursky
	October 6–15	Voyage to America aboard liner *New York*; Vienna residence maintained by Adele
	October–December	I.A.S. lectures on the consistency of the axiom of choice and the generalized continuum hypothesis
	November 9	Set-theoretic consistency results (*1938*) communicated to the National Academy of Sciences
	December 28	Lecture to 45th annual meeting of the A.M.S., Williamsburg: "The consistency of the generalized continuum hypothesis"
1939	January–June	In residence at Notre Dame as visiting faculty member; presentation of a lecture course on his consistency results; joint seminar with Menger on elementary logic
	March (?)	General abolition, following *Anschluss*, of the title of *Privatdozent*
	June 14–20	Return voyage to Europe aboard German liner *Bremen*; subsequent receipt of notice to report for military physical examination
	September 25	Submission of application to become *Dozent neuer Ordnung*
	November 27	Letter to Veblen from Gödel, noting his classification by Nazi military authorities as "fit for garrison duty" and mentioning his quest for a U.S. non-quota immigrant visa

	December 15	Lecture at Göttingen on the continuum problem, in the midst of frantic efforts to obtain exit permits and transit visas
	December 19	Issuance of German exit permits to Kurt and Adele Gödel
1940		Publication of the monograph *1940*
	January 8	Issuance of U.S. non-quota immigrant visas
	January 12	Issuance in Berlin of Russian transit visas
	January 18–March 4	Journey to America via trans-Siberian railway and U.S ship *President Cleveland* (Yokohama to San Francisco, 5 February–4 March)
	Spring	Appointment as ordinary member of the I.A.S.; membership renewed annually until 1946
	June 28	Belated approval of application to become *Dozent neuer Ordnung*; later listed officially as "Dozent für Grundlagen der Mathematik und Logik, 1941–1945"
	November 15	Lecture at Brown University: "Consistency of Cantor's continuum hypothesis"
1941	April 15	Lecture at Yale University: "In what sense is intuitionistic logic constructive?"
	Spring	Course of lectures at Princeton on the *Dialectica* interpretation
1942	Summer	Reputed proof of the independence of the axiom of choice in finite type theory, obtained while on vacation at Blue Hill House, Hancock County, Maine***
1943–44		Transition to work in philosophy, with publication of an essay (*1944*) on Russell's logic
1945	(ca.)	Treatment of life-threatening duodenal ulcer by blood transfusions
1946		Beginning of membership in the Association for Symbolic Logic
	July 1	Promotion to permanent member of the I.A.S.
	December 17	Address to Princeton bicentennial conference on problems of mathematics (*1946*)
1947	December	Publication of expository essay on Cantor's continuum problem (*1947*)
1948	April 2	Granting of U.S. citizenship to Kurt and Adele Gödel
1949	May	I.A.S. lectures on rotating universes in general relativity theory

	July	First publication of results in relativity theory (*1949*)
	August 3	Purchase of home, 129 Linden Lane, Princeton (later renumbered as 145)
	December	Publication of *1949a*, an essay on Kant and relativity theory
1950	August 31	Address to International Congress of Mathematicians, Cambridge, U.S.A.: "Rotating universes in general relativity theory"
1951	March 14	First Einstein Award (jointly to Kurt Gödel and Julian Schwinger)
	June	Honorary D.Litt., Yale University
	December 26	A.M.S. Gibbs Lecture at Brown University: "Some basic theorems on the foundations of mathematics and their implications"
1952	June	Honorary Sc.D., Harvard University
1953		Election to the National Academy of Sciences
	July 1	Promotion to professor at the I.A.S.
1955	April 18	Death of Albert Einstein, I.A.S. colleague, friend and walking companion
1957		Election as Fellow of the American Academy of Arts and Sciences
	February 8	Death of John von Neumann, I.A.S. colleague and friend
1958		Appearance of *1958* in *Dialectica* (last published paper apart from revisions of earlier works)
1959		First study of Husserl's philosophy (according to *Wang 1981*)
1961		Election to American Philosophical Society
	July 29	Premature death at the I.A.S. of Clifford Spector; his *1962* published posthumously, with postscript by Gödel
1963		Proof by Paul J. Cohen that the axiom of choice and the generalized continuum hypothesis are independent of Zermelo–Fraenkel set theory (submitted by Gödel to *Proceedings of the National Academy of Sciences*)
1964		Publication of *1964*, a revised and expanded version of *1947*
1966	April 21–23	60th birthday symposium at Ohio State University, organized by the Ohio Academy of Sciences; invitation to attend declined by Gödel

	July 23	Death, in Vienna, of Marianne (Handschuh) Gödel (mother), born on 31 August 1879
	August	Refusal by Gödel of honorary membership in Austrian Academy of Sciences
1967		Election as honorary member of London Mathematical Society
	June	Honorary Sc.D., Amherst College
1968		Election as Foreign Member of the Royal Society (U.K.)
1970		Private circulation of "Scales of functions" paper; withdrawn after errors discovered
1972		Election as corresponding Member of the Institut de France (Académie des sciences morales et politiques)
	June	Honorary Sc.D., Rockefeller University
1974	April 11	Death of Abraham Robinson, colleague and friend
1975	September 18	Award of National Medal of Science by President Ford (accepted on Gödel's behalf by Saunders Mac Lane)
1976	July 1	Retirement from I.A.S., as professor emeritus
1977	July	Hospitalization of Adele Gödel for major surgery
	July 26	Death, in Princeton, of Oskar Morgenstern, Gödel's closest personal friend
	September 18	Death of Paul Bernays, long-time friend and colleague
	Mid-December	Release of Adele from hospital
	December 29	Hospitalization of Gödel himself, at Adele's urging
1978	January 14	Death of Kurt Gödel, at Princeton Hospital, due to "malnutrition and inanition"
	March 3	I.A.S. memorial service (speeches by André Weil, Hao Wang, Simon Kochen, and Hassler Whitney)
1981	February 4	Death of Adele Gödel, born on 4 November 1899; bequest of Gödel's *Nachlass* to the I.A.S.

John W. Dawson, Jr.

Introductory note to *1929*, *1930* and *1930a*

1. Background

If from Frege's formal system presented in *1879* we excise quantification over functions, we obtain a set of formal axioms and rules of inference that is complete with respect to quantificational validity. In the opening paragraph of his *1930* (below, page 103) Gödel writes that, when a formal system is introduced, "the question at once arises whether the initially postulated system of axioms and principles of inference is complete, that is, whether it actually suffices for the derivation of *every* logico-mathematical proposition". Frege, however, never saw completeness as a problem, and indeed almost fifty years elapsed between the publication of *Frege 1879* and that of *Hilbert and Ackermann 1928*, where the question of the completeness of quantification theory was raised explicitly in print for the first time. Why? Because neither in the tradition in logic that stemmed from Frege through Russell and Whitehead, that is, logicism, nor in the tradition that stemmed from Boole through Peirce and Schröder, that is, algebra of logic, could the question of the completeness of a formal system arise.

For Frege, and then for Russell and Whitehead, logic was universal: within each explicit formulation of logic all deductive reasoning, including all of classical analysis and much of Cantorian set theory, was to be formalized. Hence not only was pure quantification theory never at the center of their attention, but metasystematic questions as such, for example the question of completeness, could not be meaningfully raised. We can give different formulations of logic, formulations that differ with respect to what logical constants are taken as primitive or what formulas are taken as formal axioms, but we have no vantage point from which we can survey a given formalism as a whole, let alone look at logic whole. In the words of *Whitehead and Russell 1910* (page 95, or *1925*, page 91),

> It is to some extent optional what ideas we take as undefined in mathematics .... We know no way of proving that such and such a system of undefined ideas contains as few as will give such and such results. Hence we can only say that such and such ideas are undefined in such and such a system, not that they are indefinable.

We are within logic and cannot look at it from outside. We are subject to what Sheffer called "the logocentric predicament" (*1926*, page 228). The only way to approach the problem of what a formal system can do

is to derive theorems. Again to quote Russell and Whitehead, "the chief reason in favor of any theory on the principles of mathematics must always be inductive, i.e., it must lie in the fact that the theory in question enables us to deduce ordinary mathematics" (*1910*, page v, or *1925*, page v). (On this point see *van Heijenoort 1967a* and *Goldfarb 1979*.)

For Peirce, Schröder and their followers, on the other hand, quantificational formulas were indeed often at the center of their attention, but now the very notion of formal system was absent. Thus in his fundamental *1915* Löwenheim deals with quantificational formulas with identity. His approach, however, is purely model-theoretic, that is, semantic. He has no formal axioms or rules of inference for quantification theory. His basic notion is that of the truth of a formula for a given interpretation in a given domain, and with that he handles validity and satisfiability. Obviously, no question of completeness of a formal system could arise here either.

To raise the question of semantic completeness the Frege–Russell–Whitehead view of logic as all embracing had to be abandoned, and Frege's notion of a formal system had to become itself an object of mathematical inquiry and be subjected to the model-theoretic analyses of the algebraists of logic. An example of what would appear to be just such an investigation, and indeed one cited by Gödel (below, page 103) as an example of what he intends to do, is *Bernays 1926* (an abbreviated version of Bernays' unpublished *Habilitationsschrift* of 1918), where, to quote Gödel, "it has been shown that every correct propositional formula does indeed follow from the axioms given in *Principia mathematica*". Gödel immediately goes on to say: "The same will be done here for a wider realm of formulas, namely those of the 'restricted functional calculus'". Bernays' result, that is, the semantic completeness of the *Principia* propositional calculus, is also contained in *Post 1921*. In this paper, the first major published work on the metamathematics of *Principia*, Post is explicit about his break with the Frege–Russell–Whitehead tradition and about the absence in the writings of the algebraists of logic of any clearly determined notion of a formal system (see the first footnote on page 169 of *Post 1921*, or footnote 7 on page 269 of *van Heijenoort 1967*). For a "general statement" of his own viewpoint Post refers his readers to *Lewis 1918*, Chapter VI, Section III (*Post 1921*, page 165, or *van Heijenoort 1967*, page 266). In this section Lewis presents what he calls a "heterodox" view of logic, opposed to the "orthodox" view of Russell and Whitehead. What Lewis had in mind was to eliminate all considerations of meaning still to be found in *Principia mathematica* and to undertake purely formal investigations: "A mathematical system is any set of strings of recognizable marks in which some of the strings are taken initially and the remainder derived from these by operations performed according to rules which are

independent of any meaning assigned to the marks" (*Lewis 1918*, page 355; in the original the passage is in italics). Or, in Post's words, "we have consistently regarded the system of *Principia* and the generalizations thereof as purely *formal developments*, and we have used whatever instruments of logic or mathematics we found useful for a study of these developments". Earlier in his paper Post had written (*1921*, pages 163–164, or *van Heijenoort 1967*, page 265):

> We here wish to emphasize that the theorems of this paper are *about* the logic of propositions but are *not included* therein. [...
> We shall show] that the postulates of *Principia* are capable of developing the complete system of the logic of propositions without ever introducing results extraneous to that system—a conclusion that could hardly have been arrived at by the particular processes used in that work.

We have said that Post showed the semantic completeness of the *Principia* formulation of the propositional calculus, and the passage just quoted strongly suggests that Post himself claimed to have done so. Such a claim, however, would seem to conflict with his assertion that he "consistently regarded the system of *Principia* ... as purely *formal developments*". For Post this conflict is resolved (or fudged) by distinguishing between intuitive interpretations or applications of the formal system and manipulations of the formal system. Thus, the truth-table mode of analysis that Post introduced (independently of the *Tractatus*) was not viewed by him as an analysis of logical truth (as it was by Wittgenstein) or as a "semantics". Rather, what he actually did is to show that the systematic manipulation of the uninterpreted truth tables yields a decision procedure for formal provability (*1921*, page 166, or *van Heijenoort 1967*, page 267). Indeed Post's interest in finding a decision procedure for provability was not confined to the propositional calculus. He writes: "Further development suggests ... this general procedure might be extended to other portions of *Principia*, and we hope at some future time to present the beginning of such an attempt" (*1921*, page 164, or *van Heijenoort 1967*, pages 265–266). It is clear from his posthumously published *1941* that in 1920–1921 Post did try to extend his "general procedure" to the *Principia* formulation of quantification theory; that is, he did try to find a "a finite method ... for determining of any enunciation [well-formed formula] in the system whether it was or was not an assertion of the system [provable in the system]" (*Davis 1965*, page 348), fully recognizing that such a method would yield an analogous method for all of *Principia* (see *Davis 1965*, page 348 and footnote 79, page 397). It was this very attempt that led Post to conjecture toward the end of 1921 that there was no such method and that

"a complete symbolic logic is impossible" (*Davis 1965*, page 416), thus foreshadowing Gödel's great incompleteness result by about a decade.

The conflict between Post's official attitude and his model-theoretic considerations is further hidden by a corollary he drew from his decision procedure: Every propositional formula becomes provable if any unprovable formula is taken as an additional primitive formula. Hence the propositional calculus satisfies a purely formal (or "syntactic") notion of completeness, often called Post completeness.

Unlike Post, Bernays (*1926*) explicitly spoke about "universally valid" ("allgemeingültige"), "universally correct" ("allgemein richtige") propositional formulas. And he stated that the system of primitive formulas of the *Principia* propositional calculus is complete "in the sense that from it by the formal rules of inference all universally correct propositional formulas can be obtained" (*1926*, page 307). However, he immediately went on to say that his result can be "sharpened", gave the statement of Post completeness and sketched a proof of it. Instead of truth tables Bernays used conjunctive normal forms, the decision procedure for validity or equivalently, as he says, derivability now being that each conjunct of a valid normal form contains both a propositional letter and its negation.

Throughout much of the 1920s it was not semantic completeness but the decision problem for quantificational validity, a problem emerging from the work of Schröder and Löwenheim, that was the dominant concern in studying quantification theory. In his survey lecture *1927* Bernays did not even mention the semantic completeness of quantification theory, but he did discuss the decision problem for that theory (and the semantic completeness of the propositional calculus). In September 1928, in his lecture at the Bologna International Congress of Mathematicians, Hilbert stated the problem of completeness for number theory in a form that was related to Post completeness for the propositional calculus: if an unprovable formula of number theory is added to the axioms of that theory, then a contradiction can be derived in the extended system (*1929*, page 140). He then said (*1929*, pages 140–141):

> The question of the completeness of the system of logical rules, put in general form, constitutes a problem of theoretical logic. Up till now we have come to the view that these rules suffice only through experiment [probieren]. We are in possession of an actual proof of this in the case of the propositional calculus. In the case of monadic predicate logic we can obtain a proof from the method for the solution of the decision problem.

The first appearance of the question in print was in *Hilbert and Ackermann 1928*. After showing that their formulation of quantification

theory is not Post complete (as no sound formulation of quantification theory can be), Hilbert and Ackermann wrote (page 68): "Whether the axiom system is complete in the sense that from it all logical formulas [logische Formeln] that are correct [richtig] for each domain of individuals can be derived is still an unsolved question. It is only known purely empirically that this axiom system suffices for all applications." (Total clarity was still not at hand. Indeed, taken literally, the question just posed was empty. The expression "logical formulas" was defined on page 54: "Among the formulas that can be proved with the help of the functional calculus those are to be distinguished which contain no individual signs and whose derivation requires as premises no formulas other than the logical primitive formulas. These formulas we shall call for short 'logical formulas'." The extension of the notions "universal validity" ("Allgemeingültigkeit") and "satisfiability" ("Erfüllbarkeit") from propositional formulas to quantificational formulas, an extension that would have permitted a non-circular posing of the question, did not appear until page 72, where the decision problem, called on page 77 "the fundamental problem of mathematical logic", was introduced. The expression "logische Formeln" and its definition disappeared in the second edition of the book, where it was replaced by "allgemeingültige Formeln" (*Hilbert and Ackermann 1938*, page 58).)

In 1929 Gödel, then a student at the University of Vienna, addressed the question put by Hilbert and Ackermann and in his doctoral dissertation (*1929*) answered it affirmatively. The degree was granted on 6 February 1930. Gödel rewrote his dissertation as an article (*1930*), received by the editorial board of *Monatshefte für Mathematik und Physik* on 22 October 1929. The note *1930a* is a short abstract.

2. The introduction to the dissertation

The introduction to the dissertation opens with a long paragraph that places the semantic completeness of quantification theory in the context of the problems and controversies surrounding the axiomatization of (parts of) mathematics. Gödel states that his completeness result provides "in a certain sense a theoretical completion of the usual method for proving consistency" (below, page 61). The usual method consists in exhibiting a model. Now, by Gödel's completeness result and its generalization to an infinite number of formulas, every first-order axiom system either is inconsistent or has a model (although not necessarily the intended model). We thus reach a strict alternative, and this is what Gödel calls the "theoretical completion" of the method.

Gödel then reminds us that for Brouwer consistency does not imply without further argument that a model can be constructed. To this

contention of Brouwer, Gödel says, some will retort that, if an axiom system is consistent, then eo ipso the notions introduced by the system are to be taken as defined and their existence to be secure (a view then generally ascribed to Hilbert, though Gödel does not mention his name, a view, of course, that renders otiose the question of semantic completeness). But, Gödel pursues, this retort can be questioned, for it assumes that no formula in the notation of the system can be proved to be undecidable (that is, neither provable nor refutable in the system). He adduces the example of an axiom system for the real numbers, for which, if undecidability of a formula were established, there would be two non-isomorphic models. On the other hand, he adds, we can prove the isomorphism of any two models of that system. Therefore the thesis that existence is to be directly equated with consistency leads to "difficulties". For

> we cannot at all exclude out of hand, however, a proof of the unsolvability of a problem if we observe that what is at issue here is only unsolvability by certain *precisely stated formal* means of inference. For, all the notions that are considered here (provable, consistent, and so on) have an exact meaning only when we have precisely delimited the means of inference that are admitted.

(These remarks of Gödel are somewhat misleading, since the distinction between first-order and second-order logic is not brought into focus. The argument for the isomorphism of any two models of an axiom system for the reals rests upon the standard interpretation of second-order logic. The distinction, once made explicit, would affect the discussion, but would not undercut Gödel's main concern, namely to suggest that certain philosophical views about the nature of mathematical existence are questionable.)

We saw that Gödel mentions Brouwer's well-known opinion on consistency. In the spring of 1928 Brouwer gave two lectures in Vienna (*1929* and *1930*). These lectures, attended by mathematicians and philosophers, had been arranged by Karl Menger. We have no direct evidence that Gödel attended them (and a letter from Gödel to George W. Corner, dated 19 January 1967, may be taken as indicating that he did not); but his relations with Menger at that time and the very content of the introduction to *1929* lead us to believe that, if he was not present at the lectures, he at least had a first-hand report about them. At the end of the first lecture, delivered on 10 March 1928, Brouwer, arguing against Hilbert, introduced a distinction between correct ("richtige") theories and consistent ("nichtkontradiktorische") theories, the second constituting a larger part of mathematics than the first (*1929*, pages 163–164; see also footnote 1, on page 2 of *1930*, as well as Heyting's comment on

that footnote in note 1 on page 599 of *Brouwer 1975*). There is, clearly, a parallelism between Brouwer's criticism of Hilbert's views and Gödel's misgivings about the sufficiency of consistency, namely, both stress that we cannot assume the solvability of every mathematical problem. But, against Hilbert (again without naming him), Gödel makes the further point that we may be able to prove the unsolvability of some mathematical problem, in the sense that a formula and its negation may be unprovable once we have, as he says, *"precisely stated formal* means of inference" (Gödel's remark is, of course, also directed against Brouwer).

These seminal ideas, adumbrating the incompleteness result, were destined to a great future; but Gödel's introductory remarks in his dissertation were left out of the printed version of the completeness proof (*1930*). From Gödel's acknowledgement of assistance (footnote 1, page 103 below) and *Wang 1981* (footnote 2, page 654) we know that some of the changes made in *1929* during the preparation of *1930* were suggested by Hans Hahn. It is a matter of speculation why the introductory remarks were omitted. Was it on Hahn's advice or because of Gödel's own caution?

3. The proof

The completeness proof is essentially the same in the dissertation and the printed version. Gödel adopts a definite system, which is a variant of the first-order fragment of *Principia* but written in the notation of *Hilbert and Ackermann 1928*, and begins the proof by adapting to this system a model-theoretic result of Skolem. In his clarification (*1920*) of *Löwenheim 1915*, Skolem showed that there is an effective way of associating with each quantificational formula Q a normalized formula F such that

(1) F has the form $\Pi\Sigma M$, where Π is a string of r universal quantifiers ($r \geq 1$), Σ is a string of s existential quantifiers ($s \geq 0$), and M is quantifier-free, and

(2) Q is satisfiable in a given domain if and only if F is.

Gödel (*1929*, Section 4, and *1930*, Theorems III and IV) replaces clause (2) by:

(2′) If F is satisfiable in a given domain, then so is Q, and, if F is refutable (in the system), then so is Q.

Hence he can restrict his attention to formulas in Skolem normal form. Let F be such a formula; consider an infinite sequence $x_0, x_1, \ldots$ of distinct individual variables foreign to F; order the r-tuples of these variables according to increasing sums of the subscripts, equal sums being ordered lexicographically. In M we replace the jth universal variable, $1 \leq j \leq r$, by the jth element of the nth r-tuple, $1 \leq n$, and replace the kth existential variable, $1 \leq k \leq s$, by $x_{(n-1)s+k}$. We thus

obtain the nth substitution instance M_n of M. Let A_n be the conjunction M_1 & M_2 & ... & M_n. We are now confronted with what we shall call here the *basic alternative*: Either, for every n, A_n, which is quantifier-free, is truth-functionally satisfiable or, for some n, it is not.

This method of analyzing a quantificational formula originated in *Löwenheim 1915*; it was refined and generalized by Skolem (*1923a* and *1929*) and further generalized by Herbrand (*1930* and *1931a*). If the first side of the alternative holds, Löwenheim (read charitably), Skolem and Gödel each show, as we shall see, that F is satisfiable in the domain of natural numbers. If the second side holds, Löwenheim and Skolem immediately conclude that F is not satisfiable. Therefore F is either $\aleph_0$-satisfiable or not satisfiable, which is (a form of) the Löwenheim–Skolem theorem. Gödel, however, supplies the simple, but for completeness essential, link to a formal system; he shows that:

(I) For each n, $F \rightarrow (P_n)A_n$ is provable in his system, where $(P_n)A_n$ is the existential closure of A_n (*1929*, page 79 below, and *1930*, Theorem VI).

Hence, if for some n, A_n is not truth-functionally satisfiable, then $\sim F$ is provable in the system. Therefore F is either $\aleph_0$-satisfiable or refutable, and we have the semantic completeness of the system (plus, because of soundness, the Löwenheim–Skolem theorem).

The relation between *Skolem 1923a* and *Gödel 1929* (and *1930*) is complex. Gödel's own assessment, more than thirty years later, is contained in three letters. On 4 October 1963 Gödel wrote to van Heijenoort:

> As far as Skolem's paper is concerned, I think I first read it about the time when I published my completeness paper. That I did not quote it must be due to the fact that either the quotations were taken over from my dissertation or that I did not see the paper before the publication of my work. At any rate I am practically sure that I did not know it when I wrote my dissertation. Otherwise I would have quoted it, since it is much closer to my work than the paper of 1920, which I did quote.

Then on 14 August 1964 Gödel wrote to van Heijenoort:

> As for Skolem, what he could justly claim, but apparently does not claim, is that, in his *1923a* paper, he implicitly proved: "Either A is provable or $\sim A$ is satisfiable" ("provable" taken in an informal sense). However, since he did not clearly formulate this result (nor, apparently, had he made it clear to himself), it seems to have remained completely unknown, as follows from the fact that Hilbert and Ackermann in 1928 do not mention it in connection with their completeness problem.

Finally on 7 December 1967 Gödel wrote to Hao Wang (*1974*, page 8):

> The completeness theorem, mathematically, is indeed an almost trivial consequence of *Skolem 1923a*. However, the fact is that, at that time, nobody (including Skolem himself) drew this conclusion (neither from *Skolem 1923a* nor, as I did, from similar considerations of his own).

Thus, according to Gödel, the only significant difference between *Skolem 1923a* and *Gödel 1929–1930* lies in the replacement of an informal notion of "provable" by a formal one, hence in the establishment of (I)—and in the explicit recognition that there is a question to be answered.

What leads to this reading of *Skolem 1923a*? In the paper Skolem often expresses the assumption that a quantificational formula F is satisfied ("erfüllt") by saying that the formula is consistent ("widerspruchsfrei", "widerspruchlos"). And "consistent" might suggest that $\sim F$ is not provable in some system left unspecified or in informal logic. But exactly the same fungibility of "satisfied" with "consistent" (and their respective cognates) runs throughout *Skolem 1920*. Skolem even gives as Löwenheim's theorem: every first-order expression is either contradictory ("widerspruchsvoll") or already satisfiable ("erfüllbar") in a denumerably infinite domain (*Skolem 1920*, page 3, second sentence; see also Theorem 2, page 6; *van Heijenoort 1967*, pages 254 and 256). Yet Gödel does not attribute to *Skolem 1920* an implicit completeness theorem. Indeed, as we have seen, in his first letter to van Heijenoort he says that *1923a* "is much closer to my work than the paper of *1920*". To see why, we must look at the crucial difference between Skolem's *1920* proof of Löwenheim's theorem and his *1923a* proof. In the former Skolem assumes that the formula F holds over an infinite domain D; he then uses the axiom of choice to replace the existential quantifiers of F by functions, and it is an easy matter to show that F holds over a denumerable subset of D. In the *1923a* proof Skolem explicitly eschews the axiom of choice, since he intends to use the theorem as part of "an investigation in the foundations of set theory" (*Skolem 1923a*, page 220; *van Heijenoort 1967*, page 293). Hence, the first side of the basic alternative comes into play, as it does also for Gödel. That is, Skolem and Gödel both intend to show the following:

(II) If, for each $n \geq 1$, there is a truth-assignment that satisfies A_n, then there is a truth-assignment T that, for each $n \geq 1$, satisfies A_n (and the formula F has a true interpretation over the natural numbers).

For Skolem the critical step in showing (II) is to show (III):

(III) The natural order of the atomic components of the A_n quickly induces an ordering $\prec$ of the truth-assignments satisfying the A_n such

that, for each $n > k \geq 1$, $T \prec T'$ only if $T \upharpoonright k \preceq T' \upharpoonright k$, where $T \upharpoonright k$ and $T' \upharpoonright k$ are, respectively, the restrictions to A_k of satisfying assignments T and T' to A_n. (*1923a*, pages 221–222; *van Heijenoort 1967*, page 294.)

Given (III), Skolem proceeds thus:

(IV) For each $n \geq 1$, there are only a finite number, say e_n, of distinct truth-assignments (with domain A_n) that satisfy A_n; let $T_{1,n}, T_{2,n}, \ldots,$ $T_{e_n,n}$ be in order these satisfying truth-assignments, and for each $j < n$, let $T_{a(j,n),j}$ be the restriction of $T_{1,n}$ to A_j, that is, $T_{a(j,n),j} = T_{1,n} \upharpoonright j$. We get at once, for each $n > k > j \geq 1$, $T_{a(j,k),j} \preceq T_{a(j,n),j}$, since $T_{1,k} \preceq T_{a(k,n),k}$ and $T_{a(k,n),k}$ is an extension of $T_{a(j,n),j}$. Whence, for each $j \geq 1$, since e_j is finite, there is an earliest $n(j)$ such that $a(j, n(j))$ becomes constant, that is, for all $n \geq n(j)$, $a(j, n(j)) = a(j, n)$, and so $T_{a(j,n(j)),j} = T_{a(j,n),j}$.

Hence, if for each $j \geq 1$ we write "T_j" for "$T_{a(j,n(j)),j}$", Skolem has shown the following:

(V) There is an infinite sequence of truth-assignments $T_1, T_2, \ldots, T_j,$ $\ldots$ such that, for each $j \geq 1$, T_j satisfies A_j and is a subpart of T_{j+1}. And the sought after truth-assignment $\mathcal{T}$ is that (infinite) assignment which, for each $j \geq 1$, agrees with T_j on A_j. (Skolem speaks of "solutions", not of "satisfying truth-assignments", writes "L" for "T", "a_j^n" for "$a(j, n)$", has no notation for "$n(j)$", does not explicitly specify $\mathcal{T}$, but says that the sequence $L_{1,1}, L_{1,2}, \ldots$ converges in the logical sense $\ldots$ and that we obtain as "limit" the satisfiability of the formula F over the domain of natural numbers (*1923a*, page 222; *van Heijenoort 1967*, page 294).)

Gödel, too, asserts (V). However, in neither *1929* nor *1930* do we find a detailed argument. His *1930* is particularly brief. After noting that, for each $n \geq 1$, (i) there exists only a finite number of satisfying truth-assignments (with domain A_n) to A_n and (ii) each satisfying assignment to A_{n+1} has a satisfying restriction to A_n, Gödel immediately continues: "it follows by familiar arguments that in this case there exists a sequence of satisfying [truth-assignments]". To what does "familiar arguments" refer? A tempting reply is König's infinity lemma (*König 1926* and *1927*). Undoubtedly, clauses (i) and (ii) fulfill the two premises of König's lemma and the first sentence of (V) is an instance of the conclusion of the lemma. But throughout *1930* Gödel has been meticulous in giving explicit references. Why assume that he made an exception for König? Perhaps then Gödel is referring not to the infinity lemma itself but to the arguments, viewed as common "folklore", that led to this lemma. The more extended discussion in *1929* seems to support such a supposition. Having stated clause (i) and a prima facie stronger form of clause (ii),

(VI) Gödel writes,

at least *one* of the satisfying systems of level 1 must be contained as a part in infinitely many satisfying systems of higher level. For *that* system, consequently, there exists a satisfying system of level 2 which contains it as a part and, moreover, is itself likewise contained as a part in infinitely many systems of higher level. The argument can be repeated indefinitely, and we show, in a familiar manner, the existence of an infinite sequence of satisfying [truth-assignments]. . . .

Here Gödel is remarkably close to König's argument in *1926* and especially to page 122 of *1927*—except that König has no phrase corresponding to "in a familiar manner". Obviously, "in a familiar manner" can refer neither to the infinity lemma nor to König's explicit arguments for this lemma. Rather, it refers to what—as König himself observed in footnotes on pages 122 and 123 of *1926*—underlies König's arguments: some form of choice, that is, in general some application of the axiom of choice or in special circumstances some ordering that obviates the need for the axiom. For Gödel's argument any uniform method of ordering the satisfying truth-assignments to the A_n will do. Indeed, in his *1929* (pages 26 and 27) Skolem explicitly uses the ordering (III) to give a proof of the Löwenheim–Skolem Theorem that has in place of his (IV) an argument exactly like Gödel's (VI). Of course, in contrast to Skolem, Gödel never challenges the use of the axiom of choice. Even more, in the third paragraph of the introduction to *1929* he says the question of completeness is not a foundational problem in the sense that consistency is; therefore, "a restriction on the means of proof does not seem to be more pressing here than for any other mathematical problem" (below, page 65). Objection to the use of classical logic would, he claims, not be justified, since the very statement of completeness involves classical notions and a thoroughly intuitionistic proof of completeness would have to do much more, namely give a decision procedure for quantification theory (below, page 65). (We have already quoted Post as saying that in metamathematical studies he has used "whatever instruments of logic or mathematics we found useful".)

If comment is a measure of interest, then the completeness of quantification theory held absolutely no interest for Skolem. There is not one reference to completeness in the fifty-one papers on logic, dating from 1913 through 1963, collected in *Skolem 1970*. (Skolem, however, in his review (*1938*, in Norwegian) of *Hilbert and Ackermann 1938*, writes that "the main advances that have been made [since the time of the first edition] are mentioned ... Chapter III [contains] proofs of the independence and the completeness of the axiom system".)

Nevertheless, Skolem comes close to a form of completeness in his *1928*. There he proposed a novel use of the basic alternative (adapted

now to prenex formulas). The search for a number n such that A_n is not truth-functionally satisfiable becomes a proof procedure for quantification theory (in the sense of *Quine 1955*, page 141), which supplants formal systems of the Frege–Hilbert type. If we complement Skolem's arguments in *1928* with some points he makes in *1923a* and *1929*, we can extract from these papers an argument that his proof procedure is indeed sound and complete.[a]

Gödel's proof should also be compared with the work of Herbrand on quantification theory, carried out during 1928–1929 (see his *1930* and *1931a*). Herbrand introduced the conjuctive expansion of a formula, which is the generalization of the conjunction A_n to an arbitrary quantificational formula, and dealt with his version of the basic alternative thus:

(1) When the first side of the alternative holds, Herbrand does not draw the conclusion that Löwenheim, Skolem and Gödel draw, namely that F is $\aleph_0$-satisfiable. Herbrand explicitly states[b] that such a conclusion could be drawn, but abstains from doing so because the conclusion would involve both a "principle of choice" and the notion of interpretation in an infinite domain, a non-finitistic principle and a non-finitistic notion that he considers to be alien to metamathematical investigations. The conclusion that Herbrand does draw, and draws finitistically, is that the negation of F is not provable, that is, F is not refutable.[c]

(2) When the second side of the alternative holds, Herbrand shows not only that the negation of F has a proof, but that it has a proof in a formal system for quantification theory in which all proofs are "cut-free" and satisfy the subformula property.

To sum up this comparison of Gödel with Skolem and Herbrand, let us consider the sequence: Frege-type formal system, proof procedure, set-theoretic validity. Skolem connects (at least implicitly) the second and third terms of the sequence, Herbrand the first and second, Gödel the first and third. (Of course, if we combine Herbrand's (2) with *Skolem 1923a* we have completeness for formulas in Skolem normal form, if with *Skolem 1929* we have it for all quantificational formulas.)

---

[a]For further exegesis of *Skolem 1928* see the following: *Feferman and Tarski 1953*; *van Heijenoort 1967*, pp. 508–512; *Wang 1970*, pp. 21–23; *Goldfarb 1971*, pp. 522–526; *Vaught 1974*, pp. 158–159; *Goldfarb 1979*, p. 363.

[b]*1930*, p. 109, or *1971*, p. 165, or *van Heijenoort 1967*, p. 552; *1931a*, p. 53, footnote 1, or *1971*, p. 256, footnote 65; *1931*, p. 4, or *1971*, p. 289, or *van Heijenoort 1967*, p. 623.

[c]On Herbrand's attitude toward completeness see *Dreben 1952* as well as Note H, written by Dreben, in *van Heijenoort 1967*, pp. 578–580, and Note N, written by Dreben and Goldfarb, in *Herbrand 1971*, pp. 265–271; on Herbrand's finitism see *van Heijenoort 1982*, pp. 77–78. On Herbrand's errors in his proof see *Dreben, Andrews and Aanderaa 1963*.

In his *1920* as well as in his *1923a*, Skolem had stated the following generalization of the Löwenheim–Skolem theorem:

A denumerable set of quantificational formulas has a denumerable model if it has a model at all.

Similarly, Gödel gives the following generalization of his completeness theorem:

A denumerable set S of quantificational formulas has a denumerable model if the negation of no finite conjunction of elements of S is provable.

In *1929* (below, page 97) Gödel merely sketches the proof of this generalization, for, as he writes, "the proof is completely analogous to the one given ... for single expressions". Most of the sketch is devoted to appropriately generalizing the construction that underlies the basic alternative and then stating the analogue of (I) above:

(VII) Let $F_1, F_2, \ldots$ be an infinite sequence of formulas in Skolem normal form; then, for each n $(n \geq 1)$,

$$F_1 \ \& \ F_2 \ \& \ \ldots \ \& \ F_n \to (P_n)T_n$$

is provable in the formal system, where $(P_n)T_n$ is the existential closure of the quantifier-free conjunction T_n that is the analogue of the quantifier-free conjunction A_n.

In *1930* (below, page 119) the generalization is labeled Theorem IX and is obtained immediately, by means of the completeness theorem, from Theorem X, which does not appear in *1929* and is known today as the compactness theorem:

A denumerable set S of quantificational formulas has a denumerable model if (and only if) every finite conjunction of elements of S has a model.

(Obviously, Theorem IX, coupled with the soundness of quantification theory, immediately yields Theorem X.) The proof sketched for Theorem X differs essentially from the *1929* sketch for the generalization (Theorem IX) only in one regard. Instead of stating (VII), Gödel writes (below, page 121): "We can easily see that $(P_n)B_n \ldots$ is a consequence of the first n expressions of the system Σ" (where the quantifier-free conjunction B_n is logically equivalent to T_n and Σ is the denumerable set of formulas $F_1, F_2, \ldots$). But, since provability in a formal system is now discarded, Gödel's argument for compactness comes very close to Skolem's (suggested) argument in *1923a* for his generalization of the Löwenheim–Skolem Theorem.

The compactness theorem attracted little attention when it first appeared; but, with the development of model theory and, later, of nonstandard analysis, it assumed more and more importance.

In *1930* (below, page 117), echoing Hilbert and Ackermann (*1928*, pages 80–81), Gödel remarks that the equivalence between validity and provability involves a kind of reduction, for the decision problem, of the nondenumerable to the denumerable "since 'valid' refers to the nondenumerable totality of functions [predicates], while 'provable' presupposes only the denumerable totality of formal proofs". (Indeed, the completeness theorem can be viewed as showing that the set of valid quantificational formulas is recursively enumerable.) Gödel's remark is the only reference to the nondenumerable in *1930*. (In *1929* (below, page 63) Gödel says that "the nondenumerable infinite, however, is not used in the main proof".) The compactness theorem is stated for denumerable sets of formulas; it was later realized that the theorem holds for nondenumerable sets (as was established by Maltsev in his *1936*), and Gödel himself had obtained a kind of compactness for nondenumerable sets of propositional formulas in his *1932c*.

In *1929* and in *1930* Gödel extends his completeness result to first-order logic with identity, a system not considered by Hilbert and Ackermann (who introduce identity as a defined notion in second-order logic), but already taken by Löwenheim as the main system in his *1915* and considered by Hilbert in his *1929a* and *1930b*.

Gödel shows, in both *1929* and *1930*, the independence of the axioms for his system of quantification theory, thus extending Bernays' independence result for the propositional calculus (*1926*) and answering a question posed by Hilbert and Ackermann for their system (*1928*, page 68). Bernays' result, obtained in 1918, was noteworthy in that it marked a sharp break with the Frege–Russell view of logic as universal (see above, page 44). Russell had explicitly questioned whether independence results were achievable for the axioms of logic (*1903*, page 15; *Whitehead and Russell 1910*, page 95, or *1925*, page 91).

Henkin gave (*1949*) a completeness proof for quantification theory that is quite different from that of Gödel. Following a train of thought initiated in part by Lindenbaum (*Tarski 1930*, page 26, or *Tarski 1956*, page 34) and Gödel himself (*1932c*), Henkin started from an arbitrary consistent set of formulas, enlarged it consistently by certain axioms eliminating existential formulas in favor of substitution instances with preassigned constants, then extended it by Lindenbaum's procedure to a maximal consistent set, for which it is easy to exhibit a model. The demonstration is relatively independent of the specific formal system adopted, using only general properties of the proof relation. Moreover, the argument remains about the same in case the initial set of formulas is not denumerable, while Gödel's proof does not extend to such a case (thus, in particular, the argument for the compactness theorem now becomes applicable to the nondenumerable case).

4. Further developments

As we saw above, Post (*1921*) and Bernays (*1926*) proved for the propositional calculus a form of syntactic completeness, namely that, if an unprovable formula is taken as an additional axiom, the resulting system is inconsistent. We also saw that, in September 1928, at the Bologna congress, Hilbert proposed a similar form of completeness for number theory. The text of his lecture, in addition to being published in the proceedings of the congress (*Hilbert 1929*), was reprinted with emendations and additions as *1929a*. In the amended text Hilbert shifts from this kind of completeness for number theory to semantic completeness for quantification theory (with identity). A system for such a logic could be obtained, he remarks, by dropping the number-theoretic axioms and introducing an arbitrary number of predicate letters. "This essentially means that we disregard the ordinal character of the number system and treat this system as any system of objects, to which predicates with one or more arguments can be attached" (*1929a*, page 9, or *1930b*, page 322; on this point see *van Heijenoort 1982*, page 66). He distinguishes formulas "that are not refutable [widerlegbar] through any definite stipulation of the suitable predicates. These formulas represent the valid logical propositions". Hilbert has now adopted the semantic viewpoint and these "not refutable" formulas are those for which there is no falsifying interpretation in any domain. He then comes to semantic completeness: "the question now arises whether all these formulas are provable through the rules of logical inference, including the so-called identity axioms; whether, in other words, the system of the usual logical rules is complete". (This was written after *Hilbert and Ackermann 1928* had already been published.)

Hilbert's goal of a Post-like completeness for number theory is, as we now know by Gödel's incompleteness result, unattainable; but a syntactic counterpart of Gödel's completeness result could perhaps be viewed as "saving" Hilbert's goal as well as emphasizing the deductive closure of quantification theory. Such a counterpart is given by Kleene: The addition to quantification theory of an unprovable quantificational formula as an axiom schema would make number theory, now based on

this enlarged logic, ω-inconsistent (Theorem 36 of *Kleene 1952*, page 395). Kleene's result is a reformulation of work of Bernays. Using the technique of arithmetization of syntax introduced by Gödel in his incompleteness paper, Bernays translates Gödel's completeness proof into number theory (*Hilbert and Bernays 1939*, pages 205–253). The crucial step is to give the number-theoretic analogue of Skolem's *1929* version of Gödel's (VI) (see page 53 above). He is thus able to show that for any consistent first-order axiom system S there is an arithmetical model, that is, a model whose predicates are expressible in first-order arithmetic (and hence the formulas of S, when interpreted by these predicates, are true in first-order arithmetic).

Kleene further shows (*1952*, pages 394–395, Theorem 35) that the arithmetic predicates interpreting S can always be taken in the class $\Sigma_2 \cap \Pi_2$ (in the hierarchy of arithmetical predicates), a result that was refined by Putnam (*1961* and *1965*): The predicates can be taken in Σ_1^*, the smallest class that contains the recursively enumerable predicates and is closed under truth functions. Earlier Kreisel (*1953*) and Mostowski (*1955*) had shown that the predicates could not be restricted to the class of recursive predicates, and Putnam (*1957*) that they could not be restricted to the class $\Sigma_1 \cup \Pi_1$.

5. Acknowledgements

We thank Warren D. Goldfarb for comments he made on a draft of this note.

Burton Dreben and Jean van Heijenoort

The translation of *1929* is by Stefan Bauer-Mengelberg and Jean van Heijenoort, that of *1930* by Stefan Bauer-Mengelberg, and that of *1930a* by John W. Dawson, Jr. The translation of *1930* was read and, after some accommodation, approved by Kurt Gödel; subsequently, minor revisions were introduced by Jean van Heijenoort.

Über die Vollständigkeit des Logikkalküls
(*1929*)

1. Einleitung

Der Hauptgegenstand der folgenden Untersuchungen ist der Beweis der Vollständigkeit des in Russell, *Principia mathematica*, P. I, Nr. 1 und Nr. 10, und ähnlich in Hilbert–Ackermann, *Grundzüge der theoretischen Logik* (zitiert als H. A.), III, § 5, angegebenen Axiomensystems des sogenannten engeren Funktionenkalküls. Dabei soll "Vollständigkeit" bedeuten, daß jede im engeren Funktionenkalkül ausdrückbare allgemein giltige Formel (allgemein giltige Zählaussage nach Löwenheim) sich durch eine endliche Reihe formaler Schlüsse aus den Axiomen deduzieren lässt. Diese Behauptung lässt sich leicht als äquivalent erkennen mit der folgenden: Jedes widerspruchslose nur aus Zählaussagen bestehende Axiomensystem[1] hat eine Realisierung. (Widerspruchslos heißt dabei, daß durch endlich viele formale Schlüsse kein Widerspruch hergeleitet werden kann.) Die zuletzt angegebene Formulierung scheint auch an sich einiges Interesse zu bieten, da die Lösung dieser Frage in gewissem Sinn eine theoretische Ergänzung der für Widerspruchslosigkeitsbeweise üblichen Methode darstellt (natürlich nur für die betrachtete spezielle Art von Axiomensystemen), denn sie würde ja eine Garantie dafür bieten, daß diese Methode in jedem Fall zum Ziele führt, d. h. daß entweder ein Widerspruch sich herstellen oder die Widerspruchslosigkeit durch ein Modell sich beweisen lassen muß.[2] Daß man aus der Widerspruchslosigkeit eines Axiomensystems nicht ohneweiters auf die Konstruierbarkeit eines Modells schließen kann, darauf hat besonders L. E. Brouwer mit Nachdruck hingewiesen. Man könnte aber vielleicht meinen, die Existenz der durch ein Axiomensystem eingeführten Begriffe sei geradezu durch ihre Widerspruchslosigkeit zu definieren und daher ein Beweis von vornehrein abzulehnen. Doch setzt diese Definition (wofern man nur die selbstverständliche Forderung stellt, daß der so eingeführte Existenzbegriff denselben Operationsregeln gehorche als der elementare) offensichtlich das Axiom von der Lösbarkeit jedes mathematischen Problems voraus. Oder genauer gesagt, sie setzt voraus, daß man von keinem Problem die Unlösbarkeit beweisen kann. Denn wäre die Unlösbarkeit irgend eines Problems (etwa aus dem Gebiet der reellen Zahlen) bewiesen, so würde daraus nach obiger Definition die Existenz zweier

---

[1] Nähere Definition folgt in einem späteren Abschnitt.

[2] Allerdings wird das Bestehen dieser Alternative nicht in intuitionistischem Sinn (d. h. durch ein Entscheidungsverfahren) bewiesen, siehe weiter unten.

On the completeness of the calculus of logic
(*1929*)

1. Introduction

The main object of the following investigations is the proof of the completeness of the axiom system for what is called the restricted functional calculus, namely the system given in *Whitehead and Russell 1910*, Part I, *1 and *10, and, in a similar way, in *Hilbert and Ackermann 1928* (hereafter cited as H. A.), III, §5. Here 'completeness' is to mean that every valid formula expressible in the restricted functional calculus (a valid *Zählaussage*, as Löwenheim would say) can be derived from the axioms by means of a finite sequence of formal inferences. This assertion can easily be seen to be equivalent to the following: Every consistent axiom system[1] consisting of only *Zählaussagen* has a realization. (Here 'consistent' means that no contradiction can be derived by means of finitely many formal inferences.) The latter formulation seems also to be of some interest in itself, since the solution of this question represents in a certain sense a theoretical completion of the usual method for proving consistency (only, of course, for the special kind of axiom systems considered here); for it would give us a guarantee that in every case this method leads to its goal, that is, that one must either be able to produce a contradiction or prove the consistency by means of a model.[2] L. E. Brouwer, in particular, has emphatically stressed that from the consistency of an axiom system we cannot conclude without further ado that a model can be constructed. But one might perhaps think that the existence of the notions introduced through an axiom system is to be defined outright by the consistency of the axioms and that, therefore, a proof has to be rejected out of hand. This definition (if only we impose the self-evident requirement that the notion of existence thus introduced obeys the same operation rules as does the elementary one), however, manifestly presupposes the axiom that every mathematical problem is solvable. Or, more precisely, it presupposes that we cannot prove the unsolvability of any problem. For, if the unsolvability of some problem (in the domain of real numbers, say) were proved, then, from the definition above, there

---

[1] A detailed definition is given in a subsequent section.

[2] To be sure, the existence of this alternative is not proved in the intuitionistic sense (that is, through a decision procedure); see below.

nicht isomorpher Realisierungen des Axiomensystems der reellen Zahlen folgen, während man anderseits die Isomorphie je zweier Realisierungen beweisen kann. Nun ist aber ein Beweis der Unlösbarkeit eines Problems durchaus nicht von vorneherein auszuschließen, wenn man bedenkt, daß es sich dabei nur um Unlösbarkeit mit gewissen *genau anzugebenden formalen* Schlußweisen handelt. Denn alle hier in Betracht kommenden Begriffe (beweisbar, widerspruchslos etc.) haben ja nur dann einen exakten Sinn, wenn man die zugelassenen Schlußweisen genau abgrenzt. Diese Überlegungen beanspruchen übrigens nur, die Schwierigkeiten, welche mit einer solchen Definition des Existenzbegriffes verbunden wären, ins rechte Licht zu setzen, ohne über ihre Möglichkeit oder Unmöglichkeit endgiltig etwas zu behaupten.

Ersetzt man den Begriff des logischen Folgens (formal beweisbar in endlich vielen Schritten) durch Implikation im Russellschen Sinn, genauer durch *formale* Implikation, wobei die Variabeln die Grundbegriffe des betreffenden Axiomensystems sind, so folgt die Existenz eines Modells für ein widerspruchsfreies (d. h. jetzt, keinen Widerspruch *implizierendes*) Axiomensystem aus der Tatsache, daß eine falsche Aussage jede andere, also auch jeden Widerspruch impliziert (daraus folgt indirekt sofort die Behauptung).[3]

Zum Schluß noch eine Bemerkung über die in der folgenden Arbeit angewandten *Beweismittel*. In diesen sind keinerlei Beschränkungen gemacht worden. Insbesondere wird vom Satz vom ausgeschlossenen Dritten für unendliche Gesamtheiten wesentlich Gebrauch gemacht (das Überabzählbare wird hingegen im Hauptbeweis nicht verwendet). Es hat vielleicht den Anschein, daß dadurch der ganze Vollständigkeitsbeweis wertlos wird. Denn was bewiesen werden soll, kann ja als eine Art Entscheidbarkeit aufgefaßt werden (jeder Ausdruck des engeren Funktionenkalküls kann entweder durch endlich viele Schlüsse als allgemein giltig erkannt oder seine Allgemeingiltigkeit durch ein Gegenbeispiel widerlegt werden). Anderseits scheint der Satz vom ausgeschlossenen Dritten nichts anderes auszusagen als die Entscheidbarkeit jedes Problems. Doch ist dagegen Folgendes einzuwenden:

1.) Wird der Satz vom ausgeschlossenen Dritten nur von intuitionistischer Seite so interpretiert.

2.) Wird, selbst wenn man diese Interpretation annimmt, damit keineswegs die Lösbarkeit mit *bestimmten* Hilfsmitteln sondern nur mit allen

---

[3]Das scheint zuerst R. Carnap in einer noch unveröffentlichten Arbeit bemerkt zu haben, die er so freundlich war, mir im Manuskript zur Verfügung zu stellen.

would follow the existence of two non-isomorphic realizations of the axiom system for the real numbers, while on the other hand we can prove the isomorphism of any two realizations. We cannot at all exclude out of hand, however, a proof of the unsolvability of a problem if we observe that what is at issue here is only unsolvability by certain *precisely stated formal* means of inference. For, all the notions that are considered here (provable, consistent, and so on) have an exact meaning only when we have precisely delimited the means of inference that are admitted. These reflections, incidentally, are intended only to properly illuminate the difficulties that would be connected with such a definition of the notion of existence, without any definitive assertion being made about its possibility or impossibility.

If we replace the notion of logical consequence (that is, of being formally provable in finitely many steps) by implication in Russell's sense, more precisely, by *formal* implication, where the ⟦functional⟧ variables are the primitive notions of the axiom system in question, then the existence of a model for a consistent axiom system (now taken to mean one that *implies* no contradiction) follows from the fact that a false proposition implies any other, hence also every contradiction (whence the assertion follows at once by indirect argument).[3]

In conclusion, let me make a remark about the *means of proof* used in what follows. Concerning them, no restriction whatsoever has been made. In particular, essential use is made of the principle of the excluded middle for infinite collections (the nondenumerable infinite, however, is not used in the main proof). It might perhaps appear that this would invalidate the entire completeness proof. For what is to be proved can, after all, be viewed as a kind of decidability (every expression of the restricted functional calculus either can be recognized as valid through finitely many inferences or its validity can be refuted by a counterexample). On the other hand, the principle of the excluded middle seems to express nothing other than the decidability of every problem. To this, however, the following objections can be made:

(1) The principle of the excluded middle is interpreted this way only by those of the intuitionistic persuasion;

(2) Even if we accept this interpretation, what is affirmed is the solvability not at all through specified means but only through all means that are

---

[3]This seems to have been noted for the first time by R. Carnap in a hitherto unpublished work, which he was kind enough to put at my disposal in a manuscript form.

überhaupt erdenklichen Hilfsmitteln behauptet,[4] während in der folgenden Arbeit gerade bewiesen wird, daß jeder allgemein giltige Ausdruck sich mit ganz *bestimmten konkret aufgezählten* Schlußregeln deduzieren lasse. Vom intuitionistischen Standpunkt aus würde das ganze Problem überhaupt ein anderes werden, weil schon der Sinn der Aussage: "Ein Relationensystem *erfüllt* einen logischen Ausdruck" (d. h. der durch Einsetzung entstehende Satz ist wahr) ein fundamental anderer wäre. Denn man müßte ja verlangen, daß die in dem Ausdruck vorkommenden Existenzialbehauptungen konstruktiv bewiesen wären. Ferner ist klar, daß ein intuitionistischer Vollständigkeitsbeweis (als Alternative: beweisbar—durch Gegenbeispiele widerlegbar) nur durch Lösung des Entscheidungsproblems der mathematischen Logik geführt werden könnte, während im Folgenden nur eine Transformation dieses Problems, nämlich seine Zurückführung auf die Frage, welche Formeln formal beweisbar sind, beabsichtigt wird. Schließlich ist auch noch zu bedenken, daß das hier behandelte Problem ja nicht erst durch den Grundlagenstreit aufgetaucht ist (wie etwa das Problem der Widerspruchslosigkeit der Mathematik), sondern, auch wenn die inhaltliche Geltung der "naiven" Mathematik niemals angezweifelt worden wäre, *innerhalb* dieser sinnvoll gestellt werden könnte (im Gegensatz z. B. zum Problem der Widerspruchslosigkeit), weshalb eine Einschränkung der Beweismittel nicht dringender zu sein scheint als bei irgend einem anderen mathematischen Problem. Soviel über den Hauptgegenstand der Arbeit.

Als Nebenuntersuchungen werden noch folgende durchgeführt werden:

1.) Wird der Vollständigkeitsbeweis auf den Fall ausgedehnt, daß man Identität als logischen Grundbegriff hinzunimmt.

2.) Wird die Frage der gegenseitigen Unabhängigkeit der Axiome für das zugrundegelegte logische Axiomensystem behandelt.

3.) Wird der Vollständigkeitssatz (jede allgemeingiltige Formel ist beweisbar) auf abzählbare Systeme von Formeln ausgedehnt.

2. Vorbereitende Bemerkungen über das zugrundegelegte logische Axiomensystem und die verwendete Terminologie

Bevor wir zum eigentlichen Thema übergehen, soll einiges über das zugrundegelegte logische Axiomensystem und die angewandten Bezeichnungen gesagt werden. Wir schließen uns dabei im wesentlichen an H.

---

[4]Ob allerdings ein so allgemeiner Lösbarkeitsbegriff und damit die in Rede stehende Interpretation vom Satz des ausgeschlossenen Dritten überhaupt einen Sinn hat, erscheint fraglich.

in any way imaginable,[4] while what is shown below is precisely that every valid expression can be derived through completely *specified, concretely enumerated* inference rules. From the intuitionistic point of view, the entire problem would be a different one, because already the meaning of the statement 'A system of relations *satisfies* a logical expression' (that is, the sentence obtained through substitution is true) would be a fundamentally different one. For we would then have to require that the existential assertions occurring in the expression be constructively proved. It is clear, moreover, that an intuitionistic completeness proof (with the alternative: provable or refutable by counterexamples) could be carried out only through the solution of the decision problem for mathematical logic, while in what follows only a transformation of that problem, namely its reduction to the question which formulas are formally provable, is intended. Finally, we must also consider that it was not the controversy regarding the foundations of mathematics that caused the problem treated here to surface (as was the case, for example, for the problem of the consistency of mathematics); rather, even if it had never been questioned that 'naive' mathematics is correct as to its content, this problem could have been meaningfully posed within this naive mathematics (unlike, for example, the problem of consistency), which is why a restriction on the means of proof does not seem to be more pressing here than for any other mathematical problem. So much for the principal topic of the present work.

The following collateral investigations will be conducted:

(1) The completeness proof will be extended to the case in which identity is added as a fundamental logical notion;

(2) The question of the mutual independence of the axioms will be treated for the logical axiom system adopted;

(3) The completeness theorem (every valid formula is provable) will be extended to denumerable systems of formulas.

2. Preliminary remarks on the logical axiom system adopted and the terminology used

Before we proceed to the subject proper, we must say a few words about the logical axiom system adopted and the notation used. Here we essentially follow H. A. The object of the investigation will be certain combina-

---

[4]It seems questionable, however, whether a notion of solvability that is so sweeping—and, consequently, the interpretation of the principle of the excluded middle that is at issue here—makes any sense at all.

A. an. Den Gegenstand der Untersuchung bilden gewisse Zeichenkombinationen (*logische Ausdrücke*). Diese bauen sich aus den Grundzeichen (*Gegenstandsvariable, Funktionsvariable*,[5] Aussagevariable, das Zeichen für oder ∨, Negation $\overline{}$, alle (x)) in der H. A., III, § 4 angegebenen Weise auf. Wir setzen voraus, daß uns für jede Variablenart abzählbar viele Zeichen zur Verfügung stehen. Es werden auch Ausdrücke, in denen das Zeichen = (in der Verwendungsweise $x = y$; x, y Individuenvariable) vorkommt, betrachtet werden. Sollen diese mitumfaßt werden, so wird das immer durch den Zusatz "im weiteren Sinn" (i. w. S.) angedeutet werden, entsprechend "im engeren Sinn" (i. e. S.). Ferner denken wir uns die Zeichen &, →, ∼, (E), in der üblichen Weise lediglich als *Abkürzungen* eingeführt. Das zugrunde gelegte logische Axiomensystem ist (im wesentlichen) das Russell'sche. D. h.: Folgende Ausdrücke sollen logische *Axiome* heißen:

$$
\begin{aligned}
&1.)\quad X \vee X \to X, &&4.)\quad (X \to Y) \to (Z \vee X \to Z \vee Y),\\
&2.)\quad X \to X \vee Y, &&5.)\quad (x)F(x) \to F(y),\\
&3.)\quad X \vee Y \to Y \vee X, &&6.)\quad (x)[X \vee F(x)] \to X \vee (x)F(x).
\end{aligned}
$$

Unter *Axiomen* i. w. S. sollen 1 bis 6, sowie

$$
7.)\quad x = x, \qquad\qquad 8.)\quad x = y \to [F(x) \to F(y)].^{6}
$$

Als Schlußregeln gelten die folgenden:

1.) Das Schlußschema.

2.) Die Einsetzungsregel für Aussage- und Funktionsvariable.

3.) Aus $A(x)$ darf $(x)A(x)$ geschlossen werden.

4.) Individuenvariable (freie oder gebundene) dürfen beliebig anders bezeichnet werden.[7]

Aus den Schlußregeln ergibt sich der Sinn von: "*beweisbar*" und "*widerlegbar*" (A ist widerlegbar soll bedeuten $\overline{A}$ ist beweisbar). Ein Ausdruck, in dem sämtliche Präfixe am Anfang stehen, soll "*Normalausdruck*" heißen. Ein Ausdruck, der keine anderen als Aussagevariable enthält, "*Aussageformel*". Die bisher eingeführten Begriffe haben das Gemeinsame, daß sie sich lediglich auf die Zeichen als räumliche Figuren beziehen. Ihnen stehen gegenüber solche, bei denen auf die *Bedeutung* der Formeln Bezug genommen wird. Sei A irgend ein logischer Ausdruck, der die Funktionsvariabeln $F_1, F_2, \ldots, F_k$, die freien Individuenvariabeln $x_1, x_2, \ldots, x_l$ und die Aussagevariabeln $X_1, X_2, \ldots, X_m$ und sonst nur gebundene Variable

---

[5]Die verschiedenstelligen Relationen sollen, wenn es die Deutlichkeit fordert, durch obere Indizes unterschieden werden.

[6]Die andern bisher aufgestellten Axiomensysteme (Frege, Bernays) unterscheiden sich vom Russellschen nur unwesentlich, so daß sich der Vollständigkeitsbeweis sofort auf sie überträgt.

[7]Bei 2.) und 4.) müssen gewisse Kautelen hinzugefügt werden, siehe H. A., III, § 5.

tions of signs (*logical expressions*). These are constructed from the primitive signs (*individual variables, functional variables,*[5] propositional variables, the signs ∨ for 'or', ⎯ for negation, (*x*) for 'all') in the manner given in H. A., III, §4. We assume that we have at our disposal denumerably many signs for each kind of variable. We shall also consider expressions in which the sign = (as used in '*x* = *y*', *x* and *y* being individual variables) occurs. Whenever such expressions are permitted, this will always be indicated by the addition 'in the extended sense' (i. t. e. s.), as opposed to 'in the restricted sense' (i. t. r. s.). Further, we consider that the signs &, →, ∼, (*E*) have been introduced in the usual way, purely as *abbreviations*. The logical axiom system adopted is (essentially) that of Russell. That is, the following expressions are to be called logical *axioms*:

(1) $X \vee X \to X$, (4) $(X \to Y) \to (Z \vee X \to Z \vee Y)$,
(2) $X \to X \vee Y$, (5) $(x)F(x) \to F(y)$,
(3) $X \vee Y \to Y \vee X$, (6) $(x)[X \vee F(x)] \to X \vee (x)F(x)$.

The *axioms* i. t. e. s. are (1) through (6), as well as

(7) $x = x$, (8) $x = y \to [F(x) \to F(y)]$.[6]

The rules of inference shall be the following:
(1) The inference schema ⟦that is, the rule of detachment⟧;
(2) The rule of substitution for propositional and functional variables;
(3) From $A(x)$, to infer $(x)A(x)$;
(4) (Free or bound) individual variables can be changed at will.[7]

The rules of inference yield the meaning of '*provable*' and '*refutable*' ('*A* is refutable' shall mean that $\overline{A}$ is provable). An expression in which all the quantifiers occur at the beginning ⟦and in which their scope is the entire formula⟧ is said to be a *normal expression*. An expression that contains no variables other than propositional variables is said to be a *propositional formula*. What the notions thus far introduced have in common is that, in the relation they bear to signs, these enter purely as figures in space. In contrast, there are those notions that depend upon the *meaning* of the formulas. Let A be any logical expression that contains the functional variables $F_1, F_2, \ldots, F_k$, the free individual variables $x_1, x_2, \ldots, x_l$, the propositional variables $X_1, X_2, \ldots, X_m$, and, otherwise, only bound variables. Let S be a system of functions, $f_1, f_2, \ldots, f_k$ (all defined in

---

[5]When clarity requires it, many-place relations are to be distinguished by superscripts.

[6]The other axiom systems set up thus far (Frege, Bernays) differ from Russell's only in inessential points, so that the completeness proof can immediately be carried over to these systems.

[7]In (2) and (4) certain provisos have to be added; see H. A., III, §5.

enthält. Wir sagen von einem System (sämtlich in demselben Denkbereich[8] definierter) Funktionen, $f_1, f_2, \ldots, f_k$, und (ebenfalls demselben Denkbereich angehörenden) Individuen, $a_1, a_2, \ldots, a_l$, sowie Aussagen, A_1, $A_2, \ldots, A_m$ —von diesem System

$$S = (f_1, f_2, \ldots, f_k; \; a_1, a_2, \ldots, a_l; \; A_1, A_2, \ldots, A_m)$$

sagen wir, daß es den logischen Ausdruck *erfülle*, wenn es in denselben eingesetzt einen (in dem betreffenden Denkbereich) wahren Satz ergibt.[9] Daraus ergibt sich ohneweiters, was unter *erfüllbar in einem bestimmten Denkbereich, erfüllbar* schlechthin (= es gibt einen Denkbereich, in dem der Ausdruck erfüllbar ist), allgemein giltig in einem bestimmten Denkbereich (= Negation nicht erfüllbar), *allgemein giltig* schlechthin verstanden werden soll.

Zur Behandlung axiomatischer Fragen sind noch einige weitere Bezeichnungen erforderlich (sie werden in der folgenden Arbeit erst von Nr. 8 an verwendet). Wir erweitern die Grundzeichen durch eine abzählbare Reihe von *Individuenkonstanten (Namen)* $a, b, c, \ldots$ und ebenso von *Funktionskonstanten (Begriffszeichen)* $f, g, h, \ldots$. Unter einem *Ausdruck* (i. e. S. und i. w. S.) soll eine mit Hilfe der bisher eingeführten Grundzeichen in bekannter Weise aufgebaute Zeichenverbindung verstanden werden. Den einen Extremfall der Ausdrücke bilden die logischen (in denen keine Konstante vorkommen, s. o.), den anderen die *Zählausdrücke* (= Ausdrücke *ohne* Funktions- und Aussagevariable, in denen sämtliche Individuenvariable gebunden sind). *Zählaxiomensystem* soll ein (endliches oder unendliches) System von Zählausdrücken heißen.[10] *Sinnvoll* für ein solches soll ein Zählausdruck heißen, in dem keine anderen Funktions- und Individualkonstanten vorkommen als in dem Axiomensystem selbst. Aus einem bestimmten Axiomensystem *deduzierbar* soll ein Ausdruck heißen, wenn er aus den logischen Axiomen *und* den das Axiomensystem konstituierenden Aussagen formal hergeleitet werden kann. Ohneweiters ist dann klar, was unter *widerspruchslos, Erfüllung (Realisierung), erfüllbar* etc., zu verstehen ist. Die meisten hier angeführten Begriffe zerfallen in solche i. w. S. und i. e. S., je nachdem ob Identität als Grundbegriff zugelassen wird oder nicht.

---

[8]Denkbereich soll stets heissen *nicht leerer* Denkbereich (sonst gilt ja Axiom 5 nicht).

[9]Man muß natürlich eine Konvention darüber treffen, in welcher Reihenfolge die f, a, A in den Ausdruck eingesetzt werden sollen.

[10]Als Beispiel kann etwa das Hilbertsche Axiomensystem der Geometrie ohne die Stetigkeitsaxiome genommen werden.

the same universal domain[8]), and of individuals (belonging to the same domain), $a_1, a_2, \ldots, a_l$, as well as propositional constants, $A_1, A_2, \ldots, A_m$. We say that this system, namely

$$(f_1, f_2, \ldots, f_k; \ a_1, a_2, \ldots, a_l; \ A_1, A_2, \ldots, A_m),$$

satisfies the logical expression if it yields a proposition that is true (in the domain in question) when it is substituted in the expression.[9] From this we see at once what we must understand by *satisfiable in a certain domain*, by *satisfiable* alone (there is a domain in which the expression is satisfiable), by *valid in a certain domain* (the negation is not satisfiable), and by *valid* alone.

In order to deal with axiomatic questions, we require some additional notational conventions (in the present work these will be used only from Section 8 onward). To the primitive signs we adjoin a denumerable sequence of *individual constants* (*names*), a, b, $c, \ldots$, and one of *functional constants* (*signs for notions*), f, g, $h, \ldots$. By an *expression* (i. t. r. s. and i. t. e. s.) we are to understand a sign sequence constructed in the well-known manner by means of the primitive signs introduced thus far. At one extreme, we have the logical expressions (in which no constants occur—see above); at the other, we have the applied expressions (that is, expressions *without* functional or propositional variables, in which all individual variables are bound). A (finite or infinite) system of applied expressions is an *applied axiom system*.[10] An applied expression in which no functional or individual constant occurs other than those in the axiom system is said to be *meaningful* for the system. An expression is said to be *derivable* from a given axiom system if it can be formally derived from the logical axioms *and* the propositions constituting the axiom system. Then, what is to be understood by *consistent, satisfaction* (*realization*), *satisfiable*, and so on, is immediately clear. Most notions introduced here can be taken i. t. e. s. or i. t. r. s., according as identity is taken as a basic notion or not.

---

[8]The universal domain must always be understood as being *non-empty* (otherwise Axiom 5 does not hold).

[9]We must, of course, make a stipulation on the order in which the f, a, A are to be substituted in the expression.

[10]Hilbert's axiom system for geometry, without the continuity axioms, can perhaps be taken as an example.

3. Zusammenstellung der im Folgenden verwendeten Sätze aus dem Funktionenkalkül

Zum Beweise des Vollständigkeitssatzes wird eine Reihe von Tatsachen aus dem Funktionenkalkül (i. e. S.) verwendet, die zum Teil schon in H. A. bewiesen sind und die hier kurz zusammengestellt seien (bez. der Beweise kann auf Hilbert–Ackermann verwiesen werden, da das dort verwendete Axiomensystem sich ohneweiters aus dem Russellschen ergibt):

1.) Jede Aussageformel ist entweder widerlegbar oder erfüllbar (denn das System der Axiome 1 bis 4 ist ja vollständig sogar im schärferen Sinn—H. A., I, § 13).

2.) Die Einsetzungsregel—H. A., III, § 7: Ist $A \sim A'$ beweisbar und geht B' aus B hervor, indem man in B A durch A' ersetzt (sei es an allen, sei es nur an einigen Stellen), so ist $B \sim B'$ beweisbar.

3.) Zu jedem logischen Ausdruck A gibt es einen Normalausdruck N, so daß $A \sim N$ eine beweisbare Formel ist—H. A., III, § 8.

4.) Jeder Ausdruck der Form

$$(Ex_1)(Ex_2)\ldots(Ex_n)A(x_1, x_2, \ldots, x_n)$$

läßt sich als äquivalent beweisen mit jedem anderen, in dem die (Ex)-Präfixe irgendwie permutiert sind.[11]

5.) Sei F ein Ausdruck, der die freien Variabeln $x_1, x_2, \ldots, x_n$, G einer, der die freien Variabeln $y_1, y_2, \ldots, y_m$ enthält, ferner soll kein x in G und kein y in F vorkommen (frei oder gebunden). Bedeutet dann (p_i) eines der Präfixe (x_i) oder (Ex_i), ebenso (q_i) eines der Präfixe (y_i) oder (Ey_i), dann ist jeder Ausdruck der Form

$$(p_1)\ldots(p_n)F(x_1, x_2, \ldots, x_n) \,\&\, (q_1)\ldots(q_m)G(y_1, y_2, \ldots, y_m)$$

als äquivalent beweisbar[12] mit jedem Ausdruck der Form

$$(P)[F(x_1, x_2, \ldots, x_n) \,\&\, G(y_1, y_2, \ldots, y_m)],$$

wobei das Präfix P aus allen p_i und allen q_i zusammengesetzt ist, die Reihenfolge der p_i untereinander und der q_i untereinander beibehalten, dagegen die Reihenfolge der p_i gegenüber den q_i ganz willkürlich festgesetzt wird. Der Beweis ergibt sich leicht durch mehrmalige Anwendung der

---

[11] Der Beweis ergibt sich durch mehrmalige Anwendung des entsprechenden Satzes für zwei Variable.

[12] A ist als äquivalent beweisbar mit B, soll immer heißen: Die Formel $A \sim B$ ist beweisbar.

3. Summary of the theorems, from the functional calculus, that will be used further on

In the proof of the completeness theorem we shall use a number of facts taken from the functional calculus (i. t. r. s.); they are in part already proved in H. A. and we give them here in a brief summary (for the proofs, we can refer the reader to H. A., since the axiom system used there results immediately from Russell's system):

(1) Every propositional formula is either refutable or satisfiable (the system consisting of Axioms 1–4 is complete even in the stricter sense; see H. A., I, §13).

(2) The replacement rule of H. A., III, §7: If $A \sim A'$ is provable and B' is obtained from B when, in B, A is replaced by A' (whether at all or only at some occurrences), then $B \sim B'$ is provable.

(3) For every logical expression A there is a normal expression N such that $A \sim N$ is a provable formula (H. A., III, §8).

(4) Every expression of the form

$$(Ex_1)(Ex_2)\ldots(Ex_n)A(x_1, x_2, \ldots, x_n)$$

can be proved to be equivalent to any other in which the existential quantifiers are arbitrarily permuted.[11]

(5) Let F be an expression containing the free variables $x_1, x_2, \ldots, x_n$, and G be an expression containing the free variables $y_1, y_2, \ldots, y_m$; moreover, no x_i is to occur in G and no y_i in F (whether free or bound). If (p_i) is one of the quantifiers (x_i) or (Ex_i), and (q_i) one of the quantifiers (y_i) or (Ey_i), then every expression of the form

$$(p_1)\ldots(p_n)F(x_1, x_2, \ldots, x_n) \,\&\, (q_1)\ldots(q_m)G(y_1, y_2, \ldots, y_m)$$

is provably equivalent[12] to every expression of the form

$$(P)[F(x_1, x_2, \ldots, x_n) \,\&\, G(y_1, y_2, \ldots, y_m)],$$

where the prefix P is built up from all the (p_i) and all the (q_i), the relative order of the (p_i) among themselves and the relative order of the (q_i) among themselves are preserved, but the relative order of the (p_i) with respect to

---

[11]This is proved by repeated application of the corresponding theorem for two variables.

[12]That A is provably equivalent to B is always to mean: The formula $A \sim B$ is provable.

Formeln

$$A \,\&\, (x)F(x) \sim (x)[A \,\&\, F(x)]$$

(entsprechend für E), im Verein mit der Einsetzungsregel, indem man den Wirkungsbereich der Präfixe p_i, q_i von außen beginnend sukzessive über den ganzen Ausdruck erweitert. Dabei hat man bei jedem Schritt die Wahl zwischen einem p_i und einem q_i, wodurch es möglich wird, alle oben umgrenzten Reihenfolgen in P zu erhalten.

6.) Als für unsere Zwecke besonders wichtig erweist sich schließlich folgender Satz: Alle Formeln der Gestalt

$$(x_1)(x_2)\ldots(x_n)F(x_1, x_2, \ldots, x_n)$$
$$\&\,(Ex_1)(Ex_2)\ldots(Ex_n)G(x_1, x_2, \ldots, x_n) \rightarrow$$
$$(Ex_1)(Ex_2)\ldots(Ex_n)[F(x_1, x_2, \ldots, x_n) \,\&\, G(x_1, x_2, \ldots, x_n)]$$

sind beweisbar.

Da der Beweis für $n > 1$ vollkommen analog verläuft, genügt es, ihn für $n = 1$ zu führen.

$$(x)F(x) \rightarrow F(x) \qquad \text{(Axiom 5)}$$
$$F(x) \rightarrow [G(x) \rightarrow F(x) \,\&\, G(x)],$$

also

$$(x)F(x) \rightarrow [G(x) \rightarrow F(x) \,\&\, G(x)]$$
$$(x)F(x) \rightarrow (x)[G(x) \rightarrow F(x) \,\&\, G(x)]$$

$$\text{(Axiom 6 und Regel 3)}$$

$$(x)[G(x) \rightarrow F(x) \,\&\, G(x)] \rightarrow \{(Ex)G(x) \rightarrow (Ex)[F(x) \,\&\, G(x)]\}$$

$$\text{(H. A., Formel 34)}.$$

Aus den beiden letzten Formeln folgt

$$(x)F(x) \rightarrow \{(Ex)G(x) \rightarrow (Ex)[F(x) \,\&\, G(x)]\},$$

was nur eine andere Schreibweise für die zu beweisende Formel ist.

4. Reduktion des Vollständigkeitssatzes auf den entsprechenden Satz für Formeln ersten Grades

Da die logischen Axiome allgemein giltig und die Schlußregeln inhaltlich richtig sind, so ist es klar, daß jede beweisbare Formel allgemein giltig ist. Der jetzt zu beweisende Vollständigkeitssatz behauptet die Umkehrung: *Jeder allgemeingiltige logische Ausdruck ist beweisbar.* Das kann offenbar auch so ausgesprochen werden: *Jeder logische Ausdruck ist entweder*

the (q_i) is set in a completely arbitrary way. The proof is readily obtained by repeated use of the formula

$$A \,\&\, (x)F(x) \sim (x)[A \,\&\, F(x)]$$

and of a similar one for the existential quantifier, together with the replacement rule, so that, beginning from the outside, one successively extends the scope of the quantifiers (p_i) and (q_i) to the whole expression. Here we have, at every step, the choice between a (p_i) and a (q_i), and we are thus able to obtain in P all the sequences allowed above.

(6) Finally, the following theorem is particularly important for our purpose: All formulas of the form

$$(x_1)(x_2)\ldots(x_n)F(x_1, x_2, \ldots, x_n)$$
$$\& \, (Ex_1)(Ex_2)\ldots(Ex_n)G(x_1, x_2, \ldots, x_n) \rightarrow$$
$$(Ex_1)(Ex_2)\ldots(Ex_n)[F(x_1, x_2, \ldots, x_n) \,\&\, G(x_1, x_2, \ldots, x_n)]$$

are provable.

Since the proof is completely similar for $n > 1$, it suffices to give it for $n = 1$:

$$(x)F(x) \rightarrow F(x) \qquad\qquad\qquad \text{(Axiom 5)}$$
$$F(x) \rightarrow [G(x) \rightarrow F(x) \,\&\, G(x)],$$

hence

$$(x)F(x) \rightarrow [G(x) \rightarrow F(x) \,\&\, G(x)]$$
$$(x)F(x) \rightarrow (x)[G(x) \rightarrow F(x) \,\&\, G(x)]$$

$$\text{(Axiom 6 and Rule 3)}$$

$$(x)[G(x) \rightarrow F(x) \,\&\, G(x)] \rightarrow \{(Ex)G(x) \rightarrow (Ex)[F(x) \,\&\, G(x)]\}$$

$$\text{(H. A., Formula 34).}$$

From the last two formulas there follows

$$(x)F(x) \rightarrow \{(Ex)G(x) \rightarrow (Ex)[F(x) \,\&\, G(x)]\},$$

which is just another way of writing the formula to be proved.

4. Reduction of the completeness theorem
to the corresponding theorem for formulas of degree 1

Since the logical axioms are valid and the rules of inference are correct in that they preserve truth, it is clear that every provable formula is valid. The completeness theorem that we must now prove states the converse: *Every*

erfüllbar oder widerlegbar, und in dieser Form soll es bewiesen werden. Der Beweis wird zunächst nur für logische Ausdrücke i. e. S. geführt und vollzieht sich so, daß der zu beweisende Satz schrittweise auf einfachere zurückgeführt wird. Zunächst ist klar, daß man sich auf Normalausdrücke beschränken kann, da ja (nach 3. Abschnitt, Satz 3) jeder logische Ausdruck A mit einem Normalausdruck N als äquivalent beweisbar ist. Ist also N widerlegbar, dann auch A; ist N erfüllbar, dann auch A (weil ja jede beweisbare Äquivalenz auch allgemein giltig ist). Ist also jeder Normalausdruck entweder widerlegbar oder erfüllbar, dann gilt dasselbe von *jedem* logischen Ausdruck. Man kann weiter voraussetzen, daß der fragliche Ausdruck keine freien Individuenvariabeln enthält. Denn enthält A etwa die freien Variabeln $x_1, x_2, \ldots, x_k$, dann enthält

$$(Ex_1)(Ex_2) \ldots (Ex_k) A(x_1, x_2, \ldots, x_k)$$

keine freien Variabeln mehr und ist *gleichzeitig* mit A sowohl erfüllbar (nach obiger Definition des Erfülltseins) als auch widerlegbar. Das letztere, weil aus

$$\overline{(Ex_1)(Ex_2) \ldots (Ex_k) A}$$

$(x_1)(x_2) \ldots (x_k)\overline{A}$ und daraus nach Axiom 5 $\overline{A}$ beweisbar ist. Man kann weiter annehmen, daß das Präfix des fraglichen Ausdrucks (ein solches muß ja vorhanden sein, wenn es sich nicht etwa um eine Aussageformel handelt) mit einem Allzeichen beginnt und einem E-Zeichen endet, denn sollte das etwa in $(P)A$ nicht der Fall sein, so genügt es, statt dessen

$$(x)(P)\{A \mathbin{\&} [F(x) \vee \overline{F(x)}]\}$$

bezw.

$$(P)(Ex)\{A \mathbin{\&} [F(x) \vee \overline{F(x)}]\}^{13}$$

zu betrachten, welche offenbar mit $(P)A$ als äquivalent beweisbar[14] und daher mit ihm *zugleich* erfüllbar oder widerlegbar sind.

Die jetzt noch in Betracht kommenden Ausdrücke sind von der Gestalt

$$(x_{1,1})(x_{1,2}) \ldots (x_{1,r_1})(Ey_{1,1})(Ey_{1,2}) \ldots (Ey_{1,s_1})$$
$$(x_{2,1})(x_{2,2}) \ldots (x_{2,r_2})(Ey_{2,1})(Ey_{2,2}) \ldots (Ey_{2,s_2})$$
$$\cdots \cdots \cdots \cdots \cdots \cdots \cdots$$
$$(x_{k,1})(x_{k,2}) \ldots (x_{k,r_k})(Ey_{k,1})(Ey_{k,2}) \ldots (Ey_{k,s_k})A(x_{l_m}, y_{l_m}).$$

k, d. h. die Anzahl der durch E-Zeichen von einander getrennten Komplexe von Allzeichen wird für den Augenblick der *Grad* des Ausdruckes genannt.

---

[13]x bedeutet dabei eine nicht in $(P)A$ vorkommende Individuenvariable.
[14]Nach 3. Abschnitt, Satz 5.

valid logical expression is provable. Clearly, this can also be expressed thus: *Every logical expression is either satisfiable or refutable,* and we shall prove it in this form. The proof will be given first only for logical expressions i. t. r. s. and will be carried out by the stepwise reduction of the theorem to simpler ones. First, it is clear that we can confine ourselves to normal expressions, since, by Theorem 3 of Section 3, every logical expression A is provably equivalent to a normal expression N. Hence, if N is refutable, so is A; if N is satisfiable, so is A (since every provable equivalence is valid). Hence, if every normal expression is either refutable or satisfiable, then so is *every* logical expression. Moreover, we can assume that the expression in question contains no free individual variables. For, if A contains, say, the free variables $x_1, x_2, \ldots, x_k$, then

$$(Ex_1)(Ex_2)\ldots(Ex_k)A(x_1, x_2, \ldots, x_k)$$

no longer contains any free variables and is satisfiable (by the definition of satisfiability given above) or refutable *according as A is.* It is refutable if A is because, from

$$\overline{(Ex_1)(Ex_2)\ldots(Ex_k)A},$$

$(x_1)(x_2)\ldots(x_k)\overline{A}$ is provable and, by Axiom 5, so is $\overline{A}$. We can further assume that the prefix of the expression in question (and, unless we are dealing with a sentential formula, there must be such a prefix) begins with a universal quantifier and ends with an existential quantifier; for, if that were not the case in $(P)A$, it would suffice to consider

$$(x)(P)\{A \,\&\, [F(x) \vee \overline{F(x)}]\}$$

or

$$(P)(Ex)\{A \,\&\, [F(x) \vee \overline{F(x)}]\}$$

instead,[13] these two formulas clearly being provably equivalent[14] to $(P)A$ and therefore being satisfiable or refutable *according as $(P)A$ is.*

The expressions that we must still consider are of the form

$$(x_{1,1})(x_{1,2})\ldots(x_{1,r_1})(Ey_{1,1})(Ey_{1,2})\ldots(Ey_{1,s_1})$$
$$(x_{2,1})(x_{2,2})\ldots(x_{2,r_2})(Ey_{2,1})(Ey_{2,2})\ldots(Ey_{2,s_2})$$

$$\cdot \ \cdot \ \cdot \ \cdot \ \cdot \ \cdot \ \cdot \ \cdot \ \cdot \ \cdot \ \cdot \ \cdot \ \cdot \ \cdot \ \cdot \ \cdot$$

$$(x_{k,1})(x_{k,2})\ldots(x_{k,r_k})(Ey_{k,1})(Ey_{k,2})\ldots(Ey_{k,s_k})A(x_{l_m}, y_{l_m}).$$

The number k, that is, the number of strings of universal quantifiers separated from one another by existential quantifiers, will for the time being be

---

[13]Here x is an individual variable not occurring in $(P)A$.

[14]By Section 3, Theorem 5.

Es soll jetzt gezeigt werden, daß, wenn jeder Ausdruck vom Grad k entweder erfüllbar oder widerlegbar ist, dasselbe auch für jeden vom Grad $k+1$ gilt, womit offenbar eine Reduktion des Vollständigkeitssatzes auf den Satz: *Jede Formel ersten Grades ist entweder erfüllbar oder widerlegbar*, erreicht sein wird. Sei also $(P)A$ ein beliebiger Ausdruck vom Grad $k + 1$; er läßt sich in der Form schreiben:

$$(x_1)(x_2)\ldots(x_r)(Ey_1)(Ey_2)\ldots(Ey_s)(P')A,$$

wobei das Präfix P' vom Grad k ist. F sei eine in A nicht vorkommende Funktionsvariable; $u_1, u_2, \ldots, u_r,\ v_1, v_2, \ldots, v_s$ nicht in A vorkommende Individuenvariable. Wir bilden den Ausdruck:[15]

$$B = (u_1)(u_2)\ldots(u_r)(Ev_1)(Ev_2)\ldots(Ev_s)F(u_1, u_2, \ldots, u_r;\ v_1, v_2, \ldots, v_s)$$
$$\&\ (x_1)(x_2)\ldots(x_r)(y_1)(y_2)\ldots(y_s)[F(x_1, \ldots, y_s) \rightarrow (P')A].$$

Da in P' die Variabeln $x_1, \ldots, y_s$ nicht vorkommen, so ist durch mehrmalige Anwendung von Axiom 6 und dem entsprechenden Satz für E beweisbar:

$$[F(x_1, \ldots, y_s) \rightarrow (P')A] \sim (P')[F(x_1, \ldots, y_s) \rightarrow A].$$

Folglich ist nach der Einsetzungsregel $B \sim B'$ beweisbar, wobei B' aus B durch Einsetzung der rechten Seite der letzten Äquivalenz an Stelle der linken entsteht. Auf B' kann aber offenbar der Satz 5, Abschnitt 3, angewendet werden. Sei

$$P' = (z_1)\ldots(z_p)(Et_1)\ldots(Et_q)(P''),$$

wo P'' vom Grade $k - 1$ ist und

$$B'' = (u_1)\ldots(u_r)(x_1)\ldots(x_r)(y_1)(y_2)\ldots(y_s)(z_1)\ldots(z_p)$$
$$(Ev_1)\ldots(Ev_s)(Et_1)\ldots(Et_q)(P'')$$
$$\{F(u_1, \ldots, u_r;\ v_1, \ldots, v_s)\ \&\ [F(x_1, \ldots, x_r;\ y_1, \ldots, y_s) \rightarrow A]\}.$$

Dann ist also nach Satz 5, Abschnitt 3, $B' \sim B''$, folglich $B \sim B''$ beweisbar. B'' hat aber den Grad k, ist also entweder erfüllbar oder widerlegbar. Ist es erfüllbar, dann auch $(P)A$, denn offensichtlich ist $B \rightarrow (P)A$ allgemein giltig (auch der formale Beweis macht keine Schwierigkeiten, wird aber hier nicht erfordert). Ist B'' widerlegbar, also $\overline{B''}$ und folglich $\overline{B}$ beweisbar, so ist auch $\overline{(P)A}$ beweisbar. Das folgt durch Einsetzung von $(P')A$ in $\overline{B}$ an Stelle von F. Dadurch geht der vor &

---

[15]Ein ähnliches Verfahren hat Th. Skolem zum Beweise des bekannten nach ihm und Löwenheim benannten Satzes verwendet.

called the *degree* of the expression. We shall show that, if every expression of degree k is either satisfiable or refutable, so is every expression of degree $k + 1$; that, clearly, yields a reduction of the completeness theorem to the theorem: *Every formula of degree 1 is either satisfiable or refutable.* Thus let $(P)A$ be any expression of degree $k + 1$; it can be written in the form

$$(x_1)(x_2)\ldots(x_k)(Ey_1)(Ey_2)\ldots(Ey_s)(P')A,$$

where the prefix P' is of degree k. Let F be a functional variable not occurring in A, and let $u_1, u_2, \ldots, u_r, v_1, v_2, \ldots, v_s$ be individual variables not occurring in A. We form the expression[15]

$$B = (u_1)(u_2)\ldots(u_r)(Ev_1)(Ev_2)\ldots(Ev_s)F(u_1, u_2, \ldots, u_r; v_1, v_2, \ldots, v_s)$$
$$\& (x_1)(x_2)\ldots(x_r)(y_1)(y_2)\ldots(y_s)[F(x_1, \ldots, y_s) \rightarrow (P')A].$$

Since the variables $x_1, \ldots, y_s$ do not occur in P', the formula

$$[F(x_1, \ldots, y_s) \rightarrow (P')A] \sim (P')[F(x_1, \ldots, y_s) \rightarrow A]$$

can be proved by repeated use of Axiom 6 and the corresponding theorem for the existential quantifier. Hence, by the replacement rule, $B \sim B'$ is provable, where B' is obtained from B through the replacement of the left side of the last equivalence by the right one. But, clearly, Theorem 5 of Section 3 can be applied to B'. Let P' be

$$(z_1)\ldots(z_p)(Et_1)\ldots(Et_q)(P''),$$

where P'' is of degree $k - 1$, and let B'' be

$$(u_1)\ldots(u_r)(x_1)\ldots(x_r)(y_1)(y_2)\ldots(y_s)(z_1)\ldots(z_p)$$
$$(Ev_1)\ldots(Ev_s)(Et_1)\ldots(Et_q)(P'')$$
$$\{F(u_1, \ldots, u_r; v_1, \ldots, v_s) \& [F(x_1, \ldots, x_r; y_1, \ldots, y_s) \rightarrow A]\}.$$

Then, by Theorem 5 of Section 3, $B' \sim B''$ is provable, and consequently so is $B \sim B''$. But B'' is of degree k, hence is either satisfiable or refutable. If it is satisfiable, so is $(P)A$, since clearly $B \rightarrow (P)A$ is valid (the formal proof itself presents no difficulties, but is not required here). If B'' is refutable, that is, if $\overline{B''}$, hence $\overline{B}$, is provable, then $\overline{(P)A}$, too, is provable. This follows by the replacement of F in $\overline{B}$ by $(P')A$. The subformula occurring before & goes therewith into $(P)A$, and the remainder (which we shall call T) becomes a tautology. Therefore, $\overline{(P)A} \& T$ and T are provable in the

---

[15]Th. Skolem has used a similar procedure for proving the well-known theorem named for him and Löwenheim.

stehende Formelteil in $(P)A$ über, der Rest (bezeichnet als T) wird tauto-logisch. Es ist also $\overline{(P)A}\,\&\,T$ und T beweisbar nach dem Aussagenkalkül, also auch $\overline{(P)A}$, d. h. $(P)A$ ist widerlegbar. Tatsächlich ist also $(P)A$ entweder erfüllbar oder widerlegbar, w. z. b. w.

5. Beweis des Vollständigkeitssatzes im engeren Sinn

Es genügt also zu zeigen, daß jede Normalformel ersten Grades entweder widerlegbar oder erfüllbar ist. Zu diesem Zweck sind einige Zwischen-betrachtungen erforderlich. Sei

$$(u_1)\ldots(u_r)(Ev_1)\ldots(Ev_s)A(u_1,\ldots,u_r;\; v_1,\ldots,v_s) = (P)A$$

irgend ein Ausdruck ersten Grades (A baut sich aus Elementarausdrücken der Form $F(u_p,u_q,\ldots,v_l,v_m)$ und Aussagevariabeln mit Hilfe von $\vee$, $\&$, $\overline{}$, $\rightarrow$, $\sim$ auf). Ferner denke man sich die aus der Reihe $x_1, x_2, \ldots$ ad infinitum entnommenen r-Tupel auf irgend eine Weise in eine Reihe geord-net. Dann soll für den Augenblick unter *"erster abgeleiteter"* von $(P)A$ der Ausdruck

$$\overset{1.)\;\ldots\quad r.)}{(Ex_1)\ldots(Ex_1)}(Ex_2)\ldots(Ex_{1+s})\,A(\overset{1.)\;\ldots\; r.)}{x_1,\ldots,x_1}, x_2,\ldots,x_{1+s})$$

verstanden werden und allgemein unter *n-ter abgeleiteter* ein Ausdruck der Form

$$(Ex_{i_{1,1}})(Ex_{i_{1,2}})\ldots(Ex_{i_{n,r+s}})$$

$$[A(x_{i_{1,1}}, x_{i_{1,2}}, \ldots, x_{i_{1,r}};\; x_{i_{1,r+1}}, \ldots, x_{i_{1,r+s}}) \qquad \text{1.)}$$

$$\&\, A(x_{i_{2,1}}, x_{i_{2,2}}, \ldots, x_{i_{2,r}};\; x_{i_{2,r+1}}, \ldots, x_{i_{2,r+s}}) \qquad \text{2.)}$$

$$\cdot\;\cdot\;\cdot\;\cdot\;\cdot\;\cdot\;\cdot\;\cdot\;\cdot\;\cdot\;\cdot\;\cdot\;\cdot\;\cdot\;\cdot\;\cdot\;\cdot\;\cdot$$

$$\&\, A(x_{i_{n,1}}, x_{i_{n,2}}, \ldots, x_{i_{n,r}};\; x_{i_{n,r+1}}, \ldots, x_{i_{n,r+s}})]. \qquad \text{n.)}$$

Er wird mit $(P_n)A_n$ bezeichnet. $(P_n)A_n$ ist also das logische Produkt von n Ausdrücken $A(\ldots)$, die sich voneinander nur durch die Bezeichnung der Variabeln unterscheiden, mit einem Präfix, in dem nur E-Zeichen vorkom-men und durch das alle in A_n vorkommenden Variabeln gebunden werden. Die Indizes sollen nach folgender Regel bestimmt werden:

1.) Die im ersten Ausdruck A vorkommenden Indizes $i_{1,1},\ldots,i_{1,r}$; $i_{1,r+1},\ldots,i_{1,r+s}$ sollen einfach die Zahlen $\overset{1)}{1},\overset{2)}{1},\ldots,\overset{r)}{1};\overset{r+1)}{2},\ldots,\overset{r+s.)}{1+s}$ sein.

2.) Die im k-ten Ausdruck A vorkommenden Indizes $i_{k,1},\ldots,i_{k,r}$ sollen das auf $i_{k-1,1},\ldots,i_{k-1,r}$ in der vorausgesetzten Anordnung folgende r-Tupel sein.

propositional calculus; hence, so is $\overline{(P)A}$, and $(P)A$ is refutable. Therefore $(P)A$ is indeed either satisfiable or refutable, q. e. d.

5. Proof of the completeness theorem in the restricted sense

It therefore suffices to show that every normal formula of degree 1 is either refutable or satisfiable. There are a number of points that we must now consider by way of preparation. Let

$$(u_1)\ldots(u_r)(Ev_1)\ldots(Ev_s)A(u_1,\ldots,u_r;\ v_1,\ldots,v_s),$$

which we shall call $(P)A$, be any expression of degree 1 (A is built up from elementary expressions of the form $F(u_p, u_q,\ldots,v_l,v_m)$ and propositional variables by means of $\vee$, $\&$, $\overline{}$, $\rightarrow$, $\sim$). Furthermore, let us imagine that the r-tuples of elements of the infinite sequence $x_1, x_2,\ldots$ are, in some way, ordered as a sequence. Then, for the time being, we shall understand by *first derived expression* of $(P)A$ the expression

$$\underbrace{(Ex_1)\ldots(Ex_1)}_{r \text{ times}}(Ex_2)\ldots(Ex_{1+s})A(\underbrace{x_1,\ldots,x_1}_{r \text{ times}}, x_2,\ldots,x_{1+s})$$

and, in general, by the *nth derived expression* an expression of the form

$$(Ex_{i_{1,1}})(Ex_{i_{1,2}})\ldots(Ex_{i_{n,r+s}})$$

$$[A(x_{i_{1,1}}, x_{i_{1,2}},\ldots,x_{i_{1,r}};\ x_{i_{1,r+1}},\ldots,x_{i_{1,r+s}}) \tag{1}$$

$$\&\ A(x_{i_{2,1}}, x_{i_{2,2}},\ldots,x_{i_{2,r}};\ x_{i_{2,r+1}},\ldots,x_{i_{2,r+s}}) \tag{2}$$

$$\cdots\cdots\cdots\cdots\cdots\cdots\cdots\cdots$$

$$\&\ A(x_{i_{n,1}}, x_{i_{n,2}},\ldots,x_{i_{n,r}};\ x_{i_{n,r+1}},\ldots,x_{i_{n,r+s}})]. \tag{n}$$

Let us call it $(P_n)A_n$. Hence $(P_n)A_n$ is the logical product of n expressions $A(\ldots)$, which differ from one another only by the subscripts of the variables, preceded by a prefix in which only existential quantifiers occur and which binds all the variables occurring in A_n. The subscripts are to be determined by the following rule:

(1) The subscripts $i_{1,1},\ldots,i_{1,r};\ i_{1,r+1},\ldots,i_{1,r+s}$ occurring in the first expression A are simply the numbers $1, 1,\ldots,1;\ 2,\ldots,1+s$, with r initial 1's.

(2) The subscripts $i_{k,1},\ldots,i_{k,r}$ occurring in the kth expression A are to be the r-tuple following $i_{k-1,1},\ldots,i_{k-1,r}$ in the assumed ordering.

3.) Die im k-ten Ausdruck vorkommenden Indizes $i_{k,r+1}, \ldots, i_{k,r+s}$ sollen irgend welche untereinander und von *sämtlichen* in den $k-1$ ersten Ausdrücken A vorkommenden Indizes sowie von sämtlichen $i_{k,1}, \ldots, i_{k,r}$ verschiedene Indizes sein. Um diese Bestimmung eindeutig zu machen wird festgesetzt, daß es von den in Betracht kommenden die mit kleinster Summe sein sollen. Man sieht ohneweiters ein, daß man die zu Anfang vorausgesetzte Anordnung der r-Tupel so einrichten kann, daß jeder der Indizes $i_{k,1}, \ldots, i_{k,r}$ schon in einem der vorhergehenden $k-1$ Ausdrücken A vorkommt (schon jede Ordnung nach steigender Summe erfüllt diese Forderung[16]).

Jetzt ergibt sich, daß für jedes k $\quad (P)A \rightarrow (P_k)A_k$ eine beweisbare Formel ist. Für $k=1$ folgt das in trivialer Weise durch Anwendung von

$$(x)F(x) \rightarrow (Ex)F(x).$$

Angenommen nun, die Behauptung sei für $k=n$ schon bewiesen, also $(P)A \rightarrow (P_n)A_n$ beweisbar. A_{n+1} läßt sich offenbar in der Form schreiben

$$A_n \,\&\, A(x_{i_{n+1,1}}, \ldots, x_{i_{n+1,r}};\ x_{i_{n+1,r+1}}, \ldots, x_{i_{n+1,r+s}}),$$

wobei die $i_{n+1,1}, \ldots, i_{n+1,r+s}$ nach obigen Regeln bestimmt sind (um die Häufung der Indizes zu vermeiden soll diese Variabelnreihe im Folgenden als $z_1, \ldots, z_r; z_{r+1}, \ldots, z_{r+s}$ geschrieben werden). Die $z_{r+1}, \ldots, z_{r+s}$ kommen in A_n, also auch in (P_n), nicht vor. Auch von $z_1, \ldots, z_r$ [17] sind sie verschieden. Dagegen kommen $z_1, z_2, \ldots, z_r$ sämtlich in A_n vor (siehe oben). Dies vorausgesetzt sind folgende Formeln beweisbar:

1.) $$(P_n)A_n \sim (Ez_1) \ldots (Ez_r)(P_n')A. \tag{1}$$

Dabei ist P_n' das Präfix, welches aus P_n durch Weglassung von $(Ez_1) \ldots (Ez_r)$ (die ja sämtlich in ihm vorkommen) entsteht. In P_n' kommt also keine der Variabeln $z_1, \ldots, z_r; z_{r+1}, \ldots, z_{r+s}$ vor. Die Beweisbarkeit von (1) ergibt sich aus Satz 4, Abschnitt 3.

2.) $$(z_1) \ldots (z_r)(Ez_{r+1}) \ldots (Ez_{r+s})A(z_1, \ldots, z_r;\ z_{r+1}, \ldots, z_{r+s}) \,\&\,$$
$$(Ez_1) \ldots (Ez_r)(P_n')A_n \rightarrow$$

---

[16]Denn jede solche Anordnung hat die Eigenschaft, daß mit dem k-ten r-Tupel höchstens eine neue Zahl und zwar die nächst folgende zu den bisherigen hinzukommt. Diese muß aber, wenn sie nicht noch früher vertreten ist, mit $i_{k-1,r+1}$ identisch sein.

[17]Es ist zu beachten, daß diese teilweise identisch sein können!

(3) The subscripts $i_{k,r+1}, \ldots, i_{k,r+s}$, occurring in the kth expression are to be distinct from one another and from *all* subscripts occurring in the first $k - 1$ expressions A, as well as from all $i_{k,1}, \ldots, i_{k,r}$. To determine the subscripts unambiguously, we stipulate that, among those that could possibly be chosen, we take those with the smallest sum. We immediately see that the ordering of the r-tuples that we initially assumed can be arranged so that each of the subscripts $i_{k,1}, \ldots, i_{k,r}$ already occurs in one of the preceding $k - 1$ expressions A (every ordering according to increasing sum already satisfies this requirement[16]).

It now turns out that, for every k, the formula $(P)A \rightarrow (P_k)A_k$ is provable. For $k = 1$, this follows trivially, by means of

$$(x)F(x) \rightarrow (Ex)F(x).$$

Let us now assume that the statement has already been proved for $k = n$, hence that $(P)A \rightarrow (P_n)A_n$ is provable. Clearly, A_{n+1} can be written in the form

$$A_n \,\&\, A(x_{i_{n+1,1}}, \ldots, x_{i_{n+1,r}};\ x_{i_{n+1,r+1}}, \ldots, x_{i_{n+1,r+s}}),$$

where the $i_{n+1,1}, \ldots, i_{n+1,r+s}$ are determined by the rules given above (to avoid the piling up of subscripts, this sequence of variables will from now on be written $z_1, \ldots, z_r;\ z_{r+1}, \ldots, z_{r+s}$). The $z_{r+1}, \ldots, z_{r+s}$ do not occur in A_n, hence not in (P_n) either. They are also different from $z_1, \ldots, z_r$.[17] On the other hand, $z_1, z_2, \ldots, z_r$ all occur in A_n (see above). This being assumed, the following formulas are provable:

1) $\qquad\qquad (P_n)A_n \sim (Ez_1) \ldots (Ez_r)(P_n')A. \qquad\qquad$ (1)

Here P_n' is the prefix that results from P_n when $(Ez_1) \ldots (Ez_r)$ (all of which occur in P_n) are omitted. Hence none of the variables $z_1, \ldots, z_r;$ $z_{r+1}, \ldots, z_{r+s}$ occurs in P_n'. The provability of (1) follows from Theorem 4 of Section 3.

2) $\quad (z_1) \ldots (z_r)(Ez_{r+1}) \ldots (Ez_{r+s})A(z_1, \ldots, z_r;\ z_{r+1}, \ldots, z_{r+s})\ \&$
$\qquad\qquad (Ez_1) \ldots (Ez_r)(P_n')A_n \rightarrow$

---

[16]For every such ordering has the property that, with the kth r-tuple, at most one new number is introduced, and indeed the one that immediately follows those that have been previously introduced. But this number, if it has not yet occurred, must be identical with $i_{k-1,r+1}$.

[17]Observe that these numbers may, in part, be identical!

$$(Ez_1)\ldots(Ez_r)[(Ez_{r+1})\ldots(Ez_{r+s})A(z_1,\ldots,z_{r+s})\,\&\,(P_n')A_n].^{18} \qquad (2)$$

Das ergibt sich aus Satz 6, Abschnitt 3, indem man

$$(Ez_{r+1})\ldots(Ez_{r+s})A(z_1,\ldots,z_r;\ z_{r+1},\ldots,z_{r+s})$$

für F und $(P_n')A_n$ für G setzt.

3.) Ist der in (2) auf das $\rightarrow$ folgende Formelteil als äquivalent erweisbar mit $(P_{n+1})A_{n+1}$. Denn auf

$$(Ez_{r+1})\ldots(Ez_{r+s})A(z_1,\ldots,z_{r+s})\,\&\,(P_n')A_n$$

kann Satz 5, 3. Abschnitt, angewendet werden, da weder die Variablen $z_{r+1},\ldots,z_{r+s}$ in $(P_n')A_n$ noch die in (P_n') enthaltenen Variabeln im ersten Teil vorkommen. Also ist der obige Ausdruck als äquivalent beweisbar mit

$$(Ez_{r+1})\ldots(Ez_{r+s})(P_n')[A(z_1,\ldots,z_{r+s})\,\&\,A_n].$$

Der ganze auf das $\rightarrow$ Zeichen in (2) folgende Formelteil ist also (nach der Einsetzungsregel) als äquivalent beweisbar mit

$$(Q_{n+1})A_n\,\&\,A(z_1,\ldots,z_{r+s}),$$

wobei sich Q_{n+1} höchstens in der Reihenfolge der E-Zeichen von P_{n+1} unterscheidet, also auch als äquivalent beweisbar mit $(P_{n+1})A_{n+1}$. Setzt man in (2) die als äquivalent gefundenen Ausdrücke ein, so sieht man, daß beweisbar ist:

$$(z_1)\ldots(z_r)(Ez_{r+1})\ldots(Ez_{r+s})A\,\&\,[(P_n)A_n\rightarrow(P_{n+1})A_{n+1}]$$

und da natürlich

$$(P)A\rightarrow(z_1)\ldots(z_r)(Ez_{r+1})\ldots(Ez_{r+s})A$$

beweisbar ist (auch wenn mehrere $z_i\,(1\le i\le r)$ identisch sind) so auch

$$(P)A\,\&\,[(P_n)A_n\rightarrow(P_{n+1})A_{n+1}].$$

Da ferner nach induktiver Annahme $(P)A\rightarrow(P_n)A_n$ beweisbar ist, so auch $(P)A\rightarrow(P_{n+1})A_{n+1}$, w. z. b. w.

---

[18]Hier und in den folgenden Formeln hat man sich in den Präfixen jedes von den verschiedenen $z_i\,(1\le i\le r)$ nur *einmal* angeschrieben zu denken.

$$(Ez_1)\ldots(Ez_r)[(Ez_{r+1})\ldots(Ez_{r+s})A(z_1,\ldots,z_{r+s})\mathbin{\&}(P_n')A_n].^{18} \qquad (2)$$

This follows from Theorem 6 of Section 3 when we put

$$(Ez_{r+1})\ldots(Ez_{r+s})A(z_1,\ldots,z_r;\ z_{r+1},\ldots,z_{r+s})$$

for F and $(P_n')A_n$ for G.

3) The subformula that, in (2), follows the sign $\rightarrow$ is provably equivalent to $(P_{n+1})A_{n+1}$. For Theorem 5 of Section 3 can be applied to

$$(Ez_{r+1})\ldots(Ez_{r+s})A(z_1,\ldots,z_{r+s})\mathbin{\&}(P_n')A_n,$$

since the variables $z_{r+1},\ldots,z_{r+s}$ do not occur in $(P_n')A_n$ and the variables contained in (P_n') do not occur in the first part. Hence the expression above is provably equivalent to

$$(Ez_{r+1})\ldots(Ez_{r+s})(P_n')[A(z_1,\ldots,z_{r+s})\mathbin{\&}A_n].$$

The entire subformula that follows the sign $\rightarrow$ in (2) is therefore, by the replacement rule, provably equivalent to

$$(Q_{n+1})A_n\mathbin{\&}A(z_1,\ldots,z_{r+s}),$$

where Q_{n+1} differs from P_{n+1} at most in the order of the existential quantifiers; hence it is also provably equivalent to $(P_{n+1})A_{n+1}$. If, in (2), expressions are replaced by those that have been found to be equivalent, we see that

$$(z_1)\ldots(z_r)(Ez_{r+1})\ldots(Ez_{r+s})A\mathbin{\&}\big[(P_n)A_n\rightarrow(P_{n+1})A_{n+1}\big]$$

is provable and, since, of course,

$$P(A)\rightarrow(z_1)\ldots(z_r)(Ez_{r+1})\ldots(Ez_{r+s})A$$

is provable (even if several z_i, $1 < i \le r$, are identical), so is

$$(P)A\mathbin{\&}\big[(P_n)A_n\rightarrow(P_{n+1})A_{n+1}\big].$$

Since further, by the induction assumption, $(P)A\rightarrow(P_n)A_n$ is provable, so is $(P)A\rightarrow(P_{n+1})A_{n+1}$, q. e. d.

---

[18] Here and in the subsequent formulas we have to imagine that each of the various z_i, $1 \le i \le r$, is written only *once*.

Mögen nun in A die Funktionsvariabeln $F_1^{i_1}, \ldots, F_p^{i_p}$ sowie die Aussage-variabeln $X_1, \ldots, X_q$ vorkommen (die Individuenvariabeln sind ja sämtlich gebunden). Wir bezeichnen als ein zu $(P)A$ gehöriges *Erfüllungssystem n-ter Stufe* ein im Bereich der natürlichen Zahlen $1 \leq x \leq 1 + ns$ definiertes System von Funktionen $f_1^{i_1}, \ldots, f_p^{i_p}$, sowie Aussagen $U_1, \ldots, U_q$, von der Eigenschaft, daß, wenn man in A_n an Stelle der Funktionsvariabeln $F_1, \ldots, F_p$ die Funktionen $f_1, \ldots, f_p$, an Stelle der Aussagevariabeln $X_1, \ldots, X_q$ die Aussagen $U_1, \ldots, U_q$ und an Stelle jedes x_i die entsprechende Ziffer i einsetzt,[19] ein wahrer Satz entsteht. Ferner bezeichnen wir als die zu $(P)A$ gehörige Aussageformel *n-ter Stufe* (C_n) die Aussageformel, welche entsteht, wenn man in A_n an Stelle der Elementarbestandteile $F_k^{i_k}(x, \ldots, x)$ Aussagevariable einsetzt und zwar an Stelle verschiedener Elementarbestandteile (sie mögen sich durch das F oder durch die x unterscheiden) verschiedene Aussagevariable, die auch sämtlich von den in A_n schon vorkommenden $X_1, \ldots, X_q$ verschieden sind. Es ist ersichtlich, daß es dann und nur dann ein Erfüllungssystem n-ter Stufe von $(P)A$ gibt, wenn die zu $(P)A$ gehörige Aussageformel n-ter Stufe erfüllbar ist. (Denn die Wahrheitswerte der Elementarbestandteile $F_l^{i_l}(k_1, \ldots, k_{i_l})$—die k bedeuten hier Ziffern—können ja von einander und von den Wahrheitswerten der X_i völlig unabhängig bestimmt werden.) Mit den jetzt bereitgestellten Hilfsmitteln kann der Beweis des Vollständigkeitssatzes (i. e. S.) zu Ende geführt werden.

Die zu $(P)A$ gehörigen Aussageformeln sind als solche entweder erfüllbar oder widerlegbar (Satz 1, Abschnitt 3) und es sind also nur zwei Fälle denkbar.

1.) Mindestens eine der zugeordneten Aussageformeln (es sei die n-ter Stufe) ist widerlegbar, d. h. $\overline{C}_n$ ist beweisbar. Daraus folgt durch Rückeinsetzung der Elementarbestandteile $F_l(x_i, \ldots, x_k)$ an Stelle der Aussage-variabeln, daß auch $\overline{A}_n$, also (nach Regel 3, Abschnitt 3) auch $(x_1) \ldots (x_{1+ns})\overline{A}_n$ beweisbar ist. Folglich auch

$$\overline{(Ex_1) \ldots (Ex_{1+ns})A_n},$$

d. h. $\overline{(P_n)A_n}$. Da aber (siehe oben) $(P)A \to (P_n)A_n$ beweisbar ist, so auch $\overline{(P)A}$. D. h. *in diesem Fall ist* $(P)A$ *widerlegbar.*

2.) Keine der zugeordneten Aussageformeln ist widerlegbar, d. h. alle sind erfüllbar. Dann gibt es zu $(P)A$ gehörige Erfüllungssysteme beliebig hoher Stufe. Nun wachsen die Denkbereiche der Erfüllungssyteme monoton mit n (sie sind ja $1 \leq x \leq 1 + ns$) und der auf $1 \leq x \leq 1 + ns$ *beschränkte*

---

[19]Man überzeugt sich leicht, daß bei der oben festgesetzten Anordnung der r-Tupel in A genau die Variabeln x_1 bis x_{1+ns} vorkommen, denn mit jedem Schritt kommen genau s neue Variable hinzu und zwar die *nächsten* s.

Assume now that the functional variables $F_1^{i_1}, \ldots, F_p^{i_p}$, as well as the propositional variables $X_1, \ldots, X_q$, occur in A (each individual variable being bound, as we know). A *satisfying system of level n* for $(P)A$ will be a system of functions, $f_1^{i_1}, \ldots, f_p^{i_p}$ (defined in the domain of the natural numbers x, with $1 \leq x \leq 1 + ns$), and of propositions, $U_1, \ldots, U_q$, such that a true proposition results when we put the functions $f_1, \ldots, f_p$ for the functional variables $F_1, \ldots, F_p$, the propositions $U_1, \ldots, U_q$ for the propositional variables $X_1, \ldots, X_q$, and, for each x_i, the corresponding numeral i.[19] Furthermore, C_n, the propositional formula of level n for $(P)A$, will be the propositional formula that results when, in A_n, the basic components $F_k^{i_k}(x, \ldots, x)$ are replaced by propositional variables, basic components that differ (whether in the F or the x's) being replaced by different propositional variables, all of which also differ from the $X_1, \ldots, X_q$ already occurring in A_n. Clearly, there exists a satisfying system of level n for $(P)A$ if and only if the propositional formula of level n associated with $(P)A$ is satisfiable. (For the truth values of the basic components $F_l^{i_l}(k_1, \ldots, k_{i_l})$, where the k's are numerals, can be determined completely independently from each other and from the truth values of the X_i.) With the means that we now have at hand, the proof of the completeness theorem (i. t. r. s.) can be brought to its conclusion.

The propositional formulas associated with $(P)A$ are, as such, either satisfiable or refutable (Theorem 1 of Section 3), and therefore only two cases are conceivable:

(1) At least one of the associated propositional formulas (the one of level n, say) is refutable, that is, $\overline{C_n}$ is provable. From that it follows, if we substitute back the basic components $F_l(x_i, \ldots, x_k)$ for the propositional variables, that $\overline{A_n}$, too, is provable, and (by Rule 3 of Section 3) so is $(x_1) \ldots (x_{1+ns})\overline{A_n}$; therefore, so is

$$\overline{(Ex_1) \ldots (Ex_{1+ns})A_n},$$

that is, $\overline{(P_n)A_n}$. Since, however, $(P)A \rightarrow (P_n)A_n$ is provable (see above), so is $\overline{(P)A}$. That is, *in this case $(P)A$ is refutable.*

(2) None of the associated propositional formulas is refutable, that is, they are all satisfiable. Then there exist satisfying systems of arbitrarily high level for $P(A)$. Now, the domains of the satisfying systems increase monotonically with n (for any element x in them, we have $1 \leq x \leq 1 + ns$) and, in a satisfying system of level higher than n, the *part*

---

[19] We can readily convince ourselves that, with the ordering of the r-tuples that was adopted above, exactly the variables $x_1, \ldots, x_{1+ns}$ occur in A, since at every step exactly s new variables are introduced, namely the *next* s variables.

Teil[20] eines Erfüllungssystems höherer als n-ter Stufe ist ein Erfüllungssystem n-ter Stufe, wie aus der Bildung von A_n durch fortgesetzte &-Verknüpfung hervorgeht. Ferner gibt es sicher nur endlich viele zu $(P)A$ gehörige Erfüllungssysteme n-ter Stufe[21] (in einem endlichen Denkbereich gibt es ja überhaupt nur endlich viele Funktionssysteme $F_1, \ldots, F_p$). Also muß mindestens *eines* der Erfüllungssysteme erster Stufe in unendlich vielen höherer Stufe als Teil enthalten sein. Zu *diesem* gibt es folglich ein Erfüllungssystem zweiter Stufe, in dem es als Teil vorkommt und zwar ein solches, das ebenfalls in unendlich vielen höherer Stufe als Teil enthalten ist. So weiter schließend zeigt man in bekannter Weise die Existenz einer Folge von Erfüllungssystemen $S_1, S_2, \ldots, S_i, \ldots$ ad infinitum, wobei $S_i \subset S_{i+1}$ und S_i von i-ter Stufe ist. (Der Definitionsbereich von S_i ist $1 \leq x \leq 1 + si$.) Wir definieren nun ein Funktionssystem S im Bereich *aller* natürlichen Zahlen—man könnte sagen—als oberen Limes der Reihe $S_1, \ldots, S_i, \ldots$—indem wir festsetzen: Die Relation $g_k^{i_k}$ (für $1 \leq k \leq p$) soll dann und nur dann zwischen den *Zahlen* $z_1, \ldots, z_{i_k}$ bestehen, wenn es in der obigen Reihe ein Erfüllungssystem S_l gibt, in dessen Definitionsbereich $z_1, \ldots, z_{i_k}$ sämtlich vorkommen und für das $f_k^{i_k}(z_1, \ldots, z_{i_k})$ besteht.[22] Ganz entsprechend soll natürlich für die Aussage X_i in S eine Aussage W_i genommen werden, für welche es ein S_l gibt, in dem die entsprechende Aussage U_i denselben Wahrheitswert hat.[23] Es ergibt sich nun ohneweiters, daß der auf $1 \leq x \leq 1 + ns$ beschränkte Teil von S mit S_n identisch, also ein Erfüllungssystem n-ter Stufe ist, daß also, wenn man in A_n an Stelle der $F_k^{i_k}$ die $g_k^{i_k}$ und an Stelle der X_i die W_i, sowie an Stelle der x_i die Zahlen i einsetzt, ein wahrer Satz entsteht (das gilt für alle n.) Daraus folgt aber weiter, daß das System $S = (g_1, \ldots, g_p; W_1, \ldots, W_q)$ den Ausdruck $(P)A$, d. h.

$$(u_1) \ldots (u_r)(Ev_1) \ldots (Ev_s)A(u_1, \ldots, v_s)$$

erfüllt. Dazu ist nur noch zu beweisen, daß es zu jedem r-Tupel von natürlichen Zahlen $k_1, \ldots, k_r$ ein s-Tupel von Zahlen $l_1, \ldots, l_s$ gibt, so daß

$$A'(k_1, \ldots, k_r;\ l_1, \ldots, l_s)$$

---

[20]Unter den auf m beschränkten Teil eines im Denkbereich M ($m \subset M$) definierten Funktionssystems S ist natürlich dasjenige in m definierte Funktionssystem zu verstehen, welches innerhalb m mit S übereinstimmt. Die im Erfüllungssystem außerdem noch vorkommenden Aussagen $U_1, \ldots, U_q$ sollen durch Aussagen gleichen Wahrheitswertes in m ersetzt werden.

[21]Erfüllungssysteme, in denen für die X_i verschiedene Aussagen aber vom gleichen Wahrheitswert eingesetzt werden, sind dabei als identisch zu betrachten.

[22]Natürlich gilt dasselbe dann für *alle* S_l, in deren Definitionsbereich $z_1, \ldots, z_{i_k}$ vorkommen.

[23]Dasselbe gilt dann für *alle* S_l.

restricted[20] by the condition $1 \le x \le 1 + ns$ is a satisfying system of level n, as is apparent from the way A_n is constructed by repeated conjunction. Furthermore, there surely exist only finitely many satisfying systems of level n for $(P)A$[21] (in a finite domain there are, after all, only finitely many systems of functions, $F_1, \ldots, F_p$). Hence at least *one* of the satisfying systems of level 1 must be contained as a part in infinitely many satisfying systems of higher level. For *that* system, consequently, there exists a satisfying system of level 2 which contains it as a part and, moreover, is itself likewise contained as a part in infinitely many systems of higher level. The argument can be repeated indefinitely, and we show, in a familiar manner, the existence of an infinite sequence of satisfying systems, $S_1, S_2, \ldots, S_i, \ldots$, where $S_i \subset S_{i+1}$ and S_i is of level i. (The domain of definition of S_i consists of the x's such that $1 \le x \le 1 + si$.) We now define a system S of functions, in the domain of *all* natural numbers, as the upper limit, so to speak, of the sequence $S_1, \ldots, S_i, \ldots$, by stipulating: the relation $g_k^{i_k}$, for $1 \le k \le p$, is to obtain between the *numbers* $z_1, \ldots, z_{i_k}$ if, in the sequence given above, there is a satisfying system S_l in whose domain of definition all of $z_1, \ldots, z_{i_k}$ occur and for which $f_k^{i_k}(z_1, \ldots, z_{i_k})$ holds.[22] Similarly, we must, of course, take for the proposition X_i in S a proposition W_i for which there exists an S_l in which the corresponding proposition U_i has the same truth value.[23] It now immediately turns out that in S the part restricted by the condition $1 \le x \le 1 + ns$ is identical with S_n, hence is a satisfying system of level n; hence, if in A_n we put the $g_k^{i_k}$ for the $F_k^{i_k}$ and the W_i for the X_i, as well as the numbers i for the x_i, we obtain a true sentence (this holds for all n). From that, however, it further follows that the system $S = (g_1, \ldots, g_p; W_1, \ldots, W_q)$ satisfies the expression $(P)A$, that is, the expression

$$(u_1) \ldots (u_r)(Ev_1) \ldots (Ev_s)A(u_1, \ldots, v_s).$$

To show this, we need merely prove that, for every r-tuple of natural numbers, $k_1, \ldots, k_r$, there exists an s-tuple of numbers, $l_1, \ldots, l_s$, such that

$$A'(k_1, \ldots, k_r;\ l_1, \ldots, l_s)$$

---

[20] By a part restricted to m of a function system S defined in a universal domain M ($m \subset M$) we must, of course, understand the function system defined in m which, within m, coincides with S. Any additional propositional variables, $U_1, \ldots, U_q$, that occur in the satisfying system are to be replaced by propositions having the same truth value in m.

[21] Here we consider as identical those satisfying systems in which different propositions having the same truth values are put for the X_i.

[22] The same then holds, of course, for *all* S_l in whose domain of definition $z, \ldots, z_{i_k}$ occur.

[23] The same then holds for *all* S_l.

eine richtige Aussage wird.[24] Das ist aber leicht einzusehen, denn das r-Tupel $k_1, \ldots, k_r$ tritt ja in der oben eingeführten Reihenfolge an irgend einer Stelle auf (etwa an der n-ten). Wir betrachten den Ausdruck A_n. Er lässt sich schreiben als

$$A_{n-1} \mathbin{\&} A(x_{k_1}, \ldots, x_{k_r};\ x_{t_1}, \ldots, x_{t_s}).^{25}$$

Durch Einsetzung von S und den Ziffern statt der Variabeln entsteht der *wahre* Satz

$$A'_{n-1} \mathbin{\&} A'(k_1, \ldots, k_r;\ t_1, \ldots, t_s).$$

Also ist auch

$$A'(k_1, \ldots, k_r;\ t_1, \ldots, t_s)$$

wahr, d. h. es gibt s Zahlen $t_1, \ldots, t_s$ von der verlangten Eigenschaft. S erfüllt $(P)A$. Tatsächlich ist also $(P)A$ entweder widerlegbar oder erfüllbar, womit der Beweis des Vollständigkeitssatzes (i. e. S.) zu Ende geführt ist. Als Korollar folgt daraus, daß jeder Ausdruck (i. e. S.) entweder widerlegbar oder im *genau* abzählbaren Denkbereich erfüllbar ist.

6. Beweis des Vollständigkeitssatzes im weiteren Sinn

Der Beweis, daß auch das logische Axiomensystem i. w. S. vollständig ist, ergibt sich aus dem Bisherigen leicht. Hier gilt allerdings nicht, daß jeder logische Ausdruck (i. w. S.) entweder widerlegbar oder im *genau* abzählbaren Denkbereich erfüllbar ist (Gegenbeispiel: $(x)(y)(x = y)$), wohl aber in einem *höchstens* abzählbaren Denkbereich. Wir beweisen den Satz hier in der Form: *Jeder logische Ausdruck im weiteren Sinn ist entweder beweisbar oder seine Negation erfüllbar.* Sei A ein logischer Ausdruck i. w. S.[26] Es mögen in ihm die Funktionsvariabeln $F_1^1, \ldots, F_{k_1}^1;\ F_1^2, \ldots, F_{k_2}^2; \ldots$ usw. und sonst keine vorkommen. Wir bilden den Ausdruck

[24]Die durch Einsetzung von S entstehenden Ausdrücke sollen mit Strichen bezeichnet werden.

[25]Mit $t_1, \ldots, t_s$ sollen der Kürze halber die nach obiger Regel bestimmten Indizes bezeichnet werden.

[26]Man kann natürlich wie oben voraussetzen, daß A keine freien Individuenvariabeln enthält.

is a correct proposition.[24] But that can readily be seen, since the r-tuple $k_1, \ldots, k_r$ occurs at some place (at the nth one, say) in the sequence introduced above. We now consider the expression A_n. It can be written as

$$A_{n-1} \,\&\, A(x_{k_1}, \ldots, x_{k_r}; \, x_{t_1}, \ldots, x_{t_s}).^{25}$$

If we substitute S and the numerals for the variables, we obtain the *true* sentence

$$A'_{n-1} \,\&\, A'(k_1, \ldots, k_r; \, t_1, \ldots, t_s).$$

Hence

$$A'(k_1, \ldots, k_r; \, t_1, \ldots, t_s)$$

is also true; that is, there are s numbers, $t_1, \ldots, t_s$, with the required property. The system S satisfies $(P)A$. Hence $(P)A$ is indeed either refutable or satisfiable, which concludes the proof of the completeness theorem (i. t. r. s.). As a corollary, it follows that every expression (i. t. r. s.) is either refutable or satisfiable in an *exactly* denumerable domain.

6. Proof of the completeness theorem in the extended sense

The proof that the logical axiom system i. t. e. s., too, is complete is readily obtained from what precedes. Here, however, it is not the case that every logical expression (i. t. e. s.) is either refutable or satisfiable in an *exactly* denumerable domain (counter-example: $(x)(y)(x = y)$); rather, this will be so in an *at most* denumerable domain. Here we prove the theorem in the form: *For every logical expression in the extended sense, either it is provable or its negation is satisfiable.* Let A be a logical expression i. t. e. s. Assume that the functional variables $F_1^1, \ldots, F_{k_1}^1; \, F_1^2, \ldots, F_{k_2}^2, \ldots$ and so on, but no others, occur in it.[26] We form the expression

---

[24]The expressions obtained from S by substitution are to be marked with the prime symbol.

[25]For the sake of brevity we shall write the subscripts determined by the rule given above as $t_1, \ldots, t_s$.

[26]We can of course assume, as above, that A contains no free individual variables.

$$(x)G(x,x) \,\&\, (x)(y)(z)\{G(x,y) \to [G(x,z) \to G(y,z)]\} \,\&\,$$
$$(x)(y)(z)\{G(x,y) \to [G(z,x) \to G(z,y)]\} \,\&\,$$
$$(x)(y)\{G(x,y) \to [F_1^1(x) \to F_1^1(y)]\} \,\&\,$$
$$(x)(y)\{G(x,y) \to [F_2^1(x) \to F_2^1(y)]\} \,\&\,$$

$$\cdots \cdots \cdots \cdots \cdots \cdots \cdots$$

$$(x)(y)(z)\{G(x,y) \to [F_1^2(x,z) \to F_1^2(y,z)]\} \,\&\,$$
$$(x)(y)(z)\{G(x,y) \to [F_1^2(z,x) \to F_1^2(z,y)]\} \,\&\,$$

$$\cdots \cdots \cdots \cdots \cdots \cdots \cdots$$

Analoge Ausdrücke für alle F_l^m.

Dieser ganze Ausdruck werde mit K' bezeichnet. Ferner bezeichne A' den Ausdruck, der aus A entsteht, wenn man überall das Gleichheitszeichen durch G ersetzt. Die Formel $K' \to A'$ enthält kein =-Zeichen mehr. Für sie gilt also, daß sie entweder beweisbar oder ihre Negation erfüllbar ist (das gilt ja sogar schon für das logische Axiomensystem im engeren Sinn).

Wenn 1.) $K' \to A'$ beweisbar ist, dann offenbar auch A. Man braucht ja nur an Stelle von G in $K' \to A'$ das =-Zeichen einzusetzen (K' möge dadurch in K, A' in A übergehen). Da K eine unmittelbare Folge von Axiom 7 und 8, $K \to A$ nach der Einsetzungsregel beweisbar ist, *ist in diesem Fall A* beweisbar.

Ist 2.) die Negation von $K' \to A'$, d. h. $K' \,\&\, \overline{A'}$ erfüllbar, dann auch $\overline{A}$, wie jetzt bewiesen werden soll. Das erfüllende Funktionssystem von $K' \,\&\, \overline{A'}$ soll mit $[g, f_l^m]$ bezeichnet werden. Offensichtlich ist g auf Grund von K' symmetrisch, reflexiv und transitiv. Der Denkbereich N von (g, f_l^m) [27] zerfällt also durch g in (höchstens abzählbar viele) Klassen $N_1, N_2, \ldots, N_k$, ... Diese Klassen sollen die Elemente eines neuen Denkbereiches (N') bilden und in diesem sollen Funktionen g', $(f_l^m)'$, erklärt werden durch die Festsetzung: $(f_l^m)'(N_{i_1}, \ldots, N_{i_m})$ soll dann und nur dann bestehen, wenn es Elemente des früheren Denkbereichs $(a_{i_1}, \ldots, a_{i_m})$ gibt, so daß $a_{i_k} \in N_{i_k}$ und $f_l^m(a_{i_1}, \ldots, a_{i_m})$ gilt. Dann gelten offenbar folgende Sätze:

1.) g' ist in N' die Identität, d. h. es gilt $g'(N_i, N_k)$ dann und nur dann, wenn N_i und N_k die selbe Klasse sind.

2.) Ersetzt man in irgend einer in N richtigen (bezw. falschen) Zählaussage i. e. S., die keine anderen Funktionskonstanten als g, f_l^m enthält und in der die Individuen $a_1, \ldots, a_n$ aus N (und keine anderen) vorkommen, die $\{g, f_l^m, a_1, \ldots, a_n\}$ durch $g', (f_l^m)'$, und diejenigen (eindeutig bestimmten) N_i, in welchen die a_i enthalten sind, so entsteht dadurch eine in N' richtige

---

[27] N soll als Menge der natürlichen Zahlen angenommen werden.

$$(x)G(x,x) \ \& \ (x)(y)(z)\{G(x,y) \rightarrow [G(x,z) \rightarrow G(y,z)]\} \ \&$$
$$(x)(y)(z)\{G(x,y) \rightarrow [G(z,x) \rightarrow G(z,y)]\} \ \&$$
$$(x)(y)\{G(x,y) \rightarrow [F_1^1(x) \rightarrow F_1^1(y)]\} \ \&$$
$$(x)(y)\{G(x,y) \rightarrow [F_2^1(x) \rightarrow F_2^1(y)]\} \ \&$$

$$\cdot \quad \cdot \quad \cdot \quad \cdot \quad \cdot \quad \cdot \quad \cdot \quad \cdot \quad \cdot \quad \cdot \quad \cdot \quad \cdot \quad \cdot$$

$$(x)(y)(z)\{G(x,y) \rightarrow [F_1^2(x,z) \rightarrow F_1^2(y,z)]\} \ \&$$
$$(x)(y)(z)\{G(x,y) \rightarrow [F_1^2(z,x) \rightarrow F_1^2(z,y)]\} \ \&$$

$$\cdot \quad \cdot \quad \cdot \quad \cdot \quad \cdot \quad \cdot \quad \cdot \quad \cdot \quad \cdot \quad \cdot \quad \cdot \quad \cdot \quad \cdot$$

(We will have similar expressions for all F_l^m.)

Let this entire expression be denoted by K'. Further, let A' be the expression that we obtain from A when we replace the identity sign everywhere by G. The formula $K' \rightarrow A'$ no longer contains the sign $=$. Hence either it is provable or its negation is satisfiable (this already holds, after all, for the logical axiom system in the restricted sense).

If (case 1) $K' \rightarrow A'$ is provable, then clearly so is A. For we need merely put the sign $=$ for G in $K' \rightarrow A'$ (then K' would go into K, A' into A). Since K is an immediate consequence of Axioms 7 and 8, and $K \rightarrow A$ is, by the substitution rule, provable, *in this case A* is provable.

If (case 2) the negation of $K' \rightarrow A'$, that is, $K' \ \& \ \overline{A'}$, is satisfiable, then so is $\overline{A}$, as we shall now prove. The function system satisfying $K' \ \& \ \overline{A'}$ will be denoted by (g, f_l^m). The expression K' being what it is, g clearly is symmetric, reflexive and transitive. Hence g partitions the domain[27] N of (g, f_l^m) into (at most denumerably many) classes, $N_1, N_2, \ldots, N_k, \ldots$. These classes are to form the elements of a new domain, N', and, in this domain, functions g', $(f_l^m)'$ are to be defined through the stipulation: $(f_l^m)'(N_{i_1}, \ldots, N_{i_m})$ is to hold if and only if in the former domain there are elements, namely $a_{i_1}, \ldots, a_{i_m}$, such that $a_{i_k} \ \epsilon \ N_{i_k}$ and $f_l^m(a_{i_1}, \ldots, a_{i_m})$ holds. Then obviously we have the following theorems:

(1) In N', g' is the identity, that is, $g'(N_i, N_k)$ holds if and only if N_i and N_k are the same class.

(2) If, in any applied formula i. t. r. s. that is correct (false) in N, that contains no functional constants other than g, f_l^m, and in which there occur the individuals $a_1, \ldots, a_n$ of N (and no others), we replace the $\{g, f_l^m, a_1, \ldots, a_n\}$ by g', $(f_l^m)'$, and the (uniquely determined) N_i in which the a_i are contained, then we obtain an applied formula that is correct (false) in N'. The statement holds by definition for sentences of the form $f_l^m(a_i, \ldots, a_m)$. But every applied formula that we need consider is

---

[27] N is to be taken as the set of natural numbers.

(bezw. falsche) Zählaussage. Die Behauptung gilt nach Definition für Aussagen der Form $f_l^m(a_1, \ldots, a_m)$. Aus solchen (und eventuell noch konstanten Aussagen) baut sich aber jede in Betracht kommende Zählaussage mit Hilfe von $\overline{}$, $\vee$, ( ) auf und man sieht leicht, daß diese Operationen die zu beweisende Eigenschaft nicht zerstören. Nun gilt die Zählaussage $\overline{A'}$ in N,[28] folglich auch die entsprechende in N'. Da aber g in N' die Identität ist, so gilt $\overline{A}$ in N,[29] $\overline{A}$ ist also erfüllbar und zwar in einem höchstens abzählbaren Denkbereich.

A ist also entweder beweisbar oder seine Negation erfüllbar, womit der Vollständigkeitssatz auch für das logische Axiomensystem i. w. S. bewiesen ist.

7. Beweis der Unabhängigkeit des logischen Axiomensystems

Jetzt soll die Frage der Unabhängigkeit und zwar für das logische Axiomensystem i. w. S. behandelt werden (für das i. e. S. folgt sie ja sofort daraus).

1.) Was die vier ersten (Aussagen) Axiome betrifft, so kann der Bernays'sche Beweis für die Unabhängigkeit der Aussageaxiome *allein* (H. A., I, §13) wörtlich wiederholt werden.[30] Man braucht nur die am angeführten Orte gegebenen arithmetischen Interpretationen auf die Formeln, welche Funktionsvariable und das =-Zeichen enthalten, auszudehnen durch die Festsetzung, daß die Klammerzeichen und die Individuenvariabeln weggelassen und in dem übrig bleibenden Formelteil die Funktionsvariabeln genau wie Aussagevariable behandelt werden sollen. (Für das =-Zeichen darf immer nur der Wert Null eingesetzt werden.) Für die Axiomensysteme, welche aus dem logischen i. w. S. durch Weglassung von Axiom 1, 3 oder 4 entstehen, gilt dann tatsächlich, daß 1.) die Axiome bei beliebiger Einsetzung den Wert 0 ergeben und 2.) diese Eigenschaft durch die Schlußregeln erhalten bleibt, während das jeweils weggelassene Axiom diese Eigenschaft nicht hat. Bei Axiom 2 ergibt sich die Schwierigkeit, daß bei beliebiger Einsetzung *nicht* alle Axiome (außer 2) den Wert Null ergeben (die Ausnahme findet für Axiome 5 und 8 bei der Einsetzung $F = 2$ statt). Doch läßt sich diese Schwierigkeit durch die Festsetzung überwinden, daß

---

[28]Genauer gesagt, es gilt in N die Aussaage, welche aus $\overline{A'}$ durch Einsetzung von $\{g, f_l^m\}$ an Stelle von $\{G, F_l^m\}$ entsteht (bezeichnet mit $\overline{A'}$).

[29]Auch hier hat man sich zunächst die Einsetzung in A vollzogen zu denken $\{(f_l^m)'\}$ an Stelle von $F_l^m\}$.

[30]Es ist ja von vornherein evident, daß keines der Aussagenaxiome durch Hinzunahme der Funktionsaxiome beweisbar werden kann.

built from such propositions (and perhaps also constant propositions) by means of &, ∨, (x), and we readily see that these operations do not destroy the property that has to be proved. Now, the applied formula $\overline{A'}$ holds in N;[28] hence so does the corresponding one in N'. But, since g is the identity in N, $\overline{A}$ holds in N.[29] Hence $\overline{A}$ is satisfiable, and it is so in an at most denumerable domain.

Thus either A is provable or its negation is satisfiable, and thereby the completeness theorem is proved for the logical axiom system i. t. e. s., too.

7. Independence proof for the logical axiom system

We shall now treat the question of independence, dealing with the logical axiom system i. t. e. s. (when its independence is established, that of the system i. t. r. s. immediately follows).

(1) As far as the first four (propositional) axioms are concerned, Bernays' independence proof for the propositional axioms *alone* (H. A., I, §13) can be repeated word for word.[30] The arithmetical interpretations given at the cited place to the formulas containing functional variables and the sign = need merely be extended by the stipulation that parentheses and individual variables are to be omitted and that, in what remains of the formula, functional variables are to be treated exactly like propositional variables. (For the sign =, only the value 0 may be substituted.) For the axiom systems that result from the axiom system i. t. e. s. when Axiom 1, 3, or 4 is omitted, it is then a fact that (1) for any substitution the axioms yield the value 0 and (2) this property is preserved by the rules of inference, while the omitted axiom does not have this property. In the case of Axiom 2 there is the difficulty that, with an arbitrary substitution, *not* all axioms (other than 2) yield the value 0 (the exception occurs for Axioms 5 and 8 with the substitution $F = 2$). We can, however, overcome this difficulty by stipulating that only the values 0 and 1 may be substituted for the functional variables. Then the axioms yield the value 0 for any (allowed) substitution, and this property is preserved when the inference rules

---

[28]More precisely, the proposition (denoted by $\overline{A'}$) that results from $\overline{A'}$ when $\{g, f_l^m\}$ is substituted for $\{G, F_l^m\}$ holds in N.

[29]Here, too, we must first imagine that, in A, $(f_l^m)'$ has been put for F_l^m.

[30]It is immediately evident that no propositional axiom can become provable when the functional axioms are added.

für Funktionsvariable nur die Werte 0 und 1 eingesetzt werden dürfen. Dann ergeben die Axiome bei beliebiger (erlaubter) Einsetzung den Wert 0 und diese Eigenschaft bleibt bei Anwendung der Schlußregeln erhalten.[31]

2.) Auch die Unabhängigkeit der Axiome 5 und 6 sowie der Schlußregeln läßt sich durch arithmetische Interpretationen beweisen. Allen diesen ist Folgendes gemeinsam:

1.) Die sowohl für Funktions- und Aussagevariable zugelassenen Werte sind $0, 1$.

2.) Die Operationen $\overline{}$ und $\vee$ haben die gewöhnliche Wirkung, d. h. $\overline{0} = 1$, $\overline{1} = 0$, $i \vee k \equiv i \times k \pmod 2$.

3.) Für das $=$ Zeichen darf nur der Wert 0 eingesetzt werden.

4.) Die Wirkung von (x), (Ex) hängt nicht ab von der Bezeichnung der darin eingesetzten Variabeln (dies erleidet nur in (d) eine Ausnahme).

Der Unabhängigkeitsbeweis erfolgt dann immer in der gleichen Weise dadurch, daß gezeigt wird: 1.) Die Axiome mit Ausnahme eventuell desjenigen, dessen Unabhängigkeit bewiesen werden soll, ergeben bei beliebiger Einsetzung den Wert 0. 2.) Diese Eigenschaft überträgt sich durch die Schlußregeln (mit Ausnahme eventuell derjenigen, deren Unabhängigkeit bewiesen werden soll). 3.) Es gibt eine im vollständigen Axiomensystem beweisbare Formel, welche diese Eigenschaft nicht hat.

a) Unabhängigkeit von Axiom 5: $(x)0 = (x)1 = 0$, Axiom 5 ergibt für $F = 1$ den Wert 1.

b) Unabhängigkeit von Axiom 6: $(x)1 = (x)0 = 1$ in Ausdrücken, welche *nicht* die Gestalt $(x)A(x)$ haben, d. h. in denen die zuletzt angewandte Operation *nicht* ( ) ist. Formeln der Gestalt $(x)A(x)$ soll bei beliebiger Einsetzung der Wert 0 zugeordnet werden. Die beweisbare Formel

$$A \vee (x)F(x) \rightarrow (x)[A \vee F(x)]$$

ergibt dann für $A = 0$, F beliebig, den Wert 1.

c) Unabhängigkeit von Schlußregel 3: $(x)0 = (x)1 = 1$. *Sämtliche* Axiome ergeben bei dieser Interpretation immer den Wert 0 und Anwendung von Regel 1, 2, 4 ändert daran nichts. Dagegen ergibt genau wie oben die beweisbare Formel

$$A \vee (x)F(x) \rightarrow (x)[A \vee F(x)]$$

---

[31]Daß dies auch bei Anwendung der Einsetzungsregel stimmt, ergibt sich daraus, daß auch jeder Ausdruck, der für eine Funktionsvariable eingesetzt werden kann, bei beliebiger Einsetzung nur die Werte 0 oder 1 annimmt. Denn enthält er ein $\vee$-Zeichen, dann folgt das daraus, daß das Resultat der $\vee$-Verknüpfung (für die betreffende Interpretation—H. A., I, §13) bei beliebiger Wertbestimmung der Glieder nur 0 oder 1 ergeben kann. Enthält er aber kein $\vee$-Zeichen, dann muß er die Gestalt $F, \overline{F}, \overline{\overline{F}}, \ldots$ usw. haben und für diese Ausdrücke ist es erst recht klar, daß sie nur die Werte 0 und 1 annehmen können.

are used.[31]

(2) The independence of Axioms 5 and 6, as well as of the inference rules, can also be proved by means of arithmetical interpretations. All these have the following in common:

(α) The admitted values for functional and propositional variables are 0 and 1;

(β) The operations $^-$ and $\vee$ have the customary effect, that is, $\overline{0} = 1$, $\overline{1} = 0$, $i \vee k \equiv i \times k \pmod 2$;

(γ) For the sign $=$ only the value 0 may be substituted;

(δ) The effect of (x) and (Ex) does not depend on the variables occurring in them (this suffers an exception only in (d)).

The independence proof then always proceeds in the same way, namely, we show: (1) The axioms, with the possible exception of the one whose independence has to be proved, yield the value 0 for any substitution; (2) this property is carried over by the rules of inference (with the exception of the one whose independence is to be proved); (3) there is a formula, provable in the complete axiom system, that does not have this property.

(a) Independence of Axiom 5: $(x)0 = (x)1 = 0$; for $F = 1$, Axiom 5 yields the value 1.

(b) Independence of Axiom 6: $(x)1 = (x)0 = 1$ in expressions that are *not* of the form $(x)A(x)$, that is, in those in which the last operation performed is *not* (x). Formulas of the form $(x)A(x)$ are, for an arbitrary substitution, to be assigned the value 0. Then the provable formula

$$A \vee (x)F(x) \rightarrow (x)[A \vee F(x)]$$

yields the value 1 for $A = 0$ and arbitrary F.

(c) Independence of Rule of inference 3: $(x)0 = (x)1 = 0$. For this interpretation *all* the axioms yield the value 0, and an application of Rule 1, 2 or 4 does not change this. But, exactly as above, the provable formula

$$A \vee (x)F(x) \rightarrow (x)[A \vee F(x)]$$

yields the value 1 for a certain substitution, hence cannot be proved without an application of Rule 3.

(d) Rule of inference 4 (renaming of variables) is, of course, indispensable only when we state the axioms with only *one definite* variable (as was

---

[31]That this is the case also when the substitution rule is used follows from the fact that every expression that can be substituted for a functional variable will, for any substitution, take only the value 0 or 1. For, if such an expression contains the sign $\vee$, this result follows from the fact that a disjunction (for the interpretation in question—H. A., I, §13) can only yield 0 or 1, whatever the values assigned to the terms may be. If, however, it does not contain the sign $\vee$, then it must have the form $F, \overline{F}, \overline{\overline{F}}, \ldots$, and for such an expression it is all the more clear that it can take on only the values 0 and 1.

für eine gewisse Einsetzung den Wert 1, sie kann also nicht ohne Anwendung von Regel 3 bewiesen werden.

d) Schlußregel 4 (Umbezeichnung der Variabeln) ist natürlich nur dann unentbehrlich, wenn man die Axiome nur *in einer bestimmten* Variabeln ausspricht (wie es oben für x geschehen ist). Im entgegengesetzten Fall müßte man unendlich viele Axiome aufstellen. Die jetzt zu gebende arithmetische Interpretation zeigt—über die Unabhängigkeit hinaus—daß Regel 4 nicht durch die schwächere ersetzt werden kann, daß nur die *freien* Variabeln umbezeichnet werden dürfen, $(x)0 = 0$, $(x)1 = 1$, dagegen für jede andere Variable z, $(z)0 = (z)1 = 0$. Schon die Formel $(x)(z)F(x, z) \rightarrow F(u, v)$ ergibt dann für $F = 1$ den Wert 1, ist also nicht ohne Hilfe von Regel 4 beweisbar und dies gilt auch für jede Formel, die daraus durch Umbezeichnen der Variabeln hervorgeht, so daß durch Weglassung von Regel 4 der Kalkül eine *wesentliche* Einbuße erleiden würde.

Daß Regel 1 und 2 nicht überflüssig sind, ist ja vollkommen trivial und die Unabhängigkeit von Axiomen 7 und 8 ergibt sich leicht daraus, daß 7 allein auch für die Allrelation, 8 allein auch für die leere Relation erfüllt ist, dagegen 7 für die leere Relation, 8 für die Allrelation (für einen Denkbereich von mindestens zwei Individuen) nicht erfüllt ist.

8. Erweiterung des Vollständigkeitssatzes auf unendliche Systeme von logischen Ausdrücken und axiomatische Anwendungen

Oben, Abschnitt 4 bis 6, wurde bewiesen, daß jeder logische Ausdruck entweder erfüllbar oder widerlegbar ist. Das gilt auch noch für jedes (höchstens abzählbare) System von logischen Ausdrücken.[32] Auch jedes solche System ist entweder erfüllbar oder widerlegbar. "Erfüllbar" soll dabei heißen: Es gibt in irgend einem Denkbereich ein System von Individuen, Funktionen und Sätzen, welche an Stelle der freien Individuenvariabeln, Funktions- und Aussagevariabeln entsprechend eingesetzt *sämtliche* Ausdrücke des Systems in wahre Sätze überführen. "Widerlegbar" soll ein System von logischen Ausdrücken heißen, wenn es ein *endliches* Teilsystem gibt, dessen logisches Produkt (&-Verknüpfung) widerlegbar ist. Der Beweis verläuft dem oben für einzelne Ausdrücke gegebenen vollkommen analog und soll deshalb nur kurz angedeutet werden. Man zeigt zunächst, daß man sich beim Beweise auf Systeme beschränken kann, in denen sämtliche Ausdrücke Normalausdrücke sind und weiter auf Systeme aus Normalausdrücken ersten Grades. Denn sei $A^1, A^2, \ldots$ irgend ein System von Normalausdrücken, die nicht sämtlich vom ersten Grad sind, dann bilde man das

---

[32]Diese Erweiterung ist besonders für axiomatische Anwendungen wichtig.

done above with x). Otherwise we would have to state infinitely many axioms. The arithmetical interpretation to be given now shows, beyond independence, that Rule 4 cannot be replaced by the weaker rule allowing the renaming of only *free* variables: $(x)0 = 0$, $(x)1 = 1$; but, for any other variable z, $(z)0 = (z)1 = 0$. Then, for $F = 1$, already the formula $(x)(z)F(x, z) \rightarrow F(u, v)$ yields the value 1, hence is not provable except by means of Rule 4, and this is also true of every formula that results from this one when variables are renamed, so that, if Rule 4 were omitted, the calculus would suffer a *substantial* loss.

That Rules 1 and 2 are not superfluous is after all completely trivial, and the independence of Axioms 7 and 8 readily results from the fact that Axiom 7, taken by itself, is satisfied by the universal relation, and Axiom 8, taken by itself, by the empty relation, while it is not the case that Axiom 7 is satisfied by the empty relation or Axiom 8 by the universal relation (for a universal domain of at least two individuals).

8. Extension of the completeness theorem to infinite systems of logical expressions; axiomatic applications

Above, in Sections 4–6, it was proved that every logical expression is either satisfiable or refutable. This holds also for every (at most denumerable) system of logical expressions.[32] Such a system, too, is either satisfiable or refutable. 'Satisfiable' here is to mean the following: In some universal domain there exists a system of individuals, functions and propositions that, when properly put at the place of the free individual variables, the functional variables and the propositional variables, turn *all* expressions of the system into true sentences. A system of logical expressions is said to be refutable if there is a *finite* subsystem whose logical product (conjunction) is refutable. The proof is completely analogous to the one given above for single expressions and will therefore only be sketched briefly. We first show that, for the proof, we can restrict ourselves to systems in which all expressions are normal expressions and, further, to systems of normal expressions of degree 1. For let $A^1, A^2, \ldots$ be any system of normal expressions that are not all of degree 1, and let us form the system $B^1, B^2, \ldots$ by repeated application of the procedure of Section 4 to the corresponding expressions A^i.[33] If now the system (B^i) is satisfiable, then so is the system

---

[32]This extension is especially important for axiomatic applications.

[33]The B^i are then normal expressions of degree 1, but will in general contain a few more functional variables than the corresponding A^i.

System $B^1, B^2, \ldots$ usw. durch mehrmalige Anwendung des Verfahrens von Abschnitt 4 auf die entsprechenden Ausdrücke A^i.[33] Ist nun das System (B^i) erfüllbar, dann auch (A^i) und zwar durch dasselbe Funktionssystem,[34] da ja $B^i \to A^i$ gilt. Ist das System (B^i) widerlegbar, dann ist (nach Definition) $\overline{B^1 \,\&\, B^2 \,\&\, \ldots \,\&\, B^n}$ für ein bestimmtes n beweisbar. Durch entsprechende Einsetzung für die in den B^i neu hinzugekommenen Variabeln folgt dann wie in Abschnitt 4, daß auch $\overline{A^1 \,\&\, A^2 \,\&\, \ldots \,\&\, A^n}$ beweisbar, d. h. das System (A^i) widerlegbar ist.

Sei jetzt $(P^1)A^1, (P^2)A^2, \ldots$ ein System von Normalausdrücken ersten Grades. Für ein solches kann man analog wie oben einen n-ten abgeleiteten Ausdruck definieren, was jetzt auseinandergesetzt werden soll. Die Leerstellen jedes A^i zerfallen in solche, die durch Allzeichen und solche, die durch E-Zeichen gebunden sind.[35] Das soll durch die Schreibweise $A^i(\ \ ;\ \ )$ angedeutet werden, wobei die Leerstellen vor dem Strichpunkt durch Allzeichen, die anderen durch E-Zeichen gebunden zu denken sind. Ferner sei bemerkt, daß im Folgenden die Zeichen z_l^i, y_l^i nicht einzelne Variable sondern k-Tupel von Variabeln bezeichnen, wobei das k sich aus dem Zusammenhang ergibt. Unter dem n-ten abgeleiteten Ausdruck des vorgelegten Systems soll der folgende verstanden werden:

$$(P_n)[A^1(z_1^1; y_1^1) \,\&\, A^1(z_2^1; y_2^1) \,\&\, \ldots \,\&\, A^1(z_{n-1}^1; y_{n-1}^1) \,\&\, A^1(z_n^1; y_n^1)$$
$$\&\, A^2(z_1^2; y_1^2) \,\&\, A^2(z_2^2; y_2^2) \&\, \ldots \,\&\, A^2(z_{n-1}^2; y_{n-1}^2)$$
$$\cdot\ \cdot\ \cdot\ \cdot\ \cdot\ \cdot\ \cdot\ \cdot\ \cdot\ \cdot\ \cdot\ \cdot\ \cdot\ \cdot\ \cdot\ \cdot\ \cdot\ \cdot\ \cdot$$
$$\&\, A^{n-1}(z_1^{n-1}; y_1^{n-1}) \,\&\, A^{n-1}(z_2^{n-1}; y_2^{n-1})$$
$$\&\, A^n(z_1^n; y_1^n)].$$

Der ganze obige Ausdruck soll mit $(P_n)T_n$ bezeichnet werden. Er baut sich in leicht ersichtlicher Weise aus der n-ten Ableitung von A^1, der $(n-1)$-Ableitung von A^2 usw. bis zur ersten Ableitung von A^n auf. Das Präfix (P_n) soll nur E-Zeichen enthalten und sämtliche Variable z, y binden. Die in den k-Tupeln z, y auftretenden Variabeln sollen sämtlich der Reihe $x_1, x_2, \ldots$ entnommen sein. Dabei sollen die Indizes der z nach folgender Regel bestimmt werden: Die in $z_1^1, z_1^2, \ldots, z_1^n$ vorkommenden Variabeln haben lauter Einsen als Indizes. z_{i+1}^l soll das auf z_i^l nächstfolgende k-Tupel sein (k bestimmt sich natürlich durch das betreffende A^l). Ferner

---

[33] Die B^i sind dann Normalausdrücke ersten Grades, enthalten i. A. aber einige Funktionsvariable mehr als die entsprechenden A^i.

[34] Natürlich mit Weglassung gewisser in den B^i nicht aber in den entsprechenden A^i vorkommenden Funktionen.

[35] Ausserdem können noch ungebundene Leerstellen vorkommen, doch stören diese die ganze folgende Betrachtung nicht, wenn man nur dafür sorgt, daß sie mit Variabeln ausgefüllt sind, welche von den x_i sämtlich verschieden sind.

(A^i), and indeed by the same systems of functions,[34] since $B^i \to A^i$ holds. If the system (B^i) is refutable, then, for a certain n, $\overline{B^1 \& B^2 \& \ldots \& B^n}$ is provable (by definition). By proper substitution for the variables that now occur in the B^i, it follows, as in Section 4, that $\overline{A^1 \& A^2 \& \ldots \& A^n}$, too, is provable; that is, the system (A^i) is refutable.

Now let $(P^1)A^1, (P^2)A^2, \ldots$ be a system of normal expressions of degree 1. For such a system we can, in a way similar to the one followed above, define an nth derived expression, as will now be explained. The argument places of each A^i divide into those that are bound by universal quantifiers and those that are bound by existential quantifiers.[35] This will be indicated by the notation $A^i(\quad ; \quad)$, where the argument places before the semi-colon are to be considered bound by universal quantifiers and the others by existential quantifiers. Furthermore, let us remark that, in what follows, the signs z_l^i, y_l^i do not denote single variables, but k-tuples of variables, where k is determined by the context. By the nth derived expression of the system adopted, we are to understand the following expression:

$$(P_n)[A^1(z_1^1; y_1^1) \& A^1(z_2^1; y_2^1) \& \ldots \& A^1(z_{n-1}^1; y_{n-1}^1) \& A^1(z_n^1; y_n^1)$$
$$\& A^2(z_1^2; y_1^2) \& A^2(z_2^2; y_2^2) \& \ldots \& A^2(z_{n-1}^2; y_{n-1}^2)$$

$$\cdots \cdots \cdots \cdots \cdots \cdots \cdots \cdots \cdots$$

$$\& A^{n-1}(z_1^{n-1}; y_1^{n-1}) \& A^{n-1}(z_2^{n-1}; y_2^{n-1})$$
$$\& A^n(z_1^n; y_1^n)].$$

The entire expression above will be denoted by $(P_n)T_n$. It is constructed, as one can readily perceive, from the nth derived form of A^1, the $(n-1)$th derived form of A^2, and so on to the first derived form of A^n. The prefix (P_n) is to contain only existential quantifiers and to bind all the variables z and y. The variables occurring in the k-tuples z and y are all to be taken from the sequence $x_1, x_2, \ldots$, and the subscripts of the z's are to be determined by the following rule: The variables occurring in $z_1^1, z_1^2, \ldots, z_1^n$ have only 1's as subscripts; z_{i+1}^l is to be the k-tuple immediately succeeding z_i^l (k is, of course, determined by the A^l in question). Further, the subscripts occurring in $y_i^1, y_{i-1}^2, \ldots, y_1^i$ are all to be different from one another. (In the schema above, these y's occupy precisely one of the diagonals that run from the upper right to the lower left.) Furthermore, they are also to be different from the subscripts occurring in the subformula that lies to the

---

[34] With the omission, of course, of certain functions that occur in the B^i, but not in the corresponding A^i.

[35] In addition there may still be free argument places; but that will not affect any of the considerations that follow, if one but sees to it that these argument places are occupied by variables all of which differ from the x_i.

sollen die in $y_i^1, y_{i-1}^2, \ldots, y_1^i$ (diese erfüllen in dem obigen Schema gerade eine der Diagonalen von rechts oben nach links unten) vorkommenden Indizes sämtlich untereinander verschieden sein. Ferner sollen sie auch von den im links oberhalb der betreffenden Diagonale stehenden Formelteil vorkommenden Indizes verschieden sein. Um diese Bestimmung eindeutig zu machen braucht man nur noch festzusetzen, daß es von den in Betracht kommenden die mit kleinster Summe sein sollen. Wie oben zeigt man dann, daß für jedes n

$$[(P^1)A^1 \& (P^2)A^2 \& \ldots \& (P^n)A^n] \to (P_n)T_n$$

beweisbar ist. Ist nun eines der $(P_n)T_n$ widerlegbar, dann ist auch

$$(P^1)A^1 \& (P^2)A^2 \& \ldots \& (P^n)A^n,$$

d. h. ein endliches Teilsystem des vorgelegten Systems, widerlegbar. Sind dagegen alle $(P_n)T_n$ (resp. die zugeordneten Aussageformeln) erfüllbar, dann ergibt sich wie oben, daß auch das ganze System von Ausdrücken $(P^i)A^i$ erfüllbar ist. Auch die Erweiterung auf Systeme, welche das =-Zeichen enthalten, ergibt sich genau wie oben.

Die Anwendung des Bisherigen auf Zählaxiomensysteme ergibt sich nun leicht. Um zu beweisen, daß jedes Zählaxiomensystem (es mag endlich oder unendlich sein) entweder ein Modell hat oder widerspruchsvoll ist, genügt es, dasjenige System von logischen Ausdrücken zu betrachten, welches aus dem Zählaxiomensystem entsteht, wenn man die Namen durch freie Variable und die Funktionskonstanten durch Funktionsvariable ersetzt (natürlich immer verschiedene durch verschiedene). Ist dann dieses System von logischen Ausdrücken erfüllbar, dann auch das Axiomensystem. Ist es widerlegbar, dann ist (nach Definition) schon das Produkt (Π) aus *endlich vielen* Ausdücken des Systems widerlegbar, also $\overline{\Pi}$ beweisbar. Also ist auch $\overline{\Pi}'$ (dieses soll durch Rückeinsetzung der Konstanten an Stelle der Variabeln aus $\overline{\Pi}$ entstehen) aus dem Axiomensystem deduzierbar (nach der Einsetzungsregel). Da aber natürlich auch Π' aus dem Axiomensystem deduzierbar ist (durch Anwendung der Formel $A \to (B \to A \& B)$), so ist in diesem Fall das Axiomensystem widerspruchsvoll. Es ergibt sich jetzt auch leicht, daß jede *richtige*, d. h. nicht durch Gegenbeispiele widerlegbare Folgerung aus einem Zählaxiomensystem mit endlich vielen formalen Schlüssen erreicht werden kann oder genauer: Jeder in einem Axiomensystem (As) sinnvolle Satz A ist entweder aus den Axiomen deduzierbar oder es gibt ein Modell für As $\& \overline{A}$. Das ergibt sich durch Betrachtung des Axiomensystems As $\& \overline{A}$, das ja entweder widerspruchsvoll oder realisierbar sein muß.

left and above the diagonal in question. In order to make this determination unique we need only specify that, of those that may be considered, we take those with the smallest sum. As above, we now show that, for every n,

$$[(P^1)A^1 \& (P^2)A^2 \& \ldots \& (P^n)A^n] \to (P_n)T_n$$

is provable. If now one of the $(P_n)T_n$ is refutable, then so is

$$(P^1)A^1 \& (P^2)A^2 \& \ldots \& (P^n)A^n;$$

that is, a finite subsystem of the system adopted is refutable. If, however, all $(P_n)T_n$ (or the associated propositional formulas) are satisfiable, then it turns out, as above, that the entire system of expressions $(P^i)A^i$ is satisfiable. The extension to systems that contain the sign $=$ is also carried out exactly as above.

It is now readily seen how what has been done so far carries over to applied first-order axiom systems. In order to prove that every applied first-order axiom system (whether finite or infinite) either has a model or is inconsistent, it suffices to consider the system of logical expressions that results from the applied axiom system when we replace the names by free variables and the functional constants by functional variables (always, of course, replacing different names or constants by different variables). Now, if this system of logical expressions is satisfiable, then so is the axiom system. If it is refutable, then (by definition) the product Π of *finitely many* expressions of the system is already refutable, hence $\overline{\Pi}$ is provable. Hence $\overline{\Pi}'$ (which we obtain from $\overline{\Pi}$ by substituting back the constants for the variables) is also derivable from the axiom system (by the substitution rule). But, since, of course, Π' is also derivable from the axiom system (by use of the formula $A \to [B \to (A \& B)]$), the axiom system is in that case inconsistent. It is now readily apparent, too, that every *correct* consequence, that is, one not refutable by counter-examples, can be reached from an applied first-order system of axioms by finitely many formal inferences, or, more precisely: For every proposition A that is meaningful in an axiom system (As), either A is derivable from the axioms or there is a model for As $\& \overline{A}$. We obtain this by considering the axiom system As $\& \overline{A}$, which, after all, must be either inconsistent or realizable.

Die Vollständigkeit der Axiome des logischen Funktionenkalküls[1]

(*1930*)

Whitehead und Russell haben bekanntlich die Logik und Mathematik so aufgebaut, daß sie gewisse evidente Sätze als Axiome an die Spitze stellten und aus diesen nach einigen genau formulierten Schlußprinzipien auf rein formalem Wege (d. h. ohne weiter von der Bedeutung der Symbole Gebrauch zu machen) die Sätze der Logik und Mathematik deduzierten. Bei einem solchen Vorgehen erhebt sich natürlich sofort die Frage, ob das an die Spitze gestellte System von Axiomen und Schlußprinzipien vollständig ist, d. h. wirklich dazu ausreicht, *jeden* logisch-mathematischen Satz zu deduzieren, oder ob vielleicht wahre (und nach anderen Prinzipien eventuell auch beweisbare) Sätze denkbar sind, welche in dem betreffenden System nicht abgeleitet werden können. Für den Bereich der logischen Aussageformeln ist diese Frage in positivem Sinn entschieden, d. h. man hat gezeigt,[2] daß tatsächlich jede richtige Aussageformel aus den in den *Principia mathematica* angegebenen Axiomen folgt. Hier soll dasselbe für einen weiteren Bereich von Formeln, nämlich für die des "engeren Funktionenkalküls",[3] geschehen, d. h. es soll gezeigt werden:

350 | Satz I: *Jede allgemeingültige[4] Formel des engeren Funktionenkalküls ist beweisbar.*

Dabei legen wir folgendes Axiomensystem[5] zugrunde:

---

[1] Einige wertvolle Ratschläge bezüglich der Durchführung verdanke ich Herrn Prof. H. Hahn.

[2] Vgl. *Bernays 1926*.

[3] In Terminologie und Symbolik schließt sich die folgende Arbeit an *Hilbert und Ackermann 1928* an. Danach gehören zum engeren Funktionenkalkül diejenigen logischen Ausdrücke, welche sich aus Aussagevariablen: $X, Y, Z \ldots$ und Funktions- (= Eigenschafts- und Relations-)variablen 1. Typs: $F(x), G(x,y), H(x,y,z) \ldots$ mittels der Operationen $\vee$ (oder), $^{-}$ (nicht), (x) (für alle), (Ex) (es gibt) aufbauen, wobei die Präfixe $(x), (Ex)$ sich *nur* auf Individuen, *nicht* auf Funktionen beziehen dürfen. Eine solche Formel heißt allgemeingültig (tautologisch), wenn bei jeder Einsetzung bestimmter Aussagen bzw. Funktionen für $X, Y, Z \ldots$ bzw. $F(x), G(x,y) \ldots$ ein wahrer Satz entsteht (z. B.: $(x)[F(x) \vee \overline{F(x)}]$).

[4] Genauer muß es heißen: "in jedem Individuenbereich allgemeingültig", was nach bekannten Sätzen dasselbe besagt wie: "im abzählbaren Individuenbereich allgemeingültig".—Bei Formeln mit freien Individuenvariablen $A(x,y,\ldots,w)$ bedeutet "allgemeingültig" die Allgemeingültigkeit von $(x)(y)\ldots(w)A(x,y,\ldots,w)$ und "erfüllbar" die Erfüllbarkeit von $(Ex)(Ey)\ldots(Ew)A$, so daß ohne Ausnahme gilt: "A ist allgemeingültig" ist gleichbedeutend mit: "$\overline{A}$ ist nicht erfüllbar".

[5] Es stimmt (bis auf das von P. Bernays als überflüssig erwiesene associative principle) mit dem in *Principia mathematica*, I, Nr. 1 und Nr. 10, gegebenen überein.

The completeness of the axioms of the functional calculus of logic[1]
(*1930*)

Whitehead and Russell, as is well known, constructed logic and mathematics by initially taking certain evident propositions as axioms and deriving the theorems of logic and mathematics from these by means of some precisely formulated principles of inference in a purely formal way (that is, without making further use of the meaning of the symbols). Of course, when such a procedure is followed the question at once arises whether the initially postulated system of axioms and principles of inference is complete, that is, whether it actually suffices for the derivation of *every* logico-mathematical proposition, or whether, perhaps, it is conceivable that there are true propositions (which may even be provable by means of other principles) that cannot be derived in the system under consideration. For the formulas of the propositional calculus the question has been settled affirmatively; that is, it has been shown[2] that every correct formula of the propositional calculus does indeed follow from the axioms given in *Principia mathematica*. The same will be done here for a wider realm of formulas, namely those of the "restricted functional calculus";[3] that is, we shall prove

Theorem I. *Every valid* [4] *formula of the restricted functional calculus is provable.*

We lay down the following system of axioms[5] as a basis:

---

[1]I am indebted to Professor H. Hahn for several valuable suggestions that were of help to me in writing this paper.

[2]See *Bernays 1926*.

[3]In terminology and symbolism this paper follows *Hilbert and Ackermann 1928*. According to that work, the restricted functional calculus contains the logical expressions that are constructed from propositional variables, $X, Y, Z, \ldots$, and functional variables (that is, variables for properties and relations) of type 1, $F(x), G(x, y), H(x, y, z), \ldots$, by means of the operations $\vee$ (or), $^-$ (not), (x) (for all), (Ex) (there exists), with the variable in the quantifiers (x) or (Ex) ranging over individuals *only, not* over functions. A formula of this kind is said to be valid (tautological) if a true proposition results from every substitution of specific propositions and functions for $X, Y, Z, \ldots$ and $F(x)$, $G(x, y), \ldots$, respectively (for example, $(x)[F(x) \vee \overline{F(x)}]$).

[4]To be more precise, we should say "valid in every domain of individuals", which, according to well-known theorems, means the same as "valid in the denumerable domain of individuals". For a formula with free individual variables, $A(x, y, \ldots, w)$, "valid" means that $(x)(y) \ldots (w)A(x, y, \ldots, w)$ is valid and "satisfiable" that $(Ex)(Ey) \ldots (Ew)$ $A(x, y, \ldots, w)$ is satisfiable, so that the following holds without exception: "A is valid" is equivalent to "$\overline{A}$ is not satisfiable".

[5]It coincides (except for the associative principle, which P. Bernays proved to be redundant) with that given in *Whitehead and Russell 1910*, *1 and *10.

Undefinierte Grundbegriffe: $\vee$, $\overline{}$, (x). (Daraus lassen sich in bekannter Weise &, $\rightarrow$, $\sim$, (Ex) definieren.)

Formale Axiome:

1. $X \vee X \rightarrow X$,
2. $X \rightarrow X \vee Y$,
3. $X \vee Y \rightarrow Y \vee X$,

4. $(X \rightarrow Y) \rightarrow (Z \vee X \rightarrow Z \vee Y)$,
5. $(x)F(x) \rightarrow F(y)$,
6. $(x)[X \vee F(x)] \rightarrow X \vee (x)F(x)$.

Schlußregeln:[6]

1. Das Schlußschema: Aus A und $A \rightarrow B$ darf B geschlossen werden.

2. Die Einsetzungsregel für Aussage- und Funktionsvariable.

3. Aus $A(x)$ darf $(x)A(x)$ geschlossen werden.

4. Individuenvariable (freie oder gebundene) dürfen durch beliebige andere ersetzt werden, soweit dadurch keine Überdeckung der Wirkungsbereiche gleichbezeichneter Variabler eintritt.

Für das Folgende ist es zweckmäßig, einige abgekürzte Bezeichnungen einzuführen.

(P), (Q), (R) etc. bedeuten irgendwie gebaute Präfixe, also endliche Zeichenreihen der Form: $(x)(Ey)$, $(y)(x)(Ez)(u)$ etc.

Kleine deutsche Buchstaben $\mathfrak{x}, \mathfrak{y}, \mathfrak{u}, \mathfrak{v}$ etc. bedeuten n-tupel von Individuenvariablen, d. h. Zeichenreihen der Form: x, y, z; x_2, x_1, x_2, x_3 etc., wobei dieselbe Variable auch mehrmals auftreten kann. Entsprechend sind die Zeichen $(\mathfrak{x}), (E\mathfrak{x})$ etc. zu verstehen. Sollte in $\mathfrak{x}$ eine Variable mehrmals vorkommen, so hat man sie natürlich in $(\mathfrak{x})$, $(E\mathfrak{x})$ nur einmal angeschrieben zu denken.

Ferner benötigen wir eine Reihe von Hilfssätzen, die hier zusammengestellt seien. Die Beweise sind nicht angeführt, da sie teils bekannt, teils leicht zu ergänzen sind:

1. Für jedes n-tupel $\mathfrak{x}$ ist beweisbar:
   (a) $(\mathfrak{x})F(\mathfrak{x}) \rightarrow (E\mathfrak{x})F(\mathfrak{x})$,
   (b) $(\mathfrak{x})F(\mathfrak{x})$ & $(E\mathfrak{x})G(\mathfrak{x}) \rightarrow (E\mathfrak{x})[F(\mathfrak{x})$ & $G(\mathfrak{x})]$,
   (c) $(\mathfrak{x})\overline{F(\mathfrak{x})} \sim \overline{(E\mathfrak{x})F(\mathfrak{x})}$.

351 | 2. Unterscheiden sich $\mathfrak{x}$ und $\mathfrak{x}'$ nur durch die Reihenfolge der Variablen, so ist beweisbar:

$$(E\mathfrak{x})F(\mathfrak{x}) \rightarrow (E\mathfrak{x}')F(\mathfrak{x}).$$

[6]Diese sind bei Russell–Whitehead nicht alle explizit formuliert, werden aber in den Deduktionen fortwährend verwendet.

Undefined primitive notions: $\vee$, $\overline{}$, and (x). (By means of these, $\&$, $\rightarrow$, $\sim$, and (Ex) can be defined in a well-known way.)

Formal axioms:

1. $X \vee X \rightarrow X,$ 4. $(X \rightarrow Y) \rightarrow (Z \vee X \rightarrow Z \vee Y),$
2. $X \rightarrow X \vee Y,$ 5. $(x)F(x) \rightarrow F(y),$
3. $X \vee Y \rightarrow Y \vee X,$ 6. $(x)[X \vee F(x)] \rightarrow X \vee (x)F(x).$

Rules of inference:[6]

1. The inferential schema: From A and $A \rightarrow B$, B may be inferred;
2. The rule of substitution for propositional and functional variables;
3. From $A(x)$, $(x)A(x)$ may be inferred;
4. Individual variables (free or bound) may be replaced by any others, so long as this does not cause overlapping of the scopes of variables denoted by the same sign.

For what follows, it will be expedient to introduce some abbreviated notations.

(P), (Q), (R), and so on stand for prefixes constructed in any way whatever, that is, finite sequences of signs of the form $(x)(Ey)$, $(y)(x)(Ez)(u)$, and the like.

Lower-case German letters, $\mathfrak{x}$, $\mathfrak{y}$, $\mathfrak{u}$, $\mathfrak{v}$, and so on, mean n-tuples of individual variables, that is, sequences of signs of the form x, y, z, or x_2, x_1, x_2, x_3, and the like, where the same variable may occur several times. The signs $(\mathfrak{x})$, $(E\mathfrak{x})$, and so on are to be understood accordingly. Should a variable occur several times in $\mathfrak{x}$, we must, of course, think of it as written only once in $(\mathfrak{x})$ or $(E\mathfrak{x})$.

Furthermore we require a number of lemmas, which are collected here. The proofs are not given, since they are in part well known, in part easy to supply.

1. For every n-tuple $\mathfrak{x}$
   (a) $(\mathfrak{x})F(\mathfrak{x}) \rightarrow (E\mathfrak{x})F(\mathfrak{x}),$
   (b) $(\mathfrak{x})F(\mathfrak{x}) \ \& \ (E\mathfrak{x})G(\mathfrak{x}) \rightarrow (E\mathfrak{x})[F(\mathfrak{x}) \ \& \ G(\mathfrak{x})],$
   (c) $(\mathfrak{x})\overline{F(\mathfrak{x})} \sim \overline{(E\mathfrak{x})F(\mathfrak{x}))}$

are provable.

2. If $\mathfrak{x}$ and $\mathfrak{x}'$ differ only in the order of the variables, then

$$(E\mathfrak{x})F(\mathfrak{x}) \rightarrow (E\mathfrak{x}')F(\mathfrak{x})$$

is provable.

---

[6]Although Whitehead and Russell use these rules throughout their derivations, they do not formulate all of them explicitly.

3. Besteht $\mathfrak{x}$ aus lauter verschiedenen Variablen und hat $\mathfrak{x}'$ dieselbe Stellenzahl wie $\mathfrak{x}$, so ist beweisbar:

$$(\mathfrak{x})F(\mathfrak{x}) \rightarrow (\mathfrak{x}')F(\mathfrak{x}')$$

auch dann, wenn in $\mathfrak{x}'$ mehrere gleiche Variable vorkommen.

4. Bedeutet (p_i) eines der Präfixe (x_i), (Ex_i) und (q_i) eines der Präfixe (y_i), (Ey_i), dann ist beweisbar:

$$(p_1)(p_2) \ldots (p_n)F(x_1, x_2, \ldots, x_n) \,\&\, (q_1)(q_2) \ldots (q_m)G(y_1, y_2, \ldots, y_m)$$
$$\sim (P)[F(x_1, x_2, \ldots, x_n) \,\&\, G(y_1, y_2, \ldots, y_m)]^{\,7}$$

für jedes Präfix (P), das sich aus den (p_i) und (q_i) zusammensetzt und der Bedingung genügt, daß für $i < k \leq n$ (p_i) vor (p_k) und für $i < k \leq m$ (q_i) vor (q_k) steht.

5. Jeder Ausdruck kann auf die Normalform gebracht werden, d. h. zu jedem Ausdruck A gibt es eine Normalformel N, so daß $A \sim N$ beweisbar ist.[8]

6. Ist $A \sim B$ beweisbar, dann auch $\mathfrak{F}(A) \sim \mathfrak{F}(B)$, wobei $\mathfrak{F}(A)$ einen beliebigen Ausdruck bedeutet, der A als Teil enthält (vgl. *Hilbert und Ackermann 1928*, III, §7).

7. Jede allgemeingültige Aussageformel ist beweisbar, d. h. die Axiome 1–4 bilden ein vollständiges Axiomensystem für den Aussagenkalkül.[9]

Wir gehen jetzt zum Beweis von Satz I über und bemerken zunächst, daß er auch in folgender Form ausgesprochen werden kann:

Satz II: *Jede Formel des engeren Funktionenkalküls ist entweder widerlegbar*[10] *oder erfüllbar* (und zwar im abzählbaren Individuenbereich).

Daß I aus II folgt, ergibt sich so: Sei A ein allgemeingültiger Ausdruck, dann ist $\overline{A}$ nicht erfüllbar, also nach II widerlegbar, d. h. $\overline{\overline{A}}$ folglich auch A ist beweisbar. Ebenso leicht sieht man die Umkehrung ein.

Wir definieren jetzt ein Klasse $\mathfrak{K}$ von Ausdrücken K durch folgende Festsetzungen:

1. K ist eine Normalformel.

2. K enthält keine freien Individuenvariablen.

352 | 3. Das Präfix von K beginnt mit einem Allzeichen und endet mit einem E-Zeichen.

Dann gilt:

---

[7] Ein analoger Satz gilt für $\vee$ statt &.

[8] Vgl. *Hilbert und Ackermann 1928*, III, §8.

[9] Vgl. *Bernays 1926*.

[10] "A ist widerlegbar" soll bedeuten: "$\overline{A}$ ist beweisbar".

3. If $\mathfrak{x}$ consists entirely of distinct variables and if $\mathfrak{x}'$ has the same number of terms as $\mathfrak{x}$, then

$$(\mathfrak{x})F(\mathfrak{x}) \rightarrow (\mathfrak{x}')F(\mathfrak{x}')$$

is provable, even when a number of identical variables occur in $\mathfrak{x}'$.

4. If (p_i) stands for one of the quantifiers (x_i) or (Ex_i) and if (q_i) stands for one of the quantifiers (y_i) or (Ey_i), then

$$(p_1)(p_2)\ldots(p_n)F(x_1, x_2, \ldots, x_n) \; \& \; (q_1)(q_2)\ldots(q_m)G(y_1, y_2, \ldots, y_m)$$
$$\sim (P)[F(x_1, x_2, \ldots, x_n) \; \& \; G(y_1, y_2, \ldots, y_m)]$$

is provable[7] for every prefix (P) that is formed from the (p_i) and the (q_i) and satisfies the condition that, for $i < k \leq n$, (p_i) precedes (p_k) and, for $i < k \leq m$, (q_i) precedes (q_k).

5. Every expression can be brought into normal form; that is, for every expression A there is a normal formula N such that $A \sim N$ is provable.[8]

6. If $A \sim B$ is provable, so is $\mathfrak{F}(A) \sim \mathfrak{F}(B)$, where $\mathfrak{F}(A)$ represents an arbitrary expression containing A as a part (see *Hilbert and Ackermann 1928*, Chapter 3, §7).

7. Every valid formula of the propositional calculus is provable; that is, Axioms 1–4 form a complete axiom system for the propositional calculus.[9]

We now proceed to the proof of Theorem I and first note that the theorem can also be stated in the following form:

Theorem II. *Every formula of the restricted functional calculus is either refutable*[10] *or satisfiable* (and, moreover, satisfiable in the denumerable domain of individuals).

That I follows from II can be seen as follows: Let A be a valid expression; then $\overline{A}$ is not satisfiable, hence according to II it is refutable; that is, $\overline{\overline{A}}$ is provable and, consequently, so is A. The converse is as apparent.

We now define a class $\mathfrak{K}$ of expressions K by means of the following stipulations:

1. K is a normal formula;

2. K contains no free individual variable;

3. The prefix of K begins with a universal quantifier and ends with an existential quantifier.

Then we have

---

[7] An analogous theorem holds with $\vee$ instead of $\&$.

[8] See *Hilbert and Ackermann 1928*, Chap. 3, §8.

[9] See *Bernays 1926*.

[10] "A is refutable" is to mean "$\overline{A}$ is provable".

Satz III: *Ist jeder $\mathfrak{K}$-Ausdruck entweder widerlegbar oder erfüllbar,*[11] *so gilt dasselbe von jedem Ausdruck.*

Beweis: Sei A ein Ausdruck, der nicht zu $\mathfrak{K}$ gehört. Er möge die freien Variablen $\mathfrak{x}$ enthalten. Wie man sofort einsieht, folgt aus der Widerlegbarkeit von A die von $(E\mathfrak{x})A$ und umgekehrt (nach Hilfssatz 1c und Schlußregel 3 bzw. Axiom 5); dasselbe gilt nach der Festsetzung in Fußnote 4 für die Erfüllbarkeit. Sei $(P)N$ die Normalform von $(E\mathfrak{x})A$, so daß

$$(E\mathfrak{x})A \sim (P)N \tag{1}$$

beweisbar ist. Ferner setze man

$$B = (x)(P)(Ey)\{N \;\&\; [F(x) \lor \overline{F(y)}]\}.^{12}$$

Dann ist

$$(P)N \sim B \tag{2}$$

beweisbar (auf Grund von Hilfssatz 4 und der Beweisbarkeit von

$$(x)(Ey)[F(x) \lor \overline{F(y)}]).$$

B gehört zu $\mathfrak{K}$, ist also nach Annahme entweder erfüllbar oder widerlegbar. Aber nach (1) und (2) zieht die Erfüllbarkeit von B die von $(E\mathfrak{x})A$, folglich auch die von A nach sich und dasselbe gilt für die Widerlegbarkeit. Auch A ist also entweder erfüllbar oder widerlegbar.

Auf Grund von Satz III genügt es also zu zeigen:

Jeder $\mathfrak{K}$-Ausdruck ist entweder erfüllbar oder widerlegbar.

Zu diesem Zweck definieren wir als Grad eines $\mathfrak{K}$-Ausdruckes[13] die Anzahl der durch E-Zeichen voneinander getrennten Komplexe von Allzeichen seines Präfixes und zeigen zunächst:

Satz IV: *Wenn jeder Ausdruck von Grad k entweder erfüllbar oder widerlegbar ist, so gilt dasselbe auch von jedem Ausdruck vom Grad $k+1$.*

Beweis: Sei $(P)A$ ein $\mathfrak{K}$-Ausdruck vom Grad $k+1$. Sei $(P) = (\mathfrak{x})(E\mathfrak{y})(Q)$ und $(Q) = (\mathfrak{u})(E\mathfrak{v})(R)$, wobei (Q) den Grad k und (R) den Grad $k-1$ hat. Sei ferner F eine nicht in A vorkommende Funktionsvariable. Setzt man dann:[14]

$$B = (\mathfrak{x}')(E\mathfrak{y}')F(\mathfrak{x}',\mathfrak{y}') \;\&\; (\mathfrak{x})(\mathfrak{y})[F(\mathfrak{x},\mathfrak{y}) \to (Q)A]$$

---

[11] "Erfüllbar" ohne Zusatz bedeutet hier und im folgenden immer: "erfüllbar im abzählbaren Individuenbereich". Dasselbe gilt für "allgemeingültig".

[12] Die Variablen x, y sollen in (P) nicht vorkommen.

[13] Im selben Sinn wird der Terminus "Grad eines Präfixes" verwendet.

[14] Ein analoges Verfahren hat Th. Skolem (*1920*) zum Beweise des Löwenheimschen Satzes verwendet.

Theorem III. *If every $\mathfrak{K}$-expression is either refutable or satisfiable,*[11] *so is every expression.*

Proof: Let A be an expression not belonging to $\mathfrak{K}$. Let it contain the free variables $\mathfrak{x}$. As is immediately obvious, the refutability of $(E\mathfrak{x})A$ follows from that of A, and conversely (by Lemma 1(c), and either Rule of inference 3 or, for the converse, Axiom 5); the same holds, according to the stipulation in footnote 4, for satisfiability. Let $(P)N$ be the normal form of $(E\mathfrak{x})A$, so that

$$(E\mathfrak{x})A \sim (P)N \tag{1}$$

is provable. Further let

$$B = (x)(P)(Ey)\{N \ \& \ [F(x) \vee \overline{F(y)}]\}.^{12}$$

Then

$$(P)N \sim B \tag{2}$$

is provable (because of Lemma 4 and the provability of

$$(x)(Ey)[F(x) \vee \overline{F(y)}]).$$

B belongs to $\mathfrak{K}$ and thus according to the assumption is either satisfiable or refutable. But, by (1) and (2), the satisfiability of B entails that of $(E\mathfrak{x})A$, hence also that of A; the same holds for refutability. Thus A, too, is either satisfiable or refutable.

Because of Theorem III, therefore, it suffices to show that

Every $\mathfrak{K}$-expression is either satisfiable or refutable.

For this purpose we define the degree of a $\mathfrak{K}$-expression[13] to be the number of blocks in its prefix that consist of universal quantifiers and are separated from each other by existential quantifiers, and we first prove

Theorem IV. *If every expression of degree k is either satisfiable or refutable, so is every expression of degree $k + 1$.*

Proof: Let $(P)A$ be a $\mathfrak{K}$-expression of degree $k+1$. Let $(P) = (\mathfrak{x})(E\mathfrak{y})(Q)$ and let $(Q) = (\mathfrak{u})(E\mathfrak{v})(R)$, where (Q) is of degree k and (R) of degree $k-1$. Further let F be a functional variable not occurring in A. If we now put[14]

$$B = (\mathfrak{x}')(E\mathfrak{y}')F(\mathfrak{x}',\mathfrak{y}') \ \& \ (\mathfrak{x})(\mathfrak{y})[F(\mathfrak{x},\mathfrak{y}) \rightarrow (Q)A]$$

and

---

[11] "Satisfiable" without additional specification here and in what follows always means "satisfiable in the denumerable domain of individuals". The same holds for "valid".

[12] The variables x and y must not occur in (P).

[13] The term "degree of a prefix" is used in the same sense.

[14] An analogous procedure was used by Skolem (*1920*) in proving Löwenheim's theorem.

und

$$C = (\mathfrak{r}')(\mathfrak{r})(\mathfrak{y})(\mathfrak{u})(E\mathfrak{y}')(E\mathfrak{v})(R)\{F(\mathfrak{r}',\mathfrak{y}') \;\&\; [F(\mathfrak{r},\mathfrak{y}) \to A]\},^{15}$$

so ergibt zweimalige Anwendung von Hilfssatz 4 im Verein mit Hilfssatz 6 die Beweisbarkeit von

$$B \sim C; \tag{3}$$

weiter ist offenbar

$$B \to (P)A \tag{4}$$

allgemeingültig. Nun hat C den Grad k, ist also nach Annahme entweder erfüllbar oder widerlegbar. Ist es erfüllbar, dann auch $(P)A$ (nach (3) und (4)). Ist es widerlegbar, dann auch B (nach (3)), d. h. dann ist $\overline{B}$ beweisbar. Setzt man in $\overline{B}$ $(Q)A$ für F ein, so ergibt sich, daß in diesem Fall beweisbar ist:

$$\overline{(\mathfrak{r}')(E\mathfrak{y}')(Q)A} \;\&\; \overline{(\mathfrak{r})(\mathfrak{y})[(Q)A \to (Q)A]}.$$

Da aber natürlich

$$(\mathfrak{r})(\mathfrak{y})[(Q)A \to (Q)A]$$

beweisbar ist, so auch $\overline{(\mathfrak{r}')(E\mathfrak{y}')(Q)A}$, d. h. in diesem Falle ist $(P)A$ widerlegbar. Tatsächlich ist also $(P)A$ entweder widerlegbar oder erfüllbar.

Es braucht jetzt nur noch gezeigt zu werden:

Satz V: *Jede Formel ersten Grades ist entweder erfüllbar oder widerlegbar.*

Zum Beweise sind einige Definitionen erforderlich. Sei $(\mathfrak{r})(E\mathfrak{y})A(\mathfrak{r};\mathfrak{y})$ (abgekürzt als $(P)A$) eine beliebige Formel ersten Grades. Dabei vertrete $\mathfrak{r}$ ein r-tupel, $\mathfrak{y}$ ein s-tupel von Variablen. Wir denken uns die der Folge $x_0, x_1, x_2, \ldots, x_i, \ldots$ entnommenen r-tupel in eine Folge:

$$\mathfrak{r}_1 = (x_0, x_0, \ldots, x_0), \; \mathfrak{r}_2 = (x_1, x_0, \ldots, x_0), \; \mathfrak{r}_3 = (x_0, x_1, x_0, \ldots, x_0) \text{ u.s.w.}$$

nach steigender Indexsumme geordenet und definieren eine Folge $\{A_n\}$ aus $(P)A$ abgeleiteter Formeln folgendermaßen:

$$A_1 = A(\mathfrak{r}_1; x_1, x_2, \ldots, x_s)$$
$$A_2 = A(\mathfrak{r}_2; x_{s+1}, x_{s+2}, \ldots, x_{2s}) \;\&\; A_1$$
$$\cdots\cdots\cdots\cdots\cdots\cdots\cdots\cdots$$
$$A_n = A(\mathfrak{r}_n; x_{(n-1)s+1}, x_{(n-1)s+2}, \ldots, x_{ns}) \;\&\; A_{n-1}$$

---

[15]Die Variablenreihen $\mathfrak{r}, \mathfrak{r}', \mathfrak{y}, \mathfrak{y}', \mathfrak{u}, \mathfrak{v}$ sind natürlich als paarweise fremd vorausgesetzt.

$$C = (\mathfrak{r}')(\mathfrak{r})(\mathfrak{n})(\mathfrak{u})(E\mathfrak{n}')(E\mathfrak{v})(R)\{F(\mathfrak{r}',\mathfrak{n}') \,\&\, [F(\mathfrak{r},\mathfrak{n}) \rightarrow A]\},^{15}$$

then a double application of Lemma 4 in combination with Lemma 6 yields the provability of

$$B \sim C; \tag{3}$$

furthermore,

$$B \rightarrow (P)A \tag{4}$$

is obviously valid. Now C is of degree k and by assumption is therefore either satisfiable or refutable. If it is satisfiable, so is $(P)A$ (by (3) and (4)). If it is refutable, so is B (by (3)); that is, $\overline{B}$ is then provable. In that case, if we substitute $(Q)A$ for F in $\overline{B}$, it follows that

$$\overline{(\mathfrak{r}')(E\mathfrak{n}')(Q)A \,\&\, (\mathfrak{r})(\mathfrak{n})[(Q)A \rightarrow (Q)A]}$$

is provable.

But since, of course,

$$(\mathfrak{r})(\mathfrak{n})[(Q)A \rightarrow (Q)A]$$

is provable, so is $\overline{(\mathfrak{r}')(E\mathfrak{n}')(Q)A}$; that is, in that case $(P)A$ is refutable. $(P)A$ is therefore indeed either refutable or satisfiable.

It now remains only to prove

Theorem V. *Every formula of degree 1 is either satisfiable or refutable.*

A few definitions are required for the proof. Let $(\mathfrak{r})(E\mathfrak{n})A(\mathfrak{r};\mathfrak{n})$ (abbreviated as $(P)A$) be any formula of degree 1. Let $\mathfrak{r}$ stand for an r-tuple and $\mathfrak{n}$ for an s-tuple of variables. We think of the r-tuples taken from the sequence $x_0, x_1, x_2, \ldots, x_i, \ldots$ as forming a sequence ordered according to increasing sum of the subscripts [and for equal sums according to some convention]:

$$\mathfrak{r}_1 = (x_0, x_0, \ldots, x_0), \quad \mathfrak{r}_2 = (x_1, x_0, \ldots, x_0), \quad \mathfrak{r}_3 = (x_0, x_1, x_0, \ldots, x_0),$$

and so forth; we now define a sequence $\{A_n\}$ of formulas derived from $(P)A$ as follows:

$$A_1 = A(\mathfrak{r}_1; x_1, x_2, \ldots, x_s),$$
$$A_2 = A(\mathfrak{r}_2; x_{s+1}, x_{s+2}, \ldots, x_{2s}) \,\&\, A_1,$$
$$\dotsc\dotsc\dotsc\dotsc\dotsc\dotsc\dotsc\dotsc\dotsc\dotsc\dotsc\dotsc\dotsc\dotsc$$
$$A_n = A(\mathfrak{r}_n; x_{(n-1)s+1}, x_{(n-1)s+2}, \ldots, x_{ns}) \,\&\, A_{n-1}.$$

---

[15] The variable-sequences $\mathfrak{r}, \mathfrak{r}', \mathfrak{n}, \mathfrak{n}', \mathfrak{u}, \mathfrak{v}$ are, of course, assumed to be pairwise disjoint.

354 | Das s-tupel $x_{(n-1)s+1}, \ldots, x_{ns}$ werde mit $\mathfrak{y}_n$ bezeichnet, so daß man
hat:

$$A_n = A(\mathfrak{x}_n; \mathfrak{y}_n) \,\&\, A_{n-1}.\tag{5}$$

Ferner definieren wir $(P_n)A_n$ durch die Festsetzung:

$$(P_n)A_n = (Ex_0)(Ex_1)\ldots(Ex_{ns})A_n.$$

Wie man sich leicht überzeugt, kommen in A_n gerade die Variablen x_0
bis x_{ns} vor, welche also sämtlich durch (P_n) gebunden werden. Ferner ist
ersichtlich, daß die Variablen des r-tupels $\mathfrak{x}_{n+1}$ schon in (P_n) vorkommen
(also insbesondere von den in $\mathfrak{y}_{n+1}$ vorkommenden verschieden sind). Was
von (P_n) übrigbleibt, wenn man die Variablen des r-tuples $\mathfrak{x}_{n+1}$ wegläßt,
werde mit (P'_n) bezeichnet, so daß, abgesehen von der Reihenfolge der
Variablen, $(E\mathfrak{x}_{n+1})(P'_n) = (P_n)$.

Diese Bezeichnungen vorausgesetzt, gilt:

Satz VI: *Für jedes n ist beweisbar:*

$$(P)A \rightarrow (P_n)A_n.$$

Zum Beweise wenden wir vollständige Induktion an:

I. $(P)A \rightarrow (P_1)A_1$ ist beweisbar, denn man hat:

$$(\mathfrak{x})(E\mathfrak{y})A(\mathfrak{x};\mathfrak{y}) \rightarrow (\mathfrak{x}_1)(E\mathfrak{y}_1)A(\mathfrak{x}_1;\mathfrak{y}_1)$$

(nach Hilfssatz 3 und Schlußregel 4) und

$$(\mathfrak{x}_1)(E\mathfrak{y}_1)A(\mathfrak{x}_1;\mathfrak{y}_1) \rightarrow (E\mathfrak{x}_1)(E\mathfrak{y}_1)A(\mathfrak{x}_1;\mathfrak{y}_1)$$

(nach Hilfssatz 1a).

II. Für jedes n ist $(P)A \,\&\, (P_n)A_n \rightarrow (P_{n+1})A_{n+1}$ beweisbar, denn man
hat:

$$(\mathfrak{x})(E\mathfrak{y})A(\mathfrak{x};\mathfrak{y}) \rightarrow (\mathfrak{x}_{n+1})(E\mathfrak{y}_{n+1})A(\mathfrak{x}_{n+1};\mathfrak{y}_{n+1})\tag{6}$$

(nach Hilfssatz 3 und Schlußregel 4) und

$$(P_n)A_n \rightarrow (E\mathfrak{x}_{n+1})(P'_n)A_n\tag{7}$$

(nach Hilfssatz 2). Ferner

$$
\begin{aligned}
(\mathfrak{x}_{n+1})(E\mathfrak{y}_{n+1})&A(\mathfrak{x}_{n+1};\,\mathfrak{y}_{n+1}) \,\&\, (E\mathfrak{x}_{n+1})(P'_n)A_n\\
&\rightarrow (E\mathfrak{x}_{n+1})[(E\mathfrak{y}_{n+1})A(\mathfrak{x}_{n+1};\mathfrak{y}_{n+1}) \,\&\, (P'_n)A_n]
\end{aligned}\tag{8}
$$

(nach Hilfssatz 1b bei der Einsetzung: $(E\mathfrak{y}_{n+1})A(\mathfrak{x}_{n+1};\mathfrak{y}_{n+1})$ für F und
$(P'_n)A_n$ für G).

Let the s-tuple $x_{(n-1)s+1}, \ldots, x_{ns}$ be denoted by $\mathfrak{y}_n$, so that we have

$$A_n = A(\mathfrak{x}_n; \mathfrak{y}_n) \ \& \ A_{n-1}. \tag{5}$$

Further we define $(P_n)A_n$ by the stipulation

$$(P_n)A_n = (Ex_0)(Ex_1) \ldots (Ex_{ns})A_n.$$

As we can easily convince ourselves, it is precisely the variables x_0 to x_{ns} that occur in A_n; hence they all are bound by (P_n). Further it is apparent that the variables of the r-tuple $\mathfrak{x}_{n+1}$ already occur in (P_n) (and therefore certainly differ from those occuring in $\mathfrak{y}_{n+1}$). Denote by (P'_n) what remains of (P_n) when the variables of the r-tuple $\mathfrak{x}_{n+1}$ are omitted, so that, except for the order of the variables, $(E\mathfrak{x}_{n+1})(P'_n) = (P_n)$.

This notation once assumed, we have

Theorem VI. *For every n*

$$(P)A \to (P_n)A_n$$

is provable.

For the proof we use mathematical induction.

I. $(P)A \to (P_1)A_1$ is provable, for we have

$$(\mathfrak{x})(E\mathfrak{y})A(\mathfrak{x}; \mathfrak{y}) \to (\mathfrak{x}_1)(E\mathfrak{y}_1)A(\mathfrak{x}_1; \mathfrak{y}_1)$$

(by Lemma 3 and Rule of inference 4) and

$$(\mathfrak{x}_1)(E\mathfrak{y}_1)A(\mathfrak{x}_1; \mathfrak{y}_1) \to (E\mathfrak{x}_1)(E\mathfrak{y}_1)A(\mathfrak{x}_1; \mathfrak{y}_1)$$

(by Lemma 1(a)).

II. For every n, $(P)A \ \& \ (P_n)A_n \to (P_{n+1})A_{n+1}$ is provable, for we have

$$(\mathfrak{x})(E\mathfrak{y})A(\mathfrak{x}; \mathfrak{y}) \to (\mathfrak{x}_{n+1})(E\mathfrak{y}_{n+1})A(\mathfrak{x}_{n+1}; \mathfrak{y}_{n+1}) \tag{6}$$

(by Lemma 3 and Rule of inference 4) and

$$(P_n)A_n \to (E\mathfrak{x}_{n+1})(P'_n)A_n \tag{7}$$

(by Lemma 2). Furthermore,

$$\begin{aligned}(\mathfrak{x}_{n+1})(E\mathfrak{y}_{n+1})A(\mathfrak{x}_{n+1}; \ \mathfrak{y}_{n+1}) \ \& \ (E\mathfrak{x}_{n+1})(P'_n)A_n \\ \to (E\mathfrak{x}_{n+1})[(E\mathfrak{y}_{n+1})A(\mathfrak{x}_{n+1}; \mathfrak{y}_{n+1}) \ \& \ (P'_n)A_n]\end{aligned} \tag{8}$$

(by Lemma 1(b) with the substitutions $(E\mathfrak{y}_{n+1})A(\mathfrak{x}_{n+1}; \mathfrak{y}_{n+1})$ for F and $(P'_n)A_n$ for G).

355 | Beachtet man, daß das Vorderglied der Implikation (8) die Konjunktion der Hinterglieder von (6) und (7) ist, so ergibt sich, daß beweisbar ist:

$$(P)A \ \& \ (P_n)A_n \rightarrow (E\mathfrak{x}_{n+1})[(E\mathfrak{y}_{n+1})A(\mathfrak{x}_{n+1};\mathfrak{y}_{n+1}) \ \& \ (P'_n)A_n]. \qquad (9)$$

Ferner folgt aus (5) und den Hilfssätzen 4, 6, 2 die Beweisbarkeit von:

$$(E\mathfrak{x}_{n+1})[(E\mathfrak{y}_{n+1})A(\mathfrak{x}_{n+1};\mathfrak{y}_{n+1}) \ \& \ (P'_n)A_n] \sim (P_{n+1})A_{n+1}. \qquad (10)$$

Aus (9) und (10) folgt II und daraus im Verein mit I Satz VI.

Mögen in A die Funktionsvariablen $F_1, F_2, \ldots, F_k$ und die Aussagevariablen $X_1, X_2, \ldots, X_l$ vorkommen. A_n baut sich dann aus Elementarbestandteilen der Form:

$$F_1(x_{p_1}, \ldots, x_{q_1}), F_2(x_{p_2}, \ldots, x_{q_2}), \ldots; \ X_1, X_2, \ldots, X_l$$

allein mittels der Operationen $\lor$ und $\overline{}$ auf. Wir ordnen jedem A_n eine Aussageformel B_n dadurch zu, daß wir die Elementarbestandteile von A_n durch Aussagevariable, und zwar verschiedene (auch wenn sie sich nur in der Bezeichnung der Individuenvariablen unterscheiden) durch verschiedene Aussagevariable, ersetzen. Ferner bezeichnen wir als "Erfüllungssystem n-ter Stufe von $(P)A$" ein im Bereich der ganzen Zahlen z ($0 \leq z \leq ns$) definiertes System von Funktionen $f_1^{(n)}, f_2^{(n)}, \ldots, f_k^{(n)}$, sowie von Wahrheitswerten $w_1^{(n)}, w_2^{(n)}, \ldots, w_l^{(n)}$ für die Aussagevariablen X_1, $X_2, \ldots, X_l$, von der Art, daß, wenn man in A_n die F_i durch die $f_i^{(n)}$, die x_i durch die Zahlen i und die X_i durch die entsprechenden Wahrheitswerte $w_i^{(n)}$ ersetzt, ein wahrer Satz entsteht. Erfüllungssysteme n-ter Stufe gibt es offenbar dann und nur dann, wenn B_n erfüllbar ist.

Jedes B_n ist als Aussageformel entweder erfüllbar oder widerlegbar (Hilfssatz 7). Es sind also nur zwei Fälle denkbar:

1. Mindestens ein B_n ist widerlegbar. Dann ist, wie man sich leicht überzeugt (Schlußregeln 2, 3; Hilfssatz 1c), auch das entsprechende $(P_n)A_n$ und folglich wegen der Beweisbarkeit von $(P)A \rightarrow (P_n)A_n$ auch $(P)A$ widerlegbar.

2. Kein B_n ist widerlegbar, also alle erfüllbar. Dann gibt es Erfüllungssysteme jeder Stufe. Da es aber für jede Stufe nur endlich viele Erfüllungssysteme gibt (wegen der Endlichkeit der zugehörigen Individuenbereiche) und da ferner jedes Erfüllungssystem $(n+1)$-ter Stufe ein solches n-ter Stufe

356 als Teil enthält[16] (was sich | sofort aus der Bildungsweise der A_n durch

---

[16]Daß ein System $\{f_1, f_2, \ldots, f_k; \ w_1, w_2, \ldots, w_l\}$ Teil eines anderen $\{g_1, g_2, \ldots, g_k; \ v_1, v_2, \ldots, v_l\}$ ist, soll bedeuten, daß:
1. der Individuenbereich der f_i Teil des Individuenbereiches der g_i ist,
2. die f_i und g_i innerhalb des engeren Bereiches übereinstimmen,
3. für jedes $i, v_i = w_i$ ist.

If we observe that the antecedent of the implication (8) is the conjunction of the consequents of (6) and (7), it is clear that

$$(P)A \mathbin{\&} (P_n)A_n \rightarrow (E\mathfrak{x}_{n+1})[(E\mathfrak{y}_{n+1})A(\mathfrak{x}_{n+1};\mathfrak{y}_{n+1}) \mathbin{\&} (P'_n)A_n] \quad (9)$$

is provable. Furthermore, from (5) and Lemmas 4, 6, and 2 the provability of

$$(E\mathfrak{x}_{n+1})[(E\mathfrak{y}_{n+1})A(\mathfrak{x}_{n+1};\mathfrak{y}_{n+1}) \mathbin{\&} (P'_n)A_n] \sim (P_{n+1})A_{n+1} \quad (10)$$

follows. II follows from (9) and (10), and from II, together with I, Theorem VI follows.

Assume that the functional variables $F_1, F_2, \ldots, F_k$ and the propositional variables $X_1, X_2, \ldots, X_l$ occur in A. Then A_n consists of elementary components of the form

$$F_1(x_{p_1}, \ldots, x_{q_1}), F_2(x_{p_2}, \ldots, x_{q_2}), \ldots, X_1, X_2, \ldots, X_l$$

compounded solely by means of the operations $\vee$ and $\overline{}$. With each A_n we associate a formula B_n of the propositional calculus by replacing the elementary components of A_n by propositional variables, making certain that different components (even if they differ only in the notation of the individual variables) are replaced by different propositional variables. Furthermore, we understand by "satisfying system of level n of $(P)A$" a system of functions $f_1^{(n)}, f_2^{(n)}, \ldots, f_k^{(n)}$, defined in the domain of integers z ($0 \leq z \leq ns$), as well as of truth values $w_1^{(n)}, w_2^{(n)}, \ldots, w_l^{(n)}$ for the propositional variables $X_1, X_2, \ldots, X_l$ such that a true proposition results if in A_n the F_i are replaced by the $f_i^{(n)}$, the x_i by the numbers i, and the X_i by the corresponding truth values $w_i^{(n)}$. Satisfying systems of level n obviously exist if and only if B_n is satisfiable.

Each B_n, being a formula of the propositional calculus, is either satisfiable or refutable (Lemma 7). Thus only two cases are conceivable:

1. At least one B_n is refutable. Then, as we can easily convince ourselves (Rules of inference 2 and 3, and Lemma 1(c)), the corresponding $(P_n)A_n$ is refutable also, and consequently, because of the provability of $(P)A \rightarrow (P_n)A_n$, so is $(P)A$.

2. No B_n is refutable; hence all are satisfiable. Then there exist satisfying systems of every level. But, since for each level there is only a finite number of satisfying systems (because the associated domains of individuals are finite) and since furthermore every satisfying system of level $n+1$

fortgesetzte &-Verknüpfung ergibt), so folgt nach bekannten Schlußweisen, daß es in diesem Fall eine Folge von Erfüllungssystemen $S_1, S_2, \ldots, S_k, \ldots$ (S_k von k-ter Stufe) gibt, deren jedes folgende das vorhergehende als Teil enthält. Wir definieren jetzt im Bereich *aller* ganzen Zahlen ≥ 0 ein System $S = \{\phi_1, \phi_2, \ldots, \phi_k; \alpha_1, \alpha_2, \ldots, \alpha_l\}$ durch die Festsetzungen:

1. $\phi_p(a_1, \ldots, a_i)$ ($1 \leq p \leq k$) soll dann und nur dann gelten, wenn für mindestens ein S_m der obigen Folge (und dann auch für alle folgenden) $f_p^{(m)}(a_1, \ldots, a_i)$ gilt.

2. $\alpha_i = w_i^{(m)}$ ($1 \leq i \leq l$) für mindestens ein (und dann auch für alle übrigen) S_m.

Dann ist ohneweiters ersichtlich, daß S die Formel $(P)A$ wahr macht. In diesem Falle ist also $(P)A$ erfüllbar, womit der Beweis der Vollständigkeit des oben angegebenen Axiomensystems zu Ende geführt ist. Es sei bemerkt, daß die nunmehr bewiesene Äquivalenz: "allgemeingültig = beweisbar" für das Entscheidungsproblem eine Reduktion des Überabzählbaren auf das Abzählbare beinhaltet, denn "allgemeingültig" bezieht sich auf die überabzählbare Gesamtheit der Funktionen, während "beweisbar" nur die abzählbare Gesamtheit der Beweisfiguren voraussetzt.

Man kann Satz I bzw. Satz II in verschiedener Richtung verallgemeinern. Zunächst ist es leicht, den Begriff der Identität (zwischen Individuen) in den Kreis der Betrachtung zu ziehen, indem man zu den obigen Axiomen 1–6 zwei weitere

$$7.\ x = x, \qquad 8.\ x = y \rightarrow [F(x) \rightarrow F(y)]$$

hinzufügt. Auch für diesen erweiterten Bereich gilt dann analog wie oben:

Satz VII: *Jede allgemeingültige (genauer: in jedem Individuenbereich allgemeingültige) Formel des erweiterten Bereiches ist beweisbar,*
und der mit VII äquivalente

Satz VIII: *Jede Formel des erweiterten Bereiches ist entweder widerlegbar oder erfüllbar* (und zwar in einem endlichen oder abzählbaren Individuenbereich).

Zum Beweise werde mit A eine beliebige Formel des erweiterten Bereiches bezeichnet. Wir bilden eine Formel B als das Produkt (&-Verknüpfung) aus: A, $(x)(x = x)$ und den sämtlichen Formeln, die aus Axiom 8 durch Einsetzung der in A vorkommenden Funktionsvariablen für F entstehen, d. h. genauer:

$$(x)(y)\,\{x = y \rightarrow [F(x) \rightarrow F(y)]\}$$

357 | für alle einstelligen Funktionsvariablen aus A,

$$(x)(y)(z)\,\{x = y \rightarrow [F(x, z) \rightarrow F(y, z)]\}\ \&$$
$$(x)(y)(z)\,\{x = y \rightarrow [F(z, x) \rightarrow F(z, y)]\}$$

contains one of level n as a part[16] (as is clear from the fact that the A_n are formed by successive conjunctions), it follows by familiar arguments that in this case there exists a sequence of satisfying systems $S_1, S_2, \ldots, S_k, \ldots$ (S_k being of level k) such that, after S_1, each contains the preceding one as a part. We now define in the domain of *all* integers ≥ 0 a system $S = \{\phi_1, \phi_2, \ldots, \phi_k; \alpha_1, \alpha_2, \ldots, \alpha_l\}$ by means of the following stipulations:

1. $\phi_p(a_1, \ldots, a_i)$ $(1 \leq p \leq k)$ holds if and only if for at least one S_m of the sequence above (and then also for all those that follow) $f_p^{(m)}(a_1, \ldots, a_i)$ holds;

2. $\alpha_i = w_i^{(m)}$ $(1 \leq i \leq l)$ for at least one S_m (and then also for all subsequent ones).

Then it is evident at once that S makes the formula $(P)A$ true. In this case, therefore, $(P)A$ is satisfiable, which concludes the proof of the completeness of the system of axioms given above. Let us note that the equivalence now proved, "valid = provable", entails, for the decision problem, a reduction of the nondenumerable to the denumerable, since "valid" refers to the nondenumerable totality of functions, while "provable" presupposes only the denumerable totality of formal proofs.

Theorem I, as well as Theorem II, can be generalized in various directions. First, it is easy to bring the notion of identity (between individuals) into consideration by adding to Axioms 1–6 above two more:

$$7. \quad x = x, \qquad 8. \quad x = y \rightarrow [F(x) \rightarrow F(y)].$$

An analogue of what we had above then holds for the extended realm of formulas too:

Theorem VII. *Every formula of the extended realm is provable if it is valid* (more precisely, if it is valid in every domain of individuals), and, equivalent to VII,

Theorem VIII. *Every formula of the extended realm is either refutable or satisfiable* (and, moreover, satisfiable in a finite or denumerable domain of individuals).

For the proof, let A denote an arbitrary formula of the extended realm. We construct a formula B as the product (conjunction) of A, $(x)(x = x)$, and all the formulas that we obtain from Axiom 8 by substituting for F the functional variables occurring in A, that is, more precisely,

$$(x)(y)\{x = y \rightarrow [F(x) \rightarrow F(y)]\}$$

---

[16]That a system $\{f_1, f_2, \ldots, f_k; w_1, w_2, \ldots, w_l\}$ is part of another, $\{g_1, g_2, \ldots, g_k; v_1, v_2, \ldots, v_l\}$, is to mean that
1. The domain of individuals of the f_i is part of the domain of individuals of the g_i;
2. The f_i and the g_i coincide within the narrower domain;
3. For every i, $v_i = w_i$.

für alle zweistelligen Funktionsvariablen aus A (inklusive "=" selbst) und entsprechenden Formeln für die drei- und mehrstelligen Funktionsvariablen aus A. Sei B' die Formel, welche aus B entsteht, wenn man darin das =-Zeichen durch eine sonst in B nicht vorkommende Funktionsvariable G ersetzt. Im Ausdruck B' kommt dann das =-Zeichen nicht mehr vor, er ist also nach dem früher Bewiesenen entweder widerlegbar oder erfüllbar. Ist er widerlegbar, dann gilt dasselbe für B, das ja aus B' durch Einsetzung von = für G entsteht. B ist aber das logische Produkt aus A und einem aus den Axiomen 7, 8 offenbar beweisbaren Formelteil. Also ist in diesem Falle auch A widerlegbar. Nehmen wir jetzt an, daß B' durch ein gewisses System von Funktionen[17] (S) im abzählbaren Individuenbereich Σ erfüllbar sei. Aus der Bildungsweise von B' ist ersichtlich, daß g (d. h. die für G einzusetzende Funktion des Systems S) eine reflexive, symmetrische und transitive Relation ist, also eine Klasseneinteilung der Elemente von Σ erzeugt, und zwar derart, daß Elemente derselben Klasse, für einander eingesetzt, an dem Bestehen oder Nichtbestehen einer im System S vorkommenden Funktion nichts ändern. Identifiziert man daher alle zu derselben Klasse gehörigen Elemente miteinander (etwa indem man die Klassen selbst als Elemente eines neuen Individuenbereiches nimmt), so geht g in die Identitätsrelation über und man hat eine Erfüllung von B, also auch von A. Tatsächlich ist also A entweder erfüllbar[18] oder widerlegbar.

Eine andere Verallgemeinerung von Satz I erhält man durch Betrachtung von abzählbar unendlichen Mengen von logischen Ausdrücken. Auch für solche gilt ein Analogon zu I und II, nämlich:

Satz IX: *Jede abzählbar unendliche Menge von Formeln des engeren Funktionenkalküls ist entweder erfüllbar* (d. h. alle Formeln des Systems sind gleichzeitig erfüllbar) *oder sie besitzt ein endliches Teilsystem, dessen logisches Produkt widerlegbar ist.*

IX ergibt sich sofort aus:

358 | Satz X: *Damit ein abzählbar unendliches System von Formeln erfüllbar sei, ist notwendig und hinreichend, daß jedes endliche Teilsystem erfüllbar ist.*

Bezüglich Satz X stellen wir zunächst fest, daß man sich bei seinem Beweise auf Systeme von Normalformeln ersten Grades beschränken kann, denn durch mehrmalige Anwendung des beim Beweise von Satz III und IV verwendeten Verfahrens auf die einzelnen Formeln kann man zu jedem Formelsystem Σ ein solches von Normalformeln ersten Grades Σ' angeben,

---

[17]Falls in A auch Aussagevariable vorkommen, muß natürlich S außer Funktionen noch Wahrheitswerte für diese Aussagevariablen enthalten.

[18]Und zwar in einem höchstens abzählbaren Denkbereich (er besteht ja aus elementfremden Klassen des abzählbaren Individuenbereichs Σ).

for all singular functional variables of A,

$$(x)(y)(z)\{x = y \to [F(x, z) \to F(y, z)]\} \, \&$$
$$(x)(y)(z)\{x = y \to [F(z, x) \to F(z, y)]\}$$

for all binary functional variables of A (including "=" itself), and corresponding formulas for the n-ary functional variables of A for which $n \geq 3$. Let B' be the formula resulting from B when the sign "=" is replaced by a functional variable G not otherwise occurring in B. Then the sign "=" no longer occurs in the expression B', which, therefore, according to what was proved above, is either refutable or satisfiable. If it is refutable, so is B, since it results from B' through the substitution of "=" for G. But B is the logical product of A and a subformula that is obviously provable from Axioms 7 and 8. In this case, therefore, A, too, is refutable. Let us now assume that B' is satisfiable in the denumerable domain Σ of individuals by a certain system S of functions.[17] From the way in which B' is formed it is clear that g (that is, the function of the system S that is to be substituted for G) is a reflexive, symmetric, and transitive relation; hence it generates a partition of the elements of Σ, in such a way, moreover, that a function occurring in the system S continues to hold, or not to hold, as the case may be, when elements of the same class are substituted for one another. If, therefore, we identify with one another all elements belonging to the same class (perhaps by taking the classes themselves as elements of a new domain of individuals), then g goes over into the identity relation and we have a satisfying system of B, hence also of A. Consequently, A is indeed either satisfiable[18] or refutable.

We obtain a different generalization of Theorem I by considering denumerably infinite sets of logical expressions. For these, too, an analogue of Theorems I and II holds, namely

Theorem IX. *Every denumerably infinite set of formulas of the restricted functional calculus either is satisfiable* (that is, all formulas of the system are simultaneously satisfiable) *or possesses a finite subsystem whose logical product is refutable.*

IX follows immediately from

Theorem X. *For a denumerably infinite system of formulas to be satisfiable it is necessary and sufficient that every finite subsystem be satisfiable.*

Concerning Theorem X we first note that in proving it we can confine ourselves to systems of normal formulas of degree 1, for, by repeated application of the procedure used in the proofs of Theorems III and IV to

---

[17]If propositional variables also occur in A, S will, of course, have to contain, besides functions, truth values for these propositional variables.

[18]And, moreover, in an at most denumerable domain (for it consists of disjoint classes of the denumerable domain Σ of individuals).

derart, daß die Erfüllbarkeit irgend eines Teilsystems von Σ mit der des entsprechenden von Σ' gleichbedeutend ist.

Sei also

$$(\mathfrak{x}_1)(E\mathfrak{y}_1)A_1(\mathfrak{x}_1;\mathfrak{y}_1), \quad (\mathfrak{x}_2)(E\mathfrak{y}_2)A_2(\mathfrak{x}_2;\mathfrak{y}_2), \quad \ldots, \quad (\mathfrak{x}_n)(E\mathfrak{y}_n)A_n(\mathfrak{x}_n;\mathfrak{y}_n), \quad \ldots$$

ein abzählbares System Σ von Normalausdrücken ersten Grades, $\mathfrak{x}_i$ sei ein r_i-tupel, $\mathfrak{y}_i$ ein s_i-tupel von Variablen. $\mathfrak{x}_1^i,\mathfrak{x}_2^i,\ldots,\mathfrak{x}_n^i,\ldots$ sei eine Folge sämtlicher aus der Folge $x_0,x_1,x_2,\ldots,x_n,\ldots$ entnommener r_i-tupel nach steigender Indexsumme, ferner sei $\mathfrak{y}_k^i$ ein s_i-tupel von Variablen der obigen Folge von der Art, daß die Variablenfolge

$$\mathfrak{y}_1^1,\mathfrak{y}_2^1,\mathfrak{y}_1^2,\mathfrak{y}_3^1,\mathfrak{y}_2^2,\mathfrak{y}_1^3,\mathfrak{y}_4^1,\ldots \text{u. s. w.,}$$

wenn man darin jedes $\mathfrak{y}_k^i$ durch das entsprechende s_i-tupel von x ersetzt, mit der Folge $x_1,x_2,\ldots,x_n,\ldots$ identisch wird. Ferner definieren wir analog wie oben eine Folge von Formeln $\{B_n\}$ durch die Festsetzungen:

$$B_1 = A_1(\mathfrak{x}_1^1;\mathfrak{y}_1^1),$$
$$B_n = B_{n-1}\ \&\ A_1(\mathfrak{x}_n^1,\mathfrak{y}_n^1)\ \&\ A_2(\mathfrak{x}_{n-1}^2;\mathfrak{y}_{n-1}^2)\ \&\ \cdots$$
$$\&\ A_{n-1}(\mathfrak{x}_2^{n-1};\mathfrak{y}_2^{n-1})\ \&\ A_n(\mathfrak{x}_1^n;\mathfrak{y}_1^n).$$

Man übersieht leicht, daß $(P_n)B_n$ (d. h. die Formel, welche aus B_n entsteht, wenn sämtliche darin vorkommende Individuenvariable durch E-Zeichen gebunden werden) eine Folgerung der ersten n Ausdrücke des obigen Systems Σ ist. Ist also jedes endliche Teilsystem von Σ erfüllbar, dann auch jedes B_n. Wenn aber jedes B_n erfüllbar ist, dann auch das ganze System Σ (was nach der beim Beweise von Satz V (vgl. S. 355) angewendeten Schlußweise folgt), womit Satz X bewiesen ist. IX und X lassen sich nach dem beim Beweise von VIII angewendeten Verfahren ohne Schwierigkeit auf Formelsysteme, welche das =-Zeichen enthalten, ausdehnen.

Man kann Satz IX noch eine etwas andere Wendung geben, wenn man sich auf Formelsysteme ohne Aussagevariable beschränkt und diese als Axiomensysteme auffaßt, deren Grundbegriffe die vor|kommenden Funktionsvariablen sind. Dann besagt Satz IX offenbar, daß jedes endliche oder abzählbare Axiomensystem, in dessen Axiomen "alle" und "es gibt" sich niemals auf Klassen oder Relationen, sondern nur auf Individuen beziehen,[19] entweder widerspruchsvoll ist, d. h. ein Widerspruch sich in endlich vielen formalen Schritten herstellen läßt, oder eine Realisierung besitzt.

---

[19]Als Beispiel kann etwa das Hilbertsche Axiomensystem der Geometrie ohne die Stetigkeitsaxiome dienen.

the individual formulas, we can specify for every system Σ of formulas a system Σ' of normal formulas of degree 1 such that the satisfiability of any subsystem of Σ is equivalent to that of the corresponding subsystem of Σ'.

Thus let

$$(\mathfrak{r}_1)(E\mathfrak{y}_1)A_1(\mathfrak{r}_1;\mathfrak{y}_1), \ (\mathfrak{r}_2)(E\mathfrak{y}_2)A_2(\mathfrak{r}_2;\mathfrak{y}_2), \ \ldots, \ (\mathfrak{r}_n)(E\mathfrak{y}_n)A_n(\mathfrak{r}_n;\mathfrak{y}_n), \ \ldots$$

be a denumerable system Σ of normal expression of degree 1, and let $\mathfrak{r}_i$ be an r_i-tuple, and $\mathfrak{y}_i$ an s_i-tuple of variables. Let $\mathfrak{r}_1^i, \mathfrak{r}_2^i, \ldots, \mathfrak{r}_n^i, \ldots$ be the sequence of all r_i-tuples taken from the sequence $x_0, x_1, x_2, \ldots, x_n, \ldots$ and ordered according to increasing sum of the subscripts [and for equal sums according to some convention]; furthermore, let $\mathfrak{y}_k^i$ be an s_i-tuple of variables, of the sequence above, such that the sequence of variables

$$\mathfrak{y}_1^1, \mathfrak{y}_2^1, \mathfrak{y}_1^2, \mathfrak{y}_3^1, \mathfrak{y}_2^2, \mathfrak{y}_1^3, \mathfrak{y}_4^1, \ldots$$

becomes identical with the sequence $x_1, x_2, \ldots, x_n, \ldots$ if every $\mathfrak{y}_k^i$ is replaced by the corresponding s_i-tuple of the x. Further we define, in a way analogous to what was done above, a sequence $\{B_n\}$ of formulas by means of the stipulations

$$B_1 = A_1(\mathfrak{r}_1^1; \mathfrak{y}_1^1),$$
$$B_n = B_{n-1} \ \& \ A_1(\mathfrak{r}_n^1, \mathfrak{y}_n^1) \ \& \ A_2(\mathfrak{r}_{n-1}^2; \mathfrak{y}_{n-1}^2) \ \& \ \cdots$$
$$\& \ A_{n-1}(\mathfrak{r}_2^{n-1}; \mathfrak{y}_2^{n-1}) \ \& \ A_n(\mathfrak{r}_1^n; \mathfrak{y}_1^n).$$

We can easily see that $(P_n)B_n$ (that is, the formula that results from B_n when all individual variables occurring in it are bound by existential quantifiers) is a consequence of the first n expressions of the system Σ given above. If, therefore, every finite subsystem of Σ is satisfiable, so is every B_n. But, if every B_n is satisfiable, so is the entire system Σ (as follows by the argument used in the proof of Theorem V (above, page 115)), and Theorem X is thus proved. Theorems IX and X can be extended without difficulty, by the procedure used in the proof of Theorem VIII, to systems of formulas containing the sign "=".

We can also give a somewhat different turn to Theorem IX if we confine ourselves to systems of formulas without propositional variables and regard them as systems of axioms whose primitive notions are the functional variables occurring in them. Then Theorem IX clearly asserts that every finite or denumerable axiom system in whose axioms "all" and "there exists" never refer to classes or relations but only to individuals[19] either is

---

[19]Hilbert's axiom system for geometry, without the axioms of continuity, can perhaps serve as an example.

Zum Schluß möge noch die Frage der Unabhängigkeit der Axiome 1–8 erörtert werden. Was die Aussagenaxiome 1–4 betrifft, so ist ja bereits von P. Bernays[20] gezeigt worden, daß keines von ihnen aus den drei anderen folgt. Daß an ihrer Unabhängigkeit auch durch Hinzunahme der Axiome 5–8 nichts geändert wird, kann durch genau dieselben Interpretationen gezeigt werden, wie sie Bernays verwendet, indem man diese auch auf Formeln, die Funktionsvariable und das =-Zeichen enthalten, ausdehnt durch die Festsetzung, daß:

1. die Präfixe und Individuenvariablen weggelassen werden,

2. in dem übrigbleibenden Formelteil die Funktionsvariablen ebenso wie Aussagevariable behandelt werden sollen,

3. für das Zeichen "=" immer nur einer der "ausgezeichneten" Werte eingesetzt werden darf.

Um die Unabhängigkeit von Axiom 5 zu zeigen, ordnen wir jeder Formel eine andere dadurch zu, daß wir die Bestandteile der Form:

$$(x)F(x), (y)F(y), \ldots; \quad (x)G(x), (y)G(y), \ldots; \ldots,^{21}$$

falls solche in ihr vorkommen, durch $X \vee \overline{X}$ ersetzen. Dadurch gehen die Axiome 1–4, 6–8 in allgemeingültige Formeln über und dasselbe gilt, wie man sich durch vollständige Induktion überzeugt, von allen aus diesen Axiomen nach Schlußregel 1–4 abgeleiteten Formeln, während Axiom 5 diese Eigenschaft nicht besitzt. In genau der gleichen Weise zeigt man die Unabhängigkeit von Axiom 6, nur muß man hier $(x)F(x), (y)F(y), \ldots$ etc. durch $X \,\&\, \overline{X}$ ersetzen. Um die Unabhängigkeit von Axiom 7 zu beweisen, bemerken wir, daß Axiom 1–6 und 8 (und daher auch alle daraus abgeleiteten Formeln) allgemeingültig bleiben, wenn man die Relation der Identität durch die leere Relation ersetzt, während das bei Axiom 7 nicht der Fall ist. Analog bleiben die aus Axiom 1–7 abgeleiteten 360 Formeln auch dann noch allgemeingültig, wenn man die Identitäts|relation durch die Allrelation ersetzt, während das für Axiom 8 (im Individuenbereich von mindestens zwei Individuen) nicht der Fall ist. Man kann sich auch leicht davon überzeugen, daß keine der Schlußregeln 1–4 überflüssig ist, worauf aber hier nicht näher eingegangen werden möge.

---

[20]Vgl. *Bernays 1926*.

[21]D. h. die einstelligen Funktionsvariablen $F, G, \ldots$ etc. mit einem vorgesetzten Alloperator, dessen Wirkungsbereich lediglich das betreffende $F, G, \ldots$ mit der zugehörigen Individuenvariablen ist.

inconsistent (that is, a contradiction can be constructed in a finite number of formal steps) or possesses a model.

Finally, let us also discuss the question of the independence of Axioms 1–8. As far as Axioms 1–4 (those of the propositional calculus) are concerned, it has already been shown by P. Bernays[20] that none of them follows from the other three. That their independence is not affected even by the addition of Axioms 5–8 can be shown by means of the very same interpretations that Bernays uses, provided that, in order to extend them to formulas containing functional variables and the sign "=", we make the following stipulations:

1. The quantifiers and individual variables are omitted;

2. In what remains of each formula the functional variables are to be treated just like propositional variables;

3. Only "distinguished" values may ever be substituted for the sign "=".

To demonstrate the independence of Axiom 5, we associate with each formula another one, which we obtain by replacing components of the form

$$(x)F(x), (y)F(y), \ldots ; (x)G(x), (y)G(y), \ldots ; \ldots ,^{21}$$

should such occur, by $X \vee \overline{X}$. Then Axioms 1–4 and 6–8 go over into valid formulas, and the same holds, as we can convince ourselves by mathematical induction, of all formulas derived from these axioms by Rules of inference 1–4; Axiom 5, however, does not possess this property. The independence of Axiom 6 can be shown in exactly the same way, except that here $(x)F(x), (y)F(y), \ldots$, and so on must be replaced by $X \& \overline{X}$. To prove the independence of Axiom 7 we note that Axioms 1–6 and 8 (and therefore also all formulas derived from them) remain valid if the identity relation is replaced by the empty relation, whereas this is not the case for Axiom 7. Similarly, the formulas derived from Axioms 1–7 remain valid when the identity relation is replaced by the universal relation, whereas this is not the case for Axiom 8 (in a domain of at least two individuals). We can also readily see that none of the Rules of inference 1–4 is redundant, but we shall not look into that more closely here.

---

[20]See *Bernays 1926*.

[21]That is, the singulary functional variables $F, G, \ldots$ preceded by a universal quantifier whose scope is just the $F, G, \ldots$ in question, along with the associated individual variable.

Über die Vollständigkeit des Logikkalküls
(*1930a*)

Bei der axiomatischen Begründung der Logik, wie sie z. B. in den *Principia Mathematica* vorliegt, taucht die Frage auf, ob die an die Spitze gestellten Axiome "vollständig" sind, d. h. wirklich dazu ausreichen, jeden richtigen Satz der Logik auf formalem Wege zu deduzieren. Dieses Problem ist bisher nur für die einfachsten logischen Sätze, nämlich die des Aussagenkalküls, gelöst. Die Antwort fällt positiv aus, d. h.: Jede richtige (allgemeingültige) Aussageformel folgt aus den Axiomen der *Principia Mathematica*. Vortragender zeigt, wie man diesen Satz auf die Formeln des engeren Funktionenkalküls (Formeln ohne gebundene Funktionsvariable) ausdehnen kann.

On the completeness of the calculus of logic
(*1930a*)

When one provides an axiomatic foundation for logic, as, for example, is done in *Principia mathematica*, the question arises whether the axioms initially adopted are "complete", that is, whether they actually suffice for the formal deduction of every correct proposition of logic. This problem has hitherto been solved only for the simplest logical propositions, namely those of the propositional calculus. The answer turns out to be positive, that is, every correct (valid) propositional formula follows from the axioms of *Principia mathematica*. The speaker shows how this theorem can be extended to the formulas of the restricted functional calculus (formulas without bound functional variables).

Introductory note to *1930b*, *1931* and *1932b*

Gödel's *1931* was undoubtedly the most exciting and the most cited article in mathematical logic and foundations to appear in the first eighty years of this century. An abstract of it, *1930b*, was presented to the Vienna Academy of Sciences by Hans Hahn on 23 October 1930. The full text of *1931* was received on 17 November 1930 for publication in *Monatshefte für Mathematik und Physik*. The note *1932b*, dated 22 January 1931, gave a more general presentation of Gödel's *1931* theorems using Peano arithmetic instead of the simple theory of types as the basic system.

In *1918* and *1922* Hilbert had addressed the crisis in the foundations of mathematics, which stemmed from the appearance around the beginning of the century of paradoxes in Cantor's theory of sets, with the following two-part proposal.

First, the structure of the language, the definitions, and all of the mathematical axioms and principles of logic, to be used in developing a suitably selected portion of the existing classical mathematics should be completely specified in terms of just the forms of the objects of the language. Briefly, the portion in question should be *formalized*, or embedded in a *formal system S*. In *S*, we will have a list chosen in advance of (*primitive*) *symbols*; certain finite sequences of (occurrences of) the symbols are defined to be *formulas*; and certain finite sequences of the formulas are to be *proofs*, the last formulas of which are *provable* or constitute (*formal*) *theorems*.

The deduction of propositions of the selected portion of mathematics, when formalized in *S*, consists simply of mechanical manipulations of the formal objects, with no reliance on their meanings (even though the meanings are what make the system of interest to us as mirroring informal mathematics). If we should be tempted to use something from the meanings or interpretation, what we use should have been put into the system *S* in the form of additional axioms or rules of inference.

In mathematics as developed informally, there is a whole range of degrees of abstraction and idealization in the concepts used, beginning with the sequence of the natural numbers $0, 1, 2, \ldots$, which we can envision very clearly as finite objects fixed by their position in the sequence, and continuing up through increasingly complicated and transcendental notions, such as those of real numbers and sets of higher and higher cardinality. If mathematicians, on introspection, had not previously admitted that they were using idealized concepts outrunning clear intuitive

meanings, they had to do so when they found that by going just a little further than before in their reasoning with idealized concepts they had fallen into contradictions (the paradoxes).

Second, Hilbert proposed that, by mathematical reasoning on the most elementary and intuitive level, it should be proved that our formal system S is (*simply*) *consistent*, i.e., that no two sequences of its formulas exist of which one is a proof in S of a formula A and the other of its negation ~A.

For, indeed, the objects of S are all finite: just symbols, finite sequences of symbols, and finite sequences of finite sequences of symbols. So no matter how abstract and transcendental the concepts in mathematics as practiced informally on its higher levels, the formal system itself can be treated as the subject matter of a new branch of mathematics (to be called "proof theory" or "metamathematics") on the same elementary level as that of the natural numbers. For metamathematical investigations in general (and in particular for a consistency proof for S), Hilbert's program called for using only the most secure methods, which in German he called "finit", usually translated as "finitary" (after *Kleene 1952*) and occasionally as "finitistic". They can be characterized as methods not using any *completed* infinity; i.e., no objects themselves infinite are to be used, and only *potentially* infinite collections of them, like the natural number sequence $0, 1, 2, \ldots$ considered as unbounded above but not as a completed collection.

After finishing his doctoral dissertation *1929*, Gödel set to work to prove the consistency of analysis, pursuant to Hilbert's program. He proposed to divide the difficulties of the problem by first reducing the consistency of analysis to that of number theory. He began by considering the model in which the set variables are interpreted as ranging over sets definable in arithmetic. He soon realized that he would need not just the consistency of number theory but also its truth. This led him to ponder Richard's paradox and the paradox of the Liar, a formal analog of which can be used to infer that truth in number theory cannot be defined in number theory (cf. *Wang 1981*, page 654). However, provability in number theory can be defined in number theory. Therefore, if the provable formulas are all true, there must be some true but unprovable formulas. Thus Gödel came to find the results he published in *1931*, which shook to its foundations Hilbert's program for foundations without quite demolishing it.

When, as Hilbert proposed, we have formalized a domain of mathematics in a formal system S, we have intended that S should encompass everything that is necessary for proving propositions belonging to that domain. So we hope that, for each formula A without free variables (each "closed formula" or "sentence"), either A or ~A will be provable in S, i.e., that S will be (*simply*) *complete*.

As the first of the two celebrated incompleteness theorems of *1931*, Gödel showed that this is not even the case when the usual elementary number theory is taken as the domain, with any of various formal systems S that come to mind for formalizing it. That is, for each of these systems, there is a proposition A of elementary number theory which is *formally undecidable* in S, i.e., such that neither the formula A expressing A nor its negation ~A is provable in S. (Cf. the title of *1931*.)

How did Gödel obtain this result?

As already indicated, in investigating a formal system S we have as the objects of our study just the symbols from a preassigned list, the finite sequences of them, and the finite sequences of finite sequences of them. So indeed we can correlate natural numbers to all these objects, distinct numbers to distinct objects. Let us fix such a correlation, calling the correlated numbers (as is now customary) *Gödel numbers* of the formal objects. This opens up the possibility that, in a formal system S embodying number theory, there will be formulas which under the interpretation talk about natural numbers, but which also, via the Gödel numbering, say things about objects of S.

Gödel ingeniously exploited this possibility in order to construct in his system S a formula A which says that every natural number x is not the Gödel number of a proof in S of a certain formula B, with B happening to be A itself. In brief, A says "I am unprovable (in S)". This is an adaptation of the ancient paradox of the Liar, in which we have a statement asserting its own falsity. For "false" Gödel simply substituted "unprovable", thereby making the statement expressible in S.

Let us suppose that S has the feature, which we normally intend it to have, that in it only true formulas are provable. Now, if Gödel's formula A were provable (in S), it would by what it says be false; so, by our supposition, it is unprovable, hence true; and so, again by our supposition, ~A is also unprovable. Thus the proposition A expressed by A is formally undecidable in S.

In Section 1 of *1931* Gödel undertakes to "sketch the main idea of the proof [of his 'first incompleteness theorem'], of course without any claim to complete precision." This corresponds to the foregoing preliminary outline; but he actually sketches the construction of the formula A (the "S" and "A" here are "PM" and "$[R(q); q]$" in his Section 1).

In the foregoing sketch, the interpretation of the formal system as formalizing a system of notions and propositions was invoked when we talked of constructing a formula A or $[R(q; q)]$ which should express its own unprovability and assumed that only true formulas are provable in S or PM. Yes, one wants this to be so, to the extent that we can be satisfied that the formulas have clear meanings. But the informal concept of truth was not commonly accepted as a definite mathematical notion, especially for systems like PM or Zermelo–Fraenkel set theory.

In Hilbert's metamathematics one is supposed to deal solely with the forms of the formulas, using only finitary reasoning.

So, near the end of Section 1, Gödel says, "The purpose of carrying out the above proof with full precision in what follows is, among other things, to replace [the assumption that every provable formula is true in the interpretation considered] by a purely formal [or metamathematical] and much weaker one."

We now proceed to articulate what Gödel does, with this aim among others, in his long Section 2.

First, Gödel specifies the symbols of the formal system P with which he will primarily deal, and lists its axioms and rules of inference. It is the system obtained from that of Whitehead and Russell's *Principia mathematica* (*1925*) by omitting the ramification of the types, taking the natural numbers as the lowest type, and adding the axioms for them which are commonly called "Peano's axioms" (*Peano 1889* and *1891*) but which Peano himself states can be found in *Dedekind 1888*. A type-2 variable is used in stating the fifth axiom (induction). In other words, it is obtained by superposing the simple theory of types (Ramsey's *1926* modification of *Principia mathematica*) on Peano's axioms for the natural numbers taken as the lowest type.

Second, Gödel establishes a specific "Gödel numbering" of the formal objects. He assigns distinct odd numbers to each of the primitive symbols; and further maps any finite sequence $n_1, n_2, \ldots, n_k$ of natural numbers onto the single natural number $2^{n_1} \cdot 3^{n_2} \cdot \ldots \cdot p_k^{n_k}$, where p_k is the kth prime number. Thus, if $n_1, n_2, \ldots, n_k$ are the Gödel numbers of the successive members of a sequence of primitive symbols, $2^{n_1} \cdot 3^{n_2} \cdot \ldots \cdot p_k^{n_k}$ is the Gödel number of the sequence; and if $n_1, n_2, \ldots, n_k$ are already the Gödel numbers of finite sequences of the symbols, $2^{n_1} \cdot 3^{n_2} \cdot \ldots \cdot p_k^{n_k}$ is the Gödel number of the sequence of those finite sequences.

Now concepts which apply to objects of the formal system, like "formula" and "proof", have corresponding concepts applying to the Gödel numbers of those objects, written "FORMULA" and "PROOF".[a] So we come face to face with the challenge of working metamathematically with the various operations and relations that have thus arisen in relation to the system.

To this end, Gödel takes a fundamental step. Actually, it is revealed in his account only later in the section. But we prefer to state it now, as the **third** step, since his next material is best understood in relation to it.

---

[a]In his original German, Gödel used italics, "*Formel*" and "*Beweis*", which in our reprinting and in the English translation are rendered by small capitals. We are using italics for emphasis, accomplished in the original German by letter spacing.

He singles out for attention number-theoretic relations $R(x_1, \ldots, x_n)$ which (in the terminology of *Kleene 1952*, page 195) are *numeralwise expressible in P* (or similarly in other systems S). These are the ones for which there is in P a formula $R(x_1, \ldots, x_n)$ (with just the free variables $x_1, \ldots, x_n$), said to *numeralwise express* $R(x_1, \ldots, x_n)$ in P, such that, for each $x_1, \ldots, x_n$,

$$R(x_1, \ldots, x_n) \rightarrow \{R(\boldsymbol{x}_1, \ldots, \boldsymbol{x}_n) \text{ is provable in } P\},$$
$$\overline{R}(x_1, \ldots, x_n) \rightarrow \{\sim R(\boldsymbol{x}_1, \ldots, \boldsymbol{x}_n) \text{ is provable in } P\},$$

where "$\overline{R}$" is read "not R", and we are not following Gödel's notation but are using "$\boldsymbol{x}_1, \ldots, \boldsymbol{x}_n$" to stand for the *numerals* expressing the natural numbers $x_1, \ldots, x_n$ in P (thus $\boldsymbol{x}_1$ is $f \ldots f0$ with x_1 f's), and "$R(\boldsymbol{x}_1, \ldots, \boldsymbol{x}_n)$" for the result of substituting $\boldsymbol{x}_1, \ldots, \boldsymbol{x}_n$ respectively for the free occurrences of the variables $x_1, \ldots, x_n$ in $R(x_1, \ldots, x_n)$. The reader should be attentive to the distinction in our notation between roman type letters used in naming formal objects (except bold face italic $\boldsymbol{x}$ is used for the numeral for the natural number x) and italic letters used in naming entities in informal number theory (except italic f is used for the formal successor symbol, and italic E in the formal existential quantifier (Ex)).[b]

The concept of "numeralwise expressibility" (and "numeralwise expressing") enters Gödel's paper implicitly in Theorem V, and explicitly a bit after he has finished proving Theorem VI, where he introduces the term "decidable (*entscheidungsdefinit*) [in P]" for it.[c]

Theorem V illustrates Gödel's propensity for speaking in terms of his numbers, rather than directly in terms of the formal objects (which we

[b]To facilitate reading this note against Gödel's paper *1931*, we retain Gödel's *logical* symbolism, both formal (introduced in his "first step") and informal or intuitive (introduced in his "fourth step"). Thus for the formal quantifiers he uses xΠ ("for all x") and (Ex) ("there exists an x"), and for the informal ones (x) ("for all x") and (Ex) ("there exists an x"). For negation ("not A") he uses $\sim$A formally and $\overline{A}$ informally; for conjunction ("A and B") A.B formally and $A \,\&\, B$ informally; for implication ("A implies B" or "if A, then B") A $\supset$ B formally and $A \rightarrow B$ informally; for disjunction ("A or B") $\lor$ both formally and informally. For equivalence, he uses A $\equiv$ B formally (the same as he uses informally beginning at his footnote 33 for definitional equality) and $A \sim B$ informally (the same as for his formal negation, except that we are printing it bigger). However, as just remarked, we are using roman letters formally and italic informally (Gödel generally uses italic for both). The formal symbolism described is that for his formal system P (for the restricted functional calculus in Theorem X it is different).

[c]To say a relation $R(x_1, \ldots, x_n)$ is "decidable" (in German, "entscheidbar"), unqualified by "in S" for a formal system S, would ordinarily be read as saying that there is an effective method or algorithm to decide as to the truth or falsity of each value of $R(x_1, \ldots, x_n)$.

prefer as being more understandable). Thus, instead of saying "a formula $R(x_1, \ldots, x_n)$ with just the free variables $x_1, \ldots, x_n$", he talks of "an n-place RELATION SIGN r (with the FREE VARIABLES $u_1, u_2, \ldots, u_n$)"; and, instead of saying the formula $R(x_1, \ldots, x_n)$ is provable, he says that its Gödel number is PROVABLE, which in the notation he has meanwhile introduced reads

$$\text{"Bew}\left[Sb\left(r^{u_1 \ldots u_n}_{Z(x_1) \ldots Z(x_n)}\right)\right]\text{".}$$

In fact, the relations $R(x_1, \ldots, x_n)$ that we will be considering in this connection will be ones that are constructively defined. For such relations, the property that a given formula $R(x_1, \ldots, x_n)$ numeralwise expresses a given relation $R(x_1, \ldots, x_n)$ can be handled in Hilbert's metamathematics. It says that all of the family of formulas are provable which are correlated in a certain definite manner to an informal constructively-defined relation $R(x_1, \ldots, x_n)$. This will be very useful in organizing our mathematical thinking about the system P.

Gödel needs to show that each of a certain stock of constructively-defined number-theoretic relations which arise, via his numbering, in studying the formal system P is numeralwise expressible in P. These are the specific relations he ultimately needs, as well as the ones that arise naturally in building toward them. He has a strategy by which he mass-produces these results. This consists in defining a class of number-theoretic functions and relations, each of which is constructively defined, and the relations among which include his desired stock and are by a general theorem each of them numeralwise expressible in P. I now list as the "fourth", "fifth" and "sixth" steps his implementation of this strategy.

Fourth, Gödel gives an exact definition of a class of number-theoretic functions $\phi(x_1, \ldots, x_n)$, which he calls "recursive". They had been used earlier (in *Dedekind 1888, Skolem 1923, Hilbert 1926* and *Ackermann 1928*), and are now called "primitive recursive" (after *Kleene 1936*).[d]

---

[d]Gödel introduces the concept of *degree* (*Stufe* in the German) of a primitive recursive function ϕ and argues by induction on the degree of ϕ in proving Theorems V and VII. This constitutes a slight blemish on his presentation. If he had really needed to use the degrees, his claims to constructiveness (following the proofs of Theorems VI and XI) would be false. For the notion of "degree" is not constructive. But in fact, for every primitive recursive function ϕ he needs to consider, he will have in hand a particular finite sequence of definitions of functions $\phi_1, \phi_2, \ldots, \phi_n$ with $\phi_n = \phi$ as described in his definition of *recursive*. We can call such a sequence a "primitive recursive definition of ϕ". These are the objects he should work with, and the length n of a given such definition can replace his degree (i.e., he does not need to use the minimum such n possible for ϕ, only the one he came up with in finding that the ϕ in question is primitive recursive). (continued on next page)

A number-theoretic relation $R(x_1, \ldots, x_n)$ $(n \geq 1)$ is *primitive recursive* if and only if there is a primitive recursive function $\phi(x_1, \ldots, x_n)$ such that, for all $x_1, \ldots, x_n,$[e]

$$R(x_1, \ldots, x_n) \sim [\phi(x_1, \ldots, x_n) = 0].$$

The primitive recursive functions and relations are constructively defined (as he remarks in his footnote 28). Theorems I–IV provide some very useful closure properties of the class of these functions and relations.

Fifth, Gödel shows that, of a list of 46 number-theoretic functions and relations which arise in discussing his system P via his numbering, the first 45 are primitive recursive. The list is constructed with great efficiency, by starting with what is needed to treat sequences of numbers primitive recursively, and proceeding step by step through each syntactic notion.

Sixth, Gödel states as his Theorem V that every primitive recursive relation is numeralwise expressible in P. The proof is given only in outline.

In fact, the proof of Theorem V is simplified if one carries it out for relations $R(x_1, \ldots, x_n, y)$ of the form $\phi(x_1, \ldots, x_n) = y$ for which there is a formula $R(x_1, \ldots, x_n, y)$ with just the free variables $x_1, \ldots, x_n, y$ (said in *Kleene 1952*, page 200, to *numeralwise represent* the function $\phi(x_1, \ldots, x_n)$ in P) such that, for each $x_1, \ldots, x_n, y$,

$$\phi(x_1, \ldots, x_n) = y \rightarrow$$
$$\{R(\boldsymbol{x_1}, \ldots, \boldsymbol{x_n}, \boldsymbol{y}) \cdot (E!y)R(\boldsymbol{x_1}, \ldots, \boldsymbol{x_n}, y) \text{ is provable in } P\},$$

where "$(E!y)$" expresses in P "there is a unique y". This is done in detail in *Kleene 1952* for his formal system (which can be described as Peano arithmetic PA).

What Gödel basically needs for his proof of Theorem VI is that, for each of his stock of constructively-defined relations $R(x_1, \ldots, x_n)$, there be a formula $R(x_1, \ldots, x_n)$ numeralwise expressing it in P. But his

---

For a reader familiar with some theory cited in the introductory note to *Gödel 1934*, it will be an easy exercise to show that there is no effective general method for finding Gödel's degree from a primitive recursive definition: If there were, applying it, for each a, to $\tau_1(a, a, x)$ as a primitive recursive function of x (τ_1 being the representing function of T_1; see footnote e), we could effectively decide whether or not $(Ex)T_1(a, a, x)$.

[e]In *1934* Gödel uses specifically the function $\phi(x_1, \ldots, x_n)$, called the *representing function* of $R(x_1, \ldots, x_n)$, whose value is 0 or 1 according as $R(x_1, \ldots, x_n)$ is true or false.

way of getting such relations $R(x_1, \ldots, x_n)$ with corresponding formulas $\mathrm{R}(\mathrm{x}_1, \ldots, \mathrm{x}_n)$ is so convenient that he focuses on it, and (at the end of his outline of his proof of Theorem V) he calls the Gödel numbers r of the formulas obtained by his procedure *recursive*. In view of the change in terminology since Gödel's paper, such formulas $\mathrm{R}(\mathrm{x}_1, \ldots, \mathrm{x}_n)$ are here called *primitive recursive*.

Using notations that entered the literature later, let us say formulas of the form $\mathrm{xIIR}(\mathrm{x})$ (respectively, of the form $\sim\mathrm{xIIR}(\mathrm{x})$), where $\mathrm{R}(\mathrm{x})$ is a primitive recursive formula, belong to the class Π_1^0 (respectively, to the class Σ_1^0).[f]

Now, **seventh**, Gödel was in a position to formulate and establish as Theorem VI, in purely metamathematical terms, the promised result (his "first incompleteness theorem"): there is a sentence A such that neither A nor $\sim$A is provable in S.

As the system S for which he establishes this, he takes any system P_κ which has the same symbols as P and comes from P by adding a class of axioms, the class κ of whose Gödel numbers is primitive recursive (i.e., κ is a primitive recursive class of FORMULAS).[g] These systems P_κ include P itself as the case when κ is the empty class. (He does not use the notation "P_κ".)

As the undecidable sentence A, he uses (in our notation) the Π_1^0 sentence $\mathrm{xIIR}(\mathrm{x}, \boldsymbol{q})$ with the R and q which we shall introduce in a moment.

As the promised metamathematical hypothesis, he requires that P_κ be "ω-consistent" in the sense we define now. Remember, a system S is *(simply) consistent* if and only if in it no two formulas A and $\sim$A are both provable. Now, a system S is *ω-consistent* if and only if in

---

[f]This is an application to metamathematical use of the notations Π_i^0 and Σ_i^0 which were introduced by *Addison 1958* and *Mostowski 1959* to describe the levels in the arithmetical hierarchy of number-theoretic relations that had previously been introduced by *Kleene 1943* and *Mostowski 1947*. For $i > 0$, the Π_i^0 relations $R(x_1, \ldots, x_n)$ (or for $n = 0$, statements) are those which can be expressed by a prefix of i alternating quantifiers $(y_1)(Ey_2)(y_3) \ldots$, with a universal quantifier (sometimes written using "Π", as Gödel does formally) first, applied to a primitive recursive relation $P(x_1, \ldots, x_n, y_1, \ldots, y_i)$; and Σ_i^0 relations similarly with an existential quantifier (sometimes written using "Σ") first (or equivalently, in the negation of the form for a Π_i^0 relation). The original definition, good for $i \geq 0$, used a "general recursive" P, but by *Kleene 1943* is equivalent for $i > 0$ to the above, and makes $\Pi_0^0 = \Sigma_0^0 = \Pi_1^0 \cap \Sigma_1^0$. (In the metamathematical application, as one finds it in *Smoryński 1977*, page 843, Π_0^0 and Σ_0^0 are just the primitive recursive formulas, so the usage for $i = 0$ is different than for the relations in the arithmetical hierarchy.)

[g]To make the proof of Theorem VI constructive, the parameter should be a primitive recursive definition of the representing function of the class κ (or a formula numeralwise expressing the relation $x \in \kappa$); see footnotes d and e.

it, for no formula S(x) containing free just the variable x, are all of the formulas A($\boldsymbol{x}$) for $x = 0, 1, 2, \ldots$ and also the formula ~xΠA(x) provable. In fact, this notion will be applied only to the Σ_1^0 formula ~xΠR(x, $\boldsymbol{q}$). This case is referred to in the current literature as 1-*consistency* (cf. *Smoryński 1977*, pages 851–852). The import of 1-consistency is simply this: every provable Σ_1^0 sentence is true. This is called the Σ_1^0-*reflection* principle. One easily sees that ω-consistency or 1-consistency implies (simple) consistency.

So Theorem VI becomes in our notation (with R and q still to be specified):

Theorem VI. *If* P_κ *is* ω-*consistent, then neither* xΠR(x, $\boldsymbol{q}$) *nor* ~xΠR(x, $\boldsymbol{q}$) *is provable in* P_κ.

Using primitive recursive relations and functions from Gödel's stock and the relation $a \,\epsilon\, \kappa$ (assumed to be primitive recursive), we can assemble a primitive recursive relation $R(x, v)$ which, *in the case that v is the Gödel number of a formula, call it* A$_v$(v), *with just the one free variable* v, *says the following:* x *is not the Gödel number of a proof in* P_κ *of* A$_v$($\boldsymbol{v}$), i.e., of the formula that comes from A$_v$(v) by substituting for the variable v the numeral $\boldsymbol{v}$ for the natural number v. By Theorem V, $R(x, v)$ is numeralwise expressed in P and hence in P_κ by a formula R(x,v). Now consider the formula xΠR(x,v), which has just the free variable v. Let its Gödel number be q (so it is A$_q$(v)). As our undecidable formula we take xΠR(x, $\boldsymbol{q}$). This says that, for all x, x is not the Gödel number of a proof in P_κ of xΠR(x, $\boldsymbol{q}$); in brief, it says "I am unprovable in P_κ."[h]

For the first half of Theorem VI (that xΠR(x, $\boldsymbol{q}$) is unprovable in P_κ), simple consistency of P_κ suffices as the hypothesis. For, if xΠR(x, $\boldsymbol{q}$) were provable (in P_κ), there would be a proof of it, say with the Gödel number x, i.e., $R(x, q)$ would be false, so ~R($\boldsymbol{x}$, $\boldsymbol{q}$) and thence ~xΠR(x, $\boldsymbol{q}$) would be provable, contradicting simple consistency.

---

[h]The diagonalization used here can be separated from the present application to give the following general "self-referential (fixed point, or diagonalization) lemma": *For each formula* B(v) *with one free variable* v, *we can find a closed formula* (or "sentence") C *with Gödel number* q *such that* C $\equiv$ B($\boldsymbol{q}$) *is provable in* P. Method of proof: Take p to be the Gödel number of $(Ey)[B(y) \cdot S(v, \boldsymbol{w}, y)]$, where w is the Gödel number of v and S(v, w, y) (given by the simplified proof of Theorem V) numeralwise represents the primitive recursive function $Sb(v^w_{Z(v)})$ composed using Gödel's Definitions 17 and 31, and $q = Sb(p^w_{Z(p)})$. For Theorem VI, we pick B(v) to express that v is not the Gödel number of a formula provable in P_κ.

The self-referential lemma is stated informally in *Gödel 1934*, §7, where he credits it to *Carnap 1934a*, page 91. It appears in more precise form in *Rosser 1939* as Lemma 1, in *Feferman 1960* as Lemma 5.1, and in *Smoryński 1977*, page 827. There is an obvious generalization to the case that B contains free variables besides v.

The other half (that, if P_κ is ω-consistent, ~xΠR(x, q) is unprovable in it) follows from the first part by Σ_1^0 reflection, since by it ~xΠR(x, q) is a Σ_1^0 sentence falsely asserting that xΠR(x, q) is provable. Or directly, by the argument of the first part (using simple consistency), none of $R(x, q)$ for $x = 0, 1, 2, \ldots$ can be false, so R(x, q) for $x = 0, 1, 2, \ldots$ are all provable, so ~xΠR(x, q) could not be provable without violating 1-consistency.

Note that, if P_κ is (simply) consistent, xΠR(x, q) is a true Π_1^0 sentence which is unprovable. P_κ is thus ω-*incomplete*, in the terminology of *Tarski 1933*, R(x, q) being an example of a formula A(x) such that A(x) for $x = 0, 1, 2, \ldots$ are all provable but xΠA(x) is not.

As a further postscript to Theorem VI, if P_κ is ω-consistent (or 1-consistent), either one of xΠR(x, q) and ~xΠR(x, q) can be adjoined to P_κ as a new axiom to give a consistent extention $P_{\kappa'}$. For, if the adjunction of one led to a contradiction, the other would be provable in P_κ, contradicting Theorem VI. Indeed, if P_κ is consistent, the adjunction of ~xΠR(x, q) gives an example of a simply consistent but ω-inconsistent (and 1-inconsistent) formal system $P_{\kappa'}$. This shows that Hilbert's proposal to prove just the simple consistency of a formal system S, if one should succeed in it, would not rule out there being in S false theorems—like ~xΠR(x, q) in $P_{\kappa'}$ if P_κ is simply consistent. Gödel stressed this in his unexpected statement *1931a* at the Königsberg meeting.

How general is Gödel's incompleteness result? In the first place, he observes that Theorem VI generalizes from a primitive recursive class κ to a class κ that is numeralwise expressible (*entscheidungsdefinit*) in P.

Furthermore, he observes that only two properties of the system P were used:

1. The class of the axioms and the relation of immediate consequence are primitive recursively definable (as soon as we replace the primitive signs in some way by natural numbers).

2. Every primitive recursive relation is numeralwise expressible in P.

In his footnote 48a he expresses the view that "the true reason for the incompleteness inherent in all formal systems of mathematics is that the formation of ever higher types can be continued into the transfinite ... while in any formal system at most denumerably many of them are available." Implicit in this remark is that the adjunction of higher types to a formal system permits one to define the notion of truth for that system, then to show that all its provable sentences are true, and hence to decide the sentence shown in Theorem VI to be undecidable in the system.

In Section 3, Theorem VII states that every primitive recursive relation is *arithmetical*, i.e., can be defined using just constant and variable natural numbers, addition $+$, multiplication $\cdot$, and the usual apparatus of elementary predicate logic with quantifiers "for all x" and equality

"=" applied only to the natural numbers. The key step is the use of the Chinese remainder theorem in the proof of its Lemma 1.[i] It follows, as Theorem VIII, that the undecidable propositions of Theorem VI are arithmetical.

In the proof of Theorem X, Gödel shows quite explicitly in a uniform way that each Π_1^0 statement is reducible to the statement of the satisfiability of a certain formula in the first-order predicate calculus (cf. footnote f). As he remarks at the end of Section 3, this argument can be formalized in P. It follows, as he states in Theorem IX, that there are formulas of the restricted predicate calculus for which neither the statement asserting (via a Gödel number of it) its validity nor the statement asserting the existence of a counterexample is provable in P_κ.

Kleene in *1936*, page 741, gave an example of a non-recursive Σ_1^0 (and hence of a non-recursive Π_1^0) class. Thence the reduction that Gödel used in proving Theorem X made immediate from Church's *1936* thesis (discussed in the note to *Gödel 1934*) the result of *Church 1936a* and *Turing 1937* that the Hilbert Entscheidungsproblem for the restricted predicate calculus is unsolvable. This was noted in *Davis 1965*, page 109. Church's examples only directly provided classes of the form $(Ex)R(a, x)$ with R shown to be general recursive, not necessarily primitive recursive (again cf. the introductory note to *Gödel 1934*). This method also applies directly to what Church in *1936a*, page 102, called the "second form" of the Entscheidungsproblem, namely to show the unsolvability of the problem of effectively deciding as to the validity or the non-validity of a formula.

In Section 4, the final Theorem XI is Gödel's famous "second incompleteness theorem". First, he observes that, for each of his systems P_κ, the proposition that P_κ is (simply) consistent is expressed (via his numbering) by a closed formula W of P, stating that there is some formula which is unprovable in P_κ. (Gödel speaks of "the SENTENTIAL FORMULA w", where w is the Gödel number of this W.) In proving the first half of Theorem VI, he showed that

(23) $\{P_\kappa \text{ is consistent}\} \rightarrow \{\text{xIIR}(\text{x}, \boldsymbol{q}) \text{ is unprovable } P_\kappa\}$.

But all of the reasoning used in proving this is on the level of elementary number theory. Therefore it is to be expected that these informal arguments can be formalized to give a proof in P and hence in P_κ of the formula expressing this informal implication. But W expresses "P_κ is

---

[i]The Chinese remainder theorem states that, if $m_1, \ldots, m_k$ are positive integers each two of which are relatively prime, then the congruences
$$x \equiv a_1 \pmod{m_1}, \ldots, x \equiv a_k \pmod{m_k}$$
have a unique solution mod $(m_1 \cdot \ldots \cdot m_k)$.

consistent", and $x\Pi R(x, \boldsymbol{q})$ expresses "$x\Pi R(x, \boldsymbol{q})$ is unprovable in P_κ".
So Gödel claims that the formula

$$(*) \qquad\qquad W \supset x\Pi R(x, \boldsymbol{q})$$

is provable in P_κ (or, as he puts it, "$w\,\mathrm{Imp}(17\,\mathrm{Gen}\,r)$ is κ-PROVABLE").
So, if W were provable in P_κ, $x\Pi R(x, \boldsymbol{q})$ would be too. Using (23), this
gives:

Theorem XI. *If P_κ is consistent, then the formula* W *which expresses
that fact in the symbolism of P is unprovable in P_κ.*

As mentioned above, Gödel attended the Königsberg meeting of
September 1930, where (in *1931a*) he presented a preliminary version of
his first incompleteness theorem (what became Theorem VI of *1931*).
John von Neumann was there, and was very much interested. At that
time Gödel only had undecidable propositions which were finitary combinatorial in nature, and von Neumann asked whether number-theoretic
undecidable propositions could also be constructed. Gödel replied that
they would have to contain concepts quite different from those occurring
in number theory like addition and multiplication. Gödel was astonished
when slightly afterward he succeeded in turning the undecidable proposition into a polynomial equation preceded by quantifiers over the natural
numbers (Theorem VIII). At the same time, but independently of this
result, Gödel derived his second incompleteness theorem (Theorem XI)
as a consequence of the first. Shortly afterward, Gödel received a letter
from von Neumann likewise suggesting the second incompleteness theorem. (Cf. *Wang 1981*, pages 654–655.)

A detailed demonstration that $(*)$ is provable in P_κ, as required for
Theorem XI, was planned to be in a sequel to *1931*, a Part II under the
same title. This was never written, in part because Gödel felt that the
results of Part I had won prompt acceptance in many quarters.[j] Certainly the idea of the argument for Theorem XI was very convincing;
but it turned out that the execution of the details required somewhat
more work and care than had been anticipated.

A demonstration that $(*)$ is provable was given in *Hilbert and Bernays
1939*, pages 283–340, for their systems Z_μ and Z formalizing elementary
number theory, whence the result readily follows for systems provably
including Z.

Elegant abstract derivability conditions were isolated in *Löb 1955* to

---

[j]That there was some resistance, however, is indicated by the letters that Zermelo
sent Gödel shortly after the Bad Elster meeting of 15 September 1931 (cf. *Grattan-
Guinness 1979* and *Moore 1980*, pages 124–129), as well as by the critical articles by
Perelman (*1936*), Kuczyński (*1938*) and Barzin (*1940*), discussed in *Ladrière 1957*,
pages 140–157. Also see *Dawson 1985* and *1985a*.

replace those used in *Hilbert and Bernays 1939* in establishing the second incompleteness theorem (Gödel's Theorem XI). These conditions are featured in the proof of the theorem in *Smoryński 1977*, page 828.

In *Feferman 1960* a large class of proof predicates was determined for which these derivability conditions were verified, and counterexamples were given to the second incompleteness theorem for proof predicates meeting some but not all of these conditions.

Theorem XI disabused the mathematical public of any hope that the consistency of a formal system like P could be proved by the use of only finitary methods selected from among the usual methods of elementary number theory. But, as Gödel remarks here, "it is conceivable that there exist finitary proofs that *cannot* be expressed in the formalism of P."

Thus Gödel's second incompleteness theorem, rather than ending efforts at finding finitary consistency proofs, pointed out the road to success for people aspiring to prove finitistically the consistency of formal systems embracing elementary number theory (and perhaps more): they must seek to use some non-elementary method which nevertheless can be construed as finitary.

Prior to 1931 there had been some modest successes in giving finitary consistency proofs for formal systems for fragments (only) of elementary number theory (or Peano arithmetic[k]) by Ackermann in *1924*, von Neumann in *1927* and Herbrand in *1931*. The Hilbert school had hoped to extend such to full elementary number theory, then to analysis, and finally to systems of set theory. (As we saw, it was Gödel's own examination of this program which led him down the road to the results in his *1931*.) The idea of what constitute "finitary" proofs had never been formulated precisely. Roughly speaking, as we said above, they were supposed to involve only constructive reasoning about finitely presented objects, making no appeal to an "actual" or "completed" infinite (thus, e.g., not applying the principle of the excluded middle to statements about all the natural numbers). All informal arguments used in carrying out Hilbert's program which had previously been accepted as being finitary could readily be formalized in systems much weaker

---

[k] In *Peano arithmetic* the symbolism is that of the first-order predicate calculus with equality, with variables only for individuals (interpreted as ranging over the natural numbers), and with an individual constant for 0 and function symbols for the successor function $+1$ (or a constant for 1) and for the sum $+$ and product $\cdot$ of two natural numbers. (Thus the formulas with free variables express the arithmetical relations as introduced by Gödel preceding his Theorem VII.) The deductive apparatus is that of the first-order predicate calculus with equality plus (as non-logical axioms) Peano's third and fourth axioms (his first two are given effect by the symbolism) and an induction axiom for each formula of the described symbolism (as instances of his fifth axiom). This system is to be distinguished from one with the "Peano axioms" of *Peano 1889* and *1891*, which are formulated in a second-order symbolism with only 0 and successor as the basic individual and function symbols.

than Peano arithmetic. At any rate, no methods had been suggested that could not readily be formalized in Peano arithmetic itself.

For this reason, Hilbert's program (at least as Hilbert had originally conceived it) might have seemed to be doomed by Gödel's second incompleteness theorem (Theorem XI). Gödel in the statement just quoted was more cautious. One can speculate that he was influenced by *Hilbert 1926*, which can be read as proposing some extension of Hilbert's original conception. However, Gödel did not indicate then what finitary methods that cannot be expressed in the formalism of P might look like; and he did not discuss this question again until his paper *1958*.

Meanwhile, Hilbert's program was continued, as Gödel had suggested it might be, by using new principles proposed as being finitary. The first such to be applied were principles of transfinite induction on certain primitive recursive well-orderings, with respect to which only the termination of finitarily constructed descending sequences needed to be assumed. Thus Gentzen, in his striking paper *1936*, proved the consistency of elementary number theory by using such a form of transfinite induction up to Cantor's ordinal ϵ_0 as his only addition to previously used finitary principles. (The ordinals less than ϵ_0 are isomorphic to a primitive recursive well-ordering by using the Cantor representation of ordinals in normal form.) Since then, Gentzen-style consistency proofs have been given for a variety of subsystems of analysis and set theory by the use of effective transfinite induction on larger and larger ordinals. The books *Takeuti 1975* and *Schütte 1977* give fairly substantial accounts of work in this direction.

Other methods, regarded as finitary in an extended sense, have been utilized for consistency proofs. One of the most striking is that of Gödel himself in his *1958*. This employs a notion of computable functions of finite type (i.e., of computable functionals), to which the process of primitive recursion can be extended in a natural way. At the beginning of *1958*, Gödel again addresses the question of how far finitary reasoning might reach. This discussion is amplified considerably in *1972*, his own revised English translation of *1958*. The essence of Gödel's later views, as represented by *1958* and *1972*, is that it is practically certain that concrete finitary methods are insufficient to prove the consistency of elementary number theory, and some abstract concepts must be used in addition. "There is nothing in the term 'finitary' which would suggest a restriction to concrete knowledge. Only Hilbert's special interpretation of it makes this restriction." (*1972*, footnote b). However, a precise definition of concrete finitary method would have to be given in order to establish with certainty the necessity of using abstract concepts. Gödel's informal efforts to delimit concrete finitary mathematics in the opening paragraphs of *1972* are of particular interest; and the connection with formal work is elaborated in his footnotes to this material.

A full critical assessment of Hilbert's program and its extensions is given in *Kreisel 1976.*

Rosser in *1936* achieved a noteworthy improvement of the first Gödel incompleteness theorem (Theorem VI) by obtaining it under the hypothesis of simple consistency instead of ω-consistency. To do this, Rosser used as his undecidable formula, call it D, a sentence which expresses that, to any Gödel number x of a proof in P_κ of D, there is a Gödel number $y \leq x$ of a proof in P_κ of ~D.[1] (Actually, W $\supset$ [D $\equiv$ xΠR(x, q)], for the R(x, q) and W of Theorems VI and XI, is provable in *P*.)

A new landmark in the expanding evolution of theory started by Gödel's *1931* is the discovery by Paris and Harrington in *1977* of a mathematically simple and interesting proposition, not depending on a numerical coding of notions from logic, which is undecidable. This proposition is a refinement of Ramsey's theorem, which is (i) a Π_2^0 statement and (ii) equivalent to the Σ_1^0-reflection principle for *PA* (Peano arithmetic), hence is independent of *PA*. Kreisel in *1980*, page 175, points out that

---

[1]It is easy to obtain such a D by the self-referential lemma (see footnote h). Or for D we can take xΠS(x, r), where $S(x, v)$ is like the $R(x, v)$ described in the first sentence following Theorem VI in this note except inserting just before the period in that sentence "*, or there exists a $y \leq x$ which is the Gödel number of a proof in P_κ of* ~$A_v(v)$", and r is the Gödel number of xΠS(x, v)".

Einige metamathematische Resultate
über Entscheidungsdefinitheit und
Widerspruchsfreiheit
(*1930b*)

Überbaut man die Peano'schen Axiome mit der Logik der *Principia mathematica*[1] (natürliche Zahlen als Individuen) samt Auswahlaxiom (für alle Typen), so entsteht ein formales System S, für welches folgende Sätze gelten:

I. Das System S ist *nicht* entscheidungsdefinit, d. h. es gibt darin Sätze A (und solche sind auch angebbar), für welche weder A noch $\overline{A}$ beweisbar

---

[1]Mit Reduzibilitätsaxiom oder ohne verzweigte Typentheorie.

it is thus also independent of the semi-formal system PA^+ consisting of PA with all the true Π_1^0 sentences added (since the sentence expressing the 1-consistency of PA is equivalent to the consistency of PA^+).[m] However, by the proof theory of PA, which extends to PA^+, these sentences have the same "provably recursive functions" (i.e., there are no new recursive functions determined by provable Π_2^0 statements). What is essentially involved is that the function determined by the Π_2^0 Paris–Harrington statement majorizes all the provably recursive functions of PA (as shown in another way by Ketonen and Solovay in *1981*).

Stephen C. Kleene

The translation of *1930b* is by Stefan Bauer-Mengelberg, and those of *1931* and *1932b* are by Jean van Heijenoort. These translations were approved by Gödel while *van Heijenoort 1967* was in preparation, being accommodated in many places to his wishes. Moreover, at that time Gödel supplied a number of short interpolations to *1931*, which are enclosed in square brackets. Subsequently, minor revisions were introduced by Jean van Heijenoort.

---

[m] PA^+ is semi-formal because the class of its axioms is not general recursive in the sense mentioned in the introductory note to *Gödel 1934*.

Some metamathematical results on completeness and consistency
(*1930b*)

If to the Peano axioms we add the logic of *Principia mathematica*[1] (with the natural numbers as the individuals) together with the axiom of choice (for all types), we obtain a formal system S, for which the following theorems hold:

I. The system S is *not* complete; that is, it contains propositions A (and we can in fact exhibit such propositions) for which neither A nor $\overline{A}$ is

---

[1] With the axiom of reducibility or without the ramified theory of types.

ist, und zwar gibt es unentscheidbare Probleme von der einfachen Struktur: $(Ex)F(x)$, wobei x über die natürlichen Zahlen läuft und F eine (sogar entscheidungsdefinite) Eigenschaft natürlicher Zahlen ist.[2]

II. Selbst wenn man alle logischen Hilfsmittel der *Principia mathematica* (insbesondere also erweiterten Funktionenkalkül[1] und Auswahlaxiom) in der Metamathematik zuläßt, gibt es *keinen Widerspruchsfreiheitsbeweis* für das System S (um so weniger, wenn man die Beweismittel irgendwie beschränkt). Ein Widerspruchsfreiheitsbeweis des Systems S kann also nur mit Hilfe von Schlußweisen geführt werden, die im System S selbst nicht formalisiert sind, und Analoges gilt auch für andere formale Systeme, etwa das Zermelo-Fränkel'sche Axiomensystem der Mengenlehre.[3]

III. Satz I läßt sich dahin verschärfen, daß auch durch Hinzufügung endlich vieler Axiome zum System S (oder unendlich vieler, die aus endlich vielen durch "Typenerhöhung" hervorgehen) *kein* entscheidungsdefinites System entsteht, sobald das erweiterte System ω-widerspruchsfrei ist. Dabei heißt ein System ω-widerspruchsfrei, wenn für keine Eigenschaft natürlicher Zahlen $F(x)$ zugleich beweisbar ist:

$$F(1), F(2), \ldots, F(n), \ldots \text{ ad infinitum}$$

215 | und

$$(Ex)\overline{F(x)}.$$

(Es gibt Erweiterungen des Systems S, welche zwar widerspruchsfrei, aber nicht ω-widerspruchsfrei sind.)

IV. Satz I gilt auch noch für alle ω-widerspruchsfreien Erweiterungen des Systems S durch *unendlich viele* Axiome, sobald die hinzugefügte Klasse von Axiomen entscheidungsdefinit ist, d. h. für jede Formel metamathematisch entscheidbar ist, ob sie ein Axiom ist oder nicht (dabei werden in der Metamathematik wieder die logischen Hilfsmittel der *Principia mathematica* vorausgesetzt).

Die Sätze I, III, IV lassen sich auch auf andere formale Systeme, z. B. das Zermelo–Fränkel'sche Axiomensystem der Mengenlehre ausdehnen, vorausgesetzt, daß die betreffenden Systeme ω-widerspruchsfrei sind.

Die Beweise dieser Sätze werden in den *Monatsheften für Mathematik und Physik* erscheinen.

---

[2]Ferner gibt es in S Formeln des engeren Funktionenkalküls, für die weder Allgemeingültigkeit noch Existenz eines Gegenbeispiels beweisbar ist.

[3]Dieses Resultat gilt insbesondere auch für das Axiomensystem der klassischen Mathematik, wie es z. B. J. von Neumann (*1927*) aufgestellt hat.

provable and, in particular, it contains (even for decidable properties F of natural numbers) undecidable problems of the simple structure $(Ex)F(x)$, where x ranges over the natural numbers.[2]

II. Even if we admit all the logical devices of *Principia mathematica* (hence in particular the extended functional calculus[1] and the axiom of choice) in metamathematics, there does *not* exist a *consistency proof* for the system S (still less so if we restrict the means of proof in any way). Hence a consistency proof for the system S can be carried out only by means of modes of inference that are not formalized in the system S itself, and analogous results hold for other formal systems as well, such as the Zermelo-Fraenkel axiom system of set theory.[3]

III. Theorem I can be sharpened to the effect that, even if we add finitely many axioms to the system S (or infinitely many that result from a finite number of them by "type elevation"), we do *not* obtain a complete system, provided the extended system is ω-consistent. Here a system is said to be ω-consistent if, for no property $F(x)$ of natural numbers,

$$F(1), F(2), \ldots, F(n), \ldots \text{ ad infinitum}$$

as well as

$$(Ex)\overline{F(x)}$$

are provable. (There are extensions of the system S that, while consistent, are not ω-consistent.)

IV. Theorem I still holds for all ω-consistent extensions of the system S that are obtained by the addition of *infinitely many* axioms, provided the added class of axioms is decidable, that is, provided for every formula it is metamathematically decidable whether it is an axiom or not (here again we suppose that in metamathematics we have at our disposal the logical devices of *Principia mathematica*).

Theorems I, III, and IV can be extended also to other formal systems, for example, to the Zermelo–Fraenkel axiom system of set theory, provided the systems in question are ω-consistent.

The proofs of these theorems will appear in *Monatshefte für Mathematik und Physik*.

---

[2]Furthermore, S contains formulas of the restricted functional calculus such that neither universal validity nor existence of a counterexample is provable for any of them.

[3]This result, in particular, holds also for the axiom system of classical mathematics, as it has been constructed, for example, by J. von Neumann (*1927*).

Über formal unentscheidbare Sätze der
Principia mathematica und verwandter Systeme I[1]
(*1931*)

1

Die Entwicklung der Mathematik in der Richtung zu größerer Exaktheit hat bekanntlich dazu geführt, daß weite Gebiete von ihr formalisiert wurden, in der Art, daß das Beweisen nach einigen wenigen mechanischen Regeln vollzogen werden kann. Die umfassendsten derzeit aufgestellten formalen Systeme sind das System der *Principia mathematica* (*PM*)[2] einerseits, das Zermelo–Fraenkelsche (von J. von Neumann weiter ausgebildete) Axiomensystem der Mengenlehre[3] andererseits. Diese beiden Systeme sind so weit, daß alle heute in der Mathematik angewendeten Beweismethoden in ihnen formalisiert, d. h. auf einige wenige Axiome und Schlußregeln zurückgeführt sind. Es liegt daher die Vermutung nahe, daß diese Axiome und Schlußregeln dazu ausreichen, *alle* mathematischen Fragen, die sich in den betreffenden Systemen überhaupt formal ausdrücken lassen, auch zu entscheiden. Im folgenden wird gezeigt, daß dies nicht der Fall ist, sondern daß es in den beiden angeführten Systemen sogar relativ einfache Probleme aus der Theorie der gewöhnlichen ganzen Zahlen gibt,[4] die sich aus den Axiomen nicht | entscheiden lassen. Dieser Umstand liegt nicht etwa an der speziellen Natur der aufgestellten Systeme, sondern gilt für eine sehr weite Klasse formaler Systeme, zu denen insbesondere alle gehören, die aus den beiden angeführten durch Hinzufügung endlich vieler Axiome entstehen,[5]

174

---

[1] Vgl. die als *1930b* erschienene Zusammenfassung der Resultate dieser Arbeit.

[2] *Whitehead und Russell 1925.* Zu den Axiomen des Systems *PM* rechnen wir insbesondere auch: Das Unendlichkeitsaxiom (in der Form: es gibt genau abzählbar viele Individuen), das Reduzibilitäts- und das Auswahlaxiom (für alle Typen).

[3] Vgl. *Fraenkel 1927, von Neumann 1925, 1928a, 1929.* Wir bemerken, daß man zu den in der angeführten Literatur gegebenen mengentheoretischen Axiomen noch die Axiome und Schlußregeln des Logikkalküls hinzufügen muß, um die Formalisierung zu vollenden.—Die nachfolgenden Überlegungen gelten auch für die in den letzten Jahren von D. Hilbert und seinen Mitarbeitern aufgestellten formalen Systeme (soweit diese bisher vorliegen). Vgl. *Hilbert 1922, 1923, 1928, Bernays 1923, von Neumann 1927, Ackermann 1924.*

[4] D. h. genauer, es gibt unentscheidbare Sätze, in denen außer den logischen Konstanten: $-$ (nicht), $\lor$ (oder), (x) (für alle), $=$ (identisch mit) keine anderen Begriffe vorkommen als $+$ (Addition), $\cdot$ (Multiplikation), beide bezogen auf natürliche Zahlen, wobei auch die Präfixe (x) sich nur auf natürliche Zahlen beziehen dürfen.

[5] Dabei werden in *PM* nur solche Axiome als verschieden gezählt, die aus einander nicht bloß durch Typenwechsel entstehen.

On formally undecidable propositions of *Principia mathematica* and related systems I[1]
(*1931*)

1

The development of mathematics toward greater precision has led, as is well known, to the formalization of large tracts of it, so that one can prove any theorem using nothing but a few mechanical rules. The most comprehensive formal systems that have been set up hitherto are the system of *Principia mathematica* (*PM*)[2] on the one hand and the Zermelo–Fraenkel axiom system of set theory (further developed by J. von Neumann)[3] on the other. These two systems are so comprehensive that in them all methods of proof today used in mathematics are formalized, that is, reduced to a few axioms and rules of inference. One might therefore conjecture that these axioms and rules of inference are sufficient to decide *any* mathematical question that can at all be formally expressed in these systems. It will be shown below that this is not the case, that on the contrary there are in the two systems mentioned relatively simple problems in the theory of integers[4] that cannot be decided on the basis of the axioms. This situation is not in any way due to the special nature of the systems that have been set up, but holds for a wide class of formal systems; among these, in particular, are all systems that result from the two just mentioned through the addition of a finite number of axioms,[5] provided no false propositions

---

[1]See a summary of the results of the present paper in *Gödel 1930b*.

[2] *Whitehead and Russell 1925*. Among the axioms of the system *PM* we include also the axiom of infinity (in this version: there are exactly denumerably many individuals), the axiom of reducibility, and the axiom of choice (for all types).

[3]See *Fraenkel 1927* and *von Neumann 1925, 1928a*, and *1929*. We note that in order to complete the formalization we must add the axioms and rules of inference of the calculus of logic to the set-theoretic axioms given in the literature cited. The considerations that follow apply also to the formal systems (so far as they are available at present) constructed in recent years by Hilbert and his collaborators. See *Hilbert 1922, 1923, 1928, Bernays 1923, von Neumann 1927*, and *Ackermann 1924*.

[4]That is, more precisely, there are undecidable propositions in which, besides the logical constants $-$ (not), $\lor$ (or), (x) (for all), and $=$ (identical with), no other notions occur but $+$ (addition) and $\cdot$ (multiplication), both for natural numbers, and in which the quantifiers (x), too, apply to natural numbers only.

[5]In *PM* only axioms that do not result from one another by mere change of type are counted as distinct.

vorausgesetzt, daß durch die hinzugefügten Axiome keine falschen Sätze von der in Fußnote 4 angegebenen Art beweisbar werden.

Wir skizzieren, bevor wir auf Details eingehen, zunächst den Hauptgedanken des Beweises, natürlich ohne auf Exaktheit Anspruch zu erheben. Die Formeln eines formalen Systems (wir beschränken uns hier auf das System PM) sind äußerlich betrachtet endliche Reihen der Grundzeichen (Variable, logische Konstante und Klammern bzw. Trennungspunkte) und man kann leicht genau präzisieren, *welche* Reihen von Grundzeichen sinnvolle Formeln sind und welche nicht.[6] Analog sind Beweise vom formalen Standpunkt nichts anderes als endliche Reihen von Formeln (mit bestimmten angebbaren Eigenschaften). Für metamathematische Betrachtungen ist es natürlich gleichgültig, welche Gegenstände man als Grundzeichen nimmt, und wir entschließen uns dazu, natürliche Zahlen[7] als solche zu verwenden. Dementsprechend ist dann eine Formel eine endliche Folge natürlicher Zahlen[8] und eine Beweisfigur eine endliche Folge von endlichen Folgen natürlicher Zahlen. Die metamathematischen Begriffe (Sätze) werden dadurch zu Begriffen (Sätzen) über natürliche Zahlen bzw. Folgen von solchen[9] und daher (wenigstens teilweise) in den Symbolen des Systems PM selbst ausdrückbar. Insbesondere kann man zeigen, daß die Begriffe "Formel", "Beweisfigur", "beweisbare Formel" innerhalb des Systems PM definierbar sind, d. h. man kann z. B. eine Formel $F(v)$ aus PM mit einer freien Variablen v (vom Typus einer Zahlenfolge) angeben,[10] so daß $F(v)$ inhaltlich interpretiert besagt: v ist eine beweisbare Formel. Nun stellen wir einen unentscheidbaren Satz des Systems PM, d. h. einen Satz A, für den weder A noch *non-A* beweisbar ist, folgendermaßen her:

175 | Eine Formel aus PM mit genau einer freien Variablen, und zwar vom Typus der natürlichen Zahlen (Klasse von Klassen) wollen wir ein *Klassenzeichen* nennen. Die Klassenzeichen denken wir uns irgendwie in eine Folge

---

[6]Wir verstehen hier und im folgenden unter "Formel aus PM" immer eine ohne Abkürzungen (d. h. ohne Verwendung von Definitionen) geschriebene Formel. Definitionen dienen ja nur der kürzeren Schreibweise und sind daher prinzipiell überflüssig.

[7]D. h. wir bilden die Grundzeichen in eineindeutiger Weise auf natürliche Zahlen ab. (Vgl. die Durchführung auf S. 179.)

[8]D. h. eine Belegung eines Abschnittes der Zahlenreihe mit natürlichen Zahlen. (Zahlen können ja nicht in räumliche Anordnung gebracht werden.)

[9]m. a. W.: Das oben beschriebene Verfahren liefert ein isomorphes Bild des Systems PM im Bereich der Arithmetik und man kann alle metamathematischen Überlegungen ebenso gut an diesem isomorphen Bild vornehmen. Dies geschieht in der folgenden Beweisskizze, d. h. unter "Formel", "Satz", "Variable" etc. *sind immer die entsprechenden Gegenstände des isomorphen Bildes zu verstehen.*

[10]Es wäre sehr leicht (nur etwas umständlich), diese Formel tatsächlich hinzuschreiben.

of the kind specified in footnote 4 become provable owing to the added axioms.

Before going into details, we shall first sketch the main idea of the proof, of course without any claim to complete precision. The formulas of a formal system (we restrict ourselves here to the system *PM*) in outward appearance are finite sequences of primitive signs (variables, logical constants, and parentheses or punctuation dots), and it is easy to state with complete precision *which* sequences of primitive signs are meaningful formulas and which are not.[6] Similarly, proofs, from a formal point of view, are nothing but finite sequences of formulas (with certain specifiable properties). Of course, for metamathematical considerations it does not matter what objects are chosen as primitive signs, and we shall assign natural numbers to this use.[7] Consequently, a formula will be a finite sequence of natural numbers,[8] and a proof array a finite sequence of finite sequences of natural numbers. The metamathematical notions (propositions) thus become notions (propositions) about natural numbers or sequences of them;[9] therefore they can (at least in part) be expressed by the symbols of the system *PM* itself. In particular, it can be shown that the notions "formula", "proof array", and "provable formula" can be defined in the system *PM*; that is, we can, for example, find a formula $F(v)$ of *PM* with one free variable v (of the type of a number sequence)[10] such that $F(v)$, interpreted according to the meaning of the terms of *PM*, says: v is a provable formula. We now construct an undecidable proposition of the system *PM*, that is, a proposition A for which neither A nor *not-A* is provable, in the following manner.

A formula of *PM* with exactly one free variable, that variable being of the type of the natural numbers (class of classes), will be called a *class sign*. We assume that the class signs have been arranged in a sequence

---

[6]Here and in what follows we always understand by "formula of *PM*" a formula written without abbreviations (that is, without the use of definitions). It is well known that [in *PM*] definitions serve only to abbreviate notations and therefore are dispensable in principle.

[7]That is, we map the primitive signs one-to-one onto some natural numbers. (See how this is done on page 157.)

[8]That is, a number-theoretic function defined on an initial segment of the natural numbers. (Numbers, of course, cannot be arranged in a spatial order.)

[9]In other words, the procedure described above yields an isomorphic image of the system *PM* in the domain of arithmetic, and all metamathematical arguments can just as well be carried out in this isomorphic image. This is what we do below when we sketch the proof; that is, by "formula", "proposition", "variable", and so on, *we must always understand the corresponding objects of the isomorphic image.*

[10]It would be very easy (although somewhat cumbersome) to actually write down this formula.

geordnet,[11] bezeichnen das n-te mit $R(n)$ und bemerken, daß sich der Begriff "Klassenzeichen" sowie die ordnende Relation R im System PM definieren lassen. Sei α ein beliebiges Klassenzeichen; mit $[\alpha; n]$ bezeichnen wir diejenige Formel, welche aus dem Klassenzeichen α dadurch entsteht, daß man die freie Variable durch das Zeichen für die natürliche Zahl n ersetzt. Auch die Tripel-Relation $x = [y; z]$ erweist sich als innerhalb PM definierbar. Nun definieren wir eine Klasse K natürlicher Zahlen folgendermaßen:

$$n \,\epsilon\, K \equiv \overline{\text{Bew}}[R(n); n] \tag{1}$$

(wobei *Bew* x bedeutet: x ist eine beweisbare Formel).[11a] Da die Begriffe, welche im Definiens vorkommen, sämtlich in PM definierbar sind, so auch der daraus zusammengesetzte Begriffe K, d. h. es gibt ein Klassenzeichen S,[12] so daß die Formel $[S; n]$ inhaltlich gedeutet besagt, daß die natürliche Zahl n zu K gehört. S ist als Klassenzeichen mit einem bestimmten $R(q)$ identisch, d. h. es gilt

$$S = R(q)$$

für eine bestimmte natürliche Zahl q. Wir zeigen nun, daß der Satz $[R(q); q]$ in PM unentscheidbar ist.[13] Denn angenommen der Satz $[R(q); q]$ wäre beweisbar, dann wäre er auch richtig, d. h. aber nach dem obigen q würde zu K gehören, d. h. nach (1) es würde $\overline{Bew}[R(q); q]$ gelten, im Widerspruch mit der Annahme. Wäre dagegen die Negation von $[R(q); q]$ beweisbar, so würde $\overline{q \,\epsilon\, K}$, d. h. $Bew[R(q); q]$ gelten. $[R(q); q]$ wäre also zugleich mit seiner Negation beweisbar, was wiederum unmöglich ist.

Die Analogie dieses Schlusses mit der Antinomie Richard springt in die Augen; auch mit dem "Lügner" besteht eine nahe Verwandtschaft,[14] denn der unentscheidbare Satz $[R(q); q]$ besagt ja, daß q zu K gehört, d. h. nach (1), daß $[R(q); q]$ nicht beweisbar ist. Wir haben also einen Satz vor uns, der

---

[11]Etwa nach steigender Gliedersumme und bei gleicher Summe lexikographisch.

[11a]Durch Überstreichen wird die Negation bezeichnet.

[12]Es macht wieder nicht die geringsten Schwierigkeiten, die Formel S tatsächlich hinzuschreiben.

[13]Man beachte, daß "$[R(q); q]$" (oder was dasselbe bedeutet "$[S; q]$") bloß eine *metamathematische Beschreibung* des unentscheidbaren Satzes ist. Doch kann man, sobald man die Formel S ermittelt hat, natürlich auch die Zahl q bestimmen und damit den unentscheidbaren Satz selbst effektiv hinschreiben.

[14]Es läßt sich überhaupt jede epistemologische Antinomie zu einem derartigen Unentscheidbarkeitsbeweis verwenden.

in some way,[11] we denote the nth one by $R(n)$, and we observe that the notion "class sign", as well as the ordering relation R, can be defined in the system *PM*. Let α be any class sign; by $[\alpha; n]$ we denote the formula that results from the class sign α when the free variable is replaced by the sign denoting the natural number n. The ternary relation $x = [y; z]$, too, is seen to be definable in *PM*. We now define a class K of natural numbers in the following way:

$$n \; \epsilon \; K \equiv \overline{Bew}[R(n); n] \tag{1}$$

(where *Bew* x means: x is a provable formula).[11a] Since the notions that occur in the definiens can all be defined in *PM*, so can the notion K formed from them; that is, there is a class sign S such that the formula $[S; n]$, interpreted according to the meaning of the terms of *PM*, states that the natural number n belongs to K.[12] Since S is a class sign, it is identical with some $R(q)$; that is, we have

$$S = R(q)$$

for a certain natural number q. We now show that the proposition $[R(q); q]$ is undecidable in *PM*.[13] For let us suppose that the proposition $[R(q); q]$ were provable; then it would also be true. But in that case, according to the definitions given above, q would belong to K, that is, by (1), $\overline{Bew}[R(q); q]$ would hold, which contradicts the assumption. If, on the other hand, the negation of $[R(q); q]$ were provable, then $\overline{q \; \epsilon \; K}$, that is, $Bew[R(q); q]$, would hold. But then $[R(q); q]$, as well as its negation, would be provable, which again is impossible.

The analogy of this argument with the Richard antinomy leaps to the eye. It is closely related to the "Liar" too;[14] for the undecidable proposition $[R(q); q]$ states that q belongs to K, that is, by (1), that $[R(q); q]$ is not provable. We therefore have before us a proposition that says about itself

---

[11]For example, by increasing sum of the finite sequence of integers that is the "class sign", and lexicographically for equal sums.

[11a]The bar denotes negation.

[12]Again, there is not the slightest difficulty in actually writing down the formula S.

[13]Note that "$[R(q); q]$" (or, which means the same, "$[S; q]$") is merely a *metamathematical description* of the undecidable proposition. But, as soon as the formula S has been obtained, we can, of course, also determine the number q and, therewith, actually write down the undecidable proposition itself. [This makes no difficulty in principle. However, in order not to run into formulas of entirely unmanageable lengths and to avoid practical difficulties in the computation of the number q, the construction of the undecidable proposition would have to be slightly modified, unless the technique of abbreviation by definition used throughout in *PM* is adopted.]

[14]Any epistemological antinomy could be used for a similar proof of the existence of undecidable propositions.

seine eigene Unbeweisbarkeit behauptet.[15] Die eben auseinandergesetzte
176 Beweismethode | läßt sich offenbar auf jedes formale System anwenden, das
erstens inhaltlich gedeutet über genügend Ausdrucksmittel verfügt, um die
in der obigen Überlegung vorkommenden Begriffe (insbesondere den Begriff
"beweisbare Formel") zu definieren, und in dem zweitens jede beweisbare
Formel auch inhaltlich richtig ist. Die nun folgende exakte Durchführung
des obigen Beweises wird unter anderem die Aufgabe haben, die zweite
der eben angeführten Voraussetzungen durch eine rein formale und weit
schwächere zu ersetzen.

Aus der Bemerkung, daß $[R(q); q]$ seine eigene Unbeweisbarkeit behauptet, folgt sofort, daß $[R(q); q]$ richtig ist, denn $[R(q); q]$ *ist* ja unbeweisbar
(weil unentscheidbar). Der *im System PM* unentscheidbare Satz wurde
also durch metamathematische Überlegungen doch entschieden. Die
genaue Analyse dieses merkwürdigen Umstandes führt zu überraschenden
Resultaten, bezüglich der Widerspruchsfreiheitsbeweise formaler Systeme,
die in Abschnitt 4 (Satz XI) näher behandelt werden.

2

Wir gehen nun an die exakte Durchführung des oben skizzierten Beweises und geben zunächst eine genaue Beschreibung des formalen Systems P, für welches wir die Existenz unentscheidbarer Sätze nachweisen
wollen. P ist im wesentlichen das System, welches man erhält, wenn man
die Peanoschen Axiome mit der Logik der PM[16] überbaut (Zahlen als
Individuen, Nachfolgerrelation als undefinierten Grundbegriff).

Die Grundzeichen des Systems P sind die folgenden:

I. Konstante: "$\sim$" (nicht), "$\vee$" (oder), "Π" (für alle), "0" (Null), "f"
(der Nachfolger von), "$($", "$)$" (Klammern).

II. Variable ersten Typs (für Individuen, d. h. natürliche Zahlen inklusive 0): "x_1", "y_1", "z_1", ...

Variable zweiten Typs (für Klassen von Individuen): "x_2", "y_2", "z_2",
....

Variable dritten Typs (für Klassen von Klassen von Individuen): "x_3",
"y_3", "z_3", ...

---

[15]Ein solcher Satz hat entgegen dem Anschein nichts Zirkelhaftes an sich, denn er
behauptet zunächst die Unbeweisbarkeit einer ganz bestimmten Formel (nämlich der
q-ten in der lexikographischen Anordnung bei einer bestimmten Einsetzung), und erst
nachträglich (gewissermaßen zufällig) stellt sich heraus, daß diese Formel gerade die ist,
in der sie selbst ausgedrückt wurde.

[16]Die Hinzufügung der Peanoschen Axiome ebenso wie alle anderen am System PM
angebrachten Abänderungen dienen lediglich zur Vereinfachung des Beweises und sind
prinzipiell entbehrlich.

that it is not provable [in *PM*].[15] The method of proof just explained can clearly be applied to any formal system that, first, when interpreted as representing a system of notions and propositions, has at its disposal sufficient means of expression to define the notions occurring in the argument above (in particular, the notion "provable formula") and in which, second, every provable formula is true in the interpretation considered. The purpose of carrying out the above proof with full precision in what follows is, among other things, to replace the second of the assumptions just mentioned by a purely formal and much weaker one.

From the remark that $[R(q); q]$ says about itself that it is not provable, it follows at once that $[R(q); q]$ is true, for $[R(q); q]$ *is* indeed unprovable (being undecidable). Thus, the proposition that is undecidable *in the system PM* still was decided by metamathematical considerations. The precise analysis of this curious situation leads to surprising results concerning consistency proofs for formal systems, results that will be discussed in more detail in Section 4 (Theorem XI).

2

We now proceed to carry out with full precision the proof sketched above. First we give a precise description of the formal system P for which we intend to prove the existence of undecidable propositions. P is essentially the system obtained when the logic of *PM* is superposed upon the Peano axioms[16] (with the numbers as individuals and the successor relation as primitive notion).

The primitive signs of the system P are the following:

I. Constants: "$\sim$" (not), "$\vee$" (or), "Π" (for all), "0" (zero), "f" (the successor of), "(", ")" (parentheses).

II. Variables of type 1 (for individuals, that is, natural numbers including 0): "x_1", "y_1", "z_1", ...

Variables of type 2 (for classes of individuals): "x_2", "y_2", "z_2", ...

Variables of type 3 (for classes of classes of individuals): "x_3", "y_3", "z_3", ...

---

[15]Contrary to appearances, such a proposition involves no faulty circularity, for initially it [only] asserts that a certain well-defined formula (namely, the one obtained from the qth formula in the lexicographic order by a certain substitution) is unprovable. Only subsequently (and so to speak by chance) does it turn out that this formula is precisely the one by which the proposition itself was expressed.

[16]The addition of the Peano axioms, as well as all other modifications introduced in the system *PM*, merely serves to simplify the proof and is dispensable in principle.

Usw. für jede natürliche Zahl als Typus.[17]

Anmerkung: Variable für zwei- und mehrstellige Funktionen (Relationen) sind als Grundzeichen überflüssig, da man Relationen als Klassen geordneter Paare definieren kann und geordnete Paare wiederum als Klassen von Klassen, z. B. das geordnete Paar a, b durch $((a), (a, b))$, wo (x, y) bzw. (x) die Klassen bedeuten, deren einzige Elemente x, y bzw. x sind.[18]

177 | Unter einem *Zeichen ersten Typs* verstehen wir eine Zeichenkombination der Form:

$$a, \ fa, \ ffa, \ fffa, \ldots \quad \text{usw.,}$$

wo a entweder 0 oder eine Variable ersten Typs ist. Im ersten Fall nennen wir ein solches Zeichen *Zahlzeichen*. Für $n > 1$ verstehen wir unter einem *Zeichen n-ten Typs* dasselbe wie *Variable n-ten Typs*. Zeichenkombinationen der Form $a(b)$, wo b ein Zeichen n-ten und a ein Zeichen $(n+1)$-ten Typs ist, nennen wir *Elementarformeln*. Die Klasse der *Formeln* definierten wir als die kleinste Klasse,[19] zu welcher sämtliche Elementarformeln gehören und zu welcher zugleich mit a, b stets auch $\sim(a)$, $(a) \lor (b)$, $x\Pi(a)$ gehören (wobei x eine beliebige Variable ist).[19a] $(a) \lor (b)$ nennen wir die *Disjunktion* aus a und $b, \sim(a)$ die *Negation* und $x\Pi(a)$ eine *Generalisation* von a. *Satzformel* heißt eine Formel, in der keine freie Variable vorkommt (*freie Variable* in der bekannten Weise definiert). Eine Formel mit genau n freien Individuenvariablen (und sonst keinen freien Variablen) nennen wir *n-stelliges Relationszeichen*, für $n = 1$ auch *Klassenzeichen*.

Unter Subst $a\binom{v}{b}$ (wo a eine Formel, v eine Variable und b ein Zeichen vom selben Typ wie v bedeutet) verstehen wir die Formel, welche aus a entsteht, wenn man darin v überall, wo es frei ist, durch b ersetzt.[20] Wir sagen, daß eine Formel a eine *Typenerhöhung* einer anderen b ist, wenn a aus b dadurch entsteht, daß man den Typus aller in b vorkommenden Variablen um die gleiche Zahl erhöht.

---

[17]Es wird vorausgesetzt, daß für jeden Variablentypus abzählbar viele Zeichen zur Verfügung stehen.

[18]Auch inhomogene Relationen können auf diese Weise definiert werden, z. B. eine Relation zwischen Individuen und Klassen als eine Klasse aus Elementen der Form: $((x_2), ((x_1), x_2))$. Alle in den *PM* über Relationen beweisbaren Sätze sind, wie eine einfache Überlegung lehrt, auch bei dieser Behandlungsweise beweisbar.

[19]Bez. dieser Definition (und analoger später vorkommender) vgl. *Lukasiewicz und Tarski 1930*.

[19a]$x\Pi(a)$ ist also auch dann eine Formel, wenn x in a nicht oder nicht frei vorkommt. In diesem Fall bedeutet $x\Pi(a)$ natürlich dasselbe wie a.

[20]Falls v in a nicht als freie Variable vorkommt, soll Subst $a\binom{v}{b} = a$ sein. Man beachte, daß "Subst" ein Zeichen der Metamathematik ist.

And so on, for every natural number as a type.[17]

Remark: Variables for functions of two or more argument places (relations) need not be included among the primitive signs, since we can define relations to be classes of ordered pairs, and ordered pairs to be classes of classes; for example, the ordered pair a, b can be defined to be $((a), (a, b))$, where (x, y) denotes the class whose sole elements are x and y, and (x) the class whose sole element is x.[18]

By a *sign of type* 1 we understand a combination of signs that has [any one of] the forms

$$a, \ fa, \ ffa, \ fffa, \ldots, \quad \text{and so on,}$$

where a is either 0 or a variable of type 1. In the first case, we call such a sign a *numeral*. For $n > 1$ we understand by a *sign of type n* the same thing as by a *variable of type n*. A combination of signs that has the form $a(b)$, where b is a sign of type n and a is a sign of type $n + 1$, will be called an *elementary formula*. We define the class of *formulas* to be the smallest class[19] containing all elementary formulas and containing $\sim(a)$, $(a) \vee (b)$, $x\Pi(a)$ (where x may be any variable)[19a] whenever it contains a and b. We call $(a) \vee (b)$ the *disjunction* of a and b, $\sim(a)$ the *negation* and $x\Pi(a)$ a *generalization* of a. A formula in which no free variable occurs (*free variable* being defined in the well-known manner) is called a *sentential formula*. A formula with exactly n free individual variables (and no other free variables) will be called an *n-place relation sign*; for $n = 1$ it will also be called a *class sign*.

By Subst $a\binom{v}{b}$ (where a stands for a formula, v for a variable, and b for a sign of the same type as v) we understand the formula that results from a if in a we replace v, wherever it is free, by b.[20] We say that a formula a is a *type elevation* of another formula b if a results from b when the type of each variable occurring in b is increased by the same number.

---

[17]It is assumed that we have denumerably many signs at our disposal for each type of variables.

[18]Nonhomogeneous relations, too, can be defined in this manner; for example, a relation between individuals and classes can be defined to be a class of elements of the form $((x_2), ((x_1), x_2))$. Every proposition about relations that is provable in *PM* is provable also when treated in this manner, as is readily seen.

[19]Concerning this definition (and similar definitions occurring below) see *Łukasiewicz and Tarski 1930*.

[19a]Hence $x\Pi(a)$ is a formula even if x does not occur in a or is not free in a. In this case, of course, $x\Pi(a)$ means the same thing as a.

[20]In case v does not occur in a as a free variable we put Subst $a\binom{v}{b} = a$. Note that "Subst" is a metamathematical sign.

Folgende Formeln (I bis V) heißen *Axiome* (sie sind mit Hilfe der in bekannter Weise definierten Abkürzungen: ., $\supset$, $\equiv$, (Ex), $=$[21] und mit Verwendung der üblichen Konventionen über das Weglassen von Klammern angeschrieben):[22]

I. 1. $\sim(fx_1 = 0)$,
　2. $fx_1 = fy_1 \supset x_1 = y_1$,
　3. $x_2(0) . x_1\Pi(x_2(x_1) \supset x_2(fx_1)) \supset x_1\Pi(x_2(x_1))$.

178 | II. Jede Formel, die aus den folgenden Schemata durch Einsetzung beliebiger Formeln für p, q, r entsteht.

1. $p \lor p \supset p$,　　　　　　3. $p \lor q \supset q \lor p$,
2. $p \supset p \lor q$,　　　　　　4. $(p \supset q) \supset (r \lor p \supset r \lor q)$.

III. Jede Formel, die aus einem der beiden Schemata

1. $v\Pi(a) \supset \text{Subst } a\binom{v}{c}$,
2. $v\Pi(b \lor a) \supset b \lor v\Pi(a)$

dadurch entsteht, daß man für a, v, b, c folgende Einsetzungen vornimmt (und in 1. die durch "Subst" angezeigte Operation ausführt):

Für a eine beliebige Formel, für v eine beliebige Variable, für b eine Formel, in der v nicht frei vorkommt, für c ein Zeichen vom selben Typ wie v, vorausgesetzt, daß c keine Variable enthält, welche in a an einer Stelle gebunden ist, an der v frei ist.[23]

IV. Jede Formel, die aus dem Schema

1. $(Eu)(v\Pi(u(v) \equiv a))$

dadurch entsteht, daß man für v bzw. u beliebige Variable vom Typ n bzw. $n+1$ und für a eine Formel, die u nicht frei enthält, einsetzt. Dieses Axiom vertritt das Reduzibilitätsaxiom (Komprehensionsaxiom der Mengenlehre).

V. Jede Formel, die aus der folgenden durch Typenerhöhung entsteht (und diese Formel selbst):

1. $x_1\Pi(x_2(x_1) \equiv y_2(x_1)) \supset x_2 = y_2$.

Dieses Axiom besagt, daß eine Klasse durch ihre Elemente vollständig bestimmt ist.

Eine Formel c heißt *unmittelbare Folge* aus a und b (bzw. aus a), wenn a die Formel $(\sim(b)) \lor (c)$ ist (bzw. wenn c die Formel $v\Pi(a)$ ist, wo v eine beliebige Variable bedeutet). Die Klasse der *beweisbaren Formeln* wird

---

[21] $x_1 = y_1$ ist, wie in *PM*, I, *13, durch $x_2\Pi(x_2(x_1) \supset x_2(y_1))$ definiert zu denken (ebenso für die höheren Typen).

[22] Um aus den angeschriebenen Schemata die Axiome zu erhalten, muß man also (in II, III, IV nach Ausführung der erlaubten Einsetzungen) noch
　1. die Abkürzungen eliminieren,
　2. die unterdrückten Klammern hinzufügen.
Man beachte, daß die so entstehenden Ausdrücke "Formeln" in obigem Sinn sein müssen. (Vgl. auch die exakten Definitionen der metamathematischen Begriffe S. 182fg.)

[23] c ist also entweder eine Variable oder 0 oder ein Zeichen der Form $f \ldots fu$, wo u entweder 0 oder eine Variable 1. Typs ist. Bezüglich des Begriffs "frei (gebunden) an einer Stelle von a" vgl. *von Neumann 1927*, I, A5.

The following formulas (I–V) are called *axioms* (we write them using the abbreviations ., $\supset$, $\equiv$, (Ex), $=$,[21] defined in the well-known manner, and observing the usual conventions about omitting parentheses):[22]

I. 1. $\sim(fx_1 = 0)$,

   2. $fx_1 = fy_1 \supset x_1 = y_1$,

   3. $x_2(0) \cdot x_1\Pi(x_2(x_1) \supset x_2(fx_1)) \supset x_1\Pi(x_2(x_1))$.

II. All formulas that result from the following schemata by substitution of any formulas whatsoever for p, q, r:

  1. $p \lor p \supset p$,   3. $p \lor q \supset q \lor p$,

  2. $p \supset p \lor q$,   4. $(p \supset q) \supset (r \lor p \supset r \lor q)$.

III. Any formula that results from either one of the two schemata

   1. $v\Pi(a) \supset \text{Subst } a\binom{v}{c}$,

   2. $v\Pi(b \lor a) \supset b \lor v\Pi(a)$

when the following substitutions are made for a, v, b, and c (and the operation indicated by "Subst" is performed in 1):

For a any formula, for v any variable, for b any formula in which v does not occur free, and for c any sign of the same type as v, provided c does not contain any variable that is bound in a at a place where v is free.[23]

IV. Every formula that results from the schema

   1. $(Eu)(v\Pi(u(v) \equiv a))$

when for v we substitute any variable of type n, for u one of type $n + 1$, and for a any formula that does not contain u free. This axiom plays the role of the axiom of reducibility (the comprehension axiom of set theory).

V. Every formula that results from

   1. $x_1\Pi(x_2(x_1) \equiv y_2(x_1)) \supset x_2 = y_2$

by type elevation (as well as this formula itself). This axiom states that a class is completely determined by its elements.

A formula c is called an *immediate consequence* of a and b if a is the formula $(\sim(b)) \lor (c)$, and it is called an *immediate consequence* of a if it is the formula $v\Pi(a)$, where v denotes any variable. The class of *provable formulas* is defined to be the smallest class of formulas that contains the

---

[21]$x_1 = y_1$ is to be regarded as defined by $x_2\Pi(x_2(x_1) \supset x_2(y_1))$, as in *PM*, I, *13 (similarly for higher types).

[22]In order to obtain the axioms from the schemata listed we must therefore

  (1) eliminate the abbreviations and

  (2) add the omitted parentheses

(in II, III, and IV after carrying out the substitutions allowed).

Note that all expressions thus obtained are "formulas" in the sense specified above. (See also the exact definitions of the metamathematical notions on pp. 163ff.)

[23]Therefore c is a variable or 0 or a sign of the form $f \ldots fu$, where u is either 0 or a variable of type 1. Concerning the notion "free (bound) at a place in a", see I, A5 in *von Neumann 1927*.

definiert als die kleinste Klasse von Formeln, welche die Axiome enthält
und gegen die Relation "unmittelbare Folge" abgeschlossen ist.[24]

Wir ordnen nun den Grundzeichen des Systems P in folgender Weise
eineindeutig natürliche Zahlen zu:

179 |
$$\text{"0" } \ldots 1 \quad \text{"} f \text{" } \ldots 3 \quad \text{"} {\sim} \text{" } \ldots 5$$
$$\text{"} \vee \text{" } \ldots 7 \quad \text{"} \Pi \text{" } \ldots 9 \quad \text{"(" } \ldots 11$$
$$\text{")" } \ldots 13,$$

ferner den Variablen n-ten Typs die Zahlen der Form p^n (wo p eine Primzahl
> 13 ist). Dadurch entspricht jeder endlichen Reihe von Grundzeichen (also
auch jeder Formel) in eineindeutiger Weise eine endliche Reihe natürlicher
Zahlen. Die endlichen Reihen natürlicher Zahlen bilden wir nun (wieder
eineindeutig) auf natürliche Zahlen ab, indem wir der Reihe $n_1, n_2, \ldots, n_k$
die Zahl $2^{n_1} \cdot 3^{n_2} \cdot \ldots \cdot p_k^{n_k}$ entsprechen lassen, wo p_k die k-te Primzahl
(der Größe nach) bedeutet. Dadurch ist nicht nur jedem Grundzeichen,
sondern auch jeder endlichen Reihe von solchen in eineindeutiger Weise
eine natürliche Zahl zugeordnet. Die dem Grundzeichen (bzw. der Grund-
zeichenreihe) a zugeordnete Zahl bezeichnen wir mit $\Phi(a)$. Sei nun irgend
eine Klasse oder Relation $R(a_1, a_2, \ldots, a_n)$ zwischen Grundzeichen oder
Reihen von solchen gegeben. Wir ordnen ihr diejenige Klasse (Relation)
$R'(x_1, x_2, \ldots, x_n)$ zwischen natürlichen Zahlen zu, welche dann und nur
dann zwischen $x_1, x_2, \ldots, x_n$ besteht, wenn es solche $a_1, a_2, \ldots, a_n$ gibt,
daß $x_i = \Phi(a_i) \, (i = 1, 2, \ldots, n)$ und $R(a_1, a_2, \ldots, a_n)$ gilt. Diejenigen
Klassen und Relationen natürlicher Zahlen, welche auf diese Weise den
bisher definierten metamathematischen Begriffen, z. B. "Variable", "Formel", "Satzformel", "Axiom", "beweisbare Formel" usw. zugeordnet sind,
bezeichnen wir mit denselben Worten in Kursivschrift [hier in Kapitäl-
chen]. Der Satz, daß es im System P unentscheidbare Probleme gibt,
lautet z. B. folgendermaßen: Es gibt SATZFORMELN a, so daß weder a
noch die NEGATION von a BEWEISBARE FORMELN sind.

Wir schalten nun eine Zwischenbetrachtung ein, die mit dem formalen
System P vorderhand nichts zu tun hat, und geben zunächst folgende Definition: Eine zahlentheoretische Funktion[25] $\phi(x_1, x_2, \ldots, x_n)$ heißt *rekursiv
definiert aus* den zahlentheoretischen Funktionen $\psi(x_1, x_2, \ldots, x_{n-1})$ und

---

[24]Die Einsetzungsregel wird dadurch überflüssig, daß wir alle möglichen Einsetzungen bereits in den Axiomen selbst vorgenommen haben (analog bei *von Neumann 1927*).

[25]D. h. ihr Definitionsbereich ist die Klasse der nicht negativen ganzen Zahlen (bzw.
der n-tupel von solchen) und ihre Werte sind nicht negative ganze Zahlen.

axioms and is closed under the relation "immediate consequence".[24]

We now assign natural numbers to the primitive signs of the system P by the following one-to-one correspondence:

$$\text{``0''} \ldots 1 \quad \text{``}f\text{''} \ldots 3 \quad \text{``}\sim\text{''} \ldots 5$$
$$\text{``}\vee\text{''} \ldots 7 \quad \text{``}\Pi\text{''} \ldots 9 \quad \text{``(''} \ldots 11$$
$$\text{``)''} \ldots 13,$$

further to the variables of type n the numbers of the form p^n (where p is a prime number > 13). Thus we have a one-to-one correspondence by which a finite sequence of natural numbers is associated with every finite sequence of primitive signs (hence also with every formula). We now map the finite sequences of natural numbers on natural numbers (again by a one-to-one correspondence), associating the number $2^{n_1} \cdot 3^{n_2} \cdot \ldots \cdot p_k^{n_k}$, where p_k denotes the kth prime number (in order of increasing magnitude), with the sequence $n_1, n_2, \ldots, n_k$. A natural number [out of a certain subset] is thus assigned one-to-one not only to every primitive sign but also to every finite sequence of such signs. We denote by $\Phi(a)$ the number assigned to the primitive sign (or to the sequence of primitive signs) a. Now let some relation (or class) $R(a_1, a_2, \ldots, a_n)$ between [or of] primitive signs or sequences of primitive signs be given. With it we associate the relation (or class) $R'(x_1, x_2, \ldots, x_n)$ between [or of] natural numbers that obtains between $x_1, x_2, \ldots, x_n$ if and only if there are some $a_1, a_2, \ldots, a_n$ such that $x_i = \Phi(a_i)$ $(i = 1, 2, \ldots, n)$ and $R(a_1, a_2, \ldots, a_n)$ holds. The relations between (or classes of) natural numbers that in this manner are associated with the metamathematical notions defined so far, for example, "variable", "formula", "sentential formula", "axiom", "provable formula", and so on, will be denoted by the same words in SMALL CAPITALS. The proposition that there are undecidable problems in the system P, for example, reads thus: There are SENTENTIAL FORMULAS a such that neither a nor the NEGATION of a is a PROVABLE FORMULA.

We now insert a parenthetic consideration that for the present has nothing to do with the formal system P. First we give the following definition: A number-theoretic function[25] $\phi(x_1, x_2, \ldots, x_n)$ is said to be *recursively defined in terms of* the number-theoretic functions $\psi(x_1, x_2, \ldots, x_{n-1})$ and

---

[24]The rule of substitution is rendered superfluous by the fact that all possible substitutions have already been carried out in the axioms themselves. (This procedure was used also in *von Neumann 1927*.)

[25]That is, its domain of definition is the class of nonnegative integers (or of n-tuples of non-negative integers) and its values are nonnegative integers.

$\mu(x_1, x_2, \ldots, x_{n+1})$, wenn für alle $x_2, \ldots, x_n, k$ [26] folgendes gilt:

$$\phi(0, x_2, \ldots, x_n) = \psi(x_2, \ldots, x_n),$$
$$\phi(k+1, x_2, \ldots, x_n) = \mu(k, \phi(k, x_2, \ldots, x_n), x_2, \ldots, x_n). \qquad (2)$$

Eine zahlentheoretische Funktion ϕ heißt *rekursiv*, wenn es eine endliche Reihe von zahlentheoretischen Funktionen $\phi_1, \phi_2, \ldots, \phi_n$ gibt, welche mit ϕ endet und die Eigenschaft hat, daß jede Funktion ϕ_k der Reihe entweder
180 aus zwei der vorhergehenden rekursiv definiert ist oder | aus irgend welchen der vorhergehenden durch Einsetzung entsteht[27] oder schließlich eine Konstante oder die Nachfolgerfunktion $x+1$ ist. Die Länge der kürzesten Reihe von ϕ_i, welche zu einer rekursiven Funktion ϕ gehört, heißt ihre *Stufe*. Eine Relation zwischen natürlichen Zahlen $R(x_1, \ldots, x_n)$ heißt *rekursiv*,[28] wenn es eine rekursive Funktion $\phi(x_1, \ldots, x_n)$ gibt, so daß für alle $x_1, x_2, \ldots, x_n$

$$R(x_1, \ldots, x_n) \sim [\phi(x_1, \ldots, x_n) = 0].\text{[29]}$$

Es gelten folgende Sätze:

I. *Jede aus rekursiven Funktionen (Relationen) durch Einsetzung rekursiver Funktionen an Stelle der Variablen entstehende Funktion (Relation) ist rekursiv; ebenso jede Funktion, die aus rekursiven Funktionen durch rekursive Definition nach dem Schema (2) entsteht.*

II. *Wenn R und S rekursive Relationen sind, dann auch $\overline{R}, R \vee S$ (daher auch $R \,\&\, S$).*

III. *Wenn die Funktionen $\phi(\mathfrak{x}), \psi(\mathfrak{y})$ rekursiv sind, dann auch die Relation: $\phi(\mathfrak{x}) = \psi(\mathfrak{y})$.*[30]

IV. *Wenn die Funktion $\phi(\mathfrak{x})$ und die Relation $R(x, \mathfrak{y})$ rekursiv sind, dann auch die Relationen S, T*

$$S(\mathfrak{x}, \mathfrak{y}) \sim (Ex)[x \leq \phi(\mathfrak{x}) \,\&\, R(x, \mathfrak{y})]$$

---

[26] Kleine lateinische Buchstaben (eventuell mit Indizes) sind im folgenden immer Variable für nicht negative ganze Zahlen (falls nicht ausdrücklich das Gegenteil bemerkt ist).

[27] Genauer: durch Einsetzung gewisser der vorhergehenden Funktionen an die Leerstellen einer der vorhergehenden, z. B. $\phi_k(x_1, x_2) = \phi_p[\phi_q(x_1, x_2), \phi_r(x_2)]$ $(p, q, r < k)$. Nicht alle Variable der linken Seite müssen auch rechts vorkommen (ebenso im Rekursionsschema (2)).

[28] Klassen rechnen wir mit zu den Relationen (einstellige Relationen). Rekursive Relationen R haben natürlich die Eigenschaft, daß man für jedes spezielle Zahlen-n-tupel entscheiden kann, ob $R(x_1, \ldots, x_n)$ gilt oder nicht.

[29] Für alle inhaltlichen (insbesondere auch die metamathematischen) Überlegungen wird die Hilbertsche Symbolik verwendet. Vgl. *Hilbert und Ackermann 1928*.

[30] Wir verwenden deutsche Buchstaben $\mathfrak{x}, \mathfrak{y}$ als abkürzende Bezeichnung für beliebige Variablen-n-tupel, z. B. $x_1, x_2, \ldots, x_n$.

$\mu(x_1, x_2, \ldots, x_{n+1})$ if

$$\phi(0, x_2, \ldots, x_n) = \psi(x_2, \ldots, x_n),$$
$$\phi(k+1, x_2, \ldots, x_n) = \mu(k, \phi(k, x_2, \ldots, x_n), x_2, \ldots, x_n) \qquad (2)$$

hold for all $x_2, \ldots, x_n, k$.[26]

A number-theoretic function ϕ is said to be *recursive* if there is a finite sequence of number-theoretic functions $\phi_1, \phi_2, \ldots, \phi_n$ that ends with ϕ and has the property that every function ϕ_k of the sequence is recursively defined in terms of two of the preceding functions, or results from any of the preceding functions by substitution,[27] or, finally, is a constant or the successor function $x + 1$. The length of the shortest sequence of ϕ_i corresponding to a recursive function ϕ is called its *degree*. A relation $R(x_1, \ldots, x_n)$ between natural numbers is said to be *recursive*[28] if there is a recursive function $\phi(x_1, \ldots, x_n)$ such that, for all $x_1, x_2, \ldots, x_n$,

$$R(x_1, \ldots, x_n) \sim [\phi(x_1, \ldots, x_n) = 0].^{[29]}$$

The following theorems hold:

I. *Every function (relation) obtained from recursive functions (relations) by substitution of recursive functions for the variables is recursive; so is every function obtained from recursive functions by recursive definition according to Schema* (2).

II. *If R and S are recursive relations, so are $\overline{R}$ and $R \vee S$ (hence also $R \,\&\, S$).*

III. *If the functions $\phi(\mathfrak{x})$ and $\psi(\mathfrak{y})$ are recursive, so is the relation $\phi(\mathfrak{x}) = \psi(\mathfrak{y})$.*[30]

IV. *If the function $\phi(\mathfrak{x})$ and the relation $R(x, \mathfrak{y})$ are recursive, so are the relations S and T defined by*

$$S(\mathfrak{x}, \mathfrak{y}) \sim (Ex)[x \leq \phi(\mathfrak{x}) \,\&\, R(x, \mathfrak{y})]$$

---

[26] In what follows, lower-case italic letters (with or without subscripts) are always variables for nonnegative integers (unless the contrary is expressly noted).

[27] More precisely, by substitution of some of the preceding functions at the argument places of one of the preceding functions, for example, $\phi_k(x_1, x_2) = \phi_p[\phi_q(x_1, x_2), \phi_r(x_2)]$ $(p, q, r < k)$. Not all variables on the left side need occur on the right side (the same applies to the recursion schema (2)).

[28] We include classes among relations (as one-place relations). Recursive relations R, of course, have the property that for every given n-tuple of numbers it can be decided whether $R(x_1, \ldots, x_n)$ holds or not.

[29] Whenever formulas are used to express a meaning (in particular, in all formulas expressing metamathematical propositions or notions), Hilbert's symbolism is employed. See *Hilbert and Ackermann 1928*.

[30] We use German letters, $\mathfrak{x}, \mathfrak{y}$, as abbreviations for arbitrary n-tuples of variables, for example, $x_1, x_2, \ldots, x_n$.

$$T(\mathfrak{x}, \mathfrak{y}) \sim (x)[x \le \phi(\mathfrak{x}) \to R(x, \mathfrak{y})]$$

sowie die Funktion ψ

$$\psi(\mathfrak{x}, \mathfrak{y}) = \varepsilon x[x \le \phi(\mathfrak{x}) \mathbin{\&} R(x, \mathfrak{y})],$$

wobei $\varepsilon x F(x)$ bedeutet: Die kleinste Zahl x, für welche $F(x)$ gilt und 0, falls es keine solche Zahl gibt.

Satz I folgt unmittelbar aus der Definition von "rekursiv". Satz II und III beruhen darauf, daß die den logischen Begriffen $\overline{}$, $\vee$, $=$ entsprechenden zahlentheoretischen Funktionen

$$\alpha(x), \quad \beta(x, y), \quad \gamma(x, y)$$

nämlich:

$$\alpha(0) = 1, \ \alpha(x) = 0 \quad \text{für} \quad x \ne 0,$$
$$\beta(0, x) = \beta(x, 0) = 0, \ \beta(x, y) = 1 \text{ wenn } x, y \text{ beide } \ne 0 \text{ sind,}$$

181 |
$$\gamma(x, y) = 0 \text{ wenn } x = y, \ \gamma(x, y) = 1 \text{ wenn } x \ne y,$$

rekursiv sind, wie man sich leicht überzeugen kann. Der Beweis für Satz IV ist kurz der folgende: Nach der Voraussetzung gibt es ein rekursives $\rho(x, \mathfrak{y})$, so daß:

$$R(x, \mathfrak{y}) \sim [\rho(x, \mathfrak{y}) = 0].$$

Wir definieren nun nach dem Rekursionsschema (2) eine Funktion $\chi(x, \mathfrak{y})$ folgendermaßen:

$$\chi(0, \mathfrak{y}) = 0$$

$$\chi(n + 1, \mathfrak{y}) = (n + 1) \cdot a + \chi(n, \mathfrak{y}) \cdot \alpha(a),^{31}$$

wobei $a = \alpha[\alpha(\rho(0, \mathfrak{y}))] \cdot \alpha[\rho(n + 1, \mathfrak{y})] \cdot \alpha[\chi(n, \mathfrak{y})]$.

$\chi(n + 1, \mathfrak{y})$ ist daher entweder $= n + 1$ (wenn $a = 1$) oder $= \chi(n, \mathfrak{y})$ (wenn $a = 0$).[32] Der erste Fall tritt offenbar dann und nur dann ein, wenn sämtliche Faktoren von a 1 sind, d. h. wenn gilt:

$$\overline{R}(0, \mathfrak{y}) \mathbin{\&} R(n + 1, \mathfrak{y}) \mathbin{\&} [\chi(n, \mathfrak{y}) = 0].$$

[31] Wir setzen als bekannt voraus, daß die Funktionen $x + y$ (Addition), $x \cdot y$ (Multiplikation) rekursiv sind.

[32] Andere Werte als 0 und 1 kann a, wie aus der Definition für α ersichtlich ist, nicht annehmen.

and

$$T(\mathfrak{x}, \mathfrak{y}) \sim (x)[x \le \phi(\mathfrak{x}) \to R(x, \mathfrak{y})],$$

as well as the function ψ defined by

$$\psi(\mathfrak{x}, \mathfrak{y}) = \varepsilon x[x \le \phi(\mathfrak{x}) \And R(x, \mathfrak{y})],$$

where $\varepsilon x F(x)$ means the least number x for which $F(x)$ holds and 0 in case there is no such number.

Theorem I follows at once from the definition of "recursive". Theorems II and III are consequences of the fact that the number-theoretic functions

$$\alpha(x), \quad \beta(x, y), \quad \gamma(x, y),$$

corresponding to the logical notions $\overline{}$, $\vee$, and $=$, namely,

$$\alpha(0) = 1, \; \alpha(x) = 0 \quad \text{for} \quad x \ne 0,$$
$$\beta(0, x) = \beta(x, 0) = 0, \; \beta(x, y) = 1 \text{ when } x \text{ and } y \text{ are both} \ne 0,$$
$$\gamma(x, y) = 0 \text{ when } x = y, \; \gamma(x, y) = 1 \text{ when } x \ne y,$$

are recursive, as we can readily see. The proof of Theorem IV is briefly as follows. By assumption there is a recursive $\rho(x, \mathfrak{y})$ such that

$$R(x, \mathfrak{y}) \sim [\rho(x, \mathfrak{y}) = 0].$$

We now define a function $\chi(x, \mathfrak{y})$ by the recursion schema (2) in the following way:

$$\chi(0, \mathfrak{y}) = 0,$$

$$\chi(n + 1, \mathfrak{y}) = (n + 1) \cdot a + \chi(n, \mathfrak{y}) \cdot \alpha(a),^{31}$$

where $a = \alpha[\alpha(\rho(0, \mathfrak{y}))] \cdot \alpha[\rho(n + 1, \mathfrak{y})] \cdot \alpha[\chi(n, \mathfrak{y})]$.

Therefore $\chi(n + 1, \mathfrak{y})$ is equal either to $n + 1$ (if $a = 1$) or to $\chi(n, \mathfrak{y})$ (if $a = 0$).[32] The first case clearly occurs if and only if all factors of a are 1, that is, if

$$\overline{R}(0, \mathfrak{y}) \And R(n + 1, \mathfrak{y}) \And [\chi(n, \mathfrak{y}) = 0]$$

holds.

---

[31] We assume familiarity with the fact that the functions $x + y$ (addition) and $x \cdot y$ (multiplication) are recursive.

[32] a cannot take values other than 0 and 1, as can be seen from the definition of α.

Daraus folgt, daß die Funktion $\chi(n,\mathfrak{y})$ (als Funktion von n betrachtet) 0 bleibt bis zum kleinsten Wert von n, für den $R(n,\mathfrak{y})$ gilt, und von da ab gleich diesem Wert ist (falls schon $R(0,\mathfrak{y})$ gilt, ist dem entsprechend $\chi(n,\mathfrak{y})$ konstant und $= 0$). Demnach gilt:

$$\psi(\mathfrak{x},\mathfrak{y}) = \chi(\phi(\mathfrak{x}),\mathfrak{y}),$$

$$S(\mathfrak{x},\mathfrak{y}) \sim R[\psi(\mathfrak{x},\mathfrak{y}),\mathfrak{y}].$$

Die Relation T läßt sich durch Negation auf einen zu S analogen Fall zurückführen, womit Satz IV bewiesen ist.

Die Funktionen $x + y$, $x \cdot y$, x^y, ferner die Relationen $x < y$, $x = y$ sind, wie man sich leicht überzeugt, rekursiv und wir definieren nun, von diesen Begriffen ausgehend, eine Reihe von Funktionen (Relationen) 1-45, deren jede aus den vorhergehenden mittels der in den Sätzen I bis IV genannten Verfahren definiert ist. Dabei sind meistens mehrere der nach Satz I bis IV erlaubten Definitionsschritte in einen zusammengefaßt. Jede der Funktionen (Relationen) 1-45, unter denen z. B. die Begriffe "FORMEL", "AXIOM", "UNMITTELBARE FOLGE" vorkommen, ist daher rekursiv.

182 | 1. $x/y \equiv (Ez)[z \le x \,\&\, x = y \cdot z]$,[33]
x ist teilbar durch y.[34]

2. $\mathrm{Prim}(x) \equiv \overline{(Ez)}[z \le x \,\&\, z \ne 1 \,\&\, z \ne x \,\&\, x/z] \,\&\, x > 1$,
x ist Primzahl.

3. $0 \,Pr\,x \equiv 0$
$(n+1)\,Pr\,x \equiv \varepsilon y[y \le x \,\&\, \mathrm{Prim}(y) \,\&\, x/y \,\&\, y > n\,Pr\,x]$,
$n\,Pr\,x$ ist die n-te (der Größe nach) in x enthaltene Primzahl.[34a]

4. $0! \equiv 1$
$(n+1)! \equiv (n+1)\cdot n!$

5. $Pr(0) \equiv 0$
$Pr(n+1) \equiv \varepsilon y[y \le \{Pr(n)\}! + 1 \,\&\, \mathrm{Prim}(y) \,\&\, y > Pr(n)]$,
$Pr(n)$ ist die n-te Primzahl (der Größe nach).

6. $n\,Gl\,x \equiv \varepsilon y[y \le x \,\&\, x/(n\,Pr\,x)^y \,\&\, \overline{x/(n\,Pr\,x)^{y+1}}]$,
$n\,Gl\,x$ ist das n-te Glied der der Zahl x zugeordneten Zahlenreihe (für $n > 0$ und n nicht größer als die Länge dieser Reihe).

7. $l(x) \equiv \varepsilon y[y \le x \,\&\, y\,Pr\,x > 0 \,\&\, (y+1)\,Pr\,x = 0]$,

---

[33]Das Zeichen $\equiv$ wird im Sinne von "Definitionsgleichheit" verwendet, vertritt also bei Definitionen entweder $=$ oder $\sim$ (im übrigen ist die Symbolik die Hilbertsche).

[34]Überall, wo in den folgenden Definitionen eines der Zeichen (x), (Ex), εx auftritt, ist es von einer Abschätzung für x gefolgt. Diese Abschätzung dient lediglich dazu, um die rekursive Natur des definierten Begriffs (vgl. Satz IV) zu sichern. Dagegen würde sich der *Umfang* der definierten Begriffe durch Weglassung dieser Abschätzung meistens nicht ändern.

[34a]Für $0 < n \le z$, wenn z die Anzahl der verschiedenen in x aufgehenden Primzahlen ist. Man beachte, daß, für $n = z + 1$, $n\,Pr\,x = 0$ ist!

From this it follows that the function $\chi(n,\mathfrak{y})$ (considered as a function of n) remains 0 up to ⟦but not including⟧ the least value of n for which $R(n,\mathfrak{y})$ holds and, from there on, is equal to that value. (Hence, in case $R(0,\mathfrak{y})$ holds, $\chi(n,\mathfrak{y})$ is constant and equal to 0.) We have, therefore,

$$\psi(\mathfrak{x},\mathfrak{y}) = \chi(\phi(\mathfrak{x}),\mathfrak{y}),$$

$$S(\mathfrak{x},\mathfrak{y}) \sim R[\psi(\mathfrak{x},\mathfrak{y}),\mathfrak{y}].$$

The relation T can, by negation, be reduced to a case analogous to that of S. Theorem IV is thus proved.

The functions $x+y$, $x \cdot y$, and x^y, as well as the relations $x < y$ and $x = y$, are recursive, as we can readily see. Starting from these notions, we now define a number of functions (relations) 1–45, each of which is defined in terms of preceding ones by the procedures given in Theorems I–IV. In most of these definitions several of the steps allowed by Theorems I–IV are condensed into one. Each of the functions (relations) 1–45, among which occur, for example, the notions "FORMULA", "AXIOM", and "IMMEDIATE CONSEQUENCE", is therefore recursive.

1. $x/y \equiv (Ez)[z \leq x \,\&\, x = y \cdot z]$,[33]
x is divisible by y.[34]

2. $\mathrm{Prim}(x) \equiv \overline{(Ez)}[z \leq x \,\&\, z \neq 1 \,\&\, z \neq x \,\&\, x/z] \,\&\, x > 1$,
x is a prime number.

3. $0 \, Pr \, x \equiv 0$,
$(n+1) \, Pr \, x \equiv \varepsilon y[y \leq x \,\&\, \mathrm{Prim}(y) \,\&\, x/y \,\&\, y > n \, Pr \, x]$,
$n \, Pr \, x$ is the nth prime number (in order of increasing magnitude) contained in x.[34a]

4. $0! \equiv 1$,
$(n+1)! \equiv (n+1) \cdot n!$.

5. $Pr(0) \equiv 0$,
$Pr(n+1) \equiv \varepsilon y[y \leq \{Pr(n)\}! + 1 \,\&\, \mathrm{Prim}(y) \,\&\, y > Pr(n)]$,
$Pr(n)$ is the nth prime number (in order of increasing magnitude).

6. $n \, Gl \, x \equiv \varepsilon y[y \leq x \,\&\, x/(n \, Pr \, x)^y \,\&\, \overline{x/(n \, Pr \, x)^{y+1}}]$,
$n \, Gl \, x$ is the nth term of the number sequence assigned to the number x (for $n > 0$ and n not greater than the length of this sequence).

7. $l(x) \equiv \varepsilon y[y \leq x \,\&\, y \, Pr \, x > 0 \,\&\, (y+1) \, Pr \, x = 0]$,

---

[33] The sign $\equiv$ is used in the sense of "equality by definition"; hence in definitions it stands for either $=$ or $\sim$ (otherwise, the symbolism is Hilbert's).

[34] Wherever one of the signs $(x), (Ex)$, or εx occurs in the definitions below, it is followed by a bound on x. This bound merely serves to ensure that the notion defined is recursive (see Theorem IV). But in most cases the *extension* of the notion defined would not change if this bound were omitted.

[34a] For $0 < n \leq z$, when z is the number of distinct prime factors of x. Note that $n \, Pr \, x = 0$ for $n = z + 1$.

$l(x)$ ist die Länge der x zugeordneten Zahlenreihe.

8. $x * y \equiv \varepsilon z \{ z \leq [Pr(l(x) + l(y))]^{x+y} \, \& $
$(n)[n \leq l(x) \rightarrow n \, Gl \, z = n \, Gl \, x] \, \& $
$(n)[0 < n \leq l(y) \rightarrow (n + l(x)) \, Gl \, z = n \, Gl \, y]\},$

$x * y$ entspricht der Operation des "Aneinanderfügens" zweier endlicher Zahlenreihen.

9. $R(x) \equiv 2^x,$

$R(x)$ entspricht der nur aus der Zahl x bestehenden Zahlenreihe (für $x > 0$).

10. $E(x) \equiv R(11) * x * R(13),$

$E(x)$ entspricht der Operation des "Einklammerns" (11 und 13 sind den Grundzeichen "(" und ")" zugeordnet).

11. $n \, \mathrm{Var} \, x \equiv (Ez)[13 < z \leq x \, \& \, \mathrm{Prim}(z) \, \& \, x = z^n] \, \& \, n \neq 0,$

x ist eine VARIABLE n-TEN TYPS.

12. $\mathrm{Var}\,(x) \equiv (En)[n \leq x \, \& \, n \, \mathrm{Var} \, x],$

x ist eine VARIABLE.

13. $\mathrm{Neg}(x) \equiv R(5) * E(x),$

$\mathrm{Neg}(x)$ ist die NEGATION von x.

183 | 14. $x \, \mathrm{Dis} \, y \equiv E(x) * R(7) * E(y),$

$x \, \mathrm{Dis} \, y$ ist die DISJUNKTION aus x und y.

15. $x \, \mathrm{Gen} \, y \equiv R(x) * R(9) * E(y),$

$x \, \mathrm{Gen} \, y$ ist die GENERALISATION von y mittels der VARIABLEN x (vorausgesetzt, daß x eine VARIABLE ist).

16. $0 \, N \, x \equiv x$
$(n + 1) \, N \, x \equiv R(3) * n \, N \, x,$

$n \, N \, x$ entspricht der Operation: "n-maliges Vorsetzen des Zeichens 'f' vor x".

17. $Z(n) \equiv n \, N \, [R(1)]$

$Z(n)$ ist das ZAHLZEICHEN für die Zahl n.

18. $\mathrm{Typ}'_1(x) \equiv (Em, n)\{m, n \leq x \, \& \, [m = 1 \lor 1 \, \mathrm{Var} \, m] \, \& $
$x = n \, N \, [R(m)]\},$[34b]

x ist ZEICHEN ERSTEN TYPS.

19. $\mathrm{Typ}_n(x) \equiv [n = 1 \, \& \, \mathrm{Typ}'_1(x)] \lor$
$[n > 1 \, \& \, (Ev)\{v \leq x \, \& \, n \, \mathrm{Var} \, v \, \& \, x = R(v)\}],$

x ist ZEICHEN n-TEN TYPS.

20. $\mathrm{Elf}(x) \equiv (Ey, z, n)[y, z, n \leq x \, \& \, \mathrm{Typ}_n(y) \, \& \, \mathrm{Typ}_{n+1}(z) \, \& $
$x = z * E(y)],$

x ist ELEMENTARFORMEL.

21. $Op(x, y, z) \equiv x = \mathrm{Neg}(y) \lor x = y \, \mathrm{Dis} \, z \lor$
$(Ev)[v \leq x \, \& \, \mathrm{Var}(v) \, \& \, x = v \, \mathrm{Gen} \, y]$

[34b]$m, n \leq x$ steht für: $m \leq x \, \& \, n \leq x$ (ebenso für mehr als zwei Variable).

$l(x)$ is the length of the number sequence assigned to x.

    8. $x * y \equiv \varepsilon z \{ z \leq [Pr(l(x) + l(y))]^{x+y} \, \& $
$$(n)[n \leq l(x) \rightarrow n \, Gl \, z = n \, Gl \, x] \, \&$$
$$(n)[0 < n \leq l(y) \rightarrow (n + l(x)) \, Gl \, z = n \, Gl \, y] \},$$

$x * y$ corresponds to the operation of "concatenating" two finite number sequences.

    9. $R(x) \equiv 2^x$,

$R(x)$ corresponds to the number sequence consisting of x alone (for $x > 0$).

    10. $E(x) \equiv R(11) * x * R(13)$,

$E(x)$ corresponds to the operation of "enclosing within parentheses" (11 and 13 are assigned to the primitive signs "(" and ")", respectively).

    11. $n \, \text{Var} \, x \equiv (Ez)[13 < z \leq x \, \& \, \text{Prim}(z) \, \& \, x = z^n] \, \& \, n \neq 0$,

x is a VARIABLE OF TYPE n.

    12. $\text{Var}\,(x) \equiv (En)[n \leq x \, \& \, n \, \text{Var} \, x]$,

x is a VARIABLE.

    13. $\text{Neg}(x) \equiv R(5) * E(x)$,

$\text{Neg}(x)$ is the NEGATION of x.

    14. $x \, \text{Dis} \, y \equiv E(x) * R(7) * E(y)$,

$x \, \text{Dis} \, y$ is the DISJUNCTION of x and y.

    15. $x \, \text{Gen} \, y \equiv R(x) * R(9) * E(y)$,

$x \, \text{Gen} \, y$ is the GENERALIZATION of y with respect to the VARIABLE x (provided x is a VARIABLE).

    16. $0 \, N \, x \equiv x$,
$$(n + 1)Nx \equiv R(3) * n \, N \, x,$$

$n \, N \, x$ corresponds to the operation of "putting the sign 'f' n times in front of x".

    17. $Z(n) \equiv n \, N \, [R(1)]$,

$Z(n)$ is the NUMERAL denoting the number n.

    18. $\text{Typ}'_1(x) \equiv (Em, n)\{m, n \leq x \, \& \, [m = 1 \vee 1 \, \text{Var} \, m] \, \&$
$$x = n \, N \, [R(m)] \},^{34b}$$

x is a SIGN OF TYPE 1.

    19. $\text{Typ}_n(x) \equiv [n = 1 \, \& \, \text{Typ}'_1(x)] \vee$
$$[n > 1 \, \& \, (Ev)\{v \leq x \, \& \, n \, \text{Var} \, v \, \& \, x = R(v)\}],$$

x is a SIGN OF TYPE n.

    20. $Elf(x) \equiv (Ey, z, n)[y, z, n \leq x \, \& \, \text{Typ}_n(y) \, \& \, \text{Typ}_{n+1}(z) \, \&$
$$x = z * E(y)],$$

x is an ELEMENTARY FORMULA.

    21. $Op(x, y, z) \equiv x = \text{Neg}(y) \vee x = y \, \text{Dis} \, z \vee$
$$(Ev)[v \leq x \, \& \, \text{Var}(v) \, \& \, x = v \, \text{Gen} \, y].$$

[34b] $m, n \leq x$ stands for $m \leq x \, \& \, n \leq x$ (similarly for more than two variables).

22. $FR(x) \equiv (n)\{0 < n \leq l(x) \rightarrow Elf(n\,Gl\,x) \lor$
$\qquad (Ep,q)[0 < p,q < n \,\&\, Op(n\,Gl\,x, p\,Gl\,x, q\,Gl\,x)]\} \,\&\,$
$\qquad l(x) > 0,$

x ist eine Reihe von FORMELN, deren jede entweder ELEMENTARFORMEL ist oder aus den vorhergehenden durch die Operationen der NEGATION, DISJUNKTION, GENERALISATION hervorgeht.

23. $\mathrm{Form}(x) \equiv (En)\{n \leq (Pr([l(x)]^2))^{x \cdot [l(x)]^2} \,\&\,$
$\qquad FR(n) \,\&\, x = [l(n)]\,Gl\,n\},^{35}$

x ist FORMEL (d. h. letztes Glied einer FORMELREIHE n).

24. $v\,\mathrm{Geb}\,n, x \equiv \mathrm{Var}(v) \,\&\, \mathrm{Form}(x) \,\&\,$
$\qquad (Ea,b,c)[a,b,c \leq x \,\&\, x = a * (v\,\mathrm{Gen}\,b) * c \,\&\,$
$\qquad \mathrm{Form}(b) \,\&\, l(a) + 1 \leq n \leq l(a) + l(v\,\mathrm{Gen}\,b)],$

die VARIABLE v ist in x an n-ter Stelle GEBUNDEN.

184 | 25. $v\,Frn, x \equiv \mathrm{Var}(v) \,\&\, \mathrm{Form}(x) \,\&\, v = n\,Gl\,x \,\&\, n \leq l(x) \,\&\, \overline{v\,\mathrm{Geb}\,n, x}$,

die VARIABLE v ist in x an n-ter Stelle FREI.

26. $v\,Frx \equiv (En)[n \leq l(x) \,\&\, v\,Frn, x],$

v kommt in x als FREIE VARIABLE vor.

27. $Su\,x\binom{n}{y} \equiv \varepsilon z\{z \leq [Pr(l(x) + l(y))]^{x+y} \,\&\, (Eu,v)[u,v \leq x \,\&\,$
$\qquad x = u * R(n\,Gl\,x) * v \,\&\, z = u * y * v \,\&\, n = l(u) + 1]\},$

$Su\,x\binom{n}{y}$ entsteht aus x, wenn man an Stelle des n-ten Gliedes von x y einsetzt (vorausgesetzt, daß $0 < n \leq l(x)$).

28. $0\,St\,v, x \equiv \varepsilon n\{n \leq l(x) \,\&\, v\,Frn, x \,\&\,$
$\qquad \overline{(Ep)}[n < p \leq l(x) \,\&\, v\,Frp, x]\}$
$(k+1)\,St\,v, x \equiv \varepsilon n\{n < k\,St\,v, x \,\&\, v\,Frn, x \,\&\,$
$\qquad \overline{(Ep)}[n < p < k\,St\,v, x \,\&\, v\,Frp, x]\},$

$k\,St\,v, x$ ist die $(k+1)$-te Stelle in x (vom Ende der FORMEL x an gezählt), an der v in x FREI ist (und 0, falls es keine solche Stelle gibt).

29. $A(v,x) \equiv \varepsilon n\{n \leq l(x) \,\&\, n\,St\,v, x = 0\},$

$A(v,x)$ ist die Anzahl der Stellen, an denen v in x FREI ist.

30. $Sb_0(x_y^v) \equiv x$
$Sb_{k+1}(x_y^v) \equiv Su[Sb_k(x_y^v)]\binom{k\,St\,v, x}{y}.$

31. $Sb(x_y^v) \equiv Sb_{A(v,x)}(x_y^v),^{36}$

$Sb(x_y^v)$ ist der oben definierte Begriff SUBST $a\binom{v}{b}.^{37}$

---

[35] Die Abschätzung $n \leq (Pr([l(x)]^2))^{x \cdot [l(x)]^2}$ erkennt man etwa so: Die Länge der kürzesten zu x gehörigen Formelreihe kann höchstens gleich der Anzahl der Teilformeln von x sein. Es gibt aber höchstens $l(x)$ Teilformeln der Länge 1, höchstens $l(x) - 1$ der Länge 2 usw., im ganzen also höchstens $l(x)(l(x)+1)/2 \leq [l(x)]^2$. Die Primzahlen aus n können also sämtlich kleiner als $Pr\{[l(x)]^2\}$ angenommen werden, ihre Anzahl $\leq [l(x)]^2$ und ihre Exponenten (welche Teilformeln von x sind) $\leq x$.

[36] Falls v keine VARIABLE oder x keine FORMEL ist, ist $Sb(x_y^v) = x$.

[37] Statt $Sb[Sb(x_y^v)_z^w]$ schreiben wir: $Sb(x_y^v \,{}_z^w)$ (analog für mehr als zwei VARIABLE).

22. $FR(x) \equiv (n)\{0 < n \leq l(x) \rightarrow Elf(n \, Gl \, x) \vee$
$\qquad (Ep, q)[0 < p, q < n \, \& \, Op(n \, Gl \, x, p \, Gl \, x, q \, Gl \, x)]\} \, \&$
$\qquad l(x) > 0,$

x is a SEQUENCE OF FORMULAS, each term of which either is an ELE-
MENTARY FORMULA or results from the preceding FORMULAS through the
operations of NEGATION, DISJUNCTION, or GENERALIZATION.

23. $\text{Form}(x) \equiv (En)\{n \leq (Pr([l(x)]^2))^{x \cdot [l(x)]^2} \, \&$
$\qquad FR(n) \, \& \, x = [l(n)] \, Gl \, n\},$[35]

x is a FORMULA (that is, the last term of a FORMULA SEQUENCE n).

24. $v \, \text{Geb} \, n, x \equiv \text{Var}(v) \, \& \, \text{Form}(x) \, \&$
$\qquad (Ea, b, c)[a, b, c \leq x \, \& \, x = a * (v \, \text{Gen} \, b) * c \, \&$
$\qquad \text{Form}(b) \, \& \, l(a) + 1 \leq n \leq l(a) + l(v \, \text{Gen} \, b)],$

the VARIABLE v is BOUND in x at the nth place.

25. $v \, Frn, x \equiv \text{Var}(v) \, \& \, \text{Form}(x) \, \& \, v = n \, Gl \, x \, \& \, n \leq l(x) \, \& \, \overline{v \, \text{Geb} \, n, \, x},$

the VARIABLE v is FREE in x at the nth place.

26. $v \, Frx \equiv (En)[n \leq l(x) \, \& \, v \, Frn, x],$

v occurs as a FREE VARIABLE in x.

27. $Su \, x\binom{n}{y} \equiv \varepsilon z \{ z \leq [Pr(l(x) + l(y))]^{x+y} \, \& \, (Eu, v)[u, v \leq x \, \&$
$\qquad x = u * R(n \, Gl \, x) * v \, \& \, z = u * y * v \, \& \, n = l(u) + 1]\},$

$Su \, x\binom{n}{y}$ results from x when we substitute y for the nth term of x (provided
that $0 < n \leq l(x)$).

28. $0 \, Stv, x \equiv \varepsilon n\{n \leq l(x) \, \& \, v \, Frn, x \, \&$
$\qquad \overline{(Ep)}[n < p \leq l(x) \, \& \, v \, Frp, x]\},$
$\qquad (k+1) \, Stv, x \equiv \varepsilon n\{n < k \, Stv, x \, \& \, v \, Frn, x \, \&$
$\qquad \overline{(Ep)}[n < p < k \, Stv, x \, \& \, v \, Frp, x]\},$

$k \, Stv, x$ is the $(k+1)$th place in x (counted from the right end of the
FORMULA x) at which v is FREE in x (and 0 in case there is no such place).

29. $A(v, x) \equiv \varepsilon n\{n \leq l(x) \, \& \, n \, Stv, x = 0\},$

$A(v, x)$ is the number of places at which v is FREE in x.

30. $Sb_0\binom{xv}{y} \equiv x,$
$\qquad Sb_{k+1}\binom{xv}{y} \equiv Su[Sb_k\binom{xv}{y}]\binom{k \, Stv, \, x}{y}.$

31. $Sb\binom{xv}{y} \equiv Sb_{A(v,x)}\binom{xv}{y},$[36]

$Sb\binom{xv}{y}$ is the notion SUBST $a\binom{v}{b}$ defined above.[37]

---

[35]That $n \leq (Pr([l(x)]^2))^{x \cdot [l(x)]^2}$ provides a bound can be seen thus: The length of
the shortest formula sequence that corresponds to x can at most be equal to the number
of subformulas of x. But there are at most $l(x)$ subformulas of length 1, at most $l(x) - 1$
of length 2, and so on, hence altogether at most $l(x)(l(x) + 1)/2 \leq [l(x)]^2$. Therefore all
prime factors of n can be assumed to be less than $Pr([l(x)]^2)$, their number $\leq [l(x)]^2$,
and their exponents (which are subformulas of x) $\leq x$.

[36]In case v is not a VARIABLE or x is not a FORMULA, $Sb\binom{xv}{y} = x$.

[37]Instead of $Sb[Sb\binom{xv}{y}\frac{w}{z}]$ we write $Sb(x\frac{v}{y}\frac{w}{z})$ (and similarly for more than two VARIA-
BLES).

32. $x \, \text{Imp} \, y \equiv [\text{Neg}(x)] \, \text{Dis} \, y$

 $x \, \text{Con} \, y \equiv \text{Neg}\{[\text{Neg}(x)] \, \text{Dis} \, [\text{Neg}(y)]\}$

 $x \, \text{Aeq} \, y \equiv (x \, \text{Imp} \, y) \, \text{Con} \, (y \, \text{Imp} \, x)$

 $v \, \text{Ex} \, y \equiv \text{Neg}\{v \, \text{Gen} \, [\text{Neg}(y)]\}$

33. $n \, Th \, x \equiv \epsilon y \{ y \leq x^{(x^n)} \, \& \, (k)[k \leq l(x) \rightarrow$

   $(k \, Gl \, x \leq 13 \, \& \, k \, Gl \, y = k \, Gl \, x) \, \vee$

   $(k \, Gl \, x > 13 \, \& \, k \, Gl \, y = k \, Gl \, x \cdot [1 \, Pr \, (k \, Gl \, x)]^n)]\},$

$n \, Th \, x$ ist die n-TE TYPENERHÖHUNG von x (falls x und $n \, Th \, x$ FORMELN sind).

Den Axiomen I, 1 bis 3, entsprechen drei bestimmte Zahlen, die wir mit z_1, z_2, z_3 bezeichnen, und wir definieren:

34. $Z\text{-}Ax(x) \equiv (x = z_1 \vee x = z_2 \vee x = z_3)$

185 | 35. $A_1\text{-}Ax(x) \equiv (Ey)[y \leq x \, \& \, \text{Form}(y) \, \& \, x = (y \, \text{Dis} \, y) \, \text{Imp} \, y],$

x ist eine durch Einsetzung in das Axiomenschema II, 1, entstehende FOR-MEL. Analog werden $A_2\text{-}Ax$, $A_3\text{-}Ax$, $A_4\text{-}Ax$ entsprechend den Axiomen II, 2 bis 4, definiert.

36. $A\text{-}Ax(x) \equiv A_1\text{-}Ax(x) \vee A_2\text{-}Ax(x) \vee A_3\text{-}Ax(x) \vee A_4\text{-}Ax(x),$

x ist eine durch Einsetzung in ein Aussagenaxiom entstehende FORMEL.

37. $Q(z, y, v) \equiv \overline{(En, m, w)}[n \leq l(y) \, \& \, m \leq l(z) \, \& \, w \leq z \, \&$

   $w = m \, Gl \, z \, \& \, w \, \text{Geb} \, n, y \, \& \, v \, Frn, y],$

z enthält keine VARIABLE, die in y an einer Stelle GEBUNDEN ist, an der v FREI ist.

38. $L_1\text{-}Ax(x) \equiv (Ev, y, z, n)\{v, y, z, n \leq x \, \& \, n \, \text{Var} \, v \, \& \, \text{Typ}_n(z) \, \&$

   $\text{Form}(y) \, \& \, Q(z, y, v) \, \& \, x = (v \, \text{Gen} \, y) \, \text{Imp} \, [Sb(y_z^v)]\},$

x ist eine aus dem Axiomenschema III, 1, durch Einsetzung entstehende FORMEL.

39. $L_2\text{-}Ax(x) \equiv (Ev, q, p)\{v, q, p \leq x \, \& \, \text{Var}(v) \, \& \, \text{Form}(p) \, \& \, \overline{v \, Frp} \, \&$

   $\text{Form}(q) \, \& \, x = [v \, \text{Gen} \, (p \, \text{Dis} \, q)] \, \text{Imp} \, [p \, \text{Dis} \, (v \, \text{Gen} \, q)]\},$

x ist eine aus dem Axiomenschema III, 2, durch Einsetzung entstehende FORMEL.

40. $R\text{-}Ax(x) \equiv (Eu, v, y, n)[u, v, y, n \leq x \, \& \, n \, \text{Var} \, v \, \&$

   $(n + 1) \, \text{Var} \, u \, \& \, \overline{u \, Fry} \, \& \, \text{Form}(y) \, \&$

   $x = u \, \text{Ex}\{v \, \text{Gen} \, [[R(u) * E(R(v))] \, \text{Aeq} \, y]\}],$

x ist eine aus dem Axiomenschema IV, 1, durch Einsetzung entstehende FORMEL.

Dem Axiom V, 1, entspricht eine bestimmte Zahl z_4 und wir definieren:

41. $M\text{-}Ax(x) \equiv (En)[n \leq x \, \& \, x = n \, Th \, z_4].$

42. $Ax(x) \equiv Z\text{-}Ax(x) \vee A\text{-}A(x) \vee L_1\text{-}Ax(x) \vee L_2\text{-}Ax(x) \vee$

   $R\text{-}Ax(x) \vee M\text{-}Ax(x),$

x ist ein AXIOM.

43. $Fl(x, y, z) \equiv y = z \, \text{Imp} \, x \vee (Ev)[v \leq x \, \& \, \text{Var}(v) \, \& \, x = v \, \text{Gen} \, y],$

x ist UNMITTELBARE FOLGE aus y und z.

32. $x \operatorname{Imp} y \equiv [\operatorname{Neg}(x)] \operatorname{Dis} y$,

$\quad x \operatorname{Con} y \equiv \operatorname{Neg}\{[\operatorname{Neg}(x)] \operatorname{Dis} [\operatorname{Neg}(y)]\}$,

$\quad x \operatorname{Aeq} y \equiv (x \operatorname{Imp} y) \operatorname{Con} (y \operatorname{Imp} x)$,

$\quad v \operatorname{Ex} y \equiv \operatorname{Neg}\{v \operatorname{Gen} [\operatorname{Neg}(y)]\}$.

33. $n \operatorname{Th} x \equiv \varepsilon y\{y \leq x^{(x^n)} \,\&\, (k)[k \leq l(x) \rightarrow$

$\quad\quad (k\, Gl\, x \leq 13 \,\&\, k\, Gl\, y = k\, Gl\, x) \vee$

$\quad\quad (k\, Gl\, x > 13 \,\&\, k\, Gl\, y = k\, Gl\, x \cdot [1 \, Pr\, (k\, Gl\, x)]^n)]\}$,

$n\, Th\, x$ is the nTH TYPE ELEVATION of x (in case x and $n\, Th\, x$ are FORMULAS).

Three specific numbers, which we denote by z_1, z_2, and z_3, correspond to the Axioms I, 1–3, and we define

34. $Z\text{-}Ax(x) \equiv (x = z_1 \vee x = z_2 \vee x = z_3)$.

35. $A_1\text{-}Ax(x) \equiv (Ey)[y \leq x \,\&\, \operatorname{Form}(y) \,\&\, x = (y \operatorname{Dis} y) \operatorname{Imp} y]$,

x is a FORMULA resulting from Axiom Schema II, 1 by substitution. Analogously, $A_2\text{-}Ax, A_3\text{-}Ax$, and $A_4\text{-}Ax$ are defined for Axioms [rather, Axiom Schemata] II, 2–4.

36. $A\text{-}A(x) \equiv A_1\text{-}Ax(x) \vee A_2\text{-}Ax(x) \vee A_3\text{-}Ax(x) \vee A_4\text{-}Ax(x)$,

x is a FORMULA resulting from a propositional axiom by substitution.

37. $Q(z,y,v) \equiv \overline{(En,m,w)}[n \leq l(y) \,\&\, m \leq l(z) \,\&\, w \leq z \,\&\,$

$\quad\quad w = m\, Gl\, z \,\&\, w \operatorname{Geb} n, y \,\&\, v\, Frn, y]$,

z does not contain any VARIABLE BOUND in y at a place at which v is FREE.

38. $L_1\text{-}Ax(x) \equiv (Ev,y,z,n)\{v,y,z,n \leq x \,\&\, n \operatorname{Var} v \,\&\, \operatorname{Typ}_n(z) \,\&\,$

$\quad\quad \operatorname{Form}(y) \,\&\, Q(z,y,v) \,\&\, x = (v \operatorname{Gen} y) \operatorname{Imp} [Sb(y_z^v)]\}$,

x is a FORMULA resulting from Axiom Schema III, 1 by substitution.

39. $L_2\text{-}Ax(x) \equiv (Ev,q,p)\{v,q,p \leq x \,\&\, \operatorname{Var}(v) \,\&\, \operatorname{Form}(p) \,\&\, \overline{v\, Frp} \,\&\,$

$\quad\quad \operatorname{Form}(q) \,\&\, x = [v \operatorname{Gen} (p \operatorname{Dis} q)] \operatorname{Imp} [p \operatorname{Dis} (v \operatorname{Gen} q)]\}$,

x is a FORMULA resulting from Axiom Schema III, 2 by substitution.

40. $R\text{-}Ax(x) \equiv (Eu,v,y,n)[u,v,y,n \leq x \,\&\, n \operatorname{Var} v \,\&\,$

$\quad\quad (n+1) \operatorname{Var} u \,\&\, \overline{u\, Fry} \,\&\, \operatorname{Form}(y) \,\&\,$

$\quad\quad x = u \operatorname{Ex} \{v \operatorname{Gen} [[R(u) * E(R(v))] \operatorname{Aeq} y]\}]$,

x is a FORMULA resulting from Axiom Schema IV, 1 by substitution.

A specific number z_4 corresponds to Axiom V, 1, and we define

41. $M\text{-}Ax(x) \equiv (En)[n \leq x \,\&\, n\, Th\, z_4]$.

42. $Ax(x) \equiv Z\text{-}Ax(x) \vee A\text{-}A(x) \vee L_1\text{-}Ax(x) \vee L_2\text{-}Ax(x) \vee$

$\quad\quad R\text{-}Ax(x) \vee M\text{-}Ax(x)$,

x is an AXIOM.

43. $Fl(x,y,z) \equiv y = z \operatorname{Imp} x \vee (Ev)[v \leq x \,\&\, \operatorname{Var}(v) \,\&\, x = v \operatorname{Gen} y]$,

x is an IMMEDIATE CONSEQUENCE of y and z.

186 | 44. $Bw(x) \equiv (n)\{0 < n \le l(x) \rightarrow Ax(n\,Gl\,x)\vee$
$(Ep,q)[0 < p,q < n \,\&\, Fl(n\,Gl\,x, p\,Gl\,x, q\,Gl\,x)]\} \,\&\,$
$l(x) > 0,$

x ist eine BEWEISFIGUR (eine endliche Folge von FORMELN, deren jede entweder AXIOM oder UNMITTELBARE FOLGE aus zwei der vorhergehenden ist).

45. $x\,B\,y \equiv Bw(x)\,\&\,[l(x)]\,Gl\,x = y,$

x ist ein BEWEIS für die FORMEL *y*.

46. $\mathrm{Bew}(x) \equiv (Ey)y\,B\,x,$

x ist eine BEWEISBARE FORMEL. ($\mathrm{Bew}(x)$ ist der einzige unter den Begriffen 1–46, von dem nicht behauptet werden kann, er sei rekursiv.)

Die Tatsache, die man vage so formulieren kann: Jede rekursive Relation ist innerhalb des Systems *P* (dieses inhaltlich gedeutet) definierbar, wird, *ohne* auf eine inhaltliche Deutung der Formeln aus *P* Bezug zu nehmen, durch folgenden Satz exakt ausgedrückt:

Satz V: *Zu jeder rekursiven Relation $R(x_1,\ldots,x_n)$ gibt es ein n-stelliges* RELATIONSZEICHEN *r (mit den* FREIEN VARIABLEN[38] *$u_1, u_2, \ldots, u_n$), so daß für alle Zahlen-n-tupel $(x_1,\ldots,x_n)$ gilt:*

$$R(x_1,\ldots,x_n) \rightarrow \mathrm{Bew}\,[Sb(r^{u_1\ \ldots\ u_n}_{Z(x_1)\ldots\,Z(x_n)})], \tag{3}$$

$$\overline{R}(x_1,\ldots,x_n) \rightarrow \mathrm{Bew}\,[\mathrm{Neg}(Sb(r^{u_1\ \ldots\ u_n}_{Z(x_1)\ldots\,Z(x_n)}))]. \tag{4}$$

Wir begnügen uns hier damit, den Beweis dieses Satzes, da er keine prinzipiellen Schwierigkeiten bietet und ziemlich umständlich ist, in Umrissen anzudeuten.[39] Wir beweisen den Satz für alle Relationen $R(x_1,\ldots,x_n)$ der Form: $x_1 = \phi(x_2,\ldots,x_n)$[40] (wo ϕ eine rekursive Funktion ist) und wenden vollständige Induktion nach der Stufe von ϕ an. Für Funktionen erster Stufe (d. h. Konstante und die Funktion $x + 1$) ist der Satz trivial. Habe also ϕ die *m*-te Stufe. Es entsteht aus Funktionen niedrigerer Stufe $\phi_1,\ldots,\phi_k$ durch die Operationen der Einsetzung oder der rekursiven Definition. Da für $\phi_1,\ldots,\phi_k$ nach induktiver Annahme bereits alles bewiesen ist, gibt es zugehörige RELATIONSZEICHEN $r_1,\ldots,r_k$, so daß (3), (4) gilt. Die Definitionsprozesse, durch die ϕ aus $\phi_1,\ldots,\phi_k$ entsteht (Einsetzung und rekursive Definition), können sämtlich im System *P* formal nachge-

187 bildet werden. Tut | man dies, so erhält man aus $r_1,\ldots,r_k$ ein neues

---

[38]Die VARIABLEN $u_1,\ldots,u_n$ können willkürlich vorgegeben werden. Es gibt z. B. immer ein *r* mit den FREIEN VARIABLEN $17, 19, 23,\ldots$ usw., für welches (3) und (4) gilt.

[39]Satz V beruht natürlich darauf, daß bei einer rekursiven Relation *R* für jedes *n*-tupel von Zahlen *aus den Axiomen des Systems P* entscheidbar ist, ob die Relation *R* besteht oder nicht.

[40]Daraus folgt sofort seine Geltung für jede rekursive Relation, da eine solche gleichbedeutend ist mit $0 = \phi(x_1,\ldots,x_n)$, wo ϕ rekursiv ist.

44. $Bw(x) \equiv (n)\{0 < n \le l(x) \to Ax(n\,Gl\,x)\,\vee$
$(Ep,q)[0 < p,q < n\;\&\;Fl(n\,Gl\,x,p\,Gl\,x,q\,Gl\,x)]\}$
$\&\;l(x) > 0,$

x is a PROOF ARRAY (a finite sequence of FORMULAS, each of which is either an AXIOM or an IMMEDIATE CONSEQUENCE of two of the preceding FORMULAS).

45. $x\,B\,y \equiv Bw(x)\;\&\;[l(x)]\,Gl\,x = y,$

x is a PROOF of the FORMULA y.

46. $Bew(x) \equiv (Ey)y\,B\,x,$

x is a PROVABLE FORMULA. ($Bew(x)$ is the only one of the notions 1–46 of which we cannot assert that it is recursive.)

The fact that can be formulated vaguely by saying that every recursive relation is definable in the system P (if the usual meaning is given to the formulas of this system) is expressed in precise language, *without* reference to any interpretation of the formulas of P, by the folllowing theorem:

Theorem V. *For every recursive relation $R(x_1,\ldots,x_n)$ there exists an n-place RELATION SIGN r (with the FREE VARIABLES[38] $u_1,u_2,\ldots,u_n$) such that for all n-tuples of numbers $(x_1,\ldots,x_n)$ we have*

$$R(x_1,\ldots,x_n) \to \mathrm{Bew}\,[Sb(r_{Z(x_1)\ldots Z(x_n)}^{u_1\ \ldots\ u_n})], \tag{3}$$

$$\overline{R}(x_1,\ldots,x_n) \to \mathrm{Bew}\,[\mathrm{Neg}(Sb(r_{Z(x_1)\ldots Z(x_n)}^{u_1\ \ldots\ u_n}))]. \tag{4}$$

We shall give only an outline of the proof of this theorem because the proof does not present any difficulty in principle and is rather long.[39] We prove the theorem for all relations $R(x_1,\ldots,x_n)$ of the form $x_1 = \phi(x_2,\ldots,x_n)$ [40] (where ϕ is a recursive function) and we use induction on the degree of ϕ. For functions of degree 1 (that is, constants and the function $x+1$) the theorem is trivial. Assume now that ϕ is of degree m. It results from functions of lower degrees, $\phi_1,\ldots,\phi_k$, through the operations of substitution or recursive definition. Since by the inductive hypothesis everything has already been proved for $\phi_1,\ldots,\phi_k$, there are corresponding RELATION SIGNS, $r_1,\ldots,r_k$, such that (3) and (4) hold. The processes of definition by which ϕ results from $\phi_1,\ldots,\phi_k$ (substitution and recursive definition) can both be formally reproduced in the system P. If this is

[38]The VARIABLES $u_1,\ldots,u_n$ can be chosen arbitrarily. For example, there always is an r with the FREE VARIABLES $17,19,23,\ldots$, and so on, for which (3) and (4) hold.

[39]Theorem V, of course, is a consequence of the fact that in the case of a recursive relation R it can, for every n-tuple of numbers, be decided *on the basis of the axioms of the system P* whether the relation R obtains or not.

[40]From this it follows at once that the theorem holds for every recursive relation, since any such relation is equivalent to $0 = \phi(x_1,\ldots,x_n)$, where ϕ is recursive.

RELATIONSZEICHEN r,[41] für welches man die Geltung von (3), (4) unter Verwendung der induktiven Annahme ohne Schwierigkeit beweisen kann. Ein RELATIONSZEICHEN r, welches auf diesem Wege einer rekursiven Relation zugeordnet ist,[42] soll *rekursiv* heißen.

Wir kommen nun ans Ziel unserer Ausführungen. Sei κ eine beliebige Klasse von FORMELN. Wir bezeichnen mit Flg(κ) (Folgerungsmenge von κ) die kleinste Menge von FORMELN, die alle FORMELN aus κ und alle AXIOME enthält und gegen die Relation "UNMITTELBARE FOLGE" abgeschlossen ist. κ heißt ω-widerspruchsfrei, wenn es kein KLASSENZEICHEN a gibt, so daß:

$$(n)[Sb(a_{Z(n)}^{v}) \; \epsilon \; \mathrm{Flg}(\kappa)] \& [\mathrm{Neg}(v \; \mathrm{Gen} \; a)] \; \epsilon \; \mathrm{Flg}(\kappa),$$

wobei v die FREIE VARIABLE des KLASSENZEICHENS a ist.

Jedes ω-widerspruchsfreie System ist selbstverständlich auch widerspruchsfrei. Es gilt aber, wie später gezeigt werden wird, nicht das Umgekehrte.

Das allgemeine Resultat über die Existenz unentscheidbarer Sätze lautet:

Satz VI: *Zu jeder ω-widerspruchsfreien rekursiven Klasse κ von* FORMELN *gibt es rekursive* KLASSENZEICHEN r, *so daß weder v* Gen r *noch* Neg(v Gen r) *zu* Flg(κ) *gehört (wobei v die* FREIE VARIABLE *aus r ist).*

Beweis: Sei κ eine beliebige rekursive ω-widerspruchsfreie Klasse von FORMELN. Wir definieren:

$$Bw_{\kappa}(x) \equiv (n)[n \leq l(x) \rightarrow Ax(n \, Gl \, x) \lor (n \, Gl \, x) \, \epsilon \, \kappa \lor$$
$$(Ep, q)\{0 < p, \; q < n \& Fl(n \, Gl \, x, \; p \, Gl \, x, \; q \, Gl \, x)\}] \& l(x) > 0 \qquad (5)$$

(vgl. den analogen Begriff 44)

$$x \, B_{\kappa} \, y \equiv Bw_{\kappa}(x) \& [l(x)] \, Gl \, x = y \qquad (6)$$
$$Bew_{\kappa}(x) \equiv (Ey)y \, B_{\kappa} \, x \qquad (6.1)$$

(vgl. die analogen Begriffe 45, 46).

Es gilt offenbar:

$$(x)[Bew_{\kappa}(x) \sim x \, \epsilon \, \mathrm{Flg}(\kappa)], \qquad (7)$$
$$(x)[Bew(x) \rightarrow Bew_{\kappa}(x)]. \qquad (8)$$

---

[41]Bei der genauen Durchführung dieses Beweises wird natürlich r nicht auf dem Umweg über die inhaltliche Deutung, sondern durch seine rein formale Beschaffenheit definiert.

[42]Welches also, inhaltlich gedeutet, das Bestehen dieser Relation ausdrückt.

done, a new RELATION SIGN r is obtained from $r_1, \ldots, r_k$,[41] and, using the inductive hypothesis, we can prove without difficulty that (3) and (4) hold for it. A RELATION SIGN r assigned to a recursive relation[42] by this procedure will be said to be *recursive*.

We now come to the goal of our discussions. Let κ be any class of FORMULAS. We denote by $\mathrm{Flg}(\kappa)$ (the set of consequences of κ) the smallest set of FORMULAS that contains all FORMULAS of κ and all AXIOMS and is closed under the relation "IMMEDIATE CONSEQUENCE". κ is said to be ω-consistent if there is no CLASS SIGN a such that

$$(n)[Sb(a_{Z(n)}^{v}) \; \epsilon \; \mathrm{Flg}(\kappa)] \, \& \, [\mathrm{Neg}(v \; \mathrm{Gen} \; a)] \; \epsilon \; \mathrm{Flg}(\kappa),$$

where v is the FREE VARIABLE of the CLASS SIGN a.

Every ω-consistent system, of course, is consistent. As will be shown later, however, the converse does not hold.

The general result about the existence of undecidable propositions reads as follows:

Theorem VI. *For every ω-consistent recursive class κ of FORMULAS there are recursive CLASS SIGNS r such that neither v Gen r nor $\mathrm{Neg}(v$ Gen $r)$ belongs to $\mathrm{Flg}(\kappa)$ (where v is the FREE VARIABLE of r).*

Proof: Let κ be any recursive ω-consistent class of FORMULAS. We define

$$Bw_\kappa(x) \equiv (n)[n \leq l(x) \rightarrow Ax(n\,Gl\,x) \vee (n\,Gl\,x) \; \epsilon \; \kappa \; \vee$$
$$(Ep,q)\{0 \; < p, \; q < n \; \& \; Fl(n\,Gl\,x, \, p\,Gl\,x, \, q\,Gl\,x)\}] \, \& \, l(x) > 0 \quad (5)$$

(see the analogous notion 44),

$$x\,B_\kappa\,y \equiv Bw_\kappa(x) \, \& \, [l(x)]\,Gl\,x = y \quad (6)$$
$$\mathrm{Bew}_\kappa(x) \equiv (Ey)y\,B_\kappa\,x \quad (6.1)$$

(see the analogous notions 45 and 46).

We obviously have

$$(x)[\mathrm{Bew}_\kappa(x) \sim x \; \epsilon \; \mathrm{Flg}(\kappa)] \quad (7)$$

and

$$(x)[\mathrm{Bew}(x) \rightarrow \mathrm{Bew}_\kappa(x)]. \quad (8)$$

---

[41] When this proof is carried out in detail, r, of course, is not defined indirectly with the help of its meaning but in terms of its purely formal structure.

[42] Which, therefore, in the usual interpretation expresses the fact that this relation holds.

188 | Nun definieren wir die Relation:

$$Q(x,y) \equiv \overline{x \, B_\kappa \, [Sb(y^{19}_{Z(y)})]}. \tag{8.1}$$

Da $x \, B_\kappa \, y$ (nach (6), (5)) und $Sb(y^{19}_{Z(y)})$ (nach Def. 17, 31) rekursiv sind, so auch $Q(x,y)$. Nach Satz V und (8) gibt es also ein RELATIONSZEICHEN q (mit den FREIEN VARIABLEN 17, 19), so daß gilt:

$$\overline{x \, B_\kappa \, [Sb(y^{19}_{Z(y)})]} \to \mathrm{Bew}_\kappa \, [Sb(q^{17}_{Z(x)}{}^{19}_{Z(y)})], \tag{9}$$

$$x \, B_\kappa \, [Sb(y^{19}_{Z(y)})] \to \mathrm{Bew}_\kappa[\mathrm{Neg}(Sb(q^{17}_{Z(x)}{}^{19}_{Z(y)}))]. \tag{10}$$

Wir setzen:

$$p = 17 \, \mathrm{Gen} \, q \tag{11}$$

(p ist ein KLASSENZEICHEN mit der FREIEN VARIABLEN 19) und

$$r = Sb(q^{19}_{Z(p)}) \tag{12}$$

(r ist ein rekursives KLASSENZEICHEN mit der FREIEN VARIABLEN 17).[43]
Dann gilt:

$$Sb(p^{19}_{Z(p)}) = Sb([17 \, \mathrm{Gen} \, q]^{19}_{Z(p)}) = 17 \, \mathrm{Gen} \, Sb(q^{19}_{Z(p)}) = 17 \, \mathrm{Gen} \, r \tag{13}$$

(wegen (11) und (12));[44] ferner:

$$Sb(q^{17}_{Z(x)}{}^{19}_{Z(p)}) = Sb(r^{17}_{Z(x)}) \tag{14}$$

(nach 12)). Setzt man nun in (9) und (10) p für y ein, so entsteht unter Berücksichtigung von (13) und (14):

$$\overline{x \, B_\kappa \, (17 \, \mathrm{Gen} \, r)} \to \mathrm{Bew}_\kappa[Sb(r^{17}_{Z(x)})] \tag{15}$$

$$x \, B_\kappa \, (17 \, \mathrm{Gen} \, r) \to \mathrm{Bew}_\kappa[\mathrm{Neg}(Sb(r^{17}_{Z(x)}))]. \tag{16}$$

189 | Daraus ergibt sich:

---

[43]r entsteht ja aus dem rekursiven RELATIONSZEICHEN q durch Ersetzen einer VARIABLEN durch eine bestimmte Zahl (p).

[44]Die Operationen Gen, Sb sind natürlich immer vertauschbar, falls sie sich auf verschiedene VARIABLE beziehen.

We now define the relation

$$Q(x, y) \equiv \overline{x \, B_\kappa \left[Sb(y^{19}_{Z(y)}) \right]}. \tag{8.1}$$

Since $x \, B_\kappa \, y$ (by (6) and (5)) and $Sb(y^{19}_{Z(y)})$ (by Definitions 17 and 31) are recursive, so is $Q(x, y)$. Therefore, by Theorem V and (8) there is a RELATION SIGN q (with the FREE VARIABLES 17 and 19) such that

$$\overline{x \, B_\kappa \left[Sb(y^{19}_{Z(y)}) \right]} \rightarrow \text{Bew}_\kappa \left[Sb(q^{17}_{Z(x)} {}^{19}_{Z(y)}) \right] \tag{9}$$

and

$$x \, B_\kappa \left[Sb(y^{19}_{Z(y)}) \right] \rightarrow \text{Bew}_\kappa [\text{Neg}(Sb(q^{17}_{Z(x)} {}^{19}_{Z(y)}))]. \tag{10}$$

We put

$$p = 17 \, \text{Gen} \, q \tag{11}$$

(p is a CLASS SIGN with the FREE VARIABLE 19) and

$$r = Sb(q^{19}_{Z(p)}) \tag{12}$$

(r is a recursive CLASS SIGN[43] with the FREE VARIABLE 17).
Then we have

$$Sb(p^{19}_{Z(p)}) = Sb([17 \, \text{Gen} \, q]^{19}_{Z(p)}) = 17 \, \text{Gen} \, Sb(q^{19}_{Z(p)}) = 17 \, \text{Gen} \, r \tag{13}$$

(by (11) and (12));[44] furthermore

$$Sb(q^{17}_{Z(x)} {}^{19}_{Z(p)}) = Sb(r^{17}_{Z(x)}) \tag{14}$$

(by (12)). If we now substitute p for y in (9) and (10), and take (13) and (14) into account, we obtain

$$\overline{x \, B_\kappa \, (17 \, \text{Gen} \, r)} \rightarrow \text{Bew}_\kappa [Sb(r^{17}_{Z(x)})], \tag{15}$$

$$x \, B_\kappa \, (17 \, \text{Gen} \, r) \rightarrow \text{Bew}_\kappa [\text{Neg}(Sb(r^{17}_{Z(x)}))]. \tag{16}$$

This yields:

---

[43]Since r is obtained from the recursive RELATION SIGN q through the replacement of a VARIABLE by a definite number p. [Precisely stated the final part of this footnote (which refers to a side remark unnecessary for the proof) would read thus: "REPLACEMENT of a VARIABLE by the NUMERAL for p".]

[44]The operations Gen and Sb, of course, can always be interchanged in case they refer to different VARIABLES.

1. 17 Gen r ist nicht κ-BEWEISBAR.[45] Denn wäre dies der Fall, so gäbe es (nach 6.1) ein n, so daß $n\,B_\kappa\,(17\;\text{Gen}\;r)$. Nach (16) gälte also:

$$\text{Bew}_\kappa[\text{Neg}(Sb(r^{17}_{Z(n)}))],$$

während andererseits aus der κ-BEWEISBARKEIT von 17 Gen r auch die von $Sb(r^{17}_{Z(n)})$ folgt. κ wäre also widerspruchsvoll (umsomehr ω-widerspruchsvoll).

2. Neg(17 Gen r) ist nicht κ-BEWEISBAR. Beweis: Wie eben bewiesen wurde, ist 17 Gen r nicht κ-BEWEISBAR, d. h. (nach 6.1) es gilt

$$(n)\,\overline{n\,B_\kappa\,(17\;\text{Gen}\;r)}.$$

Daraus folgt nach (15)

$$(n)\text{Bew}_\kappa[Sb(r^{17}_{Z(n)})],$$

was zusammen mit

$$\text{Bew}_\kappa[\text{Neg}(17\;\text{Gen}\;r)]$$

gegen die ω-Widerspruchsfreiheit von κ verstoßen würde.

17 Gen r ist also aus κ unentscheidbar, womit Satz VI bewiesen ist.

Man kann sich leicht überzeugen, daß der eben geführte Beweis konstruktiv ist,[45a] d. h. es ist intuitionistisch einwandfrei folgendes bewiesen: Sei eine beliebige rekursiv definierte Klasse κ von FORMELN vorgelegt. Wenn dann eine formale Entscheidung (aus κ) für die (effektiv aufweisbare) SATZFORMEL 17 Gen r vorgelegt ist, so kann man effektiv angeben:

1. Einen BEWEIS für Neg(17 Gen r).

2. Für jedes beliebige n einen BEWEIS für $Sb(r^{17}_{Z(n)})$, d. h. eine formale Entscheidung von 17 Gen r würde die effektive Aufweisbarkeit eines ω-Widerspruches zur Folge haben.

Wir wollen eine Relation (Klasse) zwischen natürlichen Zahlen $R(x_1,\ldots, x_n)$ *entscheidungsdefinit* nennen, wenn es ein n-stelliges RELATIONSZEICHEN r gibt, so daß (3) und (4) (vgl. Satz V) gilt. Insbesondere ist also nach Satz V jede rekursive Relation entscheidungsdefinit. Analog soll ein RELATIONSZEICHEN *entscheidungsdefinit* heißen, wenn es auf diese Weise einer entscheidungsdefiniten Relation zugeordnet ist. Es genügt nun für die Existenz von aus κ unentscheidbarer Sätze, von der Klasse κ vorauszusetzen, daß sie ω-widerspruchsfrei und entscheidungsdefinit ist. Denn die

---

[45]x ist κ-BEWEISBAR, soll bedeuten: $x\;\epsilon\;\text{Flg}(\kappa)$, was nach (7) dasselbe besagt wie: $\text{Bew}_\kappa(x)$.

[45a]Denn alle im Beweise vorkommenden Existentialbehauptungen beruhen auf Satz V, der, wie leicht zu sehen, intuitionistisch einwandfrei ist.

1. 17 Gen r is not κ-PROVABLE.[45] For, if it were, there would (by (6.1)) be an n such that $n\, B_\kappa$ (17 Gen r). Hence by (16) we would have

$$\text{Bew}_\kappa[\text{Neg}(Sb(r^{17}_{Z(n)}))],$$

while, on the other hand, from the κ-PROVABILITY of 17 Gen r that of $Sb(r^{17}_{Z(n)})$ follows. Hence, κ would be inconsistent (and a fortiori ω-inconsistent).

2. Neg(17 Gen r) is not κ-PROVABLE. Proof: As has just been proved, 17 Gen r is not κ-PROVABLE; that is (by (6.1)),

$$(n)\overline{n\, B_\kappa\,(17 \text{ Gen } r)}$$

holds. From this,

$$(n)\text{Bew}_\kappa[Sb(r^{17}_{Z(n)})]$$

follows by (15), and that, in conjunction with

$$\text{Bew}_\kappa[\text{Neg}(17 \text{ Gen } r)],$$

is incompatible with the ω-consistency of κ.

17 Gen r is therefore undecidable on the basis of κ, which proves Theorem VI.

We can readily see that the proof just given is constructive;[45a] that is, the following has been proved in an intuitionistically unobjectionable manner: Let an arbitrary recursively defined class κ of FORMULAS be given. Then, if a formal decision (on the basis of κ) of the SENTENTIAL FORMULA 17 Gen r (which [for each κ] can actually be exhibited) is presented to us, we can actually give

1. a PROOF of Neg(17 Gen r),
2. for any given n, a PROOF of $Sb(r^{17}_{Z(n)})$.

That is, a formal decision of 17 Gen r would have the consequence that we could actually exhibit an ω-inconsistency.

We shall say that a relation between (or a class of) natural numbers $R(x_1, \ldots, x_n)$ is *decidable* if there exists an n-place RELATION SIGN r such that (3) and (4) (see Theorem V) hold. In particular, therefore, by Theorem V every recursive relation is decidable. Similarly, a RELATION SIGN will be said to be *decidable* if it corresponds in this way to a decidable relation. Now it suffices for the existence of propositions undecidable on the basis of κ that the class κ be ω-consistent and decidable. For the decidability

---

[45]By "x is κ-PROVABLE" we mean $x\ \epsilon\ \text{Flg}(\kappa)$, which, by (7), means the same thing as $\text{Bew}_\kappa(x)$.

[45a]Since all existential statements occurring in the proof are based upon Theorem V, which, as is easily seen, is unobjectionable from the intuitionistic point of view.

Entscheidungsdefinitheit überträgt sich von κ auf $x\,B_\kappa\,y$ (vgl. (5), (6)) und

190   auf $Q(x,y)$ (vgl. | (8.1)) und nur dies wurde in obigem Beweise verwendet. Der unentscheidbare Satz hat in diesem Fall die Gestalt v Gen r, wo r ein entscheidungsdefinites KLASSENZEICHEN ist (es genügt übrigens sogar, daß κ in dem durch κ erweiterten System entscheidungsdefinit ist).

Setzt man von κ statt ω-Widerspruchsfreiheit, bloß Widerspruchsfreiheit voraus, so folgt zwar nicht die Existenz eines unentscheidbaren Satzes, wohl aber die Existenz einer Eigenschaft (r), für die weder ein Gegenbeispiel *angebbar*, noch beweisbar ist, daß sie allen Zahlen zukommt. Denn zum Beweise, daß 17 Gen r nicht κ-BEWEISBAR ist, wurde nur die Widerspruchsfreiheit von κ verwendet (vgl. S. 189) und aus $\overline{\text{Bew}}_\kappa(17$ Gen $r)$ folgt nach (15), daß für jede Zahl x, $Sb(r^{17}_{Z(x)})$, folglich für keine Zahl $\text{Neg}(Sb(r^{17}_{Z(x)}))$ κ-BEWEISBAR ist.

Adjungiert man $\text{Neg}(17$ Gen $r)$ zu κ, so erhält man eine widerspruchsfreie aber nicht ω-widerspruchsfreie FORMELKLASSE κ'. κ' ist widerspruchsfrei, denn sonst wäre 17 Gen r κ-BEWEISBAR. κ' ist aber nicht ω-widerspruchsfrei, denn wegen $\overline{\text{Bew}}_\kappa(17$ Gen $r)$ und (15) gilt:

$$(x)\text{Bew}_\kappa(Sb(r^{17}_{Z(x)})),$$

umsomehr also:

$$(x)\,\text{Bew}_{\kappa'}(Sb(r^{17}_{Z(x)}))$$

und anderseits gilt natürlich:

$$\text{Bew}_{\kappa'}(\text{Neg}(17 \text{ Gen } r)).^{46}$$

Ein Spezialfall von Satz VI ist der, daß die Klasse κ aus endlich vielen FORMELN (und eventuell den daraus durch TYPENERHÖHUNG entstehenden) besteht. Jede endliche Klasse κ ist natürlich rekursiv. Sei a die größte in κ enthaltene Zahl. Dann gilt in diesem Fall für κ:

$$x \,\epsilon\, \kappa \sim (Em,n)[m \leq x \,\&\, n \leq a \,\&\, n \,\epsilon\, \kappa \,\&\, x = m \, Th \, n].$$

κ ist also rekursiv. Das erlaubt z. B. zu schließen, daß auch mit Hilfe des Auswahlaxioms (für alle Typen) oder der verallgemeinerten Kontinuumhypothese nicht alle Sätze entscheidbar sind, vorausgesetzt, daß diese Hypothesen ω-widerspruchsfrei sind.

---

[46]Die Existenz widerspruchsfreier und nicht ω-widerspruchsfreier κ ist damit natürlich nur unter der Voraussetzung bewiesen, daß es überhaupt widerspruchsfreie κ gibt (d. h. daß P widerspruchsfrei ist).

carries over from κ to $x \, B_\kappa \, y$ (see (5) and (6)) and to $Q(x,y)$ (see (8.1)), and only this was used in the proof given above. In this case the undecidable proposition has the form v Gen r, where r is a decidable CLASS SIGN. (Note that it even suffices that κ be decidable in the system enlarged by κ.)

If, instead of assuming that κ is ω-consistent, we assume only that it is consistent, then, although the existence of an undecidable proposition does not follow [by the argument given above], it does follow that there exists a property (r) for which it is possible neither to give a counterexample nor to prove that it holds of all numbers. For in the proof that $17 \, Gen \, r$ is not κ-PROVABLE only the consistency of κ was used (above, page 177). Moreover from $\overline{\text{Bew}}_\kappa(17 \text{ Gen } r)$ it follows by (15) that, for every number x, $Sb(r_{Z(x)}^{17})$ is κ-PROVABLE and consequently that $\text{Neg}(Sb(r_{Z(x)}^{17}))$ is not κ-PROVABLE for any number.

If we adjoin $\text{Neg}(17 \text{ Gen } r)$ to κ, we obtain a class of FORMULAS κ' that is consistent but not ω-consistent. κ' is consistent, since otherwise 17 Gen r would be κ-PROVABLE. However, κ' is not ω-consistent, because, by $\overline{\text{Bew}}_\kappa(17 \text{ Gen } r)$ and (15),

$$(x)\text{Bew}_\kappa(Sb(r_{Z(x)}^{17}))$$

and, a fortiori,

$$(x)\text{Bew}_{\kappa'}(Sb(r_{Z(x)}^{17}))$$

hold, while on the other hand, of course,

$$\text{Bew}_{\kappa'}(\text{Neg}(17 \text{ Gen } r))$$

holds.[46]

We have a special case of Theorem VI when the class κ consists of a finite number of FORMULAS (and, if we so desire, of those resulting from them by TYPE ELEVATION). Every finite class κ is, of course, recursive. Let a be the greatest number contained in κ. Then we have for κ

$$x \, \epsilon \, \kappa \sim (Em, n)[m \leq x \, \& \, n \leq a \, \& \, n \, \epsilon \, \kappa \, \& \, x = m \, Th \, n].$$

Hence κ is recursive. This allows us to conclude, for example, that, even with the help of the axiom of choice (for all types) or the generalized continuum hypothesis, not all propositions are decidable, provided these hypotheses are ω-consistent.

---

[46]Of course, the existence of classes κ that are consistent but not ω-consistent is thus proved only on the assumption that there exists some consistent κ (that is, that P is consistent).

Beim Beweise von Satz VI wurden keine anderen Eigenschaften des Systems P verwendet als die folgenden:

1. Die Klasse der Axiome und die Schlußregeln (d. h. die Relation "unmittelbare Folge") sind rekursiv definierbar (sobald man die Grundzeichen in irgend einer Weise durch natürliche Zahlen ersetzt).

2. Jede rekursive Relation ist innerhalb des Systems P definierbar (im Sinn von Satz V).

Daher gibt es in jedem formalen System, das den Voraussetzungen 1, 2 genügt und ω-widerspruchsfrei ist, unentscheidbare Sätze der Form $(x)F(x)$, wo F eine rekursiv definierte Eigenschaft natürlicher Zahlen ist, 191 und ebenso in jeder Erweiterung eines solchen | Systems durch eine rekursiv definierbare ω-widerspruchsfreie Klasse von Axiomen. Zu den Systemen, welche die Voraussetzungen 1, 2 erfüllen, gehören, wie man leicht bestätigen kann, das Zermelo–Fraenkelsche und das von Neumannsche Axiomensystem der Mengenlehre,[47] ferner das Axiomensystem der Zahlentheorie, welches aus den Peanoschen Axiomen, der rekursiven Definition (nach Schema (2)) und den logischen Regeln besteht.[48] Die Voraussetzung 1. erfüllt überhaupt jedes System, dessen Schlußregeln die gewöhnlichen sind und dessen Axiome (analog wie in P) durch Einsetzung aus endlich vielen Schemata entstehen.[48a]

3

Wir ziehen nun aus Satz VI weitere Folgerungen und geben zu diesem Zweck folgende Definition:

Eine Relation (Klasse) heißt *arithmetisch*, wenn sie sich allein mittels der Begriffe $+$, $\cdot$ (Addition and Multiplikation, bezogen auf natürliche Zahlen)[49] und den logischen Konstanten $\vee$, $\overline{}$, (x), $=$ definieren läßt, wobei (x) und $=$ sich nur auf natürliche Zahlen beziehen dürfen.[50] Entsprechend wird der Begriff "arithmetischer Satz" definiert. Insbesondere sind z. B.

[47]Der Beweis von Voraussetzung 1. gestaltet sich hier sogar einfacher als im Falle des Systems P, da es nur eine Art von Grundvariablen gibt (bzw. zwei bei J. von Neumann).

[48]Vgl. Problem III in *Hilbert 1929a*.

[48a]Der wahre Grund für die Unvollständigkeit, welche allen formalen Systemen der Mathematik anhaftet, liegt, wie im II. Teil dieser Abhandlung gezeigt werden wird, darin, daß die Bildung immer höherer Typen sich ins Transfinite fortsetzen läßt (vgl. *Hilbert 1926*), während in jedem formalen System höchstens abzählbar viele vorhanden sind. Man kann nämlich zeigen, daß die hier aufgestellten unentscheidbaren Sätze durch Adjunktion passender höherer Typen (z. B. des Typus ω zum System P) immer entscheidbar werden. Analoges gilt auch für das Axiomensystem der Mengenlehre.

[49]Die Null wird hier und im folgenden immer mit zu den natürlichen Zahlen gerechnet.

[50]Das Definiens eines solchen Begriffes muß sich also allein mittels der angeführten

In the proof of Theorem VI no properties of the system P were used besides the following:

1. The class of axioms and the rules of inference (that is, the relation "immediate consequence") are recursively definable (as soon as we replace the primitive signs in some way by natural numbers).

2. Every recursive relation is definable (in the sense of Theorem V) in the system P.

Therefore, in every formal system that satisfies the assumptions 1 and 2 and is ω-consistent, there are undecidable propositions of the form $(x)F(x)$, where F is a recursively defined property of natural numbers, and likewise in every extension of such a system by a recursively definable ω-consistent class of axioms. As can easily be verified, included among the systems satisfying the assumptions 1 and 2 are the Zermelo–Fraenkel and the von Neumann axiom systems of set theory,[47] as well as the axiom system of number theory consisting of the Peano axioms, recursive definition (by Schema (2)), and the rules of logic.[48] Assumption 1 is satisfied by any system that has the usual rules of inference and whose axioms (like those of P) result from a finite number of schemata by substitution.[48a]

3

We shall now deduce some consequences from Theorem VI, and to this end we give the following definition:

A relation (class) is said to be *arithmetical* if it can be defined in terms of the notions $+$ and $\cdot$ (addition and multiplication for natural numbers)[49] and the logical constants $\vee$, $\overline{}$, (x), and $=$, where (x) and $=$ apply to natural numbers only.[50] The notion "arithmetical proposition" is defined

---

[47]The proof of assumption 1 turns out to be even simpler here than for the system P, since there is just one kind of primitive variables (or two in von Neumann's system).

[48]See Problem III in *Hilbert 1929a*.

[48a]As will be shown in Part II of this paper, the true reason for the incompleteness inherent in all formal systems of mathematics is that the formation of ever higher types can be continued into the transfinite (see *Hilbert 1926*, page 184), while in any formal system at most denumerably many of them are available. For it can be shown that the undecidable propositions constructed here become decidable whenever appropriate higher types are added (for example, the type ω to the system P). An analogous situation prevails for the axiom system of set theory.

[49]Here and in what follows, zero is always included among the natural numbers.

[50]The definiens of such a notion, therefore, must consist exclusively of the signs listed, variables for natural numbers, $x, y, \ldots$, and the signs 0 and 1 (variables for functions and sets are not permitted to occur). Instead of x any other number variable, of course, may occur in the quantifiers.

die Relationen "größer" und "kongruent nach einem Modul" arithmetisch, denn es gilt:

$$x > y \sim \overline{(Ez)}[y = x + z],$$
$$x \equiv y \,(\mathrm{mod}\, n) \sim (Ez)[x = y + z \cdot n \lor y = x + z \cdot n].$$

Es gilt der

Satz VII: *Jede rekursive Relation ist arithmetisch.*

Wir beweisen den Satz in der Gestalt: Jede Relation der Form $x_0 = \phi(x_1, \ldots, x_n)$, wo ϕ rekursiv ist, ist arithmetisch, und wenden vollständige Induktion nach der Stufe von ϕ an. ϕ habe die s-te Stufe ($s > 1$). Dann gilt entweder:

192 | 1. $\phi(x_1, \ldots, x_n) = \rho[\chi_1(x_1, \ldots, x_n), \chi_2(x_1, \ldots, x_n), \ldots,$
$$\chi_m(x_1, \ldots, x_n)]$$

(wo ρ und sämtliche χ_i kleinere Stufe haben als s)[51] oder:

    2. $\phi(0, x_2, \ldots, x_n) = \psi(x_2, \ldots, x_n)$
$$\phi(k + 1, x_2, \ldots, x_n) = \mu[k, \phi(k, x_2, \ldots, x_n), x_2, \ldots, x_n]$$

(wo ψ, μ niedrigere Stufe als s haben).

Im ersten Falle gilt:

$$x_0 = \phi(x_1, \ldots, x_n) \sim (Ey_1, \ldots, y_m)[R(x_0, y_1, \ldots, y_m) \,\&$$
$$S_1(y_1, x_1, \ldots, x_n) \,\& \, \ldots \,\& \, S_m(y_m, x_1, \ldots, x_n)],$$

wo R bzw. S_i die nach induktiver Annahme existierenden mit $x_0 = \rho(y_1, \ldots, y_m)$ bzw. $y = \chi_i(x_1, \ldots, x_n)$ äquivalenten arithmetischen Relationen sind. Daher ist $x_0 = \phi(x_1, \ldots, x_n)$ in diesem Fall arithmetisch.

Im zweiten Fall wenden wir folgendes Verfahren an: Man kann die Relation $x_0 = \phi(x_1, \ldots, x_n)$ mit Hilfe des Begriffes "Folge von Zahlen" (f)[52] folgendermaßen ausdrücken:

$$x_0 = \phi(x_1, \ldots, x_n) \sim (Ef)\{f_0 = \psi(x_2, \ldots, x_n) \,\& \, (k)[k < x_1 \to$$
$$f_{k+1} = \mu(k, f_k, x_2, \ldots, x_n)] \,\& \, x_0 = f_{x_1}\}.$$

Wenn $S(y, x_2, \ldots, x_n)$ bzw. $T(z, x_1, \ldots, x_{n+1})$ die nach induktiver Annahme existierenden mit $y = \psi(x_2, \ldots, x_n)$ bzw. $z = \mu(x_1, \ldots, x_{n+1})$

Zeichen, Variablen für natürliche Zahlen $x, y, \ldots$ und den Zeichen 0, 1 aufbauen (Funktions- und Mengenvariable dürfen nicht vorkommen). (In den Präfixen darf statt x natürlich auch jede andere Zahlvariable stehen.)

[51]Es brauchen natürlich nicht alle $x_1, \ldots, x_n$ in den χ_i tatsächlich vorzukommen (vgl. das Beispiel in Fußnote 27).

[52]f bedeutet hier eine Variable, deren Wertbereich die Folgen natürlicher Zahlen sind. Mit f_k wird das $(k+1)$-te Glied einer Folge f bezeichnet (mit f_0 das erste).

accordingly. The relations "greater than" and "congruent modulo n", for example, are arithmetical because we have

$$x > y \sim \overline{(Ez)}[y = x + z],$$
$$x \equiv y \,(\mathrm{mod}\,n) \sim (Ez)[x = y + z \cdot n \lor y = x + z \cdot n].$$

We now have

Theorem VII. *Every recursive relation is arithmetical.*

We shall prove the following version of this theorem: Every relation of the form $x_0 = \phi(x_1, \ldots, x_n)$, where ϕ is recursive, is arithmetical, and we shall use induction on the degree of ϕ. Let ϕ be of degree s ($s > 1$). Then we have either

1. $\phi(x_1, \ldots, x_n) = \rho[\chi_1(x_1, \ldots, x_n), \chi_2(x_1, \ldots, x_n), \ldots,$
    $$\chi_m(x_1, \ldots, x_n)]$$

(where ρ and all χ_1 are of degrees less than s)[51] or

2. $\phi(0, x_2, \ldots, x_n) = \psi(x_2, \ldots, x_n),$
    $$\phi(k + 1, x_2, \ldots, x_n) = \mu[k, \phi(k, x_2, \ldots, x_n), x_2, \ldots, x_n]$$

(where ψ and μ are of degrees less than s).

In the first case we have

$$x_0 = \phi(x_1, \ldots, x_n) \sim (Ey_1, \ldots, y_m)[R(x_0, y_1, \ldots, y_m) \,\&$$
$$S_1(y_1, x_1, \ldots, x_n) \,\& \,\ldots \,\& \,S_m(y_m, x_1, \ldots, x_n)],$$

where R and S_i are the arithmetical relations, existing by the inductive hypothesis, that are equivalent to $x_0 = \rho(y_1, \ldots, y_m)$ and $y = \chi_i(x_1, \ldots, x_n)$, respectively. Hence in this case $x_0 = \phi(x_1, \ldots, x_n)$ is arithmetical.

In the second case we use the following method. We can express the relation $x_0 = \phi(x_1, \ldots, x_n)$ with the help of the notion "sequence of numbers" (f) [52] in the following way:

$$x_0 = \phi(x_1, \ldots, x_n) \sim (Ef)\{f_0 = \psi(x_2, \ldots, x_n) \,\& \,(k)[k < x_1 \to$$
$$f_{k+1} = \mu(k, f_k, x_2, \ldots, x_n)] \,\& \,x_0 = f_{x_1}\}.$$

If $S(y, x_2, \ldots, x_n)$ and $T(z, x_1, \ldots, x_{n+1})$ are the arithmetical relations, existing by the inductive hypothesis, that are equivalent to $y = \psi(x_2, \ldots, x_n)$ and $z = \mu(x_1, \ldots, x_{n+1})$, respectively, then

$$x_0 = \phi(x_1, \ldots, x_n) \sim (Ef)\{S(f_0, x_2, \ldots, x_n) \,\& \,(k)[k < x_1 \to$$
$$T(f_{k+1}, k, f_k, x_2, \ldots, x_n)] \,\& \,x_0 = f_{x_1}\}. \tag{17}$$

---

[51] Of course, not all $x_1, \ldots, x_n$ need occur in the χ_i (see the example in footnote 27).

[52] f here is a variable with the [infinite] sequences of natural numbers as its range of values. f_k denotes the $(k + 1)$th term of a sequence f (f_0 denoting the first).

äquivalenten arithmetische Relationen sind, gilt daher:

$$x_0 = \phi(x_1, \ldots, x_n) \sim (Ef)\{S(f_0, x_2, \ldots, x_n) \,\&\, (k)[k < x_1 \rightarrow$$
$$T(f_{k+1}, k, f_k, x_2, \ldots, x_n)] \,\&\, x_0 = f_{x_1}\}. \tag{17}$$

Nun ersetzen wir den Begriff "Folge von Zahlen" durch "Paar von Zahlen", indem wir dem Zahlenpaar n, d die Zahlenfolge $f^{(n,d)}$ ($f_k^{(n,d)} = [n]_{1+(k+1)d}$) zuordnen, wobei $[n]_p$ den kleinsten nicht negativen Rest von n modulo p bedeutet.

Es gilt dann der

Hilfssatz 1: Ist f eine beliebige Folge natürlicher Zahlen und k eine beliebige natürliche Zahl, so gibt es ein Paar von natürlichen Zahlen n, d, so daß $f^{(n,\,d)}$ und f in den ersten k Gliedern übereinstimmen.

Beweis: Sei l die größte der Zahlen $k, f_0, f_1, \ldots, f_{k-1}$. Man bestimme n so, daß:

$$n \equiv f_i \, [\mathrm{mod}\,(1 + (i+1)l!)] \quad \text{für} \quad i = 0, 1, \ldots, k-1,$$

193 | was möglich ist, da je zwei der Zahlen $1 + (i+1)l!$ $(i = 0, 1, \ldots, k-1)$ relativ prim sind. Denn eine in zwei von diesen Zahlen enthaltene Primzahl müßte auch in der Differenz $(i_1 - i_2)l!$ und daher wegen $|i_1 - i_2| < l$ in $l!$ enthalten sein, was unmöglich ist. Das Zahlenpaar $n, l!$ leistet dann das Verlangte.

Da die Relation $x = [n]_p$ durch:

$$x \equiv n \,(\mathrm{mod}\,p) \,\&\, x < p$$

definiert und daher arithmetisch ist, so ist auch die folgendermaßen definierte Relation $P(x_0, x_1, \ldots, x_n)$:

$$P(x_0, \ldots, x_n) \equiv (En, d)\{S([n]_{d+1}, x_2, \ldots, x_n) \,\&\, (k)[k < x_1 \rightarrow$$
$$T([n]_{1+d(k+2)}, k, [n]_{1+d(k+1)}, x_2, \ldots, x_n)] \,\&\, x_0 = [n]_{1+d(x_1+1)}\}$$

arithmetisch, welche nach (17) und Hilfssatz 1 mit: $x_0 = \phi(x_1, \ldots, x_n)$ äquivalent ist (es kommt bei der Folge f in (17) nur auf ihren Verlauf bis zum $(x_1 + 1)$-ten Glied an). Damit ist Satz VII bewiesen.

Gemäß Satz VII gibt es zu jedem Problem der Form $(x)F(x)$ (F rekursiv) ein äquivalentes arithmetisches Problem und da der ganze Beweis von Satz VII sich (für jedes spezielle F) innerhalb des Systems P formalisieren läßt, ist diese Äquivalenz in P beweisbar. Daher gilt:

Satz VIII: *In jedem der in Satz VI genannten formalen Systeme*[53] *gibt es unentscheidbare arithmetische Sätze.*

---

[53]Das sind diejenigen ω-widerspruchsfreien Systeme, welche aus P durch Hinzufügung einer rekursiv definierbaren Klasse von Axiomen entstehen.

We now replace the notion "sequence of numbers" by "pair of numbers", assigning to the number pair n, d the number sequence $f^{(n,d)}$ ($f_k^{(n,d)} = [n]_{1+(k+1)d}$, where $[n]_p$ denotes the least nonnegative remainder of n modulo p).

We then have

Lemma 1. If f is any sequence of natural numbers and k any natural number, there exists a pair of natural numbers, n, d, such that $f^{(n,d)}$ and f agree in the first k terms.

Proof: Let l be the maximum of the numbers $k, f_0, f_1, \ldots, f_{k-1}$. Let us determine an n such that

$$n \equiv f_i \, [\mathrm{mod}(1 + (i+1)l!)] \quad \text{for} \quad i = 0, 1, \ldots, k-1,$$

which is possible, since any two of the numbers $1 + (i+1)l!$ ($i = 0, 1, \ldots, k-1$) are relatively prime. For a prime number contained in two of these numbers would also be contained in the difference $(i_1 - i_2)l!$ and therefore, since $|i_1 - i_2| < l$, in $l!$; but this is impossible. The number pair $n, l!$ then has the desired property.

Since the relation $x = [n]_p$ is defined by

$$x \equiv n \pmod{p} \,\&\, x < p$$

and is therefore arithmetical, the relation $P(x_0, x_1, \ldots, x_n)$, defined by

$$P(x_0, \ldots, x_n) \equiv (En, d)\{S([n]_{d+1}, x_2, \ldots, x_n) \,\&\, (k)[k < x_1 \rightarrow$$
$$T([n]_{1+d(k+2)}, k, [n]_{1+d(k+1)}, x_2, \ldots, x_n)] \,\&\, x_0 = [n]_{1+d(x_1+1)}\},$$

is also arithmetical. But by (17) and Lemma 1 it is equivalent to $x_0 = \phi(x_1, \ldots, x_n)$ (the sequence f enters in (17) only through its first $x_1 + 1$ terms). Theorem VII is thus proved.

By Theorem VII, for every problem of the form $(x)F(x)$ (with recursive F) there is an equivalent arithmetical problem. Moreover, since the entire proof of Theorem VII (for every particular F) can be formalized in the system P, this equivalence is provable in P. Hence we have

Theorem VIII. *In any of the formal systems mentioned in Theorem VI,*[53] *there are undecidable arithmetical propositions.*

By the remark on page 181 above, the same holds for the axiom system of set theory and its extensions by ω-consistent recursive classes of axioms.

Finally, we derive the following result:

---

[53]These are the ω-consistent systems that result from P when recursively definable classes of axioms are added.

Dasselbe gilt (nach der Bemerkung auf Seite 190) für das Axiomensystem der Mengenlehre und dessen Erweiterungen durch ω-widerspruchsfreie rekursive Klassen von Axiomen.

Wir leiten schließlich noch folgendes Resultat her:

Satz IX: *In allen in Satz* VI *genannten formalen Systemen*[53] *gibt es unentscheidbare Probleme des engeren Funktionenkalküls*[54] (d. h. Formeln des engeren Funktionenkalküls, für die weder Allgemeingültigkeit noch Existenz eines Gegenbeispiels beweisbar ist).[55]

194 | Dies beruht auf:

Satz X: *Jedes Problem der Form* $(x)F(x)$ $(F$ *rekursiv) läßt sich zurückführen auf die Frage nach der Erfüllbarkeit einer Formel des engeren Funktionenkalküls* (d. h. zu jedem rekursiven F kann man eine Formel des engeren Funktionenkalküls angeben, deren Erfüllbarkeit mit der Richtigkeit von $(x)F(x)$ äquivalent ist).

Zum engeren Funktionenkalkül (e. F.) rechnen wir diejenigen Formeln, welche sich aus den Grundzeichen: $\overline{}$, $\vee$, (x), $=$, $x, y, \ldots$ (Individuenvariable), $F(x), G(x,y), H(x,y,z), \ldots$ (Eigenschafts- und Relationsvariable) aufbauen,[56] wobei (x) und $=$ sich nur auf Individuen beziehen dürfen. Wir fügen zu diesen Zeichen noch eine dritte Art von Variablen $\phi(x)$, $\psi(x,y)$, $\chi(x,y,z)$ etc. hinzu, die Gegenstandsfunktionen vertreten (d. h. $\phi(x)$, $\psi(x,y)$ etc. bezeichnen eindeutige Funktionen, deren Argumente und Werte Individuen sind).[57] Eine Formel, die außer den zuerst angeführten Zeichen des e. F. noch Variable dritter Art $(\phi(x), \psi(x,y), \ldots$ etc.) enthält, soll eine Formel im weiteren Sinne (i. w. S.) heißen.[58] Die Begriffe "erfüllbar", "allgemeingültig" übertragen sich ohneweiters auf Formeln i. w. S und es gilt der Satz, daß man zu jeder Formel i. w. S. A eine gewöhnliche Formel des e. F. B angeben kann, so daß die Erfüllbarkeit von A mit der von B äquivalent ist. B erhält man aus A, indem man die in A vorkommenden Variablen dritter Art $\phi(x), \psi(x,y), \ldots$ durch Ausdrücke der Form:

---

[54]Vgl. *Hilbert und Ackermann 1928*. Im System P sind unter Formeln des engeren Funktionenkalküls diejenigen zu verstehen, welche aus den Formeln des engeren Funktionenkalküls der PM durch die auf S. 176 angedeutete Ersetzung der Relationen durch Klassen höheren Typs entstehen.

[55]In meiner Arbeit *1930* habe ich gezeigt, daß jede Formel des engeren Funktionenkalküls entweder als allgemeingültig nachweisbar ist oder ein Gegenbeispiel existiert; die Existenz dieses Gegenbeispiels ist aber nach Satz IX *nicht* immer nachweisbar (in der angeführten formalen Systemen).

[56]D. Hilbert und W. Ackermann rechnen in dem eben zitierten Buch (*1928*) das Zeichen $=$ nicht zum engeren Funktionenkalkül. Es gibt aber zu jeder Formel, in der das Zeichen $=$ vorkommt, eine solche ohne dieses Zeichen, die mit der ursprünglichen gleichzeitig erfüllbar ist (vgl. meiner Arbeit *1930*).

[57]Und zwar soll der Definitionsbereich immer der *ganze* Individuenbereich sein.

[58]Variable dritter Art dürfen dabei an allen Leerstellen für Individuenvariable stehen, z. B.: $y = \phi(x)$, $F(x, \phi(y))$, $G[\psi(x,(\phi(y)), x]$ usw.

Theorem IX.   *In any of the formal systems mentioned in Theorem VI,*[53] *there are undecidable problems of the restricted functional calculus*[54] (that is, formulas of the restricted functional calculus for which neither validity nor the existence of a counterexample is provable).[55]

This is a consequence of

Theorem X.   *Every problem of the form* $(x)F(x)$ *(with recursive F) can be reduced to the question whether a certain formula of the restricted functional calculus is satisfiable* (that is, for every recursive F, we can find a formula of the restricted functional calculus that is satisfiable if and only if $(x)F(x)$ is true).

By formulas of the restricted functional calculus (r. f. c.) we understand expressions formed from the primitive signs $\overline{}$, $\vee$, (x), $=$, x, $y, \ldots$ (individual variables), $F(x), G(x, y), H(x, y, z), \ldots$ (predicate and relation variables), where (x) and $=$ apply to individuals only.[56] To these signs we add a third kind of variables, $\phi(x)$, $\psi(x, y)$, $\chi(x, y, z)$, and so on, which stand for functions of individuals (that is, $\phi(x)$, $\psi(x, y)$, and so on denote single-valued functions whose arguments and values are individuals).[57] A formula that contains variables of the third kind in addition to the signs of the r. f. c. first mentioned will be called a formula in the extended sense (i. e. s.).[58] The notions "satisfiable" and "valid" carry over immediately to formulas i. e. s., and we have the theorem that, for any formula A i. e. s., we can find a formula B of the r. f. c. proper such that A is satisfiable if and only if B is. We obtain B from A by replacing the variables of the third kind, $\phi(x), \psi(x, y), \ldots$, that occur in A with expressions of the form $(\imath z)F(z, x), (\imath z)G(z, x, y), \ldots$, by eliminating the "descriptive" functions by the method used in PM (I, $*14$), and by logically multiplying[59] the formula thus obtained by an expression stating about each $F, G, \ldots$ put in place of

---

[54]See *Hilbert and Ackermann 1928*. In the system P we must understand by formulas of the restricted functional calculus those that result from the formulas of the restricted functional calculus of *PM* when relations are replaced by classes of higher types as indicated on page 153 above.

[55]In *1930* I showed that every formula of the restricted functional calculus either can be proved to be valid or has a counterexample. However, by Theorem IX the existence of this counterexample is *not* always provable (in the formal systems we have been considering).

[56]Hilbert and Ackermann (*1928*) do not include the sign $=$ in the restricted functional calculus. But for every formula in which the sign $=$ occurs there exists a formula that does not contain this sign and is satisfiable if and only if the original formula is (see *Gödel 1930*).

[57]Moreover, the domain of definition is always supposed to be the *entire* domain of individuals.

[58]Variables of the third kind may occur at all argument places occupied by individual variables, for example, $y = \phi(x)$, $F(x, \phi(y))$, $G(\psi(x, \phi(y)), x)$, and the like.

[59]That is, by forming the conjunction.

$(\imath z)F(z,x), (\imath z)G(z,x,y), \ldots$ ersetzt, die "beschreibenden" Funktionen im Sinne der *PM*, I, *14, auflöst und die so erhaltene Formel mit einem Ausdruck logisch multipliziert,[59] der besagt, daß sämtliche an Stelle der $\phi, \psi, \ldots$ gesetzte $F, G, \ldots$ hinsichtlich der ersten Leerstelle genau eindeutig sind.

Wir zeigen nun, daß es zu jedem Problem der Form $(x)F(x)$ (F rekursive) ein äquivalentes betreffend die Erfüllbarkeit einer Formel i. w. S. gibt, woraus nach der eben gemachten Bemerkung Satz X folgt.

Da F rekursiv ist, gibt es eine rekursive Funktion $\Phi(x)$, so daß

$$F(x) \sim [\Phi(x) = 0],$$

und für Φ gibt es eine Reihe von Funktionen $\Phi_1, \Phi_2, \ldots, \Phi_n$, so daß: $\Phi_n = \Phi$, $\Phi_1(x) = x + 1$ und für jedes Φ_k $(1 < k \leq n)$ entweder:

1. $$(x_2, \ldots, x_m)[\Phi_k(0, x_2, \ldots, x_m) = \Phi_p(x_2, \ldots, x_m)],$$
$$(x, x_2, \ldots, x_m)\{\Phi_k[\Phi_1(x), x_2, \ldots, x_m] =$$
$$\Phi_q[x, \Phi_k(x, x_2, \ldots, x_m), x_2, \ldots, x_m]\},$$
$$p, q < k, \tag{18}$$

195 | oder:

2. $$(x_1, \ldots, x_m)[\Phi_k(x_1, \ldots, x_m) = \Phi_r(\Phi_{i_1}(\mathfrak{x}_1), \ldots, \Phi_{i_s}(\mathfrak{x}_s))],^{60}$$
$$r < k, \quad i_v < k \quad (\text{für } v = 1, 2, \ldots, s), \tag{19}$$

oder:

3. $$(x_1, \ldots, x_m)[\Phi_k(x_1, \ldots, x_m) = \Phi_1(\Phi_1(\ldots(\Phi_1(0))\ldots))]. \tag{20}$$

Ferner bilden wir die Sätze:

$$(x)\overline{\Phi_1(x) = 0} \,\&\, (x,y)[\Phi_1(x) = \Phi_1(y) \rightarrow x = y], \tag{21}$$
$$(x)[\Phi_n(x) = 0]. \tag{22}$$

Wir ersetzen nun in allen Formeln (18), (19), (20) (für $k = 2, 3, \ldots, n$) und in (21), (22) die Funktionen Φ_i durch Funktionsvariable ϕ_i, die Zahl 0 durch eine sonst nicht vorkommende Individuenvariable x_0 und bilden die Konjunktion C sämtlicher so erhaltener Formeln.

Die Formel $(Ex_0)C$ hat dann die verlangte Eigenschaft, d. h.:

---

[59]D. h. die Konjunktion bildet.

[60]$\mathfrak{x}_i$ $(i = 1, \ldots, s)$ vertreten irgend welche Komplexe der Variablen $x_1, x_2, \ldots, x_m$; z. B.: x_1, x_3, x_2.

some $\phi, \psi, \ldots$ that it holds for a unique value of the first argument [for any choice of values for the other arguments].

We now show that, for every problem of the form $(x)F(x)$ (with recursive F), there is an equivalent problem concerning the satisfiability of a formula i. e. s., so that, on account of the remark just made, Theorem X follows.

Since F is recursive, there is a recursive function $\Phi(x)$ such that

$$F(x) \sim [\Phi(x) = 0],$$

and for Φ there is a sequence of functions, $\Phi_1, \Phi_2, \ldots, \Phi_n$, such that $\Phi_n = \Phi$, $\Phi_1(x) = x + 1$, and for every Φ_k $(1 < k \leq n)$ we have either

1. $$(x_2, \ldots, x_m)[\Phi_k(0, x_2, \ldots, x_m) = \Phi_p(x_2, \ldots, x_m)],$$
    $$(x, x_2, \ldots, x_m)\{\Phi_k[\Phi_1(x), x_2, \ldots, x_m] =$$
    $$\Phi_q[x, \Phi_k(x, x_2, \ldots, x_m), x_2, \ldots, x_m]\},$$
    with $p, q < k$,[59a]   (18)

or

2. $$(x_1, \ldots, x_m)[\Phi_k(x_1, \ldots, x_m) = \Phi_r(\Phi_{i_1}(\mathfrak{r}_1), \ldots, \Phi_{i_s}(\mathfrak{r}_s))],$$
    with $r < k$, $i_v < k$ (for $v = 1, 2, \ldots, s$),[60]   (19)

or

3. $$(x_1, \ldots, x_m)[\Phi_k(x_1, \ldots, x_m) = \Phi_1(\Phi_1(\ldots(\Phi_1(0))\ldots))].$$ (20)

We then form the propositions

$$(x)\overline{\Phi_1(x) = 0} \,\&\, (x, y)[\Phi_1(x) = \Phi_1(y) \to x = y],$$ (21)

$$(x)[\Phi_n(x) = 0].$$ (22)

In all of the formulas (18), (19), (20) (for $k = 2, 3, \ldots, n$) and in (21) and (22) we now replace the functions Φ_i by function variables ϕ_i and the number 0 by an individual variable x_0 not used so far, and we form the conjunction C of all the formulas thus obtained.

The formula $(Ex_0)C$ then has the required property, that is,

---

[59a][The last clause of footnote 27 was not taken into account in the formulas (18). But an explicit formulation of the cases with fewer variables on the right side is actually necessary here for the formal correctness of the proof, unless the identity function, $I(x) = x$, is added to the initial functions.]

[60]The $\mathfrak{r}_i$ $(i = 1, \ldots, s)$ stand for finite sequences of the variables $x_1, x_2, \ldots, x_m$; for example, x_1, x_3, x_2.

1. Wenn $(x)[\Phi(x) = 0]$ gilt, ist $(Ex_0)C$ erfüllbar, denn die Funktionen $\Phi_1, \Phi_2, \ldots, \Phi_n$ ergeben dann offenbar in $(Ex_0)C$ für $\phi_1, \phi_2, \ldots, \phi_n$ eingesetzt einen richtigen Satz.

2. Wenn $(Ex_0)C$ erfüllbar ist, gilt $(x)[\Phi(x) = 0]$.

Beweis: Seien $\Psi_1, \Psi_2, \ldots, \Psi_n$ die nach Voraussetzung existierenden Funktionen, welche in $(Ex_0)C$ für $\phi_1, \phi_2, \ldots, \phi_n$ eingesetzt einen richtigen Satz liefern. Ihr Individuenbereich sei $\mathfrak{J}$. Wegen der Richtigkeit von $(Ex_0)C$ für die Funktionen Ψ_i gibt es ein Individuum a (aus $\mathfrak{J}$), so daß sämtliche Formeln (18) bis (22) bei Ersetzung der Φ_i durch Ψ_i und von 0 durch a in richtige Sätze (18′) bis (22′) übergehen. Wir bilden nun die kleinste Teilklasse von $\mathfrak{J}$, welche a enthält und gegen die Operation $\Psi_1(x)$ abgeschlossen ist. Diese Teilklasse ($\mathfrak{J}'$) hat die Eigenschaft, daß jede der Funktionen Ψ_i auf Elemente aus $\mathfrak{J}'$ angewendet wieder Elemente aus $\mathfrak{J}'$ ergibt. Denn für Ψ_1 gilt dies nach Definition von $\mathfrak{J}'$ und wegen (18′), (19′), (20′) überträgt sich diese Eigenschaft von Ψ_i mit niedrigerem Index auf solche mit höherem. Die Funktionen, welche aus Ψ_i durch Beschränkung auf den Individuenbereich $\mathfrak{J}'$ entstehen, nennen wir Ψ_i'. Auch für diese Funktion gelten sämtliche Formeln (18) bis (22) (bei der Ersetzung von 0 durch a und Φ_i durch Ψ_i').

Wegen der Richtigkeit von (21) für Ψ_1' und a kann man die Individuen aus $\mathfrak{J}'$ eineindeutig auf die natürlichen Zahlen abbilden und zwar so, daß a in 0 und die Funktion Ψ_1' in die Nachfolgerfunktion Φ_1 übergeht. Durch diese Abbildung gehen aber sämtliche Funktionen Ψ_i' in die Funktionen Φ_i über und wegen der Richtigkeit von (22) | für Ψ_n' und a gilt

$$(x)[\Phi_n(x) = 0]$$

oder $(x)[\Phi(x) = 0]$, was zu beweisen war.[61]

Da man die Überlegungen, welche zu Satz X führen (für jedes spezielle F), auch innerhalb des Systems P durchführen kann, so ist die Äquivalenz zwischen einem Satz der Form $(x)F(x)$ (F rekursiv) und der Erfüllbarkeit der entsprechenden Formel des e. F. in P beweisbar und daher folgt aus der Unentscheidbarkeit des einen die des anderen, womit Satz IX bewiesen ist.[62]

---

[61] Aus Satz X folgt z. B., daß das Fermatsche und das Goldbachsche Problem lösbar wären, wenn man das Entscheidungsproblem des e. F. gelöst hätte.

[62] Satz IX gilt natürlich auch für das Axiomensystem der Mengenlehre und dessen Erweiterungen durch rekursiv definierbare ω-widerspruchsfreie Klassen von Axiomen, da es ja auch in diesen Systemen unentscheidbare Sätze der Form $(x)F(x)$ (F rekursiv) gibt.

1. If $(x)[\Phi(x) = 0]$ holds, $(Ex_0)C$ is satisfiable. For the functions $\Phi_1, \Phi_2, \ldots, \Phi_n$ obviously yield a true proposition when substituted for $\phi_1, \phi_2, \ldots, \phi_n$ in $(Ex_0)C$.

2. If $(Ex_0)C$ is satisfiable, $(x)[\Phi(x) = 0]$ holds.

Proof: Let $\Psi_1, \Psi_2, \ldots, \Psi_n$ be the functions (which exist by assumption) that yield a true proposition when substituted for $\phi_1, \phi_2, \ldots, \phi_n$ in $(Ex_0)C$. Let $\mathfrak{J}$ be their domain of individuals. Since $(Ex_0)C$ holds for the functions Ψ_i, there is an individual a (in $\mathfrak{J}$) such that all of the formulas (18)–(22) go over into true propositions, (18′)–(22′), when the Φ_i are replaced by the Ψ_i and 0 by a. We now form the smallest subclass of $\mathfrak{J}$ that contains a and is closed under the operation $\Psi_1(x)$. This subclass $(\mathfrak{J}')$ has the property that every function Ψ_i, when applied to elements of $\mathfrak{J}'$, again yields elements of $\mathfrak{J}'$. For this holds of Ψ_1 by the definition of $\mathfrak{J}'$, and by (18′), (19′), and (20′) it carries over from Ψ_i with smaller subscripts to Ψ_i with larger ones. The functions that result from the Ψ_i when these are restricted to the domain $\mathfrak{J}'$ of individuals will be denoted by Ψ_i'. All of the formulas (18)–(22) hold for these functions also (when we replace 0 by a and Φ_i by Ψ_i').

Because (21) holds for Ψ_1' and a, we can map the individuals of $\mathfrak{J}'$ one-to-one onto the natural numbers in such a manner that a goes over into 0 and the function Ψ_1' into the successor function Φ_1. But by this mapping the functions Ψ_i' go over into the functions Φ_i; and, since (22) holds for Ψ_n' and a,

$$(x)[\Phi_n(x) = 0],$$

that is, $(x)[\Phi(x) = 0]$, holds, which was to be proved.[61]

Since (for each particular F) the argument leading to Theorem X can be carried out in the system P, it follows that any proposition of the form $(x)F(x)$ (with recursive F) can in P be proved equivalent to the proposition that states about the corresponding formula of the r. f. c. that it is satisfiable. Hence the undecidability of one implies that of the other, which proves Theorem IX.[62]

4

The results of Section 2 have a surprising consequence concerning a consistency proof for the system P (and its extensions), which can be stated as follows:

---

[61]Theorem X implies, for example, that Fermat's problem and Goldbach's problem could be solved if the decision problem for the r. f. c. were solved.

[62]Theorem IX, of course, also holds for the axiom system of set theory and for its extensions by recursively definable ω-consistent classes of axioms, since there are undecidable propositions of the form $(x)F(x)$ (with recursive F) in these systems too.

4

Aus den Ergebnissen von Abschnitt 2 folgt ein merkwürdiges Resultat, bezüglich eines Widerspruchslosigkeitsbeweises des Systems P (und seiner Erweiterungen), das durch folgenden Satz ausgesprochen wird:

Satz XI: *Sei κ eine beliebige rekursive widerspruchsfreie*[63] *Klasse von* FORMELN, *dann gilt: Die* SATZFORMEL, *welche besagt, daß κ widerspruchsfrei ist, ist nicht κ-*BEWEISBAR; insbesondere ist die Widerspruchsfreiheit von P in P unbeweisbar,[64] vorausgesetzt, daß P widerspruchsfrei ist (im entgegengesetzten Fall ist natürlich jede Aussage beweisbar).

Der Beweis ist (in Umrissen skizziert) der folgende: Sei κ eine beliebige für die folgenden Betrachtungen ein für allemal gewählte rekursive Klasse von FORMELN (im einfachsten Falle die leere Klasse). Zum Beweise der Tatsache, daß 17 Gen r nicht κ-BEWEISBAR ist,[65] wurde, wie aus 1., Seite 189, hervorgeht, nur die Widerspruchsfreiheit von κ benutzt, d. h. es gilt:

$$\mathrm{Wid}(\kappa) \to \overline{\mathrm{Bew}_\kappa}(17 \text{ Gen } r), \tag{23}$$

d. h. nach (6.1):

$$\mathrm{Wid}(\kappa) \to (x)\overline{x \, B_\kappa \, (17 \text{ Gen } r)}.$$

Nach (13) ist 17 Gen $r = Sb(p_{Z(p)}^{19})$ und daher:

$$\mathrm{Wid}(\kappa) \to (x)\overline{x \, B_\kappa \, (Sb(p_{Z(p)}^{19}))},$$

d. h. nach (8.1):

$$\mathrm{Wid}(\kappa) \to (x)Q(x,p). \tag{24}$$

Wir stellen nun folgendes fest: Sämtliche in Abschnitt 2[66] und Abschnitt 4 bisher definierte Begriffe (bzw. bewiesene Behauptungen) sind auch in P ausdrückbar (bzw. beweisbar). Denn es wurden überall nur die gewöhnlichen Definitions- und Beweismethoden der klassischen Mathematik verwendet, wie sie im System P formalisiert sind. Insbesondere ist κ (wie jede rekursive Klasse) in P definierbar. Sei w die SATZFORMEL, durch welche in P Wid (κ) ausgedrückt wird. Die Relation $Q(x,y)$ wird gemäß (8.1), (9), (10) durch das RELATIONSZEICHEN q ausgedrückt, folglich $Q(x,p)$ durch r (da nach (12) $r = Sb(q_{Z(p)}^{19})$) und der Satz $(x)Q(x,p)$ durch 17 Gen r.

---

[63]κ ist widerspruchsfrei (abgekürzt als Wid(κ)) wird folgendermaßen definiert: Wid(κ) $\equiv (Ex)[\mathrm{Form}(x) \, \& \, \overline{\mathrm{Bew}_\kappa}(x)]$.

[64]Dies folgt, wenn man für κ die leere Klasse von FORMELN einsetzt.

[65]r hängt natürlich (ebenso wie p) von κ ab.

[66]Von der Definition für "rekursiv" auf Seite 179 bis zum Beweis von Satz VI inklusiv.

Theorem XI. *Let κ be any recursive consistent*[63] *class of* FORMULAS; *then the* SENTENTIAL FORMULA *stating that κ is consistent is not κ-*PROVABLE; in particular, the consistency of P is not provable in P,[64] provided P is consistent (in the opposite case, of course, every proposition is provable [in P]).

The proof (briefly outlined) is as follows: Let κ be some recursive class of FORMULAS chosen once and for all for the following discussion (in the simplest case it is the empty class). As appears from 1, page 177 above, only the consistency of κ was used in proving that 17 Gen r is not κ-PROVABLE;[65] that is, we have

$$\mathrm{Wid}(\kappa) \to \overline{\mathrm{Bew}_\kappa}(17 \text{ Gen } r), \tag{23}$$

that is, by (6.1),

$$\mathrm{Wid}(\kappa) \to (x)\overline{x\,B_\kappa\,(17 \text{ Gen } r)}.$$

By (13), we have

$$17 \text{ Gen } r = Sb(p^{19}_{Z(p)}),$$

hence

$$\mathrm{Wid}(\kappa) \to (x)\overline{x\,B_\kappa\,Sb(p^{19}_{Z(p)})},$$

that is, by (8.1),

$$\mathrm{Wid}(\kappa) \to (x)Q(x,p). \tag{24}$$

We now observe the following: all notions defined (or statements proved) in Section 2,[66] and in Section 4 up to this point, are also expressible (or provable) in P. For throughout we have used only the methods of definition and proof that are customary in classical mathematics, as they are formalized in the system P. In particular, κ (like every recursive class) is definable in P. Let w be the SENTENTIAL FORMULA by which $\mathrm{Wid}(\kappa)$ is expressed in P. According to (8.1), (9), and (10), the relation $Q(x,y)$ is expressed by the RELATION SIGN q, hence $Q(x,p)$ by r (since, by (12), $r = Sb(q^{19}_{Z(p)})$), and the proposition $(x)Q(x,p)$ by 17 Gen r.

Therefore, by (24), $w\,\mathrm{Imp}\,(17 \text{ Gen } r)$ is provable in P (and a fortiori κ-PROVABLE).[67] If now w were κ-PROVABLE, then 17 Gen r would also be κ-PROVABLE, and from this it would follow, by (23), that κ is not consistent.

---

[63] "κ is consistent" (abbreviated by "$\mathrm{Wid}(\kappa)$") is defined thus: $\mathrm{Wid}(\kappa) \equiv (Ex)(\mathrm{Form}(x)\,\&\,\overline{\mathrm{Bew}_\kappa}(x))$.

[64] This follows if we substitute the empty class of FORMULAS for κ.

[65] Of course, r (like p) depends on κ.

[66] From the definition of "recursive" on page 159 above to the proof of Theorem VI inclusive.

[67] That the truth of $w\,\mathrm{Imp}\,(17 \text{ Gen } r)$ can be inferred from (23) is simply due to the fact that the undecidable proposition 17 Gen r asserts its own unprovability, as was noted at the very beginning.

Wegen (24) ist also w Imp (17 Gen r) in P beweisbar[67] (um so mehr
κ-BEWEISBAR). Wäre nun w κ-BEWEISBAR, so wäre auch 17 Gen r κ-
BEWEISBAR und daraus würde nach (23) folgen, daß κ nicht widerspruchs-
frei ist.

Es sei bemerkt, daß auch dieser Beweis konstruktiv ist, d. h. er gestattet,
falls ein BEWEIS aus κ für w vorgelegt ist, einen Widerspruch aus κ effektiv
herzuleiten. Der ganze Beweis für Satz XI läßt sich wörtlich auch auf das
Axiomensystem der Mengenlehre M und der klassischen Mathematik[68] A
übertragen und liefert auch hier das Resultat: Es gibt keinen Widerspruchs-
losigkeitsbeweis für M bzw. A, der innerhalb von M bzw. A formalisiert
werden könnte, vorausgesetzt daß M bzw. A widerspruchsfrei ist. Es sei
ausdrücklich bemerkt, daß Satz XI (und die entsprechenden Resultate über
M, A) in keinem Widerspruch zum Hilbertschen formalistischen Standpunkt stehen. Denn dieser setzt nur die Existenz eines mit finiten Mitteln
geführten Widerspruchsfreiheitsbeweises voraus und es wäre denkbar, daß
es finite Beweise gibt, die sich in P (bzw. M, A) *nicht* darstellen lassen.

Da, für jede widerspruchsfreie Klasse κ, w nicht κ-BEWEISBAR ist, so
gibt es schon immer dann (aus κ) unentscheidbare Sätze (nämlich w), wenn
Neg(w) nicht κ-BEWEISBAR ist; m. a. W. man kann in Satz VI | die Vor-
aussetzung der ω-Widerspruchsfreiheit ersetzen durch die folgende: Die
Aussage "κ ist widerspruchsvoll" ist nicht κ-BEWEISBAR. (Man beachte,
daß es widerspruchsfreie κ gibt, für die diese Aussage κ-BEWEISBAR ist.)

Wir haben uns in dieser Arbeit im wesentlichen auf das System P
beschränkt und die Anwendungen auf andere Systeme nur angedeutet. In
voller Allgemeinheit werden die Resultate in einer demnächst erscheinen-
den Fortsetzung ausgesprochen und bewiesen werden. In dieser Arbeit
wird auch der nur skizzenhaft geführte Beweis von Satz XI ausführlich
dargestellt werden.

----

[67]Daß aus (23) auf die Richtigkeit von w Imp (17 Gen r) geschlossen werden kann,
beruht einfach darauf, daß der unentscheidbare Satz 17 Gen r, wie gleich zu Anfang
bemerkt, seine eigene Unbeweisbarkeit behauptet.

[68]Vgl. *von Neumann 1927*.

Let us observe that this proof, too, is constructive; that is, it allows us to actually derive a contradiction from κ, once a PROOF of w from κ is given. The entire proof of Theorem XI carries over word for word to the axiom system of set theory, M, and to that of classical mathematics,[68] A, and here, too, it yields the result: There is no consistency proof for M, or for A, that could be formalized in M, or A, respectively, provided M, or A, is consistent. I wish to note expressly that Theorem XI (and the corresponding results for M and A) do not contradict Hilbert's formalistic viewpoint. For this viewpoint presupposes only the existence of a consistency proof in which nothing but finitary means of proof is used, and it is conceivable that there exist finitary proofs that *cannot* be expressed in the formalism of P (or of M or A).

Since, for any consistent class κ, w is not κ-PROVABLE, there always are propositions (namely w) that are undecidable (on the basis of κ) as soon as $\text{Neg}(w)$ is not κ-PROVABLE; in other words, we can, in Theorem VI, replace the assumption of ω-consistency by the following: The proposition "κ is inconsistent" is not κ-PROVABLE. (Note that there are consistent κ for which this proposition is κ-PROVABLE.)

In the present paper we have on the whole restricted ourselves to the system P, and we have only indicated the applications to other systems. The results will be stated and proved in full generality in a sequel to be published soon.[68a] In that paper, also, the proof of Theorem XI, only sketched here, will be given in detail.

Note added 28 August 1963. In consequence of later advances, in particular of the fact that due to A. M. Turing's work[69] a precise and unquestionably adequate definition of the general notion of formal system[70] can now be given, a completely general version of Theorems VI and XI is now possible. That is, it can be proved rigorously that in *every* consistent formal system that contains a certain amount of finitary number theory there exist undecidable arithmetic propositions and that, moreover, the consistency of any such system cannot be proved in the system.

---

[68]See *von Neumann 1927.*

[68a][This explains the "I" in the title of the paper. The author's intention was to publish this sequel in the next volume of the *Monatshefte.* The prompt acceptance of his results was one of the reasons that made him change his plan.]

[69]See *Turing 1937,* page 249.

[70]In my opinion the term "formal system" or "formalism" should never be used for anything but this notion. In a lecture [*1946*] at Princeton (mentioned in *Princeton University 1947,* p. 11) I suggested certain transfinite generalizations of formalisms; but these are something radically different from formal systems in the proper sense of the term, whose characteristic property is that reasoning in them, in principle, can be completely replaced by mechanical devices.

Introductory note to *1931a, 1932e, f* and *g*

The items considered in this note all concern events that took place during the second Conference on Epistemology of the Exact Sciences, a meeting organized by the *Gesellschaft für empirische Philosophie* (Society for Empirical Philosophy) and held in Königsberg on 5–7 September 1930. Most interesting among them is *1931a*, which contains Gödel's unexpected announcement, during a discussion session at the conference, of the existence of formally undecidable propositions.

The program for the sessions was published in advance in *Erkenntnis* (volume 1, page 80), the journal edited by Hans Reichenbach and Rudolf Carnap on behalf of the *Gesellschaft für empirische Philosophie* and the *Verein Ernst Mach*, organizations distinct from, but closely allied with, the Vienna Circle. According to that program, the first day of the conference was devoted to hour addresses by Carnap, Arend Heyting and John von Neumann, entitled, respectively, "Die Grundgedanken des Logizismus", "Die intuitionistische Begründung der Mathematik" and "Die axiomatische Begründung der Mathematik";[a] in effect, the lectures were manifestoes for the competing philosophies of logicism, intuitionism and formalism.[b] Texts of the lectures, with slightly altered titles, were published in *Erkenntnis* the following year (*Carnap 1931, Heyting 1931* and *von Neumann 1931*) and later reviewed by Gödel for *Zentralblatt* in his *1932e, 1932f* and *1932g*; reportorial in character, the reviews themselves require no further discussion here.

The conference continued on 6 September with hour addresses by Reichenbach, Werner Heisenberg and Otto Neugebauer, followed by three twenty-minute talks, including a presentation by Gödel of the results of his dissertation. The content of Gödel's talk is summarized in the abstract *1930a* (see the introductory note to *1929*), which appeared shortly after the conference. Here we may note in passing the absence in *1930a* of any reference either to the compactness theorem or to the prior appearance of *Gödel 1930*. (That *1930* had not, or had only just,

---

[a]That is, "The basic ideas of logicism", "The intuitionistic grounding of mathematics" and "The axiomatic grounding of mathematics".

[b]Wittgenstein's views on mathematics were also presented, in a lecture by Friedrich Waismann entitled "Das Wesen der Mathematik: Der Standpunkt Wittgensteins" ("The essence of mathematics: Wittgenstein's position"). Waismann's talk, however, was added to the program without prior announcement, and only fragments of it have appeared in print. See *Waismann 1967*, pp. 19–21, 102–107, and 164, and the appendix to *Grassl 1982*.

appeared at the time of the Königsberg conference may be inferred from a receipt in Gödel's *Nachlass*, which shows that Gödel's offprints of *1930* were not shipped to him until 19 September.)

On Sunday, 7 September, the conference concluded with a roundtable discussion of the problem of providing a foundation for mathematics. Chaired by Hans Hahn, the discussion was intended as an adjunct to the addresses of Carnap, Heyting and von Neumann. A stenographic transcript was made, from which an abridgement of the discussion was prepared for publication with the three lecture texts (*Hahn et alii 1931*). In that abridgement, the participants, in addition to those already mentioned, are Gödel, Arnold Scholz and Kurt Reidemeister; however, a single page of a more extensive transcript, preserved in Gödel's *Nachlass*, indicates that there were other participants as well, including Emmy Noether and someone referred to as v. Geilen (perhaps Vitalis Geilen).

The published discussion was supplemented by a Postscript by Gödel, solicited after the conference by the editors of *Erkenntnis*, and by an extensive bibliography on the foundations of mathematics. The extract reprinted here (*Gödel 1931a*) comprises just Gödel's remarks and Postscript. For an English translation of the entire published discussion, including further commentary, see *Dawson 1984*; an English translation of Hahn's long opening statement has also appeared in *Hahn 1980*, pages 31–38.

At least in its published form, the discussion at Königsberg was dominated by Hahn and Carnap, each of whom argued in favor of the logicist position. Their remarks were overshadowed, however, by Gödel's seemingly casual but dramatically timed announcement, late in the discussion, of his discovery of formally undecidable propositions. In the published transcript, Gödel's remarks are split into two segments, separated by a one-sentence interjection by von Neumann. The tone of the first segment is speculative, echoing the views Gödel expressed in the introduction to *1929* (concerning which, see the introductory note to *1929*): contrary to Carnap, Gödel argues against adopting consistency as a criterion of adequacy for formal theories, asserting that "it remains conceivable" that one could perceive, through contentual but finitary considerations, that a statement provable within some consistent formal system is nonetheless false. At that point, as if to reinforce or clarify Gödel's remarks, von Neumann interposed that it remained unsettled whether all intuitionistically acceptable means of proof were formally representable. Then abruptly, as though emboldened and spurred by von Neumann's leader, Gödel shifts from subjunctive to indicative and forthrightly asserts that one can give examples of contentually true propositions that are unprovable within the formalized framework of classical mathematics.

Several reasons might be adduced to account for the tentative character of Gödel's initial remarks. As a newcomer, virtually unknown outside Vienna, he may understandably have advanced his views somewhat hesitantly, unsure what reaction to expect—especially in the context of a conference oriented toward the views of the Vienna Circle.[c] On other occasions, too, he seemingly exercised caution to avoid possible controversy (see the discussion in *Feferman 1984a*), and in a letter to Hao Wang of 7 December 1967 (excerpted in *Wang 1974*, pages 8–9) Gödel speaks of the "prejudice, or whatever you may call it" of logicians of the time against such transfinite concepts as that of " 'objective mathematical truth' ", prejudice that Gödel took pains to circumvent by his rigorous syntactic treatment in *1931*. Always fastidious about his published work, Gödel was undoubtedly aware of the care necessary to avoid misstating his incompleteness result; and, as the Postscript makes clear, at the time of the Königsberg conference he had not yet submitted his discovery for publication. He may well have regarded his results then as still in provisional form. In particular, the second theorem, on the unprovability of consistency, definitely came later, and Wang (*1981*, pages 645–655) has reported that at the time of his announcement Gödel had not yet cast his canonical undecidable proposition in a simple arithmetical form.

It is a remarkable historical fact that Gödel first announced *both* his completeness and incompleteness theorems to an international mathematical audience within the span of two days. (He had, however, previously spoken on the results of his dissertation at the fifteenth meeting of Karl Menger's mathematical colloquium, on 14 May 1930, in Vienna. See *Ergebnisse eines mathematischen Kolloquiums 2*, page 17.) The impact of his back-to-back announcements can well be imagined. In the words of Quine (*1979*, page 81), "Completeness was expected ... , but

---

[c]Gödel's disagreement with, and eventual alienation from, the Vienna Circle is well documented. In that regard, it is worth noting that *1931* did not appear in *Erkenntnis*, ostensibly because of its technical mathematical character; but Gödel's attitude toward the Circle may also have played a role.

an actual proof of completeness was less expected, and a notable accomplishment. It came as a welcome reassurance .... On the other hand the incompletability of elementary number theory came as an upset of firm preconceptions and a crisis in the philosophy of mathematics." Not everyone, however, immediately appreciated the significance of Gödel's remarks. Certainly von Neumann did; he is reported to have pressed Gödel for further details at the end of the session (*Wang 1981*, pages 654–655), and on 20 November, after his return to Berlin, he wrote Gödel to announce his own discovery of the unprovability of consistency.[d] On the other hand, there is evidence in Carnap's *Nachlass* that Gödel had told him of his incompleteness discovery *before* the Königsberg meeting, but that, even so, Carnap still found the result difficult to understand months later. Certainly his remarks at Königsberg suggest no prior understanding. It is notable that no further discussion of Gödel's announcement is recorded in the Königsberg transcript—his remarks are followed only by a few summarizing queries by Reidemeister —and, although the editors of *Erkenntnis* thought Gödel's result momentous enough to solicit the Postscript from him, Reichenbach failed to mention Gödel at all in his summary of the conference published shortly afterward (*Die Naturwissenschaften 18*, pages 1093–1094). See *Dawson 1985* for a detailed examination of contemporary reactions to the incompleteness theorems.

John W. Dawson, Jr.

[In the text of *Gödel 1931a*, a row of asterisks is used to indicate that a passage of remarks by other participants in the discussion has been omitted.]

---

[d]Three days earlier, the manuscript of *Gödel 1931*, containing both incompleteness results, had been received by the editors of *Monatshefte für Mathematik und Physik*; both results were also announced in Gödel's abstract *1930b*, communicated to the Vienna Academy of Sciences by Hahn on 23 October.

Diskussion zur
Grundlegung der Mathematik
(*1931a*)

* * *

Nach formalistischer Auffassung fügt man zu den sinnvollen Sätzen der
Mathematik transfinite (Schein-)Aussagen hinzu, welche an sich keinen
Sinn haben, sondern nur dazu dienen, das System zu einem abgerundeten zu machen, ebenso wie man in der Geometrie durch Einführung der
unendlich fernen Punkte zu einem abgerundeten System gelangt. Diese
Auffassung setzt voraus, daß, wenn man zum System S der sinnvollen
Sätze das System T der transfiniten Sätze und Axiome hinzufügt und dann
einen Satz aus S auf dem Umweg über Sätze aus T beweist, dieser Satz
auch inhaltlich richtig ist, daß also durch Hinzufügung der transfiniten
Axiome keine inhaltlich falschen Sätze beweisbar werden. Diese Forderung
pflegt man durch die der Widerspruchsfreiheit zu ersetzen. Ich möchte
nun darauf hinweisen, daß diese beiden Forderungen keinesfalls ohne wei-
teres als äquivalent angesehen werden dürfen. Denn wenn in einem widerspruchsfreien formalen System A (etwa dem der klassischen Mathematik)
ein sinnvoller Satz p mit Hilfe der transfiniten Axiome beweisbar ist, so
folgt aus der Widerspruchsfreiheit von A bloß, daß *non-p innerhalb* des
Systems A formal nicht beweisbar ist. Trotzdem bleibt es denkbar, daß
man *non-p* durch irgendwelche inhaltliche (intuitionistische) Überlegungen
einsehen könnte, die sich in A *nicht* formal darstellen lassen. In diesem
Falle wäre trotz der Widerspruchsfreiheit von A in A ein Satz beweisbar, dessen Falschheit man durch finite Betrachtungen einsehen könnte.
Sobald man den Begriff "sinnvoller Satz" hinreichend eng faßt (z. B. auf
die finiten Zahlgleichungen beschränkt), kann etwas Derartiges allerdings
nicht eintreten. Hingegen wäre es z. B. durchaus möglich, daß man einen
Satz der Form $(Ex)F(x)$, wo F | eine finite Eigenschaft natürlicher Zahlen
ist (die Negation des Goldbachschen Satzes hat z. B. diese Form), mit
den transfiniten Mitteln der klassischen Mathematik beweisen und andererseits durch inhaltliche Überlegungen einsehen könnte, daß alle Zahlen die
Eigenschaft *non-F* haben, und zwar bleibt dies, worauf ich eben hinweisen
möchte, auch dann noch möglich, wenn man die Widerspruchsfreiheit des
formalen Systems der klassischen Mathematik nachgewiesen hätte. Denn
man kann von keinem formalen System mit Sicherheit behaupten, daß alle
inhaltlichen Überlegungen in ihm darstellbar sind.

* * *

Discussion on
providing a foundation for mathematics
(*1931a*)

* * *

According to the formalist view one adjoins to the meaningful proposi-tions of mathematics transfinite (pseudo-)assertions, which in themselves have no meaning, but serve only to round out the system, just as in ge-ometry one rounds out a system by the introduction of points at infinity. This view presupposes that, if one adjoins to the system S of meaningful propositions the system T of transfinite propositions and axioms and then proves a theorem of S by making a detour through theorems of T, this theorem is also contentually correct, hence that through the adjunction of the transfinite axioms no contentually false theorems become provable. This requirement is customarily replaced by that of consistency. Now I would like to point out that one cannot, without further ado, regard these two demands as equivalent. For, if in a consistent formal system A (say that of classical mathematics) a meaningful proposition p is provable with the help of the transfinite axioms, there follows from the consistency of A only that not-p is not formally provable *within* the system A. Nonethe-less it remains conceivable that one could ascertain not-p through some sort of contentual (intuitionistic) considerations that are *not* formally rep-resentable in A. In that case, despite the consistency of A, there would be provable in A a proposition whose falsity one could ascertain through finitary considerations. To be sure, as soon as one interprets the notion "meaningful proposition" sufficiently narrowly (for example, as restricted to finitary numerical equations), something of that kind cannot happen. However, it is quite possible, for example, that one could prove a statement of the form $(Ex)F(x)$, where F is a finitary property of natural numbers (the negation of Goldbach's conjecture, for example, has this form), by the transfinite means of classical mathematics, and on the other hand could ascertain by means of contentual considerations that all numbers have the property not-F; indeed, and here is precisely my point, this would still be possible even if one had demonstrated the consistency of the formal sys-tem of classical mathematics. For of no formal system can one affirm with certainty that all contentual considerations are representable within it.

* * *

Man kann (unter Voraussetzung der Widerspruchsfreiheit der klassischen Mathematik) sogar Beispiele für Sätze (und zwar solche von der Art des Goldbachschen oder Fermatschen) angeben, die zwar inhaltlich richtig, aber im formalen System der klassischen Mathematik unbeweisbar sind. Fügt man daher die Negation eines solchen Satzes zu den Axiomen der klassischen Mathematik hinzu, so erhält man ein widerspruchsfreies System, in dem ein inhaltlich falscher Satz beweisbar ist.

* * *

149 | Nachtrag

Von den Herausgebern der *Erkenntnis* werde ich aufgefordert, eine Zusammenfassung der Resultate meiner jüngst in den *Monatshefte für Mathematik und Physik 38* erschienenen Abhandlung, "Über formal unentscheidbare Sätze der *Principia mathematica* und verwandter Systeme" [*1931*] zu geben, die auf der Königsberger Tagung noch nicht vorlag. Es handelt sich in dieser Arbeit um Probleme von zweierlei Art, nämlich 1. um die Frage der Vollständigkeit (Entscheidungsdefinitheit) formaler Systeme der Mathematik, 2. um die Frage der Widerspruchsfreiheitsbeweise für solche Systeme. Ein formales System heißt vollständig, wenn jeder in seinen Symbolen ausdrückbare Satz aus den Axiomen formal entscheidbar ist, d. h. wenn für jeden solchen Satz A eine nach den Regeln des Logikkalküls verlaufende endliche Schlußkette existiert, die mit irgendwelchen Axiomen beginnt und mit dem Satz A oder dem Satz *non-A* endet. Ein System $\mathfrak{S}$ heißt vollständig hinsichtlich einer gewissen Klasse 150 | von Sätzen $\mathfrak{R}$, wenn wenigstens jeder Satz von $\mathfrak{R}$ aus den Axiomen von $\mathfrak{S}$ entscheidbar ist. Was in der obigen Arbeit gezeigt wird, ist, daß es kein System mit endlich vielen Axiomen gibt, welches auch nur hinsichtlich der arithmetischen Sätze vollständig wäre.[1] Unter "arithmetischen Sätzen" sind dabei diejenigen Sätze zu verstehen, in denen keine anderen Begriffe vorkommen als $+$, $\cdot$, $=$ (Addition, Multiplikation, Identität, und zwar bezogen auf natürliche Zahlen), ferner die logischen Verknüpfungen des Aussagenkalküls und schließlich das All- und Existenzzeichen, aber nur bezogen auf Variable, deren Laufbereich die natürlichen Zahlen sind (in arithmetischen Sätzen kommen daher überhaupt keine anderen Variablen vor als solche für natürliche Zahlen). Sogar für Systeme, welche unendlich viele Axiome haben, gibt es immer unentscheidbare arithmetische Sätze, wenn nur die "Axiomenregel" gewissen (sehr allgemeinen) Voraussetzungen genügt. Insbesondere ergibt sich aus dem

---

[1] Vorausgesetzt, daß keine falschen (d. h. inhaltlich widerlegbaren) arithmetischen Sätze aus den Axiomen des betreffenden Systems beweisbar sind.

(Assuming the consistency of classical mathematics) one can even give examples of propositions (and in fact of those of the type of Goldbach or Fermat) that, while contentually true, are unprovable in the formal system of classical mathematics. Therefore, if one adjoins the negation of such a proposition to the axioms of classical mathematics, one obtains a consistent system in which a contentually false proposition is provable.

$$* \quad * \quad *$$

Postscript

I have been invited by the editors of *Erkenntnis* to give a synopsis of the results of my *1931*, which has recently appeared in *Monatshefte für Mathematik und Physik 38*, but was not yet available at the time of the Königsberg conference. That paper deals with problems of two kinds, namely: (1) the question of the completeness (decidability) of formal systems of mathematics; (2) the question of consistency proofs for such systems. A formal system is said to be complete if every proposition expressible by means of its symbols is formally decidable from the axioms, that is, if for each such proposition A there exists a finite chain of inferences, proceeding according to the rules of the logical calculus, that begins with some of the axioms and ends with the proposition A or the proposition not-A. A system $\mathfrak{S}$ is said to be complete with respect to a certain class of propositions $\mathfrak{R}$ if at least every statement of $\mathfrak{R}$ is decidable from the axioms of $\mathfrak{S}$. What is shown in the work cited above is that there is no system with finitely many axioms that is complete even with respect only to arithmetical propositions.[1] Here by "arithmetical propositions" are to be understood those propositions in which no notions occur other than $+$, $\cdot$, $=$ (addition, multiplication, identity, with respect to just the natural numbers), as well as the logical connectives of the propositional calculus and, finally, the universal and existential quantifiers, restricted, however, to variables whose domains are the natural numbers. (In arithmetical propositions, therefore, no variables other than those for natural numbers can occur at all.) Even in systems that have infinitely many axioms, there are always undecidable arithmetical propositions if only the "axiom scheme" satisfies certain (very general) assumptions. In particular, it follows from what has just been said that in all the well-known formal systems of mathematics—for example, *Principia mathematica* (together with the axioms of reducibility, choice and infinity), the Zermelo–Fraenkel and von Neumann axiom systems for set theory, and the formal systems

---

[1] Under the assumption that no false (that is, contentually refutable) arithmetical propositions are derivable from the axioms of the system in question.

Gesagten, daß es in allen bekannten formalen Systemen der Mathematik—
z. B. *Principia mathematica* (samt Reduzibilitäts-, Auswahl- und Unendlichkeitsaxiom), Zermelo–Fränkelsches und von Neumannsches Axiomensystem der Mengenlehre, formale Systeme der Hilbertschen Schule—unentscheidbare arithmetische Sätze gibt. Bezüglich der Resultate über die
Widerspruchsfreiheitsbeweise ist zunächst zu beachten, daß es sich hier um
Widerspruchsfreiheit in formalem (Hilbertschen) Sinn handelt, d. h. die
Widerspruchsfreiheit wird als rein kombinatorische Eigenschaft gewisser
Zeichensysteme und der für sie geltenden "Spielregeln" aufgefaßt. Kombinatorische Tatsachen kann man aber in den Symbolen der mathematischen
Systeme (etwa der *Principia mathematica*) zum Ausdruck bringen. Daher
wird die Aussage, daß ein gewisses formales System $\mathfrak{S}$ widerspruchsfrei ist,
häufig in den Symbolen dieses Systems selbst ausdrückbar sein (insbesondere gilt dies für alle oben angeführten Systeme). Was gezeigt wird, ist nun
das folgende: Für alle formalen Systeme, für welche oben die Existenz unentscheidbarer arithmetischer Sätze behauptet wurde, gehört insbesondere
die Aussage der Widerspruchsfreiheit des betreffenden Systems zu den in
diesem System unentscheidbaren Sätzen. D. h. ein Widerspruchsfreiheitsbeweis für eines dieser Systeme $\mathfrak{S}$ kann nur mit Hilfe von Schlußweisen
geführt werden, die in $\mathfrak{S}$ selbst nicht formalisiert sind. Für ein System, in
151　dem alle finiten (d. h. intuitionistisch einwandfreien) | Beweisformen for-
malisiert sind, wäre also ein finiter Widerspruchsfreiheitsbeweis, wie ihn die
Formalisten suchen, überhaupt unmöglich. Ob eines der bisher aufgestellten Systeme, etwa die *Principia mathematica*, so umfassend ist (bzw. ob es
überhaupt ein so umfassendes System gibt), erscheint allerdings fraglich.

Besprechung von *Neder 1931*:
Über den Aufbau der Arithmetik
(*1931b*)

Der Verfasser sucht zunächst die Peanoschen Axiome in zwei Punkten zu
verbessern: 1. ersetzt er die von Peano zugrunde gelegte Gegenstandsfunk-
tion "der Nachfolger von" durch eine Relation "x folgt unmittelbar auf y",
um die im Begriff "der Nachfolger" implizit enthaltene Existenz und Ein-
zigkeit voneinander trennen zu können; 2. beweist er die Unabhängigkeit
seiner Axiome. Bemerkenswert ist ferner, daß die Reflexivität, Symmetrie
und Transitivität der Gleichheitsbeziehung (welche als zweiter Grundbegriff

of Hilbert's school—there are undecidable arithmetical propositions. As to the results on consistency proofs, it is to be noted first of all that they have to do with consistency in the formal (Hilbertian) sense, that is, consistency is conceived as a purely combinatorial property of certain systems of signs and the "rules of the game" governing them. Combinatorial facts can, however, be expressed in the symbols of mathematical systems (say *Principia mathematica*). Therefore the assertion that a certain formal system $\mathfrak{S}$ is consistent will often be expressible in the symbols of that system itself (in particular, this is the case for all of the systems mentioned above). What is shown, now, is the following: For all formal systems for which the existence of undecidable arithmetical propositions was asserted above, the assertion of the consistency of the system in question itself belongs to the propositions undecidable in that system. That is, a consistency proof for one of these systems $\mathfrak{S}$ can be carried out only by means of methods of inference that are not formalized in $\mathfrak{S}$ itself. For a system in which all finitary (that is, intuitionistically unobjectionable) forms of proof are formalized, a finitary consistency proof, such as the formalists seek, would thus be altogether impossible. However, it seems questionable whether one of the systems hitherto set up, say *Principia mathematica*, is so all-embracing (or whether there is a system so all-embracing at all).

Review of *Neder 1931*:
On the construction of arithmetic
(*1931b*)

The author first seeks to improve the Peano axioms in two respects: (1) he replaces the function of individuals "the successor of", taken as primitive by Peano, by a relation "x follows immediately upon y", in order to be able to disentangle the existence and the uniqueness that are implicitly contained in the notion "the successor"; (2) he proves the independence of his axioms. Further, it is noteworthy that the reflexivity, symmetry, and transitivity of the relation of identity (which occurs as the second primitive notion) are not expressly postulated, but follow from the other axioms. With respect to the kinds of numbers of a higher level (fractions, negative numbers, and so on), two theses (already advocated in part by B. Russell) are presented and justified, namely: (1) it is expedient to undertake the extensions not by adjoining "new" numbers to the "old" ones that are already available, but by constructing an entirely new domain of individ-

auftritt) nicht eigens postuliert wird, sondern aus den anderen Axiomen
6   folgt. Be|züglich der höheren Zahlarten (Brüche, negative Zahlen usw.)
werden zwei (zum Teil schon von B. Russell vertretene) Thesen aufgestellt und begründet, nämlich: 1. Es ist zweckmäßig, die Erweiterungen nicht durch Adjunktion "neuer" Zahlen zu den schon vorhandenen "alten" vorzunehmen, sondern in der Weise, daß man einen ganz neuen Dingbereich konstruiert, der einen mit dem früheren isomorphen Teil enthält.
2. Es ist zweckmäßig, die negativen Zahlen erst nach den gebrochenen und
irrationalen einzuführen (unter anderem wegen des einfachen Aufbaus der
Schnittheorien im Bereich der positiven Brüche). Hinsichtlich der Einführung der reellen Zahlen weist der Verfasser auf die Vorteile einer zuerst von
A. Capelli [*1897*] angegebenen Methode hin, nach welcher reelle Zahlen
definiert werden als Paare von nicht leeren Mengen M_1/M_2 rationaler
Zahlen, die den Bedingungen genügen, daß : 1. jede Zahl aus $M_1 \leq$ ist
jeder Zahl aus M_2, 2. zu jedem positiven δ zwei Zahlen a, b existeren, so
daß $a \,\epsilon\, M_1$, $b \,\epsilon\, M_2$, $b - a < \delta$. Die Ableitung der Rechengesetze für nicht
negative reelle Zahlen auf Grund der genannten Definition wird in allen
Einzelheiten durchgeführt.

uals that contains a part isomorphic to the earlier one; (2) it is expedient to introduce the negative numbers only after the fractions and irrational numbers have been introduced (among other reasons, because of the simple structure of the theories of cuts in the domain of positive fractions). With regard to the introduction of the real numbers, the author points to the advantages of a method first presented by A. Capelli [*1897*], according to which real numbers are defined as pairs of non-empty sets M_1/M_2 of rational numbers satisfying the following conditions: (1) every number in M_1 is $\leq$ every number in M_2; (2) for each positive δ, there exist two numbers a and b such that $a \epsilon M_1$, $b \epsilon M_2$, and $b - a < \delta$. The derivation of the rules of calculation for non-negative real numbers on the basis of the aforementioned definition is carried out in all particulars.

Introductory note to *1931c*

In the paper here reviewed by Gödel, Hilbert added a new rule of a semi-infinitary character to a formal system of arithmetic with the aim of obtaining a complete and consistent system. Written at the end of his academic career, this paper was one of the last of Hilbert's individual contributions to the foundations of mathematics. It is natural to ask whether the pursuit of this novel extension of arithmetic came in response to Gödel's incompleteness theorem of *1931* or whether it had an independent origin. The question is not readily answered, since there is no mention of Gödel or of the incompleteness results in *Hilbert 1931*, nor in its successor *1931a*, where the ideas are elaborated further. The other available evidence, to be discussed below, is also equivocal in this respect.

While Hilbert had long emphasized the program of obtaining finitary consistency proofs for formal systems of arithmetic and analysis, questions of formal completeness had not been explicitly raised by him until 1928. To be sure, Hilbert's early conviction as to the solvability of every mathematical problem—so forcefully expressed in his famous list of 23 problems (*1900*)—is an informal reflection of a belief in completeness. In his paper *1929* for the International Congress of Mathematicians in Bologna which took place in 1928, Hilbert stated two problems of completeness, one for the first-order predicate calculus (completeness with respect to validity in all interpretations) and the second for a system of elementary number theory (formal completeness, in the sense of maximal consistency). The first of these problems, posed also in *Hilbert and Ackermann 1928*, was solved positively by Gödel in his dissertation *1929*, while the second was excluded by his incompleteness results of *1931*.

The timing of Gödel's and Hilbert's work is close: as concerns the former see the introductory notes in this volume to *1931a* by Dawson and to *1931* by Kleene. The relevant facts are that Gödel made a brief public announcement (*1931a*) of his incompleteness results at a special meeting on the foundations of mathematics held in Königsberg 5–7 September 1930, just prior to the meeting of the 91st Assembly of German Scientists and Physicians. Von Neumann, who was one of the principal participants in the meeting on foundations, was by all accounts the first member of Hilbert's circle to appreciate the significance of Gödel's results for Hilbert's program. According to *Wang 1981*, von Neumann enthusiastically took up Gödel's ideas and pursued them on the spot.[a]

---

[a]See *Wang 1981*, pp. 654–655, where Gödel is given as the source for this report.

It happens that Hilbert gave his famous lecture *Naturerkennen und Logik* (*1930*) as the opening address for the main assembly directly following. We can only speculate whether von Neumann had sufficient time to communicate Gödel's new results to Hilbert in those days, and have them be absorbed. At any rate, von Neumann could well have written him about this in the following months, after Hilbert returned to Göttingen.

During this same period in Göttingen, Bernays was acting as Hilbert's close assistant and collaborator on matters concerning logic and the foundations of mathematics.[b] Bernays wrote Gödel on 24 December 1930 (from Berlin, where he was spending the holidays with his family), requesting a copy of galley proofs for *1931*, saying that Courant and Schur had told him that it contained important and surprising results. (Courant was also prominent in Hilbert's circle.) Gödel forwarded the proof sheets immediately and Bernays quickly responded with a lengthy letter from Göttingen on 18 January 1931 in which, among other things, he described Gödel's results as a considerable step forward in the investigation of foundational problems.[c] Offprints of *1931* were sent a few months later by Gödel to both Bernays and Hilbert, as acknowledged by Bernays in a letter of 20 April 1931. However Gödel himself never corresponded directly with Hilbert, nor did they ever meet (*Reid 1970*, page 217).[d]

Hilbert's publication *1931*, at issue here, was based on a talk he gave to the Philosophical Society of Hamburg in December 1930. Its contents were described to Gödel by Bernays in his letter of 18 January 1931, but no statement one way or the other was made there concerning the genesis of the ideas involved. The elaboration in *Hilbert 1931a* was based on a further presentation to the Göttingen Academy of Sciences at its meeting of 17 July 1931. Hilbert was surely apprised by then of the significance of *Gödel 1931*, but he still made no reference to that work.

---

[b] According to Bernays' own short biography (*1976b*), his work as assistant to Hilbert began in 1917. Bernays was later appointed associate professor without tenure [*nichtbeamteter ausserordentlicher Professor*] in 1922. He was deprived of that position in 1933 as a "non-Aryan" and thereupon left for Zurich. The principal published results of the collaboration are the volumes *Hilbert and Bernays 1934* and *1939*.

[c] Four letters from Bernays to Gödel during the period 1930–1931 were found in Gödel's *Nachlass*; unfortunately, we do not have Gödel's replies and can only infer their contents. All further correspondence between the two dates from 1935. A considerable amount of this has been preserved both in the original and in copies, in the *Nachlass* of each of them. (See *Dawson 1985*.)

[d] This is not surprising, in view of the great asymmetry of their positions at that point. It should be mentioned that Hilbert had retired at the beginning of 1930, when he turned 68 (*Reid 1970*, pp. 190ff).

In his survey article on Hilbert's foundational contributions for Hilbert's collected works, Bernays (*1935a*, page 215) says that even before Gödel's incompleteness results were known ("bekannt"), Hilbert had already given up the original form of his completeness problem and, in its place, had taken up the ideas of *Hilbert 1931*. However, in a letter to Constance Reid, dated 3 August 1966, Bernays reports that he had expressed prior doubts to Hilbert about the conjectured completeness of number theory, and that Hilbert was very "angry" about this suggestion, and was then also angry at Gödel's results.[e,f] But *then* (according to Bernays in this same letter), Hilbert went on to deal with this obstacle in a positive way, as evidenced by introduction of the new rule of inference in *1931* and *1931a*. The reports of Bernays in 1935 and 1966 do not quite jibe, and are subject matter for speculation.

In all, then, the available evidence leaves us with no clear-cut answer as to the possible influence of Gödel's work on that of Hilbert.

We now turn to an examination of the content of *Hilbert 1931* in relation to that of *Gödel 1931*. Hilbert's system is a first-order form of Peano arithmetic, with the usual axioms for zero and successor, the scheme of (primitive) recursive definition, and the axiom (scheme) of complete induction (*1931*, pages 489–490); it is here denoted 'Z'. From his *1929*, Hilbert takes a solution of the completeness problem for a system (S) to consist in demonstrating the following:

(C1) Every statement $\mathfrak{S}$ shown to be consistent (with S) is provable (in S); and

(C2) if a statement $\mathfrak{S}$ is shown to be consistent (with S) then the same cannot be done for its negation $\overline{\mathfrak{S}}$.

Of course, implicit throughout is the general requirement of Hilbert's program that all metamathematical results are to be established by finitary methods.

Hilbert's new rule of *1931* (page 491) is as follows. Whenever $A(x)$ is a quantifier-free formula for which it can be shown (finitarily) that $A(z)$

---

[e]This information from Bernays is reported in *Reid 1970*, pp. 198–199. In connection with Hilbert's response here, Reid drew my attention to p. 173, op. cit., where Bernays told about the "violent arguments" that he and Hilbert would have concerning the foundations of mathematics. Bernays attributed the emotional character of these arguments "to a fundamental 'opposition' [*Gegensatz*] in Hilbert's feelings about mathematics ... namely his resistance to Kronecker's tendency to restrict mathematical methods ..., particularly, set theory [on the one hand]...[and his] thought that Kronecker had probably been right .... It became his goal ... to do battle with Kronecker with his own weapon of finiteness ...."

[f]Curiously (as W. Sieg brought to my attention), Bernays himself conjectured the completeness of arithmetic in a paper "Die Philosophie der Mathematik und die Hilbertischen Beweistheorie" published in 1930. See p. 59 of the collection *Bernays 1976*, where the 1930 paper is reprinted.

is a correct numerical formula for each particular numerical instance z, then its universal generalization $(x)A(x)$ may be taken as a new premise (*Ausgangsformel*) in all further proofs. The (semi-formal) system with this new (informal) rule is here denoted 'Z^*'. Hilbert claims to have a finitary proof of the consistency of Z^* by an extension of a proof of the consistency of Z due to Ackermann (*1924*) and (in revised form) von Neumann (*1927*). However, the latter proofs accomplished much less than was considered at the time; the matter is discussed below. In any case, Hilbert supplies no details for his own extension of the earlier work.

Assuming the consistency of Z^*, Hilbert goes on to establish the completeness conditions (C1) and (C2) for $\mathfrak{S}$ of the form $(x)A(x)$ with A quantifier-free, by the following simple line of argument. Every correct numerical statement is provable in Z. Hence, if $(x)A(x)$ is shown consistent with Z^* then each instance $A(z)$ is correct; for otherwise $\overline{A(z)}$ would be correct, so provable in Z and this would yield an inconsistency of Z^* with $(x)A(x)$. By the new rule, it follows that $(x)A(x)$ is provable in Z^* and (C1) is thus shown for $\mathfrak{S}$ of the form $(x)A(x)$. Using the consistency of Z^*, it is then immediate that (C2) also holds for such $\mathfrak{S}$. It is of course trivial that (C2) holds as well for $\mathfrak{S}'$ of the form $(Ex)A(x)$, but Hilbert observes that this does not tell us (C1) holds for such $\mathfrak{S}'$. (Indeed, this will fail by Gödel's incompleteness result of *1931* if the semi-formal system Z^* is contained in an ω-consistent formal system S.)

As it happens, the arguments of Ackermann and von Neumann referred to above only succeeded in establishing the consistency by finitary means of a weak subsystem Z_o of Z, obtained by considerable restriction of the induction scheme. This was pointed out by Bernays (*1935a*, page 213) in his discussion of Hilbert's foundational program.[g] (The reasons for the necessity to restrict induction are elaborated in *Hilbert and Bernays 1939*, pages 123–125). Hilbert's argument can be carried over, mutatis mutandis, to a system Z_o^* obtained from Z_o by closing under his new rule, but the interest of Hilbert's result is thereby correspondingly diminished. In any case, the above shows that Hilbert's arguments come relatively "cheap" once consistency of such Z_o^* is accepted.

At the time Bernays corresponded with Gödel in 1930–1931, the restricted scope of the Ackermann–von Neumann consistency proofs was apparently not yet realized, and was thus a cause for puzzlement. On the one hand, it was believed that a finitary consistency proof had been given for Z, so it apparently followed from Gödel's second incompleteness theorem that not all finitary arguments can be formalized in Z. On the other hand, von Neumann was already making a case (as Bernays

[g]See also *Bernays 1935a*, p. 211, footnote 2.

wrote on 18 January 1931) that every finitary proof can be formalized in
the system P of *Gödel 1931*. It was natural to ask what, among all that
had previously been presented as finitary consistency proofs, could not
already be formalized in Z. (Gödel himself was cautious on the scope of
finitary proofs in general, at the end of *1931*.[h])

The main defect of *Hilbert 1931* is that the precisely defined formal
system Z is replaced by an imprecisely defined system Z*, using the
vague and general concept of finitary proof in an essential way. Gödel
himself told Carnap that he viewed this as a step compromising Hilbert's
program;[i] after all, that program had rested entirely on treating mathematics as formalized in combinatorially described, precisely delimited
languages and axiom systems. In his review, though, Gödel is again
(properly) cautious: he describes *Hilbert 1931* as a "substantial supplement to the formal steps taken thus far ... ".

The vagueness just mentioned needs also to be kept in mind when
Hilbert is credited (as is often done) with the introduction of the rule
of "infinite induction", or what is nowadays called the ω-rule: this allows the inference of $(x)A(x)$ for any formula A for which $A(z)$ has been
proved for each numeral z. Unlike Hilbert's rule, no restriction is here
made on the form of the formula A or the manner in which the numer-

---

[h]In his introduction to *Hilbert and Bernays 1934*, Hilbert stood firm in defense of
his program, saying that Gödel's incompleteness results only showed that sharper
methods must be utilized in order to obtain more far-reaching finitary consistency
proofs.

[i]According to a memorandum by Carnap of 21 May 1931; see *Dawson 1985*, footnote 13.

Besprechung von *Hilbert 1931*:
Die Grundlegung der elementaren Zahlenlehre
(*1931c*)

Der Vortrag bringt außer einer kurzen Darstellung des formalistischen
Standpunktes und einer Reihe philosophischer und historischer Bemerkungen auch eine wesentliche Ergänzung der bisherigen formalen Ansätze
zur Begründung der Zahlentheorie. In dem Vortrag "Probleme der Grundlegung der Mathematik" (*1929a*), hatte Hilbert unter anderem die Aufgabe
gestellt, für den Formalismus der Zahlentheorie folgende beiden die Vollständigkeit des Systems betreffenden Sätze zu beweisen: 1. Wenn eine

ical instances $A(z)$ are to be proved. Apparently, the first to consider such a frankly infinitary rule was Tarski in a lecture he gave in 1927,[j] although his first publication in which this was stated is *Tarski 1933*. It was then formulated again (as the rule DC2) in *Carnap 1934a*, §14 (= *1937*, §14) and in *1935*, leading to its designation as "Carnap's rule" in *Rosser 1937*, where it was taken up for systematic treatment.

A syntactically finite formal rule closer to Hilbert's own is that which is sometimes called the "formalized ω-rule". Using Gödel's representation of the proof relation for Z within itself, this permits the inference of $(x)A(x)$, for any formula $A(x)$ for which it has been proved, *in* Z, that every numerical instance $A(z)$ is provable in Z.[k] It is readily seen that the system Z' thus obtained has a primitive recursive axiomatization, and Gödel's incompleteness theorem applies (granting the consistency of Z'). This was established in *Rosser 1937*, page 135, where incompleteness was also obtained for arbitrary finite iterations of the rule. It was later shown in *Feferman 1962* and *Feferman and Spector 1962* that the procedure by which Z' is obtained from Z can be iterated into the recursive transfinite, with both completeness and incompleteness results, depending on the "path" taken for that iteration.

Solomon Feferman

[j]See *Tarski 1933*, footnote 2 (appearing in tranlation on p. 279 in *Tarski 1956*). The rule happens also to have been formulated in Bernays' letter to Gödel of 18 January 1931, though it may be that Bernays intended a formalized version in the sense discussed below.

[k]The first published discussion of this rule is to be found in *Rosser 1937*, pp. 134–135, where the idea for it is credited to Kleene; see also *Feferman 1962*, p. 296.

Review of *Hilbert 1931*:
The foundation of elementary number theory
(*1931c*)

Besides a short presentation of the formalist viewpoint and a series of philosophical and historical remarks, this lecture offers a substantial supplement to the formal steps taken thus far toward laying a foundation for number theory. In his lecture *1929a* Hilbert had, among other things, posed the problem of proving, for the formalism of number theory, the following two theorems, which concern the completeness of the system:

Aussage als widerspruchsfrei erwiesen werden kann, so ist sie auch beweisbar. 2. Wenn für einen Satz 𝕾 die Widerspruchsfreiheit mit den Axiomen der Zahlentheorie nachgewiesen werden kann, so kann nicht auch für 𝕾̄ die Widerspruchsfreiheit mit jenen Axiomen nachgewiesen werden. Um zum Beweise dieser beiden Sätze, wenigstens für gewisse Spezialfälle zu gelangen, wird das formale System der Zahlentheorie durch folgende, ihrer Struktur nach ganz neuartige Schlußregel erweitert: Falls nachgewiesen ist, daß die Formel $A(z)$ allemal, wenn z eine vorgelegte Ziffer ist, eine richtige numerische Formel wird, so darf die Formel $(x)A(x)$ als Ausgangsformel angesetzt werden. Der Widerspruchfreiheitsbeweis für die neue Regel wird skizziert. Außerdem ergibt sich unmittelbar, daß in dem erweiterten System für Aussagen der Form $(x)A(x)$ (wobei $A(x)$ außer x keine Variable mehr enthält) Satz 1. und 2. und für Aussagen der Form $(Ex)A(x)$ wenigstens Satz 2. gilt.

Besprechung von *Betsch 1926*: *Fiktionen in der Mathematik* (*1931d*)

Die von der Wiener Akademie preisgekrönte umfangreiche Arbeit greift über das Thema "Fiktionen in der Mathematik" nach allen Richtungen weit hinaus. Es werden einerseits die Fiktionen im allgemeinen und im Zusammenhang damit das philosophische Wirklichkeits- und das Wahrheitsproblem besprochen, andererseits die Grundlagenprobleme der Mathematik ausführlich behandelt. Die Darstellungsweise ist in der Hauptsache historisch referierend. Was das eigentliche Thema betrifft, d. h. die Frage, ob und in welchem Sinn es Fiktionen in der Mathematik gibt, so ist die Stellungnahme des Autors vorwiegend negativ. Hinsichtlich der echten Fiktionen (widerspruchsvolle aber zweckmäßige Annahmen) fällt die Antwort durchwegs verneinend aus und alle diesbezüglichen Vaihingerschen Behauptungen werden als Irrtümer nachgewiesen. Ob man unechte Fiktionen (mit der Wirklichkeit nicht übereinstimmende Annahmen) in der Mathematik zugeben kann, hängt, wie der Autor zu zeigen sucht, von der Stellungnahme in den Grundlagenfragen ab, wobei Konventionalismus und Empirismus (im Sinn von Mill) den Fiktionen noch am meisten Raum geben. Zum Schluß wird darauf hingewiesen, daß Fiktionen in einem ganz anderen Sinn (als versuchsweise Annahmen, provisorische Begriffsbildungen) in der historischen Entwicklung der Mathematik eine große Rolle spielen.

(1) if a proposition can be shown to be consistent, then it is also provable; (2) if consistency with the axioms of number theory can be demonstrated for a sentence $\mathfrak{S}$, then consistency with those axioms cannot also be demonstrated for its negation $\overline{\mathfrak{S}}$. In order to make possible the proof of these two theorems, at least for certain special cases, the formal system of number theory is extended by the following rule of inference, which, structurally, is of an entirely new kind: in case it is demonstrated that, whenever z is a particular numeral, the formula $A(z)$ is a correct numerical formula, then the formula $(x)A(x)$ may be taken as a premise. A consistency proof for the new rule is sketched. In addition, it follows immediately that in the extended system Theorems 1 and 2 hold for propositions of the form $(x)A(x)$ (where $A(x)$ contains no variable other than x) and that at least Theorem 2 holds for propositions of the form $(Ex)A(x)$.

Review of *Betsch 1926*: *Fictions in mathematics* (*1931d*)

This comprehensive work, which was awarded a prize by the Vienna Academy, reaches in all directions far beyond the theme "fictions in mathematics". On the one hand, it discusses fictions in general and, in connection with them, the philosophical problems of reality and of truth; on the other, it treats in detail the problems of the foundations of mathematics. The method of presentation is chiefly historical. As concerns the theme proper, that is, the question whether and in what sense there are fictions in mathematics, the author's attitude is predominantly negative. With respect to genuine fictions (contradictory but expedient assumptions), his answer always comes out negative, and all of Vaihinger's assertions in this regard are shown to be erroneous. Whether unreal fictions (assumptions not in accord with reality) can be admitted to occur in mathematics depends, as the author seeks to show, upon one's attitude toward foundational questions, conventionalism and empiricism (in Mill's sense) in the end allowing such fictions the widest scope. In conclusion it is pointed out that fictions in a quite different sense (as tentative assumptions or provisional definitions) play a large role in the historical development of mathematics.

Besprechung von *Becker 1930*:
Zur Logik der Modalitäten
(*1931e*)

6 | Der Verfasser knüpft an C. I. Lewis "System of strict implication" (ein Aussagenkalkül mit den Grundbegriffen: "und", "nicht", "unmöglich") an und bemerkt, daß man in diesem System durch iterierte Anwendung der Prädikate "nicht", "unmöglich" unendlich viele unreduzierbare Modalitäten der Aussage erzeugen kann. Diese Modalitäten könne man nicht einmal ihrer logischen Stärke nach in eine lineare Rangordnung bringen, in der Weise, daß von zwei bejahenden (bzw. verneinenden) immer eine von beiden die andere implizieren würde. (Strenge Beweise für diese Behauptungen werden allerdings nicht gegeben.) Um diesem Mangel des Lewisschen Systems abzuhelfen, schlägt Verfasser verschiedene zusätzliche Axiome vor und sucht dann ein möglichst voraussetzungsarmes System zu bestimmen, in dem noch lineare Rangordnung besteht. Im ganzen ergeben sich drei verschiedene Arten des Modalitätenkalküls, die, wie Verfasser meint, bei verschiedenen Arten von "Notwendigkeit" Anwendung finden können. Was die rein formale Seite betrifft, so wird man wohl kaum etwas einwenden können, aber es bleiben noch wesentliche Lücken auszufüllen, auf die Verfasser zum Teil selbst hinweist. Vor allem ist nirgends gezeigt, daß die drei aufgestellten Systeme tatsächlich untereinander und vom Lewisschen verschieden sind (d. h. ob die Zusatzaxiome nicht etwa äquivalent sind oder aus den Lewisschen folgen); ferner auch, ob die erhaltenen sechs bzw. zehn Grundmodalitäten nicht noch weiter reduziert werden können. Zum Schluß bespricht Verfasser die Zusammenhänge, welche nach seiner Meinung zwischen Modalitätenlogik und Brouwer–Heytingscher intuitionistischer Logik sowohl vom formalen als vom phänomenologischen Standpunkt bestehen. Der diesbezügliche formale Ansatz dürfte allerdings kaum zum Ziele führen.

Review of *Becker 1930*:
On modal logic
(*1931e*)

The author's starting point is C. I. Lewis' "system of strict implication" (a propositional calculus with the primitive notions "and", "not", "impossible"), and he remarks that in this system we can generate, through iterated application of the predicates "not" and "impossible", infinitely many irreducible modalities for a proposition. These modalities cannot even be linearly ordered according to their logical strength in the sense that, of any two affirming modalities, one will imply the other, and similarly for negating ones. (To be sure, strict proofs of these assertions are not given.) To remedy this defect of Lewis' system, the author proposes various additional axioms and then seeks to specify a system, with as few assumptions as possible, for which a linear ordering still exists. All in all, three different kinds of the calculus of modalities emerge, which, so the author believes, can be applied when different kinds of "necessity" are considered. As far as the purely formal side is concerned, one can hardly take exception to anything here, but there remain essential gaps to be filled in, some of which the author himself points out. Above all, it is nowhere shown that the three systems set up really differ from one another and from Lewis' system (in other words, that the additional axioms are not in fact equivalent and do not follow from Lewis'); nor, furthermore, that the six, or ten, basic modalities obtained cannot be still further reduced. In conclusion, the author discusses, from a formal as well as a phenomenological standpoint, the connections that in his opinion obtain between modal logic and the intuitionistic logic of Brouwer and Heyting. It seems doubtful, however, that the steps here taken to deal with this problem on a formal plane will lead to success.

Besprechung von *Hasse und Scholz 1928*: *Die Grundlagenkrisis der griechischen Mathematik* (*1931f*)

Dieses anregend geschriebene Büchlein gibt eine sehr interessante Darstellung der Entwicklung der Lehre vom Irrationalen bei den Griechen. Nach der Auffassung der Autoren haben die Pythagoräer die auftretenden Schwierigkeiten zunächst durch Einführung unendlich kleiner Größen zu überwinden versucht und erst Zenon hat durch seine bekannten Paradoxien (fliegender Pfeil etc.) die Unhaltbarkeit dieses Standpunktes dargetan und dadurch eine Grundlagenkrisis hervorgerufen, welche ihrerseits der Ausgangspunkt für den exakten Aufbau durch Eudoxos und andere gewesen ist. Diese Hypothese wirft ein ganz neues Licht auf Zenon, der als Vorkämpfer der exakten Methoden in der Mathematik erscheint und in diesem Sinn mit Weierstraß in Parallele gesetzt wird. Das Buch enthält noch eine Reihe anderer interessanter historischer Bemerkungen sowie in einem Anhang eine eingehende Kritik der von Oswald Spengler vertretenen Ansicht über das Verhältnis der Griechen zum Problem des Irrationalen. Besonders hervorzuheben ist auch die Klarheit des Stiles und die Schärfe und Prägnanz der Formulierungen.

Besprechung von *von Juhos 1930*: *Das Problem der mathematischen Wahrscheinlichkeit* (*1931g*)

Der Zweck dieser Schrift ist eine Analyse der der mathematischen Wahrscheinlichkeitsrechnung zugrunde liegenden Begriffe und Grundsätze, insbesondere ihre Reinigung von allen empirischen Beimengungen. Um dieses Ziel (d. h. einen rein logischen Aufbau) zu erreichen, werden die in der Wahrscheinlichkeitsrechnung betrachteten physikalischen Ereignisse (Kugelziehungen etc.) durch "willkürliche Auswahlakte" ersetzt gedacht, welche "absolut zufällige, raumzeitlich nicht bedingte (deshalb unwirkliche) Ereignisse" und daher "rein logische Konstruktionen" sein sollen. Die mathematische Wahrscheinlichkeit wird schließlich definiert als ein gewis-

Review of *Hasse and Scholz 1928*:
The foundational crisis in Greek mathematics
(*1931f*)

This stimulating little book depicts in a very interesting way how the doctrine of the irrational developed among the Greeks. In the authors' view, the Pythagoreans at first attempted to overcome the emerging difficulties by introducing infinitely small magnitudes, and only Zeno, through his well-known paradoxes (the flying arrow, etc.), demonstrated the untenability of this standpoint, thereby precipitating a foundational crisis, which, for its part, was the point of departure for the rigorous construction by Eudoxus and others. This hypothesis casts an entirely new light on Zeno, who appears as an early champion of rigorous methods in mathematics, and in that sense is compared to Weierstrass. In addition, the book contains a series of other interesting historical remarks, as well as, in an appendix, a penetrating criticism of the view advocated by Oswald Spengler concerning the attitude of the Greeks toward the problem of the irrational. The clarity of style as well as the sharpness and terse suggestiveness of the formulations are also especially notable.

Review of *von Juhos 1930*:
The problem of
mathematical probability
(*1931g*)

The object of this monograph is to present an analysis of the notions and fundamental principles underlying the calculus of mathematical probabilities, and in particular to purge them of all empirical impurities. To attain this goal (that is, a purely logical construction), the physical events (drawing of balls, etc.) considered in the calculus of probabilities are thought of as being replaced by "arbitrary acts of choice", which are supposed to be "absolutely random events that are not spatio-temporally conditioned (hence unreal)" and therefore "purely logical constructs". Mathematical probability is ultimately defined as a fraction associated with certain propo-

sen Aussagefunktionen zugeordneter Bruch, nämlich das Verhältnis der Anzahl derjenigen Argumente, welche die betreffende Aussagefunktion zu einem wahren Satz machen, zur Anzahl derjenigen, welche sie zu einem sinnvollen Satz machen. In einem Anhang wird auf das Verhältnis zum physikalischen Wahrscheinlichkeitsbegriff eingegangen.

sitional functions, namely, the ratio of the number of those arguments that turn the propositional function in question into a true statement to the number of those that turn it into a meaningful statement. The relation to the physical notion of probability is considered in an appendix.

Introductory note to *1932*

In this short note on intuitionistic propositional logic (**H**),[a] Gödel shows that

(1)   **H** cannot be viewed as a system of many-valued logic[b]

(that is to say, we cannot find a finite set M of truth values, with a subset $D \subset M$ of designated values, plus an interpretation of $\rightarrow$, $\wedge$, $\vee$ by binary operations on M and an interpretation of $\neg$ by a unary operation on M, such that $\mathbf{H} \vdash A$ if and only if, for all valuations ϕ in M, $\phi(A) \in D$) and that

(2)   there is an infinite descending chain of logics intermediate in strength between **A** (classical propositional logic) and **H**.

From Gödel's argument one sees that one can take for this chain

$$\mathbf{A} = \mathbf{L}_2 \supset \mathbf{L}_3 \supset \mathbf{L}_4 \supset \dots,$$

where $\mathbf{L}_n$ is the set of propositional formulas identically valid on the

[a]For the formal systems designated by A and H in Gödel's text we use bold face **A** and **H**, respectively, for greater typographical clarity.

[b]For an introduction to many-valued logics, see, e.g., *Rautenberg 1979*, Chapter III.

Zum intuitionistischen Aussagenkalkül
(*1932*)

[In Beantwortung einer von Hahn aufgeworfenen Frage] für das von A. Heyting[1] aufgestellte System H des intuitionistischen Aussagenkalküls gelten folgende Sätze:

[1] *Heyting 1930*.

n-element linearly ordered Heyting algebra (pseudo-Boolean algebra). A finite axiomatization of $\mathbf{L}_n$ ($n \geq 2$) was first given by I. Thomas (*1962*), based on Dummett's (*1959*) axiomatization of the logic **LC** of tautologies on the linear Heyting algebra of order type ω. **LC** can be axiomatized by adding to **H** the axiom $(P \rightarrow Q) \vee (Q \rightarrow P)$, or, equivalently, the following characterization of $\vee$:

$$(A \vee B) \leftrightarrow [((A \rightarrow B) \rightarrow B) \wedge ((B \rightarrow A) \rightarrow A)].$$

$\mathbf{L}_n$ is then axiomatizable as $\mathbf{LC} + F_{n+1}$, where F_{n+1} is as defined in Gödel's note.

It is to be noted that (1) is in fact a consequence of (2), since it is not difficult to show that any propositional logic characterized by a finite set of truth values (in the sense indicated above) and containing **H** has only finitely many proper strengthenings.

Gödel's second result may be regarded as the first contribution to the topic of intermediate propositional logic.[c]

In the last line of his note Gödel stated the disjunction property for **H**; a proof was given by Gentzen in his *1935*.

A. S. Troelstra

[c]There is now an extensive literature on the subject. A survey of the literature up till 1970 may be found in *Hosoi and Ono 1973*. *Minari 1983* presents an extensive bibliography with historical comments. The reasons for studying intermediate logics are mainly technical; for example, intermediate logics give rise to interesting algebraic theories.

On the intuitionistic propositional calculus
(*1932*)

[Answer to a question posed by Hahn:] For the system H, set up by Heyting,[1] of the intuitionistic propositional calculus the following theorems hold:

[1] *Heyting 1930.*

I. *Es gibt keine Realisierung mit* endlich *vielen Elementen (Wahrheits-werten), für welche die und nur die in* H *beweisbaren Formeln erfüllt sind (d. h. bei beliebiger Einsetzung ausgezeichnete Werte ergeben).*

II. *Zwischen* H *und dem System* A *des gewöhnlichen Aussagenkalküls liegen unendlich viele Systeme, d. h. es gibt eine monoton abnehmende Folge von Systemen, welche sämtlich* H *umfassen und in* A *enthalten sind.*

Der Beweis ergibt sich aus folgenden Tatsachen: Sei F_n die Formel:

$$\sum_{1 \leq i < k \leq n} (a_i \supset\subset a_k)$$

wobei Σ die iterierte $\vee$-Verknüpfung bedeutet und die a_i Aussagevariable sind. F_n ist erfüllt für jede Realisierung mit weniger als n Elementen, für welche alle in H beweisbaren Formeln erfüllt sind. Denn bei jeder Einsetzung wird in mindestens einem Summanden von F_n a_i und a_k durch dasselbe Element e ersetzt und $e \supset\subset e . \vee b$ ergibt bei beliebigem b einen ausgezeichneten Wert, weil die Formel $a \supset\subset a . \vee b$ in H beweisbar ist. Sei ferner S_n die folgende Realisierung:

Elemente: $\{1, 2, \ldots, n\}$, ausgezeichnetes Element: 1;

$a \vee b = \min(a, b);$ $a \wedge b = \max(a, b);$

$a \supset b = 1$ für $a \geq b;$ $a \supset b = b$ für $a < b;$

$\neg a = n$ für $a \neq n,$ $\neg n = 1.$

Dann sind für S_n sämtliche Formeln aus H und die Formel F_{n+1} sowie alle F_i mit größerem Index erfüllt, dagegen F_n sowie | alle F_i mit kleinerem Index *nicht* erfüllt. Insbesonders ergibt sich daraus, daß *kein* F_n in H beweisbar ist. Es gilt übrigens ganz allgemein, daß eine Formel der Gestalt $A \vee B$ in H nur dann beweisbar sein kann, wenn entweder A oder B in H beweisbar ist.

I. *There is no realization with* finitely *many elements* (*truth values*) *for which the formulas provable in H, and only those, are satisfied* (*that is, yield designated values for an arbitrary assignment*).

II. *Infinitely many systems lie between H and the system A of the ordinary propositional calculus, that is, there is a monotonically decreasing sequence of systems all of which include H as a subset and are included in A as subsets.*

The proof results from the following facts: Let F_n be the formula

$$\sum_{1 \leq i < k \leq n} (a_i \supset\subset a_k),$$

where Σ denotes the iterated $\vee$-connective and the a_i are propositional variables. F_n is satisfied in every realization with fewer than n elements in which all formulas provable in H are satisfied. For, with every assignment, a_i and a_k are replaced by the same element e in at least one of the summands of F_n, and $e \supset\subset e . \vee b$ yields a designated value for arbitrary b, since the formula $a \supset\subset a . \vee b$ is provable in H. Further, let S_n be the following realization:

Elements: $\{1, 2, \ldots, n\}$; designated element: 1;

$a \vee b = \min(a, b); \quad a \wedge b = \max(a, b);$

$a \supset b = 1$ for $a \geq b; \quad a \supset b = b$ for $a < b;$

$\neg a = n$ for $a \neq n, \quad \neg n = 1.$

Then, for S_n, all formulas of H and the formula F_{n+1}, as well as all F_i with greater subscript, are satisfied, while F_n, as well as all F_i with smaller subscript, are *not* satisfied. In particular, it follows that *no* F_n is provable in H. Besides, the following holds with full generality: a formula of the form $A \vee B$ can only be provable in H if either A or B is provable in H.

Introductory note to *1932a, 1933i* and *l*

This note discusses Gödel's published writings on decision problems for quantification theory.[a] A class of quantificational formulas is said to be *decidable* if and only if there is an effective procedure that determines, for each formula in the class, whether or not it is satisfiable. Prior to the rigorous explications of "effective" in the mid-1930s, various classes were shown decidable by exhibiting decision procedures that are clearly effective in any intuitive sense of the word. In this period, though, there could be no question of undecidability proofs. Instead, various classes were shown to be *reduction classes*. That is, the decision problem for all of quantification theory was reduced to the problem for these classes. The undecidability of reduction classes could be inferred once the recursive undecidability of all of quantification theory was shown (*Church 1936a, Turing 1937*).

The classes Gödel considers are *prefix classes*, that is, classes of closed, prenex formulas demarcated by the form of the quantifier prefix. Moreover, with one exception discussed below, the formulas are in pure quantification theory, or, as Gödel puts it, the restricted functional calculus: they contain predicate letters of arbitrary degree, but lack constants, function signs, and the identity predicate.

In *1928a* Ackermann showed the $\exists \ldots \exists \forall \exists \ldots \exists$ class to be decidable, and in *1920* Skolem showed the $\forall \ldots \forall \exists \ldots \exists$ class to be a reduction class.[b] In *1932a* and *1933i* Gödel strengthens these results by showing the $\exists \ldots \exists \forall \forall \exists \ldots \exists$ class to be decidable and the $\forall \forall \forall \exists \ldots \exists$ class to be a reduction class. A reasonably sharp boundary between decidable and undecidable is thereby exhibited: two contiguous universal quantifiers are not powerful enough to yield undecidability, whereas three are.

Gödel's reduction proof is relatively straightforward. Starting with an $\forall \ldots \forall \exists \ldots \exists$ formula, he reduces the number of universal quantifiers by stages. At each stage, new clauses are conjoined to the formula to encode the existence of a one-one correlation of elements of the universe with pairs of elements. Once this is done, the work of two universal quantifiers can be done by one quantifier ranging over elements that are correlated with pairs; hence the number of universal quantifiers can

---

[a] Aside from the writings discussed here, Gödel published two reviews (*1932l, 1934a*) of articles in this area. Both are short summaries of the results obtained in the articles under review (*Kalmár 1932, Skolem 1933*).

[b] The notation used here for prefix classes should be clear. For example, the prefixes of formulas in the $\exists \ldots \exists \forall \exists \ldots \exists$ class consist of any number of existential quantifiers followed by one universal quantifier followed by any number of existential quantifiers.

be reduced. But since three universal quantifiers are used to encode the required information about the correlation, the number of universal quantifiers cannot be reduced below three.

An encoding of a correlation between pairs of elements and elements was in fact first used by Löwenheim (*1915*) to show that the class of formulas containing only dyadic predicate letters is a reduction class. Gödel's proof amounts to an adaption of Löwenheim's, with special care taken to minimize the number of new quantifiers needed for the encoding.

Refinements of this sort of reduction can serve to pare the number of existential quantifiers as well. Efforts along these lines culminated in *Surányi 1950*, where it is shown that the $\forall\forall\exists$ class is undecidable. A rather different approach, however, was used by Kahr, Moore and Wang in their *1962* to obtain the undecidability of the $\forall\exists\forall$ class.

The two undecidability results just mentioned, together with the decidability of the $\exists\ldots\exists\forall\forall\exists\ldots\exists$ class proved by Gödel and of the $\exists\ldots\exists\forall\ldots\forall$ class proved earlier by Bernays and Schönfinkel (*1928*), yield an exhaustive classification of the prefix classes into decidable and undecidable: a prefix class if undecidable if and only if it allows either at least two universal quantifiers separated by an existential quantifier, or else at least three universal quantifiers governing an existential quantifier.

Gödel begins his decidability proof for the $\exists\ldots\exists\forall\forall\exists\ldots\exists$ class in *1932a* by reducing that class to the $\forall\forall\exists\ldots\exists$ class, using a technique for eliminating initial existential quantifiers due to Ackerman (*1928a*). He then provides an effective criterion for the satisfiability of any $\forall\forall\exists\ldots\exists$ formula. This criterion is in essence the same as that formulated independently by Kalmár (*1933*) and by Schütte (*1934*), and is described succinctly by Gödel in his review (*1933l*) of Kalmár's paper. We shall briefly indicate its form. Let $F = \forall x \forall y \exists z_1 \ldots \exists z_m H$ be an $\forall\forall\exists\ldots\exists$ formula, where H is quantifier-free. For $k > 0$, a *k-table* is an interpretation of the predicate letters of F over the universe $\{1, \ldots, k\}$. The Gödel–Kalmár–Schütte Criterion requires the existence of a nonempty set $\mathbf{T}$ of $(m+2)$-tables such that, first, $T \models H[1, 2, 3, \ldots, m+2]$ for each $T \in \mathbf{T}$ and, second, $\mathbf{T}$ fulfills certain closure conditions. The Criterion is effective, since there are only finitely many $(m + 2)$-tables. Gödel, Kalmár, and Schütte each show that such a set $\mathbf{T}$ can be extracted from any model for F and, conversely, given such a set a model for F can be inductively constructed. Hence fulfillment of the Criterion is equivalent to satisfiability.

As Gödel notes in *1933i*, his argument in *1932a* contains a lacuna: in the inductive construction of the model, he overlooks the case in which the two universal variables take the same value. Kalmár handles this case by incorporating into the Criterion a closure condition that

Gödel lacks; but in *1933i* Gödel fills the lacuna in a different way. Let H' be obtained from the matrix H of F by replacing y with x, and let $G = \forall x \forall y \exists z_1 \ldots \exists z_n J$ be a prenex form of $\forall x \forall y (\exists z_1 \ldots \exists z_m H \,\&\, \exists z_1 \ldots \exists z_m H')$. Then G is equivalent to F; moreover, a structure $\mathfrak{A}$ of cardinality ≥ 2 will be a model for G just in case $\mathfrak{A} \models \exists z_1 \ldots \exists z_n J(a,b)$ for all *distinct* members a and b of the universe. Thus the case overlooked in *1932a* need not be treated at all; applied to G, the proof of the sufficiency of the Criterion as formulated by Gödel goes through.

The bulk of *1933i* is devoted to a question beyond that of decidability. Call a class of quantificational formulas *finitely controllable* if and only if every satisfiable formula in it has a finite model. A syntactically specified class is decidable if it is finitely controllable, but the converse does not hold in general. Gödel proves the finite controllability of the $\forall\forall\exists\ldots\exists$ class. Since Ackermann's technique for eliminating initial existential quantifiers preserves satisfiability over finite universes, the finite controllability of the $\exists\ldots\exists\forall\forall\exists\ldots\exists$ class may be inferred. This shows that every decidable prefix class is in fact finitely controllable.

Gödel's proof, although concise, is rather subtle. Let $G = \forall x \forall y \exists z_1 \ldots \exists z_n J$ be as above, and suppose G is satisfiable. Thus there is a set $\mathbf{T}$ of $(n+2)$-tables that fulfills the Gödel–Kalmár–Schütte Criterion. Gödel then constructs a finite universe $\mathfrak{J}$ and a correlation of a number of distinct n-tuples with each pair of distinct elements of $\mathfrak{J}$; each such n-tuple is, moreover, associated with a member of $\mathbf{T}$. Two ingenious lemmas of finite combinatorics (Lemmas 1 and 2, pages 311ff. below) allow this correlation to be made so that no n-tuple is correlated with both a pair $\langle a, b \rangle$ and a pair $\langle c, a \rangle$; and if $\langle c_1, \ldots, c_n \rangle$ is correlated with $\langle a, b \rangle$ then neither a nor b occurs in any n-tuple correlated with a pair $\langle c_i, d \rangle$ or $\langle d, c_i \rangle$. These properties, in turn, allow a univocal definition of interpretations M of the predicate letters over the universe $\mathfrak{J}$ so that, if $\langle c_1, \ldots, c_n \rangle$ is correlated with $\langle a, b \rangle$, then $c_1, \ldots, c_n$ are "potential values" of the variables $z_1, \ldots, z_n$ for values a and b of x and y. That is, the relations between the c_i and a, between the c_i and b, and the relations among the c_i match—in an appropriate sense—the relations given by whatever $(n+2)$-table in $\mathbf{T}$ is associated with $\langle c_1, \ldots, c_n \rangle$. The closure properties of $\mathbf{T}$ given in the Gödel–Kalmár–Schütte Criterion then insure that for all distinct a and b there will be some T in $\mathbf{T}$ and some associated n-tuple $\langle c_1, \ldots, c_n \rangle$ such that *all* relations among $a, b, c_1, \ldots, c_n$ match those given by T. This yields $M \models J(a, b, c_1, \ldots, c_n)$. Thus, for all distinct a and b in $\mathfrak{J}$, $M \models \exists z_1 \ldots \exists z_n J(a, b)$, so that M is a finite model for G.

Subsequent to Gödel's work, other proofs of the finite controllability of the $\forall\forall\exists\ldots\exists$ class were devised: that of *Schütte 1934a* independently, and those of *Ackermann 1954*, *Dreben 1962*, and *Dreben and Goldfarb 1979* in the light of Gödel's. These proofs are all, at bottom, similar; they

rely on at least some of the combinatorial devices Gödel originated, in particular his Lemma 2. Only recently has a proof that dispenses with these devices been formulated (*Gurevich and Shelah 1983*). Gödel's combinatorial lemmas have been generalized: Lemma 1 in *Aanderaa and Goldfarb 1974* and Lemma 2 in *Erdös 1963* (simplified proofs are in *Dreben and Goldfarb 1979*, Appendix §2). Such generalizations have been applied in finite controllability proofs for other classes of quantificational formulas (*Aanderaa and Goldfarb 1974*; *Dreben and Goldfarb 1979*, Chapters 3 and 4).

In the last sentence of *1933i* Gödel claims, with no substantiation, that the same method as used for the ∀∀∃...∃ class suffices to show the finite controllability of the ∀∀∃...∃ class extended by the inclusion of the identity sign "=" in the language. In the mid-1960s, however, Stål Aanderaa devised some examples that showed the Gödel–Kalmár–Schütte Criterion not to be sufficient for the satisfiability of ∀∀∃...∃ formulas that contain "=". Thus it became unclear how Gödel might have intended his method to be applied, and on 24 May 1966 Burton Dreben wrote Gödel to inquire. Gödel responded on 19 July 1966,

> I am sorry I don't have any notes about the exact procedure for proving Theorem 1 of my paper [[*1933i*]] in case the formula contains "=". However, I remember that the idea was to formulate the auxiliary concepts and the lemmas under the assumption that in addition to the relations F_i an equivalence relation leaving the F_i invariant is given. No difficulty arose in carrying this through.

Gödel let the matter rest there until 1970. Then, on 3 April 1970, Dana Scott wrote to Dreben and to Hao Wang:

> In a recent telephone conversation [[Gödel]] mentioned that he was able to recall the method by which he dealt with the presence of *equality* in the matrix of the formula in his solvable prefix class. Since both of you have thought about this question, he asked me to write you the idea.

The idea Scott presents is fairly simple. Let, for example, $F = \forall x \forall y \exists z_1 H$; then one can easily find a formula G equivalent to F over universes of cardinality ≥ 4 that has the form

$$(*) \qquad \forall x \forall y [x \neq y \to \exists z_1 \exists z_2 (z_1 \neq y \ \& \ z_2 \neq x \ \& \ z_2 \neq y \ \& \ J)],$$

where J is a quantifier-free formula that lacks the identity sign. If we use special quantifiers (called "sharp" by Scott) that demand that the value of each quantified variable be distinct from the values of the previous quantified variables, then G can be written

$(**)$ $\qquad\qquad\qquad\qquad \forall^\# x \forall^\# y \exists^\# z_1 \exists^\# z_2 J.$

Scott continues, "Finally, Gödel claims that his original method applies unchanged to sharp formulas."

Unfortunately, this assertion is false. For, if we apply the Gödel–Kalmár–Schütte Criterion to G by treating J as though it were the matrix of G, then the Criterion fails to be necessary for satisfiability. If, on the other hand, we apply the Criterion to G and take the matrix to be that of a prenex form of $(*)$, which does contain "$=$", then the Criterion is not sufficient for satisfiability.

By the early 1970s it had become widely recognized that the decision problem for the $\forall\forall\exists\ldots\exists$ class with identity was open. It was settled in 1983 by the present author. Contrary to Gödel's claim, this class is undecidable; indeed, even the $\forall\forall\exists$ class with identity is undecidable (*Goldfarb 1984, 1984a*). The heart of the undecidability proof is the construction of a satisfiable formula F in the $\forall\forall\exists$ class with identity with the following property: for some dyadic predicate letter S of F, every model for F contains an ω-sequence $a_0, a_1, a_2, \ldots$ of distinct elements on which the interpretation of S must be that of "successor". This shows at once that the $\forall\forall\exists$ class with identity is not finitely controllable. Moreover, the formula F may be exploited to obtain encodings, by $\forall\forall\exists$ formulas with identity, of computational processes whose halting

---

Ein Spezialfall des Entscheidungsproblems der theoretischen Logik
(*1932a*)

Im Anschluß an die für den Beweis der Vollständigkeit des Funktionenkalküls verwendete Methode (*1930*) kann man ein Verfahren entwickeln, welches für jede Formel mit einer Normalform der Gestalt:

$$(Ex_1, \ldots, x_n)(y_1, y_2)(Ez_1, \ldots, z_m)A(x_i, y_i, z_i)$$

zu entscheiden gestattet, ob sie erfüllbar ist. Man führt das Problem zunächst in bekannter Weise[1] auf Formeln der Gestalt:

$$(x_1, x_2)(Ex_3, \ldots, x_n)F(x_1, \ldots, x_n) \tag{1}$$

---

[1] Vgl. *Ackermann 1928a*, S. 647.

problem is undecidable.

Gödel's error may have been engendered by the manner in which he fills the lacuna of his *1932a* argument. Recall that to do this he moves from, for example, a formula $F = \forall x \forall y \exists z_1 H$ to a prenex form $G = \forall x \forall y \exists z_1 \exists z_2 J$ of $\forall x \forall y (\exists z_1 H \wedge \exists z_1 H')$, where H' comes from H by replacement of y with x. He infers that

(1) if F is satisfiable, then the Gödel–Kalmár–Schütte Criterion holds for G.

This is justified since F is equivalent to G. He also infers that

(2) if the Criterion holds for G, then F is satisfiable.

This is true since F is also equivalent to $\forall x \forall y (x \neq y \rightarrow \exists z_1 \exists z_2 J)$. Now, if the original formula F contains "=", then in order to eliminate "=" from J (so that the Criterion will be sufficient for satisfiability), one must replace all occurrences of "$x = y$" by a sign for falsity. But, once J is transformed in this manner, F is no longer equivalent to $\forall x \forall y \exists z_1 \exists z_2 J$; indeed, the latter may be unsatisfiable even when F is satisfiable. Thus (1) fails. In view both of the fact that Gödel did not argue for (1) in the case without "="—no doubt thinking it obvious—and of the suggestion contained in Scott's letter, it seems likely that in 1933 Gödel simply overlooked the difference that the presence of "=" makes at this juncture.

Warren D. Goldfarb

A special case of the decision problem for theoretical logic
(*1932a*)

In connection with the method used to prove the completeness of the functional calculus (*1930*), one can develop a procedure that, for each formula with a normal form of the form

$$(Ex_1, \ldots, x_n)(y_1, y_2)(Ez_1, \ldots, z_m)A(x_i, y_i, z_i),$$

permits us to decide whether it is satisfiable. First, one reduces the problem in a well known way[1] to formulas of the form

$$(x_1, x_2)(Ex_3, \ldots, x_n)F(x_1, \ldots, x_n). \tag{1}$$

---

[1] See *Ackermann 1928a*, p. 647.

zurück. Wir setzen zunächst voraus, daß in F nur zweistellige Funktions-variable, und zwar $\phi_1, \phi_2, \ldots, \phi_s$ vorkommen. Sei A die Menge der Formeln $\phi_i(x_1, x_1)$, $i = 1, 2, \ldots, s$; B die Menge der Formeln $\phi_i(x_p, x_q)$, $i = 1, 2, \ldots, s$; $p = 1, 2$; $q = 1, 2$. Es gilt $A \subset B$. Eine eindeutige Zuordnung von Wahrheitswerten $+, -$ zu den Formeln aus A (bzw. B) bezeichnen wir allgemein $\mathfrak{A}_i$ ($i = 1, 2, \ldots$ usw.) bzw. | $\mathfrak{B}_i$ ($i = 1, 2, \ldots$ usw.)[2] (es gibt nur endlichviele $\mathfrak{A}_i$ und $\mathfrak{B}_i$). Wenn eine Realisierung der ϕ_i in einem Individuenbereich $\mathfrak{J}$ gegeben ist, so liefert jedes Element $a \; \epsilon \; \mathfrak{J}$ bzw. je zwei Elemente $a, b \; \epsilon \; \mathfrak{J}$ ein bestimmtes $\mathfrak{A}_i$ bzw. $\mathfrak{B}_j$, wenn man a mit x_1 und b mit x_2 identifiziert. Dieses $\mathfrak{A}_i, \mathfrak{B}_j$ bezeichnen wir auch mit $\mathfrak{A}_a, \mathfrak{B}_{a,b}$. Falls ein $\mathfrak{A}_k$ mit einem $\mathfrak{B}_i$ für die Menge A übereinstimmt, schreiben wir $\mathfrak{A}_k \subset \mathfrak{B}_i$. Mit $\mathfrak{B}'_i$ bezeichnen wir dasjenige $\mathfrak{B}_j$, welches aus $\mathfrak{B}_i$ durch Vertauschung von x_1 mit x_2 hervorgeht.

Sei ein System von Funktionen $\psi_1, \ldots, \psi_s$ in einem Individuenbereich $\mathfrak{J}$ gegeben, welches die Formel (1) erfüllt. Wir bilden die Menge $\mathfrak{P}$ bzw. $\mathfrak{Q}$ derjenigen $\mathfrak{A}_i$ bzw. $\mathfrak{B}_j$, welche man erhält, wenn man in $\mathfrak{A}_a$ (bzw. $\mathfrak{B}_{a,b}$) a, b alle Individuen aus $\mathfrak{J}$ durchlaufen läßt. Die (endlichen und nicht leeren) Mengen $\mathfrak{P}, \mathfrak{Q}$ genügen offenbar folgenden Bedingungen:

I. Wenn $\mathfrak{A}_i \; \epsilon \; \mathfrak{P}$ und $\mathfrak{A}_k \; \epsilon \; \mathfrak{P}$, so gibt es ein $\mathfrak{B}_j \; \epsilon \; \mathfrak{Q}$, so daß $\mathfrak{A}_i \subset \mathfrak{B}_j$ und $\mathfrak{A}_k \subset \mathfrak{B}'_j$.

II. Wenn $\mathfrak{B}_j \; \epsilon \; \mathfrak{Q}$, so gibt es ein für den Bereich von n Individuen $a_1, a_2, \ldots, a_n$ definiertes System von Funktionen $\chi_1, \chi_2, \ldots, \chi_s$, so daß:

    1. $\mathfrak{B}_j = \mathfrak{B}_{a_1, a_2}$.

    2. $F(x_1, \ldots, x_n)$ den Wahrheitswert $+$ erhält, wenn man darin die ϕ_i durch die χ_i und die x_i durch die a_i ersetzt.

    3. $\mathfrak{B}_{a_i, a_k} \; \epsilon \; \mathfrak{Q}$ und $\mathfrak{A}_{a_i} \; \epsilon \; \mathfrak{P}$ für alle $i, k = 1, 2, \ldots, n$.

Ob es für eine vorgegebene Formel der Gestalt (1) ein Paar $\mathfrak{P}, \mathfrak{Q}$ nicht leerer Mengen gibt, welches den Bedingungen I, II genügt, kann man offenbar immer in endlichvielen Schritten entscheiden (weil es nur endlichviele $\mathfrak{A}_i, \mathfrak{B}_k$ gibt). Existiert ein solches Paar *nicht*, so ist die vorgegebene Formel *nicht* erfüllbar. Existiert ein solches Paar, dann ist die Formel erfüllbar. Denn es gilt folgendes: Wenn $\mathfrak{P}, \mathfrak{Q}$ den Bedingungen I, II genügen und wenn für die in einem endlichen Bereich $\mathfrak{R}$ definierten Funktionen $\phi_i, \ldots, \phi_s$ $\mathfrak{A}_x \; \epsilon \; \mathfrak{P}$ und $\mathfrak{B}_{x,y} \; \epsilon \; \mathfrak{Q}$ für beliebige $x, y \; \epsilon \; \mathfrak{R}$, dann kann man, wenn zwei beliebige Individuen a, b aus $\mathfrak{R}$ vorgegeben sind, die Funktionen ϕ_i auf einen durch eine Menge $\mathfrak{R}'$ von $n - 2$ Individuen $(r'_1, \ldots, r'_{n-2})$ vermehrten Bereich $\mathfrak{R} + \mathfrak{R}'$ so erweitern, daß $F(a, b, r'_1, \ldots, r'_{n-2})$ richtig ist und daß auch für den erweiterten Bereich immer $\mathfrak{A}_x \; \epsilon \; \mathfrak{P}$ und $\mathfrak{B}_{x,y} \; \epsilon \; \mathfrak{Q}$.

Denn man kann die ϕ_i zunächst wegen II für die Elementepaare der Menge $\{a, b\} + \mathfrak{R}'$ und dann wegen I für die Paare (x, y) mit $x \; \epsilon \; \mathfrak{R} - \{a, b\}$, $y \; \epsilon \; \mathfrak{R}'$ so definieren, daß die geforderten Bedingungen erfüllt sind.

---

[2] D. h. wir numerieren die obigen Zuordnungen irgendwie und bezeichnen die i-te mit $\mathfrak{A}_i$ (bzw. $\mathfrak{B}_i$).

We initially assume that only two-place functional variables occur in F, namely $\phi_1, \phi_2, \ldots, \phi_s$. Let A be the set of formulas $\phi_i(x_1, x_1)$, where $i = 1, 2, \ldots, s$; and let B be the set of formulas $\phi_i(x_p, x_q)$, where $i = 1, 2, \ldots, s$ and $p = 1, 2$ and $q = 1, 2$. Then $A \subset B$ holds. A single-valued assignment of truth values $+, -$ to the formulas of A (or B) will be generally denoted by $\mathfrak{A}_i$ ($i = 1, 2, \ldots$, etc.) (or $\mathfrak{B}_i$ ($i = 1, 2, \ldots$, etc.), respectively)[2] (there being only finitely many $\mathfrak{A}_i$ and $\mathfrak{B}_i$). If a realization of the ϕ_i in a domain $\mathfrak{J}$ of individuals is given, then each element $a \in \mathfrak{J}$ yields a certain $\mathfrak{A}_i$ and any two elements $a, b \in \mathfrak{J}$ yield a certain $\mathfrak{B}_j$ if a is identified with x_1 and b with x_2. We also denote this pair $\mathfrak{A}_i, \mathfrak{B}_j$ by $\mathfrak{A}_a, \mathfrak{B}_{a,b}$. In case an $\mathfrak{A}_k$ coincides with a $\mathfrak{B}_i$ on the set A, we write $\mathfrak{A}_k \subset \mathfrak{B}_i$. We denote by $\mathfrak{B}_i'$ that $\mathfrak{B}_j$ which results from $\mathfrak{B}_i$ when x_1 and x_2 are interchanged.

Let a system $\psi_1, \ldots, \psi_s$ of functions on a domain $\mathfrak{J}$ of individuals be given which satisfy formula (1). We form the set $\mathfrak{P}$ (or $\mathfrak{Q}$, respectively) of those $\mathfrak{A}_i$ (or $\mathfrak{B}_j$, respectively) that one obtains if in $\mathfrak{A}_a$ (or $\mathfrak{B}_{a,b}$, respectively) one allows a and b to run through all individuals of $\mathfrak{J}$. The (finite and non-empty) sets $\mathfrak{P}$ and $\mathfrak{Q}$ obviously satisfy the following conditions:

I. If $\mathfrak{A}_i \in \mathfrak{P}$ and $\mathfrak{A}_k \in \mathfrak{P}$, then there is a $\mathfrak{B}_j \in \mathfrak{Q}$ such that $\mathfrak{A}_i \subset \mathfrak{B}_j$ and $\mathfrak{A}_k \subset \mathfrak{B}_j'$.

II. If $\mathfrak{B}_j \in \mathfrak{Q}$, then there is a system of functions $\chi_1, \chi_2, \ldots, \chi_s$, defined on the domain of n individuals $a_1, a_2, \ldots, a_n$, such that:

    1. $\mathfrak{B}_j = \mathfrak{B}_{a_1, a_2}$;

    2. $F(x_1, \ldots, x_n)$ receives the truth-value $+$ if in it we replace the ϕ_i by the χ_i and the x_i by the a_i;

    3. $\mathfrak{B}_{a_i, a_k} \in \mathfrak{Q}$ and $\mathfrak{A}_{a_i} \in \mathfrak{P}$ for all $i, k = 1, 2, \ldots, n$.

Since there are only finitely many $\mathfrak{A}_i, \mathfrak{B}_k$, one can obviously always decide in finitely many steps whether, for a given formula of the form (1), there is a pair $\mathfrak{P}, \mathfrak{Q}$ of non-empty sets that satisfies conditions I and II. If such a pair does *not* exist, the given formula is *not* satisfiable. If such a pair does exist, then the formula is satisfiable. For the following holds: If $\mathfrak{P}$ and $\mathfrak{Q}$ satisfy conditions I and II, and if, for the functions $\phi_1, \ldots, \phi_s$ defined on a finite domain $\mathfrak{R}$, $\mathfrak{A}_x \in \mathfrak{P}$ and $\mathfrak{B}_{x,y} \in \mathfrak{Q}$ for arbitrary $x, y \in \mathfrak{R}$, then, given two arbitrary individuals a and b of $\mathfrak{R}$, one can extend the functions ϕ_i to a domain $\mathfrak{R} + \mathfrak{R}'$ (which is $\mathfrak{R}$ augmented by a set $\mathfrak{R}'$ of $n - 2$ individuals $r_1', \ldots, r_{n-2}'$) so that $F(a, b, r_1', \ldots, r_{n-2}')$ is correct and so that $\mathfrak{A}_x \in \mathfrak{P}$ and $\mathfrak{B}_{x,y} \in \mathfrak{Q}$ always hold for the extended domain as well.

For one can define the ϕ_i first, because of II, for pairs of elements of the set $\{a, b\} + \mathfrak{R}'$, and then, because of I, for the pairs (x, y) with $x \in \mathfrak{R} - \{a, b\}$ and $y \in \mathfrak{R}'$, so that the required conditions are satisfied.

---

[2]That is, we number the assignments above in some way and denote the ith one by $\mathfrak{A}_i$ (or $\mathfrak{B}_i$, respectively).

Falls in F mehr als zweistellige Funktionen vorkommen, hat man unter B (bzw. A) die Menge der Formeln zu verstehen, welche man aus den ϕ_i erhält, indem man auf alle möglichen Arten die Variablen x_1, x_2 (bzw. x_1) an die Leerstellen einsetzt, und kann dann ebenso schließen.

Über Vollständigkeit und Widerspruchsfreiheit
(*1932b*)

Sei Z das formale System, welches man erhält, indem man die Peanoschen Axiome durch das Schema der rekursiven Definition (nach einer Variablen) und die logischen Regeln des *engeren* Funktionenkalküls ergänzt. Z soll also keine anderen Variablen als solche für Individuen (d. h. natürliche Zahlen) enthalten und das Prinzip der vollständigen Induktion muß daher als Schlußregel formuliert werden. Dann gilt:

1. Jedes Z umfassende[1] formale System S mit endlich vielen Axiomen und der Einsetzungs- und Implikationsregel als einzigen Schlußprinzipien ist unvollständig, d. h. es gibt darin Sätze (und zwar Sätze aus Z), die aus den Axiomen von S unentscheidbar sind, vorausgesetzt, daß S ω-widerspruchsfrei ist. Dabei heiße ein System ω-widerspruchsfrei, wenn für keine Eigenschaft F natürlicher Zahlen zugleich $(Ex)\overline{F(x)}$ und sämtliche Formeln $F(i)$, $i = 1, 2, \ldots$ usw. beweisbar sind.

2. In jedem solchen System S ist insbesondere die Aussage, daß S widerspruchsfrei ist (genauer die mit ihr äquivalente arithmetische Aussage, welche man erhält, indem man die Formeln ein-eindeutig auf natürliche Zahlen abbildet), unbeweisbar.

Satz 1 und 2 gelten auch für Systeme mit unendlich vielen Axiomen und anderen Schlußprinzipien als den genannten, vorausgesetzt, daß bei einer Durchnumerierung der Formeln (nach wachsender Länge und bei gleicher Länge lexikographisch) die Klasse der den Axiomen zugeordneten Nummern, sowie die folgende Relation $R(x_1, x_2, \ldots, x_n)$ zwischen natürlichen Zahlen: "Die Formel mit der Nummer x_1 folgt aus den Formeln mit den Nummern $x_2, \ldots, x_n$ durch einmalige Anwendung einer der Schlußregeln" im System Z enthalten und entscheidungsdefinit sind. Dabei heißt eine Relation (Klasse) $R(x_1, \ldots, x_n)$ entscheidungsdefinit in Z, wenn für jedes n-tupel von natürlichen Zahlen $(k_1, \ldots, k_n)$ entweder $R(k_1, \ldots, k_n)$ oder $\overline{R}(k_1, \ldots, k_n)$ in Z beweisbar ist (bisher ist keine entscheidungsdefinite

---

[1] Ein formales System S umfaßt ein anderes T, soll bedeuten: Jeder in T ausdrückbare (beweisbare) Satz ist auch in S ausdrückbar (beweisbar).

In case functions of more than two arguments occur in F, one must take B (or A, respectively) to be the set of formulas that one obtains from the ϕ_i by inserting the variables x_1, x_2 (or x_1, respectively) into the argument places in all possible ways, and one can then reason just as before.

On completeness and consistency
(*1932b*)

[The introductory note to *Gödel 1932b*, as well as to related items, can be found on page 126, immediately preceding *1930b*.]

Let Z be the formal system that we obtain by supplementing the Peano axioms with the schema of definition by recursion (on one variable) and the logical rules of the *restricted* functional calculus. Hence Z is to contain no variables other than variables for individuals (that is, natural numbers), and the principle of mathematical induction must therefore be formulated as a rule of inference. Then the following hold:

1. Given any formal system S in which there are finitely many axioms and in which the sole principles of inference are the rule of substitution and the rule of implication, if S contains[1] Z, S is incomplete, that is, there are

---

[1] That a formal system S contains another formal system T means that every proposition expressible (provable) in T is expressible (provable) also in S.

[Remark by the author, 18 May 1966:] [This definition is not precise, and, if made precise in the straightforward manner, it does not yield a sufficient condition for the nondemonstrability in S of the consistency of S. A sufficient condition is obtained if one uses the following definition: "S contains T if and only if every meaningful formula (or axiom or rule (of inference, of definition, or of construction of axioms)) of T *is* a meaningful formula (or axiom, and so forth) of S, that is, if S is an extension of T."

Under the weaker hypothesis that Z is recursively one-to-one translatable into S, with demonstrability preserved in this direction, the consistency, even of very strong systems S, *may* be provable in S and even in primitive recursive number theory. However, what can be shown to be unprovable in S is the fact that the rules of the equational calculus applied to equations, between primitive recursive terms, demonstrable in S yield only correct numerical equations (provided that S possesses the property that is asserted to be unprovable). Note that it is necessary to prove this "outer" consistency of S (which for the usual systems is trivially equivalent with consistency) in order to "justify", in the sense of Hilbert's program, the transfinite axioms of a system S. ("Rules of the equational calculus" in the foregoing means the two rules of substituting primitive recursive terms for variables and substituting one such term for another to which it has been proved equal.)

The last-mentioned theorem and Theorem 1 of the paper remain valid for much weaker systems than Z, in particular for primitive recursive number theory, that is, what remains of Z if quantifiers are omitted. With insignificant changes in the wording of the conclusions of the two theorems they even hold for any recursive translation into S of the equations between primitive recursive terms, under the sole hypothesis of ω-consistency (or outer consistency) of S in this translation.]

Zahlrelation bekannt, die nicht schon in Z enthalten und entscheidungsdefinit wäre).

Denkt man sich das System Z sukzessive dadurch erweitert, daß man
Variable für Klassen von Zahlen, Klassen von Klassen von Zahlen usw.
sowie die zugehörigen Komprehensionsaxiome einführt, so erhält man eine
(ins Transfinite fortsetzbare) Reihe von formalen Systemen, die den obigen
Voraussetzungen genügen, und es zeigt sich, daß die Widerspruchsfreiheit
(ω-Widerspruchsfreiheit) der vorhergehenden innerhalb der folgenden beweisbar ist. Auch die zum Beweise von Satz 1 konstruierten unentscheid-
baren Sätze werden | durch Hinzunahme höherer Typen und der zugehöri-
gen Axiome entscheidbar; doch kann man in den höheren Systemen nach
demselben Verfahren andere unentscheidbare Sätze konstruieren usf. Alle
so konstruierten Sätze sind zwar in Z ausdrückbar (also zahlentheoretische
Sätze), aber nicht in Z, sondern erst in höheren Systemen, z. B. dem der
Analysis entscheidbar. Falls man die Mathematik typenlos aufbaut, wie
es im Axiomensystem der Mengenlehre geschieht, so treten an Stelle der
Typenerweiterungen Mächtigkeitsaxiome (d. h. Axiome, welche die Existenz von Mengen immer höherer Mächtigkeit fordern) und es ergibt sich
daraus, daß gewisse in Z unentscheidbare arithmetische Sätze durch Mächtigkeitsaxiome entscheidbar werden, z. B. durch das Axiom, daß es Mengen
gibt, deren Mächtigkeit größer ist als jedes α_n, wenn $\alpha_0 = \aleph_0$, $\alpha_{n+1} = 2^{\alpha_n}$.

in S propositions (in particular, propositions of Z) that are undecidable on the basis of the axioms of S, provided that S is ω-consistent.

Here a system is said to be ω-consistent if, for no property F of natural numbers, $(Ex)\overline{F(x)}$ as well as all the formulas $F(i)$, $i = 1, 2, \ldots$ are provable.

2. In particular, in every system S of the kind just mentioned the proposition that S is consistent (more precisely, the equivalent arithmetic proposition that we obtain by mapping the formulas one-to-one on natural numbers) is unprovable.

Theorems 1 and 2 hold also for systems in which there are infinitely many axioms and in which there are other principles of inference than those mentioned above, provided that when we enumerate the formulas (in order of increasing length and, for equal length, in lexicographical order) the class of numbers assigned to the axioms is definable and decidable in the system Z, and that the same holds of the following relation $R(x_1, x_2, \ldots, x_n)$ between natural numbers: "the formula with number x_1 follows from the formulas with numbers $x_2, \ldots, x_n$ by a single application of one of the rules of inference". Here a relation (class) $R(x_1, x_2, \ldots, x_n)$ is said to be decidable in Z if for every n-tuple $(k_1, k_2, \ldots, k_n)$ of natural numbers either $R(k_1, k_2, \ldots, k_n)$ or $\overline{R}(k_1, k_2, \ldots, k_n)$ is provable in Z. (At present no decidable number-theoretic relation is known that is not definable and decidable already in Z.)

If we imagine that the system Z is successively enlarged by the introduction of variables for classes of numbers, classes of classes of numbers, and so forth, together with the corresponding comprehension axioms, we obtain a sequence (continuable into the transfinite) of formal systems that satisfy the assumptions mentioned above, and it turns out that the consistency (ω-consistency) of any of those systems is provable in all subsequent systems. Also, the undecidable propositions constructed for the proof of Theorem 1 become decidable by the adjunction of higher types and the corresponding axioms; however, in the higher systems we can construct other undecidable propositions by the same procedure, and so forth. To be sure, all the propositions thus constructed are expressible in Z (hence are number-theoretic propositions); they are, however, not decidable in Z, but only in higher systems, for example, in that of analysis. In case we adopt a type-free construction of mathematics, as is done in the axiom system of set theory, axioms of cardinality (that is, axioms postulating the existence of sets of ever higher cardinality) take the place of the type extensions, and it follows that certain arithmetic propositions that are undecidable in Z become decidable by axioms of cardinality, for example, by the axiom that there exist sets whose cardinality is greater than every α_n, where $\alpha_0 = \aleph_0, \alpha_{n+1} = 2^{\alpha_n}$.

Introductory note to *1932c*

We easily impose a true-false dichotomy on the truth-functional formulas of propositional logic by assigning arbitrary truth values to all
the propositional letters and then letting the truth tables determine
the truth values of all compound formulas. But now let us imagine,
instead of the familiar formulas, a well-ordered but perhaps indenumerable supply. It is still to be closed under the operations '~' and '⊃', but
there need no longer be minimal units such as letters. Can we specify a
true-false dichotomy for these formulas, conformable still to the truth-
functional constraints on '~' and '⊃'? Gödel does so by a transfinite
recursion based on Lukasiewicz's three axioms (*Lukasiewicz and Tarski
1930*) and modus ponens.

Eine Eigenschaft der Realisierungen
des Aussagenkalküls
(*1932c*)

In Beantwortung einer mir von Menger brieflich gestellten Frage kann
man folgenden Satz beweisen:

Es sei irgend eine Menge $\mathfrak{S}$ von Dingen p, q, r,... gegeben, in der eine
einstellige Operation $\sim p$ und eine zweistellige Operation $p \supset q$ definiert
sind, die nicht aus der Menge $\mathfrak{S}$ herausführen. Es sei ferner $\mathfrak{T}$ eine echte
Teilmenge von $\mathfrak{S}$, die folgenden Bedingungen genügt:

21 | I. Falls p, q, r beliebige Dinge aus $\mathfrak{S}$ sind, gehören immer die folgenden
drei Dinge zu $\mathfrak{T}$:

(a) $$(\sim p \supset p) \supset p,$$

(b) $$p \supset (\sim p \supset q),$$

(c) $$(p \supset q) \supset [(q \supset r) \supset (p \supset r)].$$

II. Wenn sowohl p als $p \supset q$ zu $\mathfrak{T}$ gehören, dann gehört auch q zu $\mathfrak{T}$.

Unter diesen Voraussetzungen gibt es immer *eine Einteilung von $\mathfrak{S}$ in
zwei elementefremde Klassen: $\mathfrak{S} = \mathfrak{W} + \mathfrak{F}$, $\mathfrak{T} \subset \mathfrak{W}$, so daß*

(a) *von den beiden Dingen p und $\sim p$ immer genau eines zu $\mathfrak{W}$ und*

Lindenbaum had shown (see *Tarski 1930*) how any consistent set of closed formulas of first-order predicate logic can be extended to a true-false dichotomy. Gödel's result, though limited to propositional logic, is stronger in applying to the case of indenumerably many formulas. His method of proof is a transfinite extension of that used by Lindenbaum.

The paper is a pioneer venture in the metalogic of indenumerable notation. Maltsev reopened this topic in *1936*, apparently unaware of Gödel's paper. The compactness theorem, proved by Gödel for the denumerable case as a corollary to his completeness proof in *1930*, was extended by Maltsev to the indenumerable and shown to have fruitful applications in algebra.

W. V. Quine

A property of the realizations of the propositional calculus
(*1932c*)

In answer to a question that Menger asked me in a letter, one can demonstrate the following theorem:

Let any set $\mathfrak{S}$ of objects $p, q, r, \ldots$ be given on which there are defined a unary operation $\sim p$ and a binary operation $p \supset q$ that do not lead outside the set $\mathfrak{S}$. Further, let $\mathfrak{T}$ be a proper subset of $\mathfrak{S}$ satisfying the following conditions:

I. In case p, q, r are arbitrary objects of $\mathfrak{S}$, the following three objects always belong to $\mathfrak{T}$:

(a) $\qquad\qquad (\sim p \supset p) \supset p,$

(b) $\qquad\qquad p \supset (\sim p \supset q),$

(c) $\qquad\qquad (p \supset q) \supset [(q \supset r) \supset (p \supset r)].$

II. If both p and $p \supset q$ belong to $\mathfrak{T}$, then q also belongs to $\mathfrak{T}$.

Under these assumptions there always exists *a partition of $\mathfrak{S}$ into two disjoint classes $\mathfrak{W}$ and $\mathfrak{F}$ ($\mathfrak{S} = \mathfrak{W} + \mathfrak{F}$), with $\mathfrak{T} \subset \mathfrak{W}$, such that*

(a) *of any two objects p and $\sim p$, exactly one belongs to $\mathfrak{W}$ and one*

eines zu $\mathfrak{F}$ gehört,

(b)   das Ding $p \supset q$ dann und nur dann zu $\mathfrak{F}$ gehört, wenn p zu $\mathfrak{W}$
und q zu $\mathfrak{F}$ gehört.

Das heißt: Wenn eine beliebige Realisierung der Axiome des Aussagenkalküls gegeben ist, so können die Elemente ("Aussagen") in zwei fremde
Klassen eingeteilt werden, welche sich ganz so wie die Klassen der wahren
und der falschen Aussagen des gewöhnlichen Aussagenkalküls verhalten.

Beweisskizze: Man denke sich $\mathfrak{S}$ wohlgeordnet; dann ordne man jeder
Ordinalzahl α eine Teilklasse $\mathfrak{T}_\alpha$ von $\mathfrak{S}$ zu durch die Festsetzungen:

1. $\mathfrak{T}_0 = \mathfrak{T}$.

2. Wenn α Limeszahl ist, $\mathfrak{T}_\alpha = \Sigma_{\mu < \alpha} \mathfrak{T}_\mu$.

3. Wenn a das erste Element aus $\mathfrak{S}$ ist, für welches weder a noch $\sim a$ zu
$\mathfrak{T}_\alpha$ gehört: $\mathfrak{T}_{\alpha+1} =$ Menge derjenigen $x \, \epsilon \, \mathfrak{S}$, für welche $(a \supset x) \, \epsilon \, \mathfrak{T}_\alpha$. Falls
ein solches a nicht existiert, sei $\mathfrak{T}_{\alpha+1} = \mathfrak{T}_\alpha$.

Durch transfinite Induktion zeigt man:

A. Jedes $\mathfrak{T}_\alpha$ genügt den beiden Forderungen I und II, die oben für $\mathfrak{T}$
aufgestellt wurden.

B. $\mathfrak{T}_\alpha \subset \mathfrak{T}_\beta$ für $\alpha \leq \beta$.

C. x und $\sim x$ gehören niemals beide zu $\mathfrak{T}_\alpha$.

Wenn ν die kleinste Ordinalzahl ist, für welche $\mathfrak{T}_\nu = \mathfrak{T}_{\nu+1}$, so hat
$\mathfrak{W} = \mathfrak{T}_\nu$, $\mathfrak{F} = \mathfrak{S} - \mathfrak{T}_\nu$ die verlangten Eigenschaften.

B zeigt man z. B. folgendermaßen: Weil für $\mathfrak{T}_\alpha$ nach induktiver Annahme
A erfüllt ist, gilt für beliebige $x \, \epsilon \, \mathfrak{S} : x \supset . a \supset x \, \epsilon \, \mathfrak{T}_\alpha$ und daher folgt
aus $x \, \epsilon \, \mathfrak{T}_\alpha$ $a \supset x \, \epsilon \, \mathfrak{T}_\alpha$, d. h. $x \, \epsilon \, \mathfrak{T}_{\alpha+1}$, also $\mathfrak{T}_\alpha \subset \mathfrak{T}_{\alpha+1}$. Daß für $\mathfrak{W} = \mathfrak{T}_\nu$
Bedingung (b) erfüllt ist, folgt daraus, daß wegen A gilt: $x \supset . y \supset x \, \epsilon \, \mathfrak{T}_\nu$,
$\sim y \supset . y \supset x \, \epsilon \, \mathfrak{T}_\nu$, $\sim x \supset [y \supset . \sim (y \supset x)] \, \epsilon \, \mathfrak{T}_\nu$ und zwar für beliebige x,
$y \, \epsilon \, \mathfrak{S}$.

Besprechung von *Skolem 1931*:
Über einige Satzfunktionen in der Arithmetik
(*1932d*)

Der Verfasser betrachtet diejenigen Aussagefunktionen und Sätze,
welche sich mittels der fünf logischen Grundoperationen (nicht, oder, und,
alle, es gibt) aus Beziehungen der Form $A = B$, $A < B$ aufbauen, wobei
A, B zahlentheoretische Funktionen sind, die durch (beliebig iterierte) Anwendung der drei Operationen der Addition, der Multiplikation mit einer
Rationalzahl und der Bildung des nächst kleineren Ganzen entstehen. Als

to $\mathfrak{F}$;

(b) *the object* $p \supset q$ *belongs to* $\mathfrak{F}$ *if and only if* p *belongs to* $\mathfrak{W}$ *and* q *belongs to* $\mathfrak{F}$.

That is, if an arbitrary realization of the axioms of the propositional calculus is given, the elements ("propositions") can be divided into two disjoint classes that behave exactly like the classes of the true and the false propositions of the ordinary propositional calculus.

Sketch of the proof: Suppose $\mathfrak{S}$ to be well-ordered; then associate with each ordinal α a subclass $\mathfrak{T}_\alpha$ of $\mathfrak{S}$ according to the following stipulations:

1. $\mathfrak{T}_0 = \mathfrak{T}$.

2. If α is a limit ordinal, $\mathfrak{T}_\alpha = \Sigma_{\mu<\alpha}\mathfrak{T}_\mu$.

3. If a is the first element of $\mathfrak{S}$ for which neither a nor $\sim a$ belongs to $\mathfrak{T}_\alpha$, let $\mathfrak{T}_{\alpha+1}$ be the set of those $x \,\epsilon\, \mathfrak{S}$ for which $(a \supset x) \,\epsilon\, \mathfrak{T}_\alpha$. In case no such a exists, let $\mathfrak{T}_{\alpha+1}$ be $\mathfrak{T}_\alpha$.

By transfinite induction one shows that:

A. Every $\mathfrak{T}_\alpha$ satisfies both of the requirements I and II that were set up above for $\mathfrak{T}$.

B. $\mathfrak{T}_\alpha \subset \mathfrak{T}_\beta$ for $\alpha \leq \beta$.

C. x and $\sim x$ never both belong to $\mathfrak{T}_\alpha$.

If ν is the smallest ordinal for which $\mathfrak{T}_\nu = \mathfrak{T}_{\nu+1}$, then ⟦the partition⟧ $\mathfrak{W} = \mathfrak{T}_\nu$, $\mathfrak{F} = \mathfrak{S} - \mathfrak{T}_\nu$ has the required properties.

B can be shown, for example, as follows: Since $\mathfrak{T}_\alpha$ satisfies A by the induction hypothesis, $x \supset .a \supset x \,\epsilon\, \mathfrak{T}_\alpha$ holds for arbitrary $x \,\epsilon\, \mathfrak{S}$, and therefore from $x \,\epsilon\, \mathfrak{T}_\alpha$ it follows that $a \supset x \,\epsilon\, \mathfrak{T}_\alpha$, that is, $x \,\epsilon\, \mathfrak{T}_{\alpha+1}$, hence $\mathfrak{T}_\alpha \subset \mathfrak{T}_{\alpha+1}$. That condition (b) is satisfied by $\mathfrak{W} = \mathfrak{T}_\nu$ follows from the fact that, on account of A, $x \supset .y \supset x \,\epsilon\, \mathfrak{T}_\nu$, $\sim y \supset .y \supset x \,\epsilon\, \mathfrak{T}_\nu$ and $\sim x \supset [y \supset .\sim(y \supset x)] \,\epsilon\, \mathfrak{T}_\nu$ hold, and this is so for arbitrary $x, y \,\epsilon\, \mathfrak{S}$.

Review of *Skolem 1931*:
On some propositional functions in arithmetic
(*1932d*)

The author considers those propositional functions and statements that are built up by means of the five primitive logical operations (not, or, and, all, there exists) from relations of the form $A = B$ and $A < B$, where A and B are number-theoretic functions that result from (arbitrarily iterated) application of the three operations of addition, multiplication by a rational number, and formation of the next smaller whole number. The domain

Laufbereich der Variablen werden die ganzen rationalen Zahlen betrachtet, doch gelten alle Überlegungen für beliebige Bereiche von Dingen, für
welche die drei genannten Operationen definiert sind und gewissen
einfachen Forderungen genügen. Für die Sätze der oben angegebenen
Form wird ein Entscheidungsverfahren entwickelt, welches darauf beruht,
daß jede Aussagefunktion der betrachteten Art mit einer "elementaren"
äquivalent ist, d. h. einer solchen, die keine All- und Es gibt-Zeichen enthält.
Dieses Resultat wird benutzt, um die Relativität der Begriffe der Wohlordnung und der induktiven Zahl an einem Beispiel zu illustrieren. Der Verfasser konstruiert nämlich für eine gewisse geordnete Menge von Individuen
einen Bereich von Funktionen, der die ordnende Relation selbst enthält,
gegen die fünf logischen Grundoperationen abgeschlossen ist und in bezug
auf den die gegebene Menge wohlgeordnet ist, d. h. jede mit Hilfe von Funktionen des Bereiches definierte Klasse enthält ein erstes Element. Andererseits kann man aber außerhalb des Bereiches liegende Klassen angeben, für
die das nicht der Fall ist. Das erwähnte Resultat wird ferner benutzt, um
ein Entscheidungsverfahren zu skizzieren für Sätze, die sich aus Bestandteilen der Form:

$$ax_1^{\alpha_1} \cdot x_2^{\alpha_2} \cdot \ldots \cdot x_n^{\alpha_n} = b$$

mittels der fünf logischen Grundoperationen aufbauen, wobei der Lauf-
bereich der Variablen sowie die Konstanten a, b, $\alpha_1, \ldots, \alpha_n$ positive ganze
Zahlen sein sollen. Auch hier gelten die Betrachtungen für allgemeinere
Dingbereiche, z. B. für die ganzen Zahlen eines beliebigen algebraischen
Zahlkörpers.

Besprechung von *Carnap 1931*:
Die logizistische Grundlegung der Mathematik
(*1932e*)

Dieser Vortrag bringt außer einer sehr klaren Darstellung des logizisti-
schen Standpunktes auch eine ausführliche Darlegung der heute noch bestehenden Schwierigkeiten und einen Versuch zu deren Überwindung. Die
These des Logizismus besagt zweierlei, nämlich 1. daß sich sämtliche mathematische *Begriffe* durch explizite Definition auf die logischen zurückführen
lassen, 2. daß sämtliche mathematischen *Sätze* aus den logischen Grundsätzen deduktiv ableitbar sind. Der Sinn dieser beiden Behauptungen
wird näher präzisiert und an Beispielen (Begriff der natürlichen und der
reellen Zahl) erläutert. Bezüglich des zweiten Punktes wird auf die Schwie-

of the variables is taken to be the totality of all rational numbers, but all considerations hold for arbitrary domains of objects for which the three operations mentioned are defined and satisfy certain simple requirements. For the statements of the form specified above, a decision procedure is developed, which depends on the fact that every propositional function of the kind considered is equivalent to an "elementary" one, that is, one that contains no universal or existential quantifier. This result is used in order to illustrate by means of an example the relativity of the notions of well-ordering and of inductive number. Namely, for a certain ordered set of individuals the author constructs a domain of functions that contains the ordering relation itself, is closed under the five primitive logical operations, and with respect to which the given set is well-ordered, that is, every class defined by means of functions of the domain contains a first element. But on the other hand one can specify classes lying outside the domain for which that is not the case. The aforementioned result is further used in order to sketch a decision procedure for statements that are built up from constituents of the form

$$ax_1^{\alpha_1} \cdot x_2^{\alpha_2} \cdot \ldots \cdot x_n^{\alpha_n} = b$$

by means of the five primitive logical operations, where the domain of the variables as well as that of the constants a, b, $\alpha_1, \ldots, \alpha_n$ is to be the positive whole numbers. Here, too, the considerations hold for more general domains of objects, for example, for the integers of an arbitrary algebraic number field.

Review of *Carnap 1931*:
The logicist's way of founding mathematics
(*1932e*)

[The introductory note to *Gödel 1932e*, as well as to related items, can be found on page 196, immediately preceding *1931a*.]

Besides giving a very clear presentation of the logicist point of view, this address also provides a detailed exposition of the difficulties still persisting today, as well as an attempt to overcome them. The thesis of logicism is twofold, namely, (1) that all mathematical *notions* are reducible to logical ones through explicit definitions and (2) that all mathematical *theorems* are

rigkeit hingewiesen, die durch das Unendlichkeits- und das Auswahlaxiom entsteht, da diese beiden Axiome wegen ihres existenzialen Charakters nicht zu den logischen Grundsätzen gehören, aber zum Beweise vieler mathematischer Sätze erforderlich sind. Doch könne man diese Schwierigkeit leicht überwinden, indem man, wie Russell es tut, die genannten Axiome in die Voraussetzungen der betreffenden Sätze aufnimmt, wodurch man rein logisch beweisbare Sätze erhält. Eine viel wesentlichere Schwierigkeit liegt, wie der Verfasser ausführt, in den sogenannten imprädikativen Begriffsbildungen. Um den bei dieser Definitionsart scheinbar auftretenden Zirkel zu vermeiden, stellte Russell seine "verzweigte" Typentheorie auf, durch welche aber eine adäquate Behandlung der reellen Zahlen unmöglich wird, wenn man nicht zu dem (heute allgemein als unzulässig erkannten) Reduzibilitätsaxiom seine Zuflucht nimmt. Die Lösung der Schwierigkeit sucht der Verfasser in der Weglassung der verzweigten Typentheorie, die deshalb überflüssig sei, weil die imprädikativen Definitionen in Wahrheit gar keinen Zirkel enthalten. Auch F. P. Ramsey habe diesen Standpunkt zu begründen versucht, doch seien seine Argumente wegen des ihnen zugrunde liegenden Begriffsabsolutismus (Platonismus) unannehmbar. Der Verfasser sucht auf einem anderen Wege die Zulässigkeit der imprädikativen Definitionen darzutun und erläutert dies an dem Begriff der induktiven Zahl. Eine Zahl heißt bekanntlich induktiv, wenn ihr jede "erbliche" Eigenschaft zukommt, die der Null zukommt. Scheinbar geht in diese Definition die Gesamtheit aller Eigenschaften von Zahlen und daher auch die zu definierende Eigenschaft selbst ein (worin eben der vermeintliche Zirkel besteht). Dieser Anschein verschwindet aber, wenn man das Wort "jede" nicht extensional, im Sinn von "*jede einzelne*" auffaßt, sondern die obige Definition so interpretiert: Eine Zahl n heißt induktiv, wenn aus der Voraussetzung, eine Eigenschaft sei erblich und komme der Null zu, *logisch folgt*, daß sie der Zahl n zukommt. Den Unterschied dieser beiden Bedeutungen von "jedes" charakterisiert der Verfasser (im Anschluß an F. Kaufmann) durch die Worte "numerische" und "spezifische" Allgemeinheit.—Was die logischen Paradoxien betrifft, die für Russell auch einer der Gründe für die Aufstellung der verzweigten Typentheorie waren, so verweist der Verfasser auf die diesbezüglichen Untersuchungen von Ramsey, der gezeigt habe, daß die in logischen Zeichen überhaupt darstellbaren Antinomien schon durch die einfache Typentheorie ausgeschlossen werden.

deductively derivable from the principles of logic. The meaning of these two assertions is made more precise and is illustrated by examples (the notions of natural number and of real number). With regard to the second point, reference is made to the difficulty that arises from the axiom of infinity and the axiom of choice, since these two axioms, on account of their existential character, do not belong to the principles of logic but are necessary for the proofs of many mathematical theorems. This difficulty, however, can easily be overcome if we add, as Russell does, the aforementioned axioms to the premises of the theorems in question, thereby obtaining theorems that are provable in a purely logical way. A much more essential difficulty, as the author explains, lies in the introduction of notions by what are called impredicative definitions. In order to avoid the circularity that seems to appear in connection with this kind of definition, Russell set up his "ramified" theory of types, in which, however, no adequate treatment of the real numbers is possible without recourse to the axiom of reducibility (today generally regarded as inadmissible). The author seeks to resolve the difficulty by doing without the ramified theory of types, which is said to be superfluous, because the impredicative definitions do not in fact involve any circularity at all. According to the text under review, F. P. Ramsey, too, tried to establish this point of view, but with arguments that are unacceptable because of the conceptual absolutism (Platonism) underlying them. The author seeks to demonstrate the admissibility of the impredicative definitions in another way, and he illustrates this by the notion of inductive number. As is well known, a number is said to be inductive if every "hereditary" property that belongs to zero belongs to it. This definition seems to involve the totality of all properties of numbers and therefore also the very property being defined (and the presumed circularity consists precisely in this). But this impression disappears if one does not interpret the word "every" extensionally, in the sense of "*every single one*", but interprets the definition above as follows: a number n is said to be inductive if, from the assumption that a property is hereditary and belongs to zero, *it logically follows* that the property belongs to the number n. Following F. Kaufmann, the author characterizes the difference between these two meanings of the word "every" by the words "numerical" and "specific" generality.

With regard to the logical paradoxes, which for Russell were another reason for setting up the ramified theory of types, the author refers to the relevant investigations of Ramsey, who is said to have shown that the antinomies that can in any way be represented in logical symbols are already excluded by the simple theory of types.

Besprechung von *Heyting 1931*:
Die intuitionistische Grundlegung der Mathematik
(*1932f*)

Der Verfasser setzt zunächst die philosophische Einstellung des Intuitionisten auseinander. Für diesen ist die Mathematik natürliche Funktion des Intellekts, Erzeugnis des menschlichen Geistes, und daher spricht er den mathematischen Gegenständen keine objektive, vom Denken unabhängige Existenz zu. Aus dieser Auffassung, daß die mathematischen Objekte nur insofern vorhanden sind, als sie vom menschlichen Denken tatsächlich erfaßt werden können, ergibt sich die Ablehnung reiner Existenzialbeweise sowie des Satzes vom ausgeschlossenen Dritten in allen den Fällen, wo eine Entscheidung der Alternative nicht tatsächlich herbeigeführt werden kann.—Der intuitionistische Aufbau der Mathematik wird an mehreren Beispielen illustriert. Zunächst wird die Brouwersche Definition der reellen

322 Zahl und der Begriff der Wahlfolge erläutert. Dieser Begriff er|setzt, wie der Verfasser zeigt, nicht so sehr die einzelne reelle Zahl als vielmehr die Gesamtheit aller möglichen reellen Zahlen, bzw. Teilmengen dieser Gesamtheit, falls man die Wahlfreiheit irgendwie gesetzmäßig einschränkt. Ferner wird die Brouwersche Mengendefinition kurz besprochen und ein Beweis skizziert für den in der intuitionistischen Mathematik geltenden Satz: Wenn jeder reellen Zahl eine Nummer zugeordnet ist, so haben sie alle dieselbe Nummer.—Am Schluß des Vortrages kommt die intuitionistische Aussagenlogik zur Sprache. Es wird unterschieden zwischen Aussagen und Behauptungen (bei Frege und Russell durch das Zeichen ⊢ angedeutet). Eine Aussage drückt eine Erwartung (Intention) aus, und zwar geht diese auf ein als möglich gedachtes Erlebnis; die entsprechende Behauptung bedeutet die Erfüllung der Intention. Z. B. bedeutet die Aussage "C ist rational" die Erwartung, man könne zwei ganze Zahlen a, b finden derart, daß $C = a/b$. Der Satz vom ausgeschlossenen Dritten drückt die Erwartung aus, man könne jeden mathematischen Satz entweder beweisen oder auf einen Widerspruch zurückführen, und ein Beweis für ihn wäre nur möglich durch Angabe einer allgemeinen Methode, welche es gestattet, diese Entscheidung immer herbeizuführen. Die Intentionen auf das Zutreffen und die Beweisbarkeit einer Aussage sind voneinander zu unterscheiden, wenn auch die zugehörigen Behauptungen gleichbedeutend sind. Falls man die beiden Aussagen negiert, sind auch die Behauptungen nicht mehr gleichbedeutend, und daraus folgert der Verfasser, daß ein Beweis für die grundsätzliche Unentscheidbarkeit einer mathematischen Frage durchaus im Bereich des Möglichen liegt.

Review of *Heyting 1931*:
The intuitionist's way of founding mathematics
(*1932f*)

[The introductory note to *Gödel 1932f*, as well as to related items, can be found on page 196, immediately preceding *1931a*.]

The author first of all expounds the philosophical attitude of the intuitionist, for whom mathematics is a natural function of the intellect, a product of the human spirit, and who therefore grants no objective existence, independent of thought, to mathematical entities. This conception—that mathematical objects exist only insofar as they can actually be comprehended by human thought—leads to the rejection of pure existence proofs, as well as of the principle of the excluded middle in all cases in which a decision among the alternatives cannot actually be made.

The intuitionistic construction of mathematics is illustrated by several examples. First, Brouwer's definition of real number and the notion of choice sequence are explained. This notion, as the author shows, does not so much replace any individual real number as it does the totality of all possible real numbers, or subsets of that totality, in case the freedom of choice is somehow restricted by a rule. Further, Brouwer's definition of set is briefly discussed, and a proof is sketched for the following theorem of intuitionistic mathematics: If a whole number is associated with each real number, then all real numbers will have the same value associated with them.

At the conclusion of the lecture intuitionistic propositional logic is discussed. A distinction is made between propositions and assertions (the latter indicated in Frege and Russell by the sign $\vdash$). A proposition expresses an expectation (intention) and, moreover, one directed toward an event regarded as possible; the corresponding assertion signifies the fulfillment of the intention. For example, the proposition "C is rational" signifies the expectation that one can find two whole numbers a and b such that $C = a/b$. The principle of the excluded middle expresses the expectation that, for every mathematical statement, one can either prove it or reduce it to a contradiction, and a proof of it would only be possible through specification of a general method that allows this decision always to be made. The intention that a proposition holds and the intention that it is provable must be distinguished from each other, even if the corresponding assertions mean the same. If one negates the two propositions, then the assertions no longer mean the same, and from this the author concludes that a proof that a mathematical question is undecidable in principle lies entirely within the realm of the possible.

247

Besprechung von *von Neumann 1931*:
Die formalistische Grundlegung der Mathematik
(*1932g*)

Das Ziel der formalistischen Grundlegung der Mathematik ist, wie der Verfasser ausführt, die Rechtfertigung der klassischen Mathematik bei Anerkennung der von den Intuitionisten gegen sie erhobenen Bedenken. In der Kritik der transfiniten (existentialen) Schlußweisen—für welche ein sehr instruktives Beispiel gegeben wird—stimmt daher der Formalist mit dem Intuitionisten überein, aber er richtet sein Augenmerk darauf, daß diese Schlußweisen, wenn sie nicht als Methoden des inhaltlichen Denkens, sondern als ein nach gewissen Konventionen verlaufendes Verfahren zur Ableitung von Formeln aufgefaßt werden, einen durchaus finiten Sinn erhalten, da ja die Formeln und die mit ihnen vollzogenen Operationen finiten Charakter haben. Die Beschreibung der Verfahren der klassischen Mathematik übernimmt der Formalismus zum großen Teil den Arbeiten der logistischen Schule, aber da er diese Verfahren nicht als inhaltlich einleuchtende Beweismethoden, sondern bloß als "Formelspiel" ansieht, so entsteht für ihn die Aufgabe, die Brauchbarkeit dieses "Spiels" für die Wissenschaft darzutun. Dies soll in der Weise geschehen, daß man zeigt: Jede in endlich vielen Schritten kontrollierbare (nachrechenbare) Zahlformel, die nach den Spielregeln der klassischen Mathematik abgeleitet werden kann, muß sich bei effektivem Nachrechnen als richtig herausstellen. Damit wäre die Verwendbarkeit der klassischen Mathematik zur abkürzenden Berechnung arithmetischer Ausdrücke dargetan, und gerade in diesem Sinn wird sie auf die Erfahrung angewandt. Der Beweis der oben bezeichneten Tatsache läuft auf den Beweis der Widerspruchsfreiheit hinaus, der daher das zentrale Problem ist. Er muß natürlich mit inhaltlich einwandfreien (d. h. intuitionistischen) Methoden geführt werden. Das für Widerspruchsfreiheitsbeweise angewendete Verfahren wird in seinen Grundzügen beschrieben und zum Schluß darauf hingewiesen, daß bisher ein Widerspruchsfreiheitsbeweis erst für gewisse beschränkte *Teilsysteme* der klassischen Mathematik geglückt ist.—Der Vortrag charakterisiert den Stand der Dinge im Zeitpunkt der Königsberger Tagung. Die seither erzielten Resultate [[*Gödel 1931*]] über die Unmöglichkeit von Widerspruchsfreiheitsbeweisen mit geringeren Beweismitteln, als sie das System selbst formalisiert enthält, sind noch nicht berücksichtigt.

Review of *von Neumann 1931*:
The formalist's way of founding mathematics
(*1932g*)

[The introductory note to *Gödel 1932g*, as well as to related items, can be found on page 196, immediately preceding *1931a*.]

The formalist's goal in providing a foundation for mathematics is, as the author explains, to vindicate classical mathematics while taking into account the doubts raised about it by the intuitionists. In criticizing transfinite (existential) modes of inference—of which a very instructive example is given—the formalist thus concurs with the intuitionist. Yet he directs his attention to the fact that these modes of inference attain a thoroughly finitary meaning if they are viewed, not as methods of contentual thinking, but as a procedure for deriving formulas according to certain conventions, since the formulas and the operations carried out with them do, after all, have a finitary character. In describing the procedures of classical mathematics, formalism in large part adopts the work of the logicist school. However, since formalism regards these procedures not as contentually evident methods of proof but merely as a "formula game", it faces the problem of demonstrating the utility of this "game" for science. This is supposed to be done by our showing that every numerical formula verifiable (calculable) in finitely many steps that can be derived according to the rules of the game by which classical mathematics is played must turn out to be correct when actually calculated. The applicability of classical mathematics to shortening the computation of arithmetical expressions would thereby be demonstrated, and it is in just this sense that classical mathematics is applied to experience. The proof of the fact described above amounts to the proof of consistency, which therefore is the central problem. It must, of course, be carried out by means of contentually unobjectionable (that is, intuitionistic) methods. The basic features of the method employed for consistency proofs are described, and in conclusion it is pointed out that up to now consistency proofs have been successfully carried out only for certain restricted *subsystems* of classical mathematics.

The lecture characterizes the state of affairs at the time of the Königsberg conference. The results subsequently obtained [[*Gödel 1931*]] concerning the impossibility of carrying out consistency proofs with means of proof weaker than those formalized in the system itself are not yet taken into account.

Besprechung von *Klein 1931*:
Zur Theorie der abstrakten Verknüpfungen
(*1932h*)

Der Verfasser untersucht Dingbereiche, in denen zwei Operationen $a \cup b$, $a \cap b$ definiert sind und den Axiomen des Klassenkalküls für Summe und Durchschnitt genügen. Das Analogon zur Komplementbildung wird eingeführt als eineindeutige symmetrische Abbildung des Bereiches in sich, durch welche die beiden Verknüpfungen $\cup$, $\cap$ miteinander vertauscht werden. Als Realisierungen für die Axiome kommen außer Klassen- und Aussagenkalkül z. B. die Operationen $\max(a, b)$, $\min(a, b)$ oder größter gemeinsamer Teiler, kleinstes gemeinsames Vielfaches in Betracht. Nach Ableitung verschiedener einfacher Sätze (z. B. der Eindeutigkeit der Zerlegung der Elemente des Bereiches in Primelemente, falls eine solche Zerlegung überhaupt existiert) wird eine sogenannte "charakteristische" Abbildung eingeführt. Darunter wird verstanden eine eindeutige Abbildung des Bereiches auf eine Abelsche Gruppe (oder eine Teilmenge einer solchen), welche folgender Bedingung genügt:

$$\{a \cup b\} = \{a\} + \{b\} - \{a \cap b\}, \tag{1}$$

(wobei mit $+, -$ die Gruppenoperation und ihre Inverse und mit $\{\ \}$ die abbildende Funktion bezeichnet ist). Ein Beispiel für eine solche Abbildung liefert der Begriff "Anzahl der Elemente" für endliche Mengen, ferner viele zahlentheoretische Funktionen (Teilersumme, Teileranzahl usw.) bei der Deutung von $\cap$, $\cup$ als größter gemeinsamer Teiler, kleinstes gemeinsames Vielfaches und der Multiplikation als Gruppenoperation. Die Relation (1) wird auf n Variable verallgemeinert und liefert in dieser Form neben anderen Anwendungen das Schema für das eratosthenische und ähnliche arithmetische Siebverfahren.

Review of *Klein 1931*:
On the theory of abstract connectives
(*1932h*)

The author investigates domains of individuals in which two operations, $a \cup b$ and $a \cap b$, are defined and satisfy the axioms of the calculus of classes for union and intersection. The analog of complementation is introduced as a one-to-one symmetric mapping of the domain onto itself, by means of which the two connectives $\cup$ and $\cap$ are interchanged. In addition to the calculi of classes and of propositions, the operations $\max(a, b)$ and $\min(a, b)$, for example, or greatest common divisor and least common multiple, are considered as realizations of the axioms. After derivation of various simple theorems (for example, the uniqueness of the decomposition of elements of the domain into prime elements, if such a decomposition exists at all), what is called a "characteristic" mapping is introduced. By this is meant a single-valued mapping of the domain onto an Abelian group (or a subset of such a group) satisfying the following condition:

$$\{a \cup b\} = \{a\} + \{b\} - \{a \cap b\}, \tag{1}$$

(where the group operation and its inverse are denoted by $+$ and $-$, respectively, and the mapping function by $\{\ \}$). The notion "number of elements" for finite sets furnishes an example of such a mapping, as do many number-theoretic functions (sum of divisors, number of divisors, and so forth), $\cap$ and $\cup$ being interpreted as greatest common divisor and least common multiple, respectively, and multiplication being taken as the group operation. The relation (1) is generalized to n variables and in this form furnishes, among other applications, the schema for the sieve of Eratosthenes and similar arithmetical sieve methods.

Besprechung von *Hoensbroech 1931*: Beziehungen zwischen Inhalt und Umfang von Begriffen
(*1932i*)

Versteht man unter Inhaltsvermehrung eines Begriffes die Hinzufügung von Bestandteilen, ohne daß die bisherige Verbindung seiner Bestandteile zerstört wird, so braucht, wie der Verfasser im Anschluß an eine Bemerkung von Bolzano feststellt, der Umfang bei einer Inhaltsvermehrung durchaus nicht abzunehmen. Z. B. hat in diesem Sinne "blau oder rot" größeren Inhalt als "blau". Der Verfasser untersucht nun die verschiedenen Arten der Inhaltsvermehrung, die unter diesem Gesichtspunkt möglich sind (konjunktive und diskonjunktive Einfügung, Negation, Einsetzung eines Namens, Hinzufügen von Quantifikatoren) bezüglich ihrer Wirkung auf den Umfang der Begriffe.

Besprechung von *Klein 1932*: Über einen Zerlegungssatz in der Theorie der abstrakten Verknüpfungen
(*1932j*)

Im Anschluß an eine frühere Arbeit (*1931*) definiert der Verfasser als "Verband" einen Dingbereich, in dem zwei kommutative und assoziative Operationen $\cup, \cap$ definiert sind, für welche überdies gilt:

$$a \cup b = a \cdot \equiv \cdot a \cap b = b \quad \text{und} \quad a \cup a = a.$$

Wenn $a \cup b = b$, heißt a Teilding von b. Unter der Voraussetzung, daß die Operation $\cap$ distributiv bez. $\cup$ ist, daß ferner ein Einheitselement e für $\cup$ existiert und daß jedes Element nur endlich viele Teildinge hat, kann ein Element a des Verbandes auf *höchstens eine* Art in der Form:

$$a = p_1 \cup p_2 \cup \cdots \cup p_n \tag{1}$$

dargestellt werden, wobei die p_i "Primärelemente" über verschiedenen

Review of *Hoensbroech 1931*:
Relations between the content and extension of notions
(*1932i*)

If by an increase in the content of a notion we understand the adjoining of constituents without the existing connection between constituents being destroyed, then, as the author establishes in conjunction with a remark of Bolzano, extension need not be at all diminished by an increase in content. For example, in this sense "blue or red" has greater content than "blue". The author then investigates the different kinds of increase in content that are possible from this point of view (conjunctive and disjunctive joining, negation, substitution of a name, adjoining of quantifiers) with regard to their effect on the extension of notions.

Review of *Klein 1932*:
On a decomposition theorem in the theory of abstract connectives
(*1932j*)

In connection with an earlier work (*1931*), the author defines "lattice" to mean a domain of individuals in which two commutative and associative operations $\cup$ and $\cap$ are defined for which, moreover,

$$a \cup b = a \cdot \equiv \cdot a \cap b = b \quad \text{and} \quad a \cup a = a$$

hold. If $a \cup b = b$, a is called a part of b. Under the assumptions that the operation $\cap$ is distributive with respect to $\cup$, that, further, a unit element e exists for $\cup$, and that every element has only finitely many parts, an element a of the lattice can be represented in *at most one* way in the form

$$a = p_1 \cup p_2 \cup \cdots \cup p_n, \tag{1}$$

where the p_i are "primary elements" over different "prime elements". Here p is called a prime element if it has no proper part other than e, and q

"Primelementen" sind. Dabei heißt p Primelement, wenn es keinen echten Teil außer e hat, und q Primärelement über p, wenn p das einzige in q enthaltene Primelement ist. Nimmt man noch an, daß jedes Nicht-Primärelement $a \neq e$ in zwei von a verschiedene Elemente zerfällt, so folgt auch die *Existenz* einer Darstellung der Form (1) für jedes Element des Verbandes.—Außer diesem Hauptergebnis werden noch verschiedene Untersuchungen insbesondere über die Unabhängigkeit der genannten Voraussetzungen durchgeführt.

is called a primary element over p if p is the only prime element included in q. If in addition it is assumed that every non-primary element $a \neq e$ can be decomposed into two elements distinct from a, the *existence* of a representation in the form (1) also follows for every element of the lattice.

In addition to this principal result, various other investigations are carried out, especially concerning the independence of the stated assumptions.

Introductory note to *1932k, 1934e* and *1936b*

Church had some quite novel ideas for "a system of mathematical logic free of some of the complications entailed by Bertrand Russell's theory of types, and [which] would at the same time avoid the well known paradoxes ... by weakening the classical principle of reductio ad absurdum." To this end he described a formal system in *1932*. Finding that he had not succeeded there in avoiding a modified form of the Russell paradox, he made some changes in *1933*.

In his system, both as in *1932* and as in *1933*, he used an operator $\lambda \mathbf{x} \mathbf{M}$ standing for the function which $\mathbf{M}$ (containing free occurrences of $\mathbf{x}$) is of $\mathbf{x}$. That is, $\lambda \mathbf{x} \mathbf{M}$ denotes the function which has the value denoted by $\mathbf{M}$ for each value of $\mathbf{x}$. At the conclusion of *1933* he proposed a way of using this to develop the theory of positive integers within his system. Kleene, in work done in 1932 and 1933 (published in *1934* and *1935*), took up this last suggestion, and pursued it using Church's λ-operator separately from the other ingredients of the system (particularly from Church's descriptive operator ι) to represent functions of positive integers.[a] Thus arose the very fruitful notion known now as "(Church–Kleene) λ-definability". For the role this has played in connection with the notion of algorithmic computability and its relations to Gödel's work, see the introductory note to *Gödel 1934*. The early history is given quite fully in the opening

[a] *Kleene 1934* made important use of *Rosser 1935–1935a*.

Besprechung von *Church 1932* :
A set of postulates for the foundation of logic
(*1932k*)

Es wird ein formales System für Logik und Mathematik angegeben, welches ohne Verwendung der Typentheorie und ohne Anlehnung an den axiomatischen Aufbau der Mengenlehre die Antinomien auf einem anderen Wege, nämlich durch Einschränkung des Satzes vom ausgeschlossenen Dritten vermeiden soll. Man kann in diesem System z. B. die Menge

pages of *Kleene 1981*. A very comprehensive treatment of the subsystem of Church's *1932* and *1933* systems called the "λ-calculus" (in which λ-definability operates), with references to the extensive literature over the intervening years, is in *Barendregt 1981* and *1984*.

The full system of *Church 1933* (like that of *Church 1932*) turned out to be inconsistent, because a form of the Richard paradox can be derived in it, as was done in *Kleene and Rosser 1935*. The full system comprised, in addition to the λ-calculus, the descriptive operator ι and an abstraction operator *A*, a restricted generality operator Π, existence Σ, conjuction & and negation ~. One can say that the generality operator Π, used in the context of λ-definability, was primarily the villain in producing the contradictions. The consistency of just the λ-calculus was proved in *Church and Rosser 1936*. Building on this consistency result, Church in *1935* considered a hierarchy of consistent systems— starting with the λ-calculus augmented by a connective δ (which expressed equality in it) and using "a sequence, extending into the second number class, of more and more inclusive definitions of the notion" of the consequences of a formula. He speculated that in the aggregate they comprise elementary number theory, "so that we obtain, in a certain sense, a metamathematical proof of the freedom from contradiction of elementary number theory." This speculation has not been pursued.

Gödel's reviews *1932k*, *1934e* and *1936b* of Church's papers *1932*, *1933* and *1935* are essentially reportorial in nature.

Stephen C. Kleene

Review of *Church 1932* :
A set of postulates for the foundation of logic
(*1932k*)

A formal system for logic and mathematics is presented that is supposed to avoid the antinomies, without use of the theory of types or reliance upon the axiomatic construction of set theory, but in another way, namely, by restricting the principle of the excluded middle. For example, while in this system the set of all sets that do not contain themselves (or the

der Mengen, die sich nicht enthalten (bzw. die zugehörige Aussagefunktion), zwar bilden, aber von ihr nicht beweisen, daß sie sich entweder selbst enthält oder nicht enthält. Auch der Satz, daß eine Annahme, aus der ein Widerspruch folgt, falsch ist, gilt nicht allgemein. Ob wirklich alle Antinomien (z. B. auch die Buralifortische) in seinem System vermieden werden und ob es zum Aufbau der Mathematik ausreicht, hält der Verfasser selbst für noch nicht feststehend, hofft aber jedenfalls durch einige Modifikationen dahin gelangen zu können. In seiner jetzigen Form enthält das System acht Grundzeichen (außer den Variablen), fünfunddreißig Axiome und fünf Schlußregeln. Entsprechend der Auffassung, daß Aussagefunktionen für gewisse Werte der Variablen weder wahr noch falsch (d. h. sinnlos) werden können, tritt der Allbegriff als zweistelliger Operator $\Pi(a, b)$ auf, mit der Bedeutung für alle x, für die $a(x)$ gilt, gilt auch $b(x)$, so daß "für alle" sich immer nur auf einen beschränkten Bereich a von Dingen beziehen kann. Der Gebrauch von freien Variablen wird vermieden. Substitutions- und Implikationsregel sind in die folgende Regel zusammengefaßt: Aus
146 | $F(a)$ und $\Pi(F, G)$ darf $G(a)$ geschlossen werden. Folgender Satz gilt, wie gezeigt wird, auch für das neue System: Ist $(\exists x)F(x)$ beweisbar und G aus $F(x)$ ableitbar, dann ist $\Pi(F, G)$, und wenn x in G nicht als freie Variable vorkommt, G selbst beweisbar.

Besprechung von *Kalmár 1932*:
Ein Beitrag zum Entscheidungsproblem
(*1932l*)

Das Entscheidungsproblem des engeren Funktionenkalküls wird auf den Fall zurückgeführt, daß der betrachtete Ausdruck in der Normalform ein Präfix der Gestalt

$$(Ex_1)(Ex_2)\ldots(Ex_{m-1})(y_1)(y_2)(Ex_m)(y_3)(y_4)\ldots(y_n)$$

besitzt, d. h. es wird gezeigt: Zu jedem Ausdruck A des engeren Funktionenkalküls kann man einen anderen B mit einem Präfix der obigen Gestalt angeben, so daß A dann und nur dann erfüllbar (bzw. in einem endlichen Individuenbereich erfüllbar) ist, wenn dasselbe für B gilt. Ist insbesondere B in einem Bereich I erfüllbar, so ist A in demselben Bereich I erfüllbar. Ist A in einem Bereich I erfüllbar, so B in einem umfassenderen I', dessen Kardinalzahl sich aus der von I leicht berechnen läßt. B enthält um einige

corresponding propositional function) can be formed, it cannot be proved that it either does or does not contain itself. Moreover, the proposition that a hypothesis from which a contradiction follows is false does not hold generally. It has not yet been established, according to the author, whether all antinomies (for example, that of Burali-Forti) are actually avoided in his system and whether it suffices for the construction of mathematics; but in any case he hopes to be able to achieve these goals by making a few modifications. In its present form the system contains eight primitive symbols (in addition to the variables), thirty-five axioms and five rules of inference. In conformity with the view that propositional functions can be neither true nor false (that is, be meaningless) for certain values of the variables, the notion of universality appears as a two-place operator $\Pi(a, b)$, with the meaning that, for all x for which $a(x)$ holds, $b(x)$ also holds, so that "for all" always relates only to a restricted domain a of individuals. The use of free variables is avoided. The rules of substitution and implication are combined in the following rule: From $F(a)$ and $\Pi(F, G)$, one may infer $G(a)$. As is shown, the following proposition, too, holds for the new system: If $(\exists x)F(x)$ is provable and G is derivable from $F(x)$, then $\Pi(F, G)$ is provable, and, if x does not occur in G as a free variable, then G itself is provable.

Review of *Kalmár 1932*:
A contribution to the decision problem
(*1932l*)

The decision problem for the restricted functional calculus is reduced to the case in which the expression under consideration possesses, in normal form, a prefix of the form

$$(Ex_1)(Ex_2)\ldots(Ex_{m-1})(y_1)(y_2)(Ex_m)(y_3)(y_4)\ldots(y_n).$$

That is, it is shown that for every expression A of the restricted functional calculus one can specify another expression B, with a prefix of the form displayed above, so that A is satisfiable (or satisfiable in a finite domain of individuals) if and only if the same is true of B. In particular, if B is satisfiable in a domain I, then A is satisfiable in that same domain I. If A is satisfiable in a domain I, then B is satisfiable in a more extensive domain I', whose cardinality may easily be computed from that of I. B will

Funktionsvariable mehr als A, man kann aber, wie gezeigt wird, immer erreichen, daß B nur binäre Funktionsvariable enthält, wenn dies für A der Fall ist.

Besprechung von *Huntington 1932* :
A new set of independent postulates for the algebra of logic with special reference to Whitehead and Russell's *Principia mathematica* (*1932m*)

Es werden zwei neue Axiomensysteme für den Klassenkalkül (Aussagen-kalkül) angegeben. Das erste verwendet Addition $(a+b)$ und Komplement-bildung (a') als Grundbegriffe und fordert neben Kommutativität und Assoziativität von $+$ nur die beiden folgenden Axiome:

$$a + a = a$$

$$(a' + b')' + (a' + b)' = a.$$

Das andere benutzt $a \supset b$ als einzige Grundoperation $(a \supset b = a' + b)$, was wesentlich kompliziertere Axiome erforderlich macht. Die Unabhängigkeit der Axiome und ihre Äquivalenz mit den vom Verfasser und anderen früher aufgestellten soll in einer späteren Abhandlung bewiesen werden. Bezüglich eines von B. A. Bernstein (*1931*) aufgestellten Systems wird bemerkt, daß man es durch ein Axiom ergänzen müsse, um den gewöhnlichen Klassen-kalkül zu erhalten.

contain a few more functional variables than A, but, as is shown, one can always ensure that B contains only two-place functional variables whenever A does.

Review of *Huntington 1932*:
A new set of independent postulates for
the algebra of logic with special reference to
Whitehead and Russell's *Principia mathematica*
(*1932m*)

Two new axiom systems for the calculus of classes (propositional calculus) are specified. The first employs addition $(a + b)$ and complementation (a') as primitive notions, and, in addition to the commutativity and associativity of $+$, requires only the following two axioms:

$$a + a = a$$

$$(a' + b')' + (a' + b)' = a.$$

The other uses $a \supset b$ as the only primitive operation $(a \supset b = a' + b)$, and this makes it necessary to adopt substantially more complicated axioms. The independence of the axioms, as well their equivalence with those put forward earlier by the author and others, is to be proved in a later article. With regard to a system set up by B. A. Bernstein (*1931*), the author remarks that it must be supplemented by one axiom if one is to obtain the ordinary calculus of classes.

Besprechung von *Skolem 1932*:
Über die symmetrisch allgemeinen Lösungen
im identischen Kalkul
(*1932n*)

Sei

$$f\{a_e, x_k\} = 0 \tag{1}$$

eine Gleichung des Klassenkalküls mit den unbestimmten Koeffizienten a_e und den Unbekannten x_k. Eine allgemeine Lösung

$$x_i = g_i\{a_e, u_k\} \tag{2}$$

von (1) heißt symmetrisch in bezug auf eine Gruppe G von Permutationen der a_e und x_i, welche (1) in sich überführen, wenn es zu jedem S aus G eine Permutation T der u_i gibt, so daß S und T, gleichzeitig auf das Gleichungssystem (2) angewandt, dieses in sich überführen. Gilt außerdem $x_i = g_i\{a_e, x_k\}$ für alle Lösungen x_i von (1) und vertauscht ferner die zu einem S gehörige Permutation T die u_i in derselben Weise wie S die x_i, so wird die Lösung normal genannt. Es wird gezeigt: 1. Eine Gleichung mit konstanten Koeffizienten (ohne a_e) ist dann und nur dann symmetrisch allgemein lösbar in bezug auf G, wenn sie lösbar bleibt bei Gleichsetzung der x_i in jedem Transitivitätssystem, in welches die x_i durch G zerfallen. 2. Kommen unbestimmte Koeffizienten a_e vor, so sei $\Sigma_A A f_A\{x_k\}$ die Boolesche Entwicklung von (1) nach den a_e (wobei A die Konstituenten sind) und G_A die Untergruppe von G, welche A in sich überführt. Die notwendige und hinreichende Bedingung für symmetrisch allgemeine Lösbarkeit von (1) in bezug auf G (wobei die Bedingung der Lösbarkeit identisch in den a_e als erfüllt vorausgesetzt wird) besteht darin, daß jedes $f_A\{x_k\} = 0$ (dies ist eine Gleichung mit konstanten Koeffizienten) in bezug auf G_A symmetrisch allgemein lösbar ist. Im allgemeinen Fall hat man, um symmetrisch allgemeine Lösbarkeit zu erreichen, für die a_e als Nebenbedingungen hinzuzufügen: 1. die Bedingung der Lösbarkeit, 2. $\Sigma_B B = 0$, wobei B diejenigen Konstituenten A durchläuft, für welche $f_A\{x_k\}$ in bezug auf G_A *nicht* symmetrisch allgemein lösbar ist. Die Bedingungen für normale Lösbarkeit sind dieselben. Es folgen Betrachtungen darüber, wann eine Gleichung symmetrisch lösbar ist mit Hilfe von Parametern u_i, die durch eine Relation $\chi\{u_i\}$ verbunden sind, und dies führt u. a. zu einer Bedingung dafür, daß eine Gleichung symmetrisch in bezug auf die Unbekannten und ihre Negate lösbar ist.

Review of *Skolem 1932*:
On symmetric general solutions
in the calculus of ⟦Boolean⟧ identities
(*1932n*)

Let

$$f\{a_e, x_k\} = 0 \tag{1}$$

be an equation in the calculus of classes with undetermined coefficients a_e and unknowns x_k. A general solution

$$x_i = g_i\{a_e, u_k\} \tag{2}$$

of (1) is said to be symmetric with respect to a group G of permutations, of the a_e and x_i, that transform (1) into itself if for each S of G there is a permutation T of the u_i such that S and T, if applied simultaneously to the system of equations (2), will transform it into itself. If, in addition, $x_i = g_i\{a_e, x_k\}$ holds for all solutions x_i of (1) and if, furthermore, the permutation T corresponding to some S permutes the u_i in the same way as that in which S permutes the x_i, then the solution is said to be normal. It is shown that: 1. An equation with constant coefficients (without a_e) has a symmetric general solution with respect to G if and only if it remains solvable when the x_i are equated within each system of transitivity ⟦orbit⟧ into which G splits the x_i. 2. If undetermined coefficients a_e occur, let $\Sigma_A A f_A\{x_k\}$ be the Boolean expansion of (1) in terms of the a_e (the A being the constituents) and let G_A be the subgroup of G that maps the A into themselves. The necessary and sufficient condition for (1) to have a symmetric general solution with respect to G (the condition of solvability being assumed to be fulfilled identically in the a_e) is that every ⟦equation⟧ $f_A\{x_k\} = 0$ (this being an equation with constant coefficients) have a symmetric general solution with respect to G_A. In the general case, to obtain a symmetric general solution, one must place these auxiliary conditions on the a_e: 1. the condition of solvability; 2. $\Sigma_B B = 0$, where B ranges over those constituents A for which $f_A\{x_k\}$ has *no* symmetric general solution with respect to G_A. The conditions for normal solvability are the same. There follow considerations as to when, for an equation, a symmetric solution can be obtained with the help of parameters u_i that are connected by a relation $\chi\{u_i\}$, and this leads, among other things, to a condition that an equation must satisfy in order to have a solution symmetric with respect to the unknowns and their complements.

Besprechung von *Dingler 1931*:
Philosophie der Logik und Arithmetik
(*1932o*)

Das Buch beschäftigt sich hauptsächlich mit dem Wesen der mathematischen Kalküle. Zur Aufstellung und Handhabung derselben sind nach Ansicht des Verfassers gewisse, nicht weiter analysierbare "Grundfähigkeiten" erforderlich, die als "aktive Logik" dem Kalkül selbst als "passiver Logik" gegenübergestellt werden (ungefähr entsprechend der Hilbertschen Unterscheidung von Metamathematik und Mathematik). Art und Verwendungsweise der Grundfähigkeiten (besonders zum Aufbau der Arithmetik) werden auseinandergesetzt und verschiedene erkenntnistheoretische Fragen von diesem Gesichtspunkte aus behandelt. Als konkrete Anwendung der allgemeinen Betrachtungen folgt eine Untersuchung über ein formales System der Arithmetik, welche den von Hilbert gesuchten Widerspruchsfreiheitsbeweis liefern soll, in Wahrheit aber bestenfalls die triviale Tatsache der Widerspruchsfreiheit der elementaren Rechenregeln (distributives Gesetz etc.) zeigen könnte.

Review of *Dingler 1931*:
Philosophy of logic and arithmetic
(*1932o*)

This book is principally concerned with the essential nature of mathematical calculi. In the author's view, certain "basic capabilities", not further analyzable, are required for the construction and manipulation of those calculi; they are opposed, as "active logic", to the "passive logic" of the calculus itself (corresponding approximately to Hilbert's distinction between metamathematics and mathematics). The nature and manner of use of these basic capabilities (especially in the construction of arithmetic) are set forth, and various epistemological questions are considered from this perspective. As a concrete application of the general discussion, there follows an investigation of a formal system of arithmetic, which is supposed to furnish the consistency proof sought by Hilbert, but which, in truth, could at best demonstrate the trivial fact that the elementary rules of computation (distributive law, and so on) are consistent.

Introductory note to *1933*

William Tuthill Parry presented a deviant propositional calculus of the sort now called relevance logic, in which a conditional formula failed to be valid if its consequent contained a new letter. Gödel's remarks are

〚 Über die Parryschen Axiome 〛
(*1933*)

"*p* impliziert *q* analytisch" kann man vielleicht so interpretieren: "*q* ist aus *p* und den logischen Axiomen ableitbar und enthält keine anderen Begriffe als *p*" und es wäre, nachdem man diese Definition genauer präzisiert hat, ein Vollständigkeitsbeweis für die Parryschen Axiome 〚*Parry 1933*〛 zu erstreben, in dem Sinn, daß alle Sätze, welche für die obige Interpretation von → gelten, ableitbar sind.

In der Diskussion wird die Frage aufgeworfen, wieviele verschiedene Wahrheitswerte es im Heytingschen Aussagenkalkül gibt, d. h. wieviele nicht äquivalente Funktionen einer Variablen ($p \lor \neg p$ z. B. ist weder mit p noch mit $\neg p$ noch mit $\neg\neg p$ äquivalent). Bisher ist nicht einmal bekannt, ob es endlich oder unendlich viele sind.

here reported. In the discussion someone (evidently Hahn) wondered how many truth values are needed in modeling Heyting's intuitionistic propositional calculus. Gödel answered the question a few months after the colloquium and thus before this publication of it; see *Gödel 1932*.

W. V. Quine

⟦ On Parry's axioms ⟧
(*1933*)

One can perhaps interpret "*p* analytically implies *q*" as follows: "*q* is derivable from *p* and the axioms of logic, and *q* contains no notions other than those in *p*"; and, after this definition has been made more precise, it would be appropriate to strive for a proof of the completeness of Parry's axioms ⟦*Parry 1933*⟧, in the sense that all statements that hold when → is interpreted as suggested above are derivable.

During the discussion the following question was raised: How many different truth values are there in Heyting's propositional calculus, that is, how many non-equivalent functions of one variable? (For example, $p \vee \neg p$ is not equivalent to p, or to $\neg p$, or to $\neg\neg p$.) At present it is not even known whether there are finitely or infinitely many.

Introductory note to *1933a*

Hahn had asked whether every independence proof for formulas of the propositional calculus could be carried out with a finite model. Gödel here presents an independence proof using an infinite model and

Über Unabhängigkeitsbeweise im Aussagenkalkül
(*1933a*)

Läßt sich jeder Unabhängigkeitsbeweis innerhalb des Aussagenkalküls mit Hilfe von *endlichen* Modellen (Matrizen[1]) führen? Diese von Hahn aufgeworfene Frage ist zu verneinen. Z. B. ist die Aussage $p \supset {\sim}{\sim}p$ von folgenden drei Axiomen:

1) $p \supset p$,   2) $p \supset {\sim}{\sim}q . \supset . p \supset q$,   3) ${\sim}{\sim}p \supset {\sim}{\sim}q . \supset . p \supset q$

mit Einsetzungs- und Implikationsregel als Schlußschemata unabhängig.

Dies zeigt folgende unendliche Matrix: Elemente seien die ganzen Zahlen ≥ 0; ausgezeichnetes Element sei 0; ${\sim}p = p + 1$; $p \supset q = 0$ für $p \geq q$; $p \supset q = 1$ für $p < q$.

Anderseits aber gilt: Für jede *endliche* normale Matrix, für welche 1), 2), 3) erfüllt sind, ist auch $p \supset {\sim}{\sim}p$ erfüllt. Beweis: Es sei $\mathfrak{A}$ die endliche Menge der Elemente, $\mathfrak{A}^* \subset \mathfrak{A}$ die der ausgezeichneten Elemente einer normalen Matrix, für welche 1), 2), 3) erfüllt sind, d. h. $\sim$ und $\supset$ sind als Operationen innerhalb $\mathfrak{A}$ gemäß folgenden Bedingungen definiert:

a) Die Axiome 1), 2), 3) ergeben für beliebige $p, q \in \mathfrak{A}$ immer Elemente aus $\mathfrak{A}^*$.

b) Aus $a \in \mathfrak{A}^*$ und $a \supset b \in \mathfrak{A}^*$ folgt immer $b \in \mathfrak{A}^*$.

Es ist zu zeigen, daß dann auch $p \supset {\sim}{\sim}p \in \mathfrak{A}^*$ für jedes $p \in \mathfrak{A}$. Da $\mathfrak{A}$ nur endlich viele Elemente enthält, muß es zwei Zahlen k, n geben, so daß

---

[1] Über den Begriff der logischen Matrix vgl. *Łukasiewicz und Tarski 1930*.

shows that no finite model would suffice. It was less than three months after this colloquium that Gödel answered a previous query of Hahn's about the intuitionistic propositional calculus by again showing the inadequacy of finite models; see *Gödel 1932.*

W. V. Quine

On independence proofs in the
propositional calculus
(*1933a*)

Can every independence proof for a formula of the propositional calculus be carried out by means of *finite* models (matrices[1])? This question, raised by Hahn, is to be answered in the negative. For example, the proposition $p \supset \sim\sim p$ is independent of the following three axioms:

$$(1) \ \ p \supset p, \quad (2) \ \ p \supset \sim\sim q . \supset . p \supset q, \quad (3) \ \sim\sim p \supset \sim\sim q . \supset . p \supset q,$$

with the rules of substitution and implication as inference schemas.

This is shown by the following infinite matrix: let its elements be the integers ≥ 0; let 0 be the distinguished element; let $\sim p = p+1$; let $p \supset q = 0$ for $p \geq q$; and let $p \supset q = 1$ for $p < q$.

But, on the other hand, the following holds: For every *finite* normal matrix for which (1), (2) and (3) are satisfied, $p \supset \sim\sim p$ is also satisfied. Proof: Let $\mathfrak{A}$ be the finite set of elements and $\mathfrak{A}^* \subset \mathfrak{A}$ the set of the distinguished elements of a normal matrix for which (1), (2) and (3) are satisfied, that is, $\sim$ and $\supset$ are defined as operations within $\mathfrak{A}$ in accordance with the following conditions:

(a) For arbitrary $p, q \in \mathfrak{A}$, axioms (1), (2) and (3) always yield elements of $\mathfrak{A}^*$.

(b) From $a \in \mathfrak{A}^*$ and $a \supset b \in \mathfrak{A}^*$, it always follows that $b \in \mathfrak{A}^*$.

It is to be shown that, in that case, also $p \supset \sim\sim p \in \mathfrak{A}^*$ for every $p \in \mathfrak{A}$. Since $\mathfrak{A}$ contains only finitely many elements, there must exist two numbers

---

[1] Concerning the notion of logical matrix, see *Łukasiewicz and Tarski 1930.*

$$\sim^{2k} p = \sim^{2n} p$$

und $k < n$. Wegen 1) gilt dann:

$$\sim^{2k} p \supset \sim^{2n} p \; \epsilon \; \mathfrak{A}^*;$$

wegen 3) gilt:

$$(\sim^{2k} p \supset \sim^{2n} p \,.\, \supset \,.\, \sim^{2k-2} p \supset \sim^{2n-2} p) \; \epsilon \; \mathfrak{A}^*$$

und daher: $\sim^{2k-2} p \supset \sim^{2n-2} p \; \epsilon \; \mathfrak{A}^*$. Durch mehrmalige Anwendung dieses Schlusses folgt schließlich $p \supset \sim^{2(n-k)} p \; \epsilon \; \mathfrak{A}^*$. Wendet man darauf $(n - k - 1)$-mal 2) an, so folgt $p \supset \sim\sim p \; \epsilon \; \mathfrak{A}^*$, w. z. b. w. In den obigen Axiomen könnte man statt $\sim\sim$ überall $\sim$ setzen. Das durch die Axiome 1), 2), 3) definierte logische System wäre aber dann kein Teilsystem der gewöhnlichen Logik (trotzdem wäre es widerspruchsfrei in dem Sinn, daß A und $\sim A$ niemals beide beweisbar sind).

Zu jeder gegenüber der Einsetzungsregel abgeschlossenen Menge M von Formeln des Aussagenkalküls gibt es eine kleinste Kardinalzahl m ($\leq \aleph_0$), so daß Modelle mit m, aber nicht mit weniger Elementen existieren. Unter einem Modell wird dabei eine Matrix verstanden, für welche alle und nur die Formeln aus M erfüllt sind. Für die Menge der aus obigen Axiomen ableitbaren Formeln ist diese Zahl $\aleph_0$; ebenso z. B. für den Heytingschen Aussagenkalkül.[2] Die Formelmengen, für die $m = 2$ ist, sind isomorph mit solchen Teilsystemen des gewöhnlichen Aussagenkalküls, welche alle und nur die Tautologien enthalten, die sich unter alleiniger Verwendung gewisser Begriffe (z. B. mit $\supset$ allein) ausdrücken lassen.

---

[2] Vgl. *Gödel 1933n* ⟦= *1932*⟧.

k and n such that

$$\sim^{2k}p = \sim^{2n}p$$

and $k < n$. On account of (1)

$$\sim^{2k}p \supset \sim^{2n}p \ \epsilon \ \mathfrak{A}^*$$

then holds; on account of (3)

$$(\sim^{2k}p \supset \sim^{2n}p . \supset . \sim^{2k-2}p \supset \sim^{2n-2}p) \ \epsilon \ \mathfrak{A}^*$$

holds, and therefore $\sim^{2k-2}p \supset \sim^{2n-2}p \ \epsilon \ \mathfrak{A}^*$. By repeated application of this inference, $p \supset \sim^{2(n-k)}p \ \epsilon \ \mathfrak{A}^*$ eventually follows. If we apply (2) to this $(n - k - 1)$ times, then $p \supset \sim\sim p \ \epsilon \ \mathfrak{A}^*$ follows, as was to be proved. In the axioms above one could put $\sim$ everywhere in place of $\sim\sim$; but then the logical system defined by axioms (1), (2) and (3) would not be a subsystem of ordinary logic (it would nevertheless be consistent in the sense that A and $\sim A$ are never both provable).

Whenever a set M of formulas of the propositional calculus is closed with respect to the rule of substitution, there is a smallest cardinal number m ($\leq \aleph_0$) such that models with m elements, but not with fewer elements, exist. Here a model must be understood to be a matrix for which all and only the formulas of M are satisfied. For the set of formulas derivable from the axioms above, this number is $\aleph_0$; likewise, for example, for Heyting's propositional calculus.[2] The sets of formulas for which $m = 2$ are isomorphic with those subsystems of the ordinary propositional calculus that contain all and only those tautologies that can be expressed by use of but one of a certain group of notions (for example, by means of $\supset$ alone).

[2] See *Gödel 1933n* $[\![= 1932]\!]$.

Introductory note to *1933b, c, d, g* and *h*

In addition to the many important logical results on which Gödel reported in Karl Menger's *Ergebnisse eines mathematischen Kolloquiums* at the University of Vienna, there are five lesser known ones on geometry. In fact, Gödel had taken courses with Menger, and these notes are best seen within the framework of a program for reforming and purifying differential and projective geometry developed by Menger in the colloquium. In order to free the notion of "curvature" from "the complicated conceptual machinery" (*Menger 1952*) of classical differential geometry, in particular from coordinates, parameterizations and differentiability assumptions, Menger proposed to study suitable n-tuples of points in compact convex metric spaces. The triangle inequality implies the existence of three points in the Euclidean plane isometric to any triple of points in such a space, and their curvature is taken to be the reciprocal of the radius of the circle circumscribed around them. Menger then defined the curvature at a point of a curve in his space as the number from which the curvature of any three sufficiently close isometric points in the Euclidean plane differs arbitrarily little. Several results were based on this definition as well as on modifications of it by Franz Alt and by Gödel himself (*Alt 1933*). For surfaces the problem of curvature is more ambiguous and difficult. One now considers quadruples of points in Menger's spaces; but points isometric to them may not exist at all in Euclidean space, and even when they do exist, the reciprocal of the radius of their circumscribed sphere is of no particular significance for the problem of curvature. Gödel (*1933b*) shows, however, in answer to a question of Laura Klanfer, that this reciprocal can be used to prove that, if an isometric Euclidean quadruple exists and is non-coplanar, then the metric quadruple is isometric, under the geodesic metric, to four points of a metricized sphere of suitable radius. In Gödel's only joint paper (*1933h*), so called "volume determinants" $D(p_1, p_2, p_3, p_4)$ of metric quadruples are introduced, and it is suggested that their strong convergence to 0 for suitable sequences of quadruples can clarify in a coordinate-free manner the local Euclidean behavior of Gaussian surfaces. Menger and Blumenthal in their *1970* discuss the relation of these ideas to Wald's eventual solution of this problem.

To define convexity for his spaces, Menger had introduced a metrically defined betweenness relation that Abraham Wald (*1931*) succeeded in characterizing by means of six postulates which any ternary relation between points must satisfy. Gödel (*1933c*) showed in turn that Wald's characterization can be transformed by means of a simple coding into a

theorem about triples of points in the Cartesian space of real numbers. The study of dimension also led Menger (*1932a*) to formulate various kinds of covering systems for compact spaces, in particular, the idea of an "unlimited covering system" for a space. Gödel (*1933g*) makes essential use of this idea to show that every one-to-one map ϕ of the real projective plane that preserves straight lines is a collineation. Specifically, he observes that the "conic-section neighborhoods" of the projective plane are such a covering system, and this allows him to prove that such a map ϕ must be continuous, from which his result follows by the fundamental theorem of projective geometry.

Menger's approach (*1928*) to projective n-space tried to avoid the disadvantages of those proposed by Oswald Veblen and David Hilbert: the latter's "point-line-plane" formulation would require a new geometrical primitive for each new dimension, while the former's definitional introduction of lines and planes bestowed an unjustifiably distinguished role on undefined points, as well as making projective geometry *too dependent on set theory*. Menger thus posited a *Feld* of undefined elements (see *Bergmann 1931*) and two algebraic operations of join and meet on them and defined an inclusion relation between them corresponding to the incidence relation. He could then define points as those elements that include no non-empty proper parts, thereby formalizing Euclid's hitherto logically unusable definition of a point as "that which has no parts". Gödel (*1933d*) remarked, however, that such definitions could not be formulated in the subsystem consisting of sentences with no existential quantifiers when put in normal form. In a letter dated 2 January 1968, Menger confessed to Gödel that he did not understand the remark, since "P is a point" could just as well be defined by saying that any element included in P is identical either with P or with the empty set. Gödel replied that on this definition any proposition containing "P is a point" as a hypothesis would still contain existential quantifiers in normal form, but he granted that *some* propositions about the defined term would be expressible in the restricted system. Gödel's suggestion to investigate systems of purely universal sentences about Menger's *Felder* appears to be a natural one, however, when we examine their subsequent historical development. In fact, it is of a piece with another suggestion (to be discussed below) made by Gödel a year later to Garrett Birkhoff, which sheds some interesting light on his thinking about completeness problems outside arithmetic.

First of all, Menger's *Felder* were not literally fields but *lattices* (the term was introduced in *Birkhoff 1933*), even though they were introduced to "obtain a far-reaching analogy with abstract algebra, where, in defining a field, one also starts with a single class of undefined elements and two undefined operations of addition and multiplication"

(*Menger 1936*). (Herbert Mehrtens (*1979*) gives a revealing account of the place of Menger's work in the emergence of lattice theory.) His axiomatic theory did not attempt to describe a single structure, but rather to formulate simple laws which held simultaneously in several domains, especially the field of finite sets and n-dimensional projective space. The lattice structure of the linear subspaces of projective geometry, in particular the failure of the distributive law for their joins and meets, had been noted already by Alwin Korselt and was later rediscovered independently by Menger and Birkhoff, who also discovered a weaker special case of this law, namely modularity,

$$\text{if } x \leq z, \text{ then } x \cup (y \cap z) = (x \cup y) \cap z,$$

which, together with the laws of idempotency, commutativity, associativity, and absorption, proved to be essential in the lattice-theoretic derivation of projective laws. The independence of the distributive law from the others governing "and" and "or" was as important for the emergence of lattice theory as that of the parallel postulate was for non-Euclidean geometry. In fact, the paper of Georg Wernick (*1929*) in which the former is rigorously established for the propositional calculus was cited not only by Fritz Klein (*1932*), where the term *Verband* is introduced, but also by Gödel (*1933d*) as a contribution to the study of his restricted system. Indeed, all the lattice laws mentioned above contain, in normal form, only universal quantifiers, so that as far as *axioms* are concerned, Gödel's suggestion would lead one to study modular lattices (the complementation axiom contains a mixed quantifier prefix).

In any case, we do have some record of Gödel's skepticism that even complemented modular lattices provide a *complete* foundation for projective geometry, and it is interesting to compare the kind of incompleteness he suspected here with that which he had recently discovered in formal arithmetic. Thus Birkhoff (*1935*)—and, independently, also Menger (*1936*)—proved a representation theorem stating that every complemented modular lattice is isomorphic to the direct product of a finite Boolean algebra and a finite number of projective geometries and, conversely, that every such direct product is a complemented modular lattice. This shows, according to Birkhoff, that his lattice axioms "in

one sense ... are complete" (*1935*, page 743), but he attributes (*ibid.*, footnote 16) to a personal suggestion of Gödel the important further question of whether "all the laws of combination of projective manifolds" can actually be derived from them "by the ordinary rules of inference". That this could not be assumed to follow from the representation theorem was "illustrated by the fact that every finite field obeys the law $ab = ba$, although this is no longer true without the restriction of finiteness" (*ibid.*). Gödel was clearly thinking here of the classic results of Hilbert (*1899*), where the connection between fields and projective "laws of combination" was clearly established and where, in particular, it was shown that Pascal's theorem is equivalent to the commutativity of field multiplication. But Hilbert also showed that Pascal's law, which was known to be equivalent to the fundamental theorem of projective geometry, is not a consequence of the projective space axioms without continuity, nor, in the plane, even of Desargues's theorem. Although it was finally shown by M. P. Schützenberger (*1945*) how to express Desargues' theorem as a lattice identity, it turns out that Pascal's theorem cannot be equivalent to any lattice identity, let alone a consequence of modularity and complementation (see *Birkhoff 1940*, page 71). It is thus clear that the fundamental "law of combination" of projective space is not a consequence of the lattice axioms. Moreover, it is precisely this law that is needed to coordinatize the projective plane, in particular, to prove that "the points on a line in our algebra of geometry behave like the numbers of a field in the algebra of numbers" (*Menger 1940*). The goal of the lattice-theoretic foundation of projective geometry had been a purely elementary algebraic one, that is, one not using continuity axioms and hence not requiring any set theory. Gödel seems pretty clearly to have anticipated the general obstacle that Hilbert's negative results would pose for this goal, though not the specific proof that Pascal's law cannot be formulated as a lattice identity. Just as he had shown that the categoricity of second-order arithmetic by no means guarantees the deductive completeness of its first-order formalization, so too the categoricity of projective geometry with continuity does not entail a complete elementary foundation for it in lattice theory.

Judson Webb

Über die metrische Einbettbarkeit der Quadrupel des R_3 in Kugelflächen

(1933b)

In Beantwortung einer im 37. Kolloquium (*Klanfer 1933*) aufgeworfenen Frage gilt: *Ein metrisches Quadrupel, das mit vier Punkten des R_3 kongruent ist, ist, wenn es nicht mit vier Punkten der Ebene kongruent ist, auch kongruent mit vier Punkten einer Kugelfläche, in welcher als Abstand je zweier Punkte die Länge des kürzesten sie auf ihr verbindenden Bogens erklärt ist.* Zum Beweise sei T ein Tetraeder im R_3, dessen sechs Kantenlängen $a_1, a_2, \ldots, a_6$ den Abständen der vier gegebenen Punkte gleich seien und von dem wir | annehmen können, daß es nicht in einer Ebene liegt. R heiße der Radius der T umgeschriebenen Kugel. Wir setzen

$$a_{i,x} = \frac{2}{x} \sin \frac{a_i x}{2};$$

dann ist also $a_{i,x}$ die Länge der Sehne eines Kreisbogens der Länge a_i auf einem Kreis vom Radius $\frac{1}{x}$; für $x = 0$ hat $a_{i,x}$ den Grenzwert a_i. Ein (eventuell ebenes) Tetraeder im R_3, dessen sechs Kanten die Längen $a_{i,x}$ besitzen, wollen wir, falls ein solches Tetraeder existiert, mit $T(x)$ bezeichnen. Dabei setzen wir $T(0) = T$. Den reziproken Radius der $T(x)$ umgeschriebenen Kugel nennen wir $f(x)$; es ist dann also $f(0) = \frac{1}{R}$; liegt $T(x)$ in einer Ebene, so ist $f(x) = 0$. Zwischenbehauptung: Heißt a die größte der sechs Zahlen a_i, so existiert im offenen Intervall $\left(0, \frac{\pi}{a}\right)$ eine Zahl x, sodaß $f(x) = x$. Beweisskizze: Es sei $\bar{x}$ die größte positive Zahl $\leq \frac{\pi}{a}$, so daß für jedes x des abgeschlossenen Intervalles $[0, \bar{x}]$ das Tetraeder $T(x)$ existiert. (Die Existenz eines solchen $\bar{x}$ ergibt sich aus Stetigkeitsgründen). $f(x)$ ist eine im Intervall $[0, \bar{x}]$ stetige Funktion. Ist erstens $\bar{x} = \frac{\pi}{a}$, dann existiert also $T\left(\frac{\pi}{a}\right)$ und die größte Kante dieses Tetraeders hat die Länge $\frac{2a}{\pi}$, so daß der Radius der Umkugel $> \frac{a}{\pi}$ und $f\left(\frac{\pi}{a}\right) < \frac{\pi}{a}$ gilt. Da andererseits $f(0) = \frac{1}{R} > 0$, so muß für einen Zwischenwert $f(x) = x$ gelten. Ist zweitens $\bar{x} < \frac{\pi}{a}$, dann muß $T(\bar{x})$ in einer Ebene liegen, denn sonst würde für jedes hinreichend wenig von $\bar{x}$ verschiedene x auch $T(x)$ existieren entgegen der Definition von $\bar{x}$; also ist $0 = f(\bar{x}) < \bar{x}$, was zusammen mit $f(0) > 0$ die

On the isometric embeddability of quadruples
of points of R_3 in the surface of a sphere
(*1933b*)

In answer to a question raised in the 37th colloquium (*Klanfer 1933*), the following holds: *A metric quadruple* ⟦*a quadruple of points in a metric space*⟧ *that is congruent to four points in R_3 is, if it is not congruent to four points of the plane, congruent to four points on the surface of a sphere on which the distance between any two points is defined to be the length of the shortest arc joining them.* For the proof, let T be a tetrahedron (which, we may assume, does not lie in a plane) in R_3 whose six edges are of lengths $a_1, a_2, \ldots, a_6$ that are equal to the distances between the four given points. Let R be the radius of the sphere circumscribed about T. We put

$$a_{i,x} = \frac{2}{x} \sin \frac{a_i x}{2};$$

accordingly, $a_{i,x}$ is then the length of the chord of a circular arc of length a_i on a circle of radius $\frac{1}{x}$; $a_{i,x}$ has the limiting value a_i for $x = 0$. Let $T(x)$ be a (possibly planar) tetrahedron in R_3 whose six edges are of lengths $a_{i,x}$, if such a tetrahedron exists. In this context we set $T(0) = T$. The reciprocal of the radius of the sphere circumscribed about $T(x)$ we call $f(x)$; thus $f(0) = \frac{1}{R}$ and, if $T(x)$ lies in a plane, then $f(x) = 0$. Intermediate assertion: If a is the largest of the six numbers a_i, then in the open interval $\left(0, \frac{\pi}{a}\right)$ there exists a number x for which $f(x) = x$. Sketch of proof: Let $\overline{x}$ be the largest positive number $\leq \frac{\pi}{a}$ such that, for every x in the closed interval $[0, \overline{x}]$, the tetrahedron $T(x)$ exists. (The existence of such an $\overline{x}$ follows by continuity.) On the interval $[0, \overline{x}]$, $f(x)$ is a continuous function. If, first, $\overline{x} = \frac{\pi}{a}$, then $T\left(\frac{\pi}{a}\right)$ exists, and the longest edge of this tetrahedron has length $\frac{2a}{\pi}$, so that the radius of the circumscribed sphere is $> \frac{a}{\pi}$, and $f\left(\frac{\pi}{a}\right) < \frac{\pi}{a}$ holds. Since, on the other hand, $f(0) = \frac{1}{R} > 0$, $f(x) = x$ must hold for some intermediate value x. If, second, $\overline{x} < \frac{\pi}{a}$, then $T(\overline{x})$ must lie in a plane since, otherwise, for every x differing from $\overline{x}$ by sufficiently little, $T(x)$ would exist, contrary to the definition of $\overline{x}$; hence $0 = f(\overline{x}) < \overline{x}$, which together with $f(0) > 0$ yields the existence of an $x < \frac{\pi}{a}$ with $f(x) = x$,

Existenz eines $x < \dfrac{\pi}{a}$ mit $f(x) = x$ ergibt, womit die Zwischenbehauptung bewiesen ist. Für die der Bedingung $f(x) = x$ gemäß bestimmte Zahl x hat das Tetraeder $T(x)$ die Kantenlängen $a_{i,x}$ und eine umgeschriebene Kugel vom Radius $\dfrac{1}{x}$. Also haben die Eckpunkte von $T(x)$ auf der Umkugel wegen $x < \dfrac{\pi}{a_i}$ die Bogenabstände a_i, so daß auf dieser Kugel ein mit den gegebenen vier Punkten kongruentes Quadrupel existiert.

Über die Waldsche Axiomatik
des Zwischenbegriffes
(*1933c*)

In seiner Axiomatik des Zwischenbegriffes (*1931* und *1932*) beweist Wald, daß drei Punkte | a, b, c in einer gewissen Bedingungen genügenden Relation aller Punktetripel aller metrischen Räume dann und nur dann stehen, wenn für ihre Abstände die Beziehung $ab + bc = ac$ gilt, d. h. wenn b im Mengerschen Sinne zwischen a und c liegt. Unter den der Relation auferlegten Bedingungen tritt dabei die Kongruenzinvarianz auf, d. h. wenn drei Punkte a, b, c in der betreffenden Relation stehen, so sollen auch je drei mit ihnen kongruente Punkte a', b', c' in der betreffenden Relation zu einander stehen. Da die Gesamtheit aller zum Punktetripel a, b, c kongruenten Tripel durch die drei Abstandszahlen ab, bc, ac gekennzeichnet ist, so läßt sich die Waldsche Axiomatik des Zwischenbegriffes als ein Satz über Tripel reeller Zahlen, mit anderen Worten über den dreidimensionalen Cartesischen Raum aussprechen. Wir erhalten denselben, indem wir jedem metrischen Punktetripel a, b, c den Punkt

$$(+) \qquad\qquad x = ab,\ \ y = bc,\ \ z = ac$$

des R_3 zuordnen. Da für a, b, c die Dreiecksungleichung gilt und Abstände nicht negativ sind, treten dabei nur Punkte des R_3 auf, für die

$$(*) \qquad x \geq 0,\ \ y \geq 0,\ \ z \geq 0,\ \ (x + y - z)(x - y + z)(-x + y + z) \geq 0$$

gilt. Es liegt b zwischen a und c dann und nur dann, wenn der dem Tripel a, b, c zugeordnete Punkt x, y, z in dem im Raumteil $(*)$ enthaltenen Teil der Ebene $x + y = z$ liegt. Der Waldschen Axiomatik entspricht dann folgender Satz:

whereby the intermediate assertion is proved. For the number x determined by the condition $f(x) = x$, the tetrahedron $T(x)$ has edges with lengths $a_{i,x}$ and a circumscribed sphere of radius $\dfrac{1}{x}$. Hence, since $x < \dfrac{\pi}{a_i}$, the vertices of $T(x)$ on the circumscribed sphere are separated by arc lengths a_i, so that on this sphere a quadruple [of points] exists that is congruent with the four given points.

On Wald's axiomatization of the notion of betweenness
(*1933c*)

In his axiomatization of the notion of betweenness (*1931* and *1932*), Wald proves that three points a, b and c stand in a relation defined for all triples of points in all metric spaces and satisfying certain conditions if and only if the relation $ab + bc = ac$ holds for the distances between them, that is, if and only if b lies between a and c in Menger's sense. Among the conditions imposed on the relation there appears that of invariance under congruence, that is, if three points a, b and c satisfy the relation in question, then any three points a', b' and c' congruent to them are also to stand in that relation to one another. Since the totality of all triples congruent to the triple of points a, b and c is characterized by the three distances ab, bc and ac, Wald's axiomatization of the notion of betweenness can be expressed as a proposition about triples of real numbers, in other words, about three-dimensional Cartesian space. We obtain that proposition by assigning to each metric triple of points a, b and c the point

$$(+) \qquad\qquad x = ab,\ y = bc,\ z = ac$$

of R_3. Since the triangle inequality holds for a, b, c and since distances are non-negative, only such points of R_3 will occur for which the inequalities

$$(*) \qquad x \geq 0,\ y \geq 0,\ z \geq 0,\ (x + y - z)(x - y + z)(-x + y + z) \geq 0$$

hold. Then b lies between a and c if and only if the point x, y, z assigned to the triple a, b, c lies in the part of the plane $x + y = z$ contained in the subspace $(*)$. To Wald's axiomatization there then corresponds the following proposition:

Es sei M eine Teilmenge des Raumteiles (∗) *des R_3 mit folgenden Eigenschaften:*

1. *Liegt der Punkt x, y, z in M, so auch der Punkt y, x, z.*

2. *Liegt der Punkt x, y, z in M, so liegt der Punkt z, y, x nicht in M.*

3. *Liegen die Punkte u, v, w und w, x, y in M, so liegen für jedes z die Punkte u, z, y und v, x, z, falls sie im Raumteil (∗) liegen, auch in M.*

4. *Für jedes $\bar{z} > 0$ existiert mindestens ein Punkt $x, y, \bar{z}$ von M.*

5. *Für jedes $\bar{z} > 0$ ist die Menge bestehend aus den Punkten $0, \bar{z}, \bar{z}$ und $\bar{z}, 0, \bar{z}$, sowie allen Punkten $x, y, \bar{z}$ von M abgeschlossen.*

Dann ist M der den Ungleichungen (∗) *genügende Teil der Ebene $x + y = z$.*

Es wird durch diesen Satz nahegelegt, allgemein für metrische Räume die durch (+) definierte Abbildung ihrer Punktetripel in den R_3 zu betrachten.

Zur Axiomatik der elementargeometrischen Verknüpfungsrelationen
(*1933d*)

Man sollte das System aller jener Aussagen über Felder untersuchen, die in der Normalform keine Existenzpräfixe enthalten. Die Begriffe Punkt und Gerade, die unter Verwendung von Existenzpräfixen definierbar sind (z. B. Punkt ist ein Element, zu dem kein nicht leeres Element existiert, das sein echter Teil ist), sind in diesem engeren System undefinierbar. Untersuchungen in dieser Richtung finden sich bei Wernick (*1929*).

Let M be a subset, of the subspace $(*)$ of R_3, with the following proper-ties:

1. If the point x, y, z lies in M, so does the point y, x, z.

2. If the point x, y, z lies in M, then the point z, y, x does not lie in M.

3. If the points u, v, w and w, x, y lie in M, then, for every z, the points u, z, y and v, x, z, in case they lie in the subspace $(*)$, also lie in M.

4. For every $\overline{z} > 0$ there exists at least one point $x, y, \overline{z}$ in M.

5. For every $\overline{z} > 0$, the set consisting of the points $0, \overline{z}, \overline{z}$ and $\overline{z}, 0, \overline{z}$, as well as all points $x, y, \overline{z}$ of M, is closed.

Then M is the part of the plane $x + y = z$ satisfying the inequalities $(*)$.

This proposition makes it plausible, when we are dealing with metric spaces, to consider quite generally the mapping defined by $(+)$ of their triples of points into R_3.

On the axiomatization of the relations of connection in elementary geometry
(*1933d*)

One should investigate the system of all those propositions about lattices that in normal form contain no existential quantifiers. The notions of point and line, which are definable by use of existential quantifiers (for example, a point is an element for which there does not exist a non-empty element that is a proper part of it), are not definable in this more restricted system. Investigations in this direction are to be found in *Wernick 1929*.

Introductory note to *1933e*

Gödel's paper *1933e*, given at Menger's colloquium in Vienna on 28 June 1932, deals with the relationship between classical and intuitionistic first-order arithmetic and shows how to interpret the former within the latter by a simple translation.

Below, we shall use $\wedge$, $\vee$, $\forall$, $\exists$, $\rightarrow$, $\neg$ for the logical operators, x, y, $z, \ldots$ for individual variables, F, G for formulas, and P for atomic formulas.[a]

Let **A** (from "Aussagenlogik") be a system of classical propositional logic and **H** (from "Heyting") the intuitionistic system of propositional logic as formulated by Heyting (*1930*).

Using a result from Glivenko (*1929*) that $\mathbf{H} \vdash \neg A$ if and only if $\mathbf{A} \vdash \neg A$, Gödel begins with a very economical proof of

$$(1) \qquad \qquad \mathbf{H} \vdash F \quad \text{if and only if } \mathbf{A} \vdash F,$$

for F in the $\wedge$, $\neg$-fragment. From this he concludes that, for the mapping $'$ defined by $P' := P$ (P atomic), $(F \wedge G)' := F' \wedge G'$, $(F \vee G)' := \neg(\neg F' \wedge \neg G')$, $(F \rightarrow G)' := \neg(F' \wedge \neg G')$, we have

$$(2) \qquad \qquad \text{if } \mathbf{A} \vdash F, \quad \text{then } \mathbf{H} \vdash F'.$$

Gödel then extends these results to arithmetic. For classical arithmetic, denoted in the present note by **Z** (from "Zahlentheorie"), he takes the system of Herbrand (*1931*), with four groups (A, B, C, D) of nonlogical axioms. As we shall see, **Z** is not quite a formal system as the term is usually understood today.

Group A consists of the usual axioms for equality and successor. Group B consists of all instances of the induction schema. Group C consists of "defining axioms" for functions, which permit us to compute the values of the functions for each set of arguments; moreover, the fact that the function value is uniquely determined by the axioms must have been established finitistically. In particular, Group C contains defining equations for all primitive recursive functions. Group D consists of all finitistically provable purely universal statements (Π_1^0-state-

---

[a]Following Herbrand and Heyting, Gödel formally distinguishes between classical $\rightarrow$, $\sim$, $\cdot$ and intuitionistic $\supset$, $\neg$, $\wedge$, but this is not necessary for the present purposes. For $\forall$, $\exists$ Gödel uses ( ), E. For the formal systems designated by A, H and H' in Gödel's text we use boldface **A**, **H** and **H'**, respectively, for greater typographical clarity.

ments).[b]

Here the term "finitistic" should be understood in the sense of Hilbert or Herbrand (e.g., *1930*); finitistic mathematics may be characterized as the mathematics of finite manipulations of finitely presented, combinatorial configurations. (N.B.: in *1931*, Herbrand uses "intuitionistic" instead of "finitistic").

H′, the intuitionistic counterpart of **Z**, is now taken to be the corresponding system based on Heyting's formalization (*1930a*) of intuitionistic mathematics (**M** for short). Disregarding some inessential syntactic details, we can now state Gödel's main result as follows. Extending the mapping by $(\forall x F)' := \forall x F'$, we have

(3) $$\text{if } \mathbf{Z} \vdash F, \quad \text{then } \mathbf{H'} \vdash F'.$$

Gödel treats $\exists$ as defined in **Z**, hence there is no translation clause for it; if $\exists$ is treated as a primitive, the obvious clause becomes $(\exists x F)' := \neg \forall x \neg F'$.

H′ is not actually a subsystem of Heyting's **M**; in particular, axioms of Groups C and D have to be added, as well as variables and quantifiers ranging over N. These modifications are obviously justified from the intuitionistic point of view.

In **Z** the natural numbers start with 0, but in **M**, and in **H′** as described by Gödel, they start with 1 instead. This causes a slight difficulty in the definition of the translation which seems to have been overlooked by Gödel. Suppose, for example, addition to have been specified by axioms in Group C :

$$x + 0 = x, \quad x + Sy = S(x + y)$$

(we use the successor symbol "S" where Herbrand writes "$+1$" and Heyting uses "seq'"). Gödel's translation would transform the first axiom into $x + 1 = x$, which is obviously false. The easiest way to repair this is probably to modify **H′** and let the natural numbers start with 0 in **H′** too.

Gödel concludes the note with the observation that his results show that intuitionistic arithmetic is only apparently weaker than classical arithmetic. In fact, as a corollary to (3) and Gödel's second lemma, we have that, if **Z** is formulated on the basis of $\wedge$, $\forall$, $\neg$ as primitives, then **Z** is a subsystem of **H′**. Thus we have a technically very simple

---

[b]Nowadays classical first-order arithmetic **PA** is usually based on A, B and defining axioms for $+$, $\cdot$, or perhaps for all primitive recursive functions; and similarly for intuitionistic first-order arithmetic **HA**. The further axioms in C, D are not essential for Gödel's result. As to the identification of Herbrand's "intuitionnistiquement" in his description of the axioms of C and D with "finitistically", see *van Heijenoort 1982*.

consistency proof of **Z** relative to intuitionistic arithmetic.

According to Bernays (*1967*), Gödel's paper showed the members of the Hilbert school that there were alternatives to finitistic reasoning as a basis for metamathematics, and also that intuitionistic principles went beyond finitism. The latter is in fact implicit in the observation just made, namely that **Z** is part of **H′**, whereas on the other hand a finitistic justification of **Z** is clearly a non-trivial enterprise. Also, it is to be noted that Heyting's explanation of the intuitionistic logical operators, as given e.g. in *Heyting 1931* (an explanation which enters into the intuitionistic justification of the formalism of *Heyting 1930a*), relies on abstract concepts such as "construction" that go well beyond the scope of finitistic reasoning.

Another way of looking at Gödel's result is to think of **H′** as consisting of **Z** extended with two new "constructive" logical operators $\vee$, $\exists$, in addition to the definable classical $\vee'$, $\exists'$. However, it should be kept in mind that the law of double negation $\neg\neg F \to F$ is not generally valid for F containing the new $\vee$, $\exists$.

Gödel's paper had been anticipated to a not inconsiderable extent by Kolmogorov (*1925*). Kolmogorov describes a translation which consists in simultaneously inserting $\neg\neg$ in front of each subformula. For the $\to$, $\neg$ fragment of propositional logic he gives full details; the treatment of the quantifiers is only sketched. There is a careful introduction by H. Wang to the English translation of Kolmogorov's paper in *van Heijenoort 1967*, although Wang perhaps slightly overstated the case for Kolmogorov's anticipation of Heyting and Gödel. Kolmogorov's paper, written in Russian, attracted little or no attention and was apparently unknown to Gödel.

The same result (3) was proved, independently of Gödel, by Gentzen and Bernays in 1933, using a slightly more elegant translation which preserves $\to$; however, Gentzen withdrew his paper from publication when he learned of Gödel's result. Gentzen later used the translation in his *1969* (page 313, footnote 13, and page 315, footnote 46).

Kuroda 1951 contains another very elegant translation: insert $\neg\neg$ after each universal quantifier and in front of the whole formula. All these translations yield intuitionistically equivalent results and thus may be regarded as variants of Gödel's ′. His translation and its variants are often referred to as "the negative translation", since the translated formulas are intuitionistically equivalent to formulas lacking the positive operators $\vee$, $\exists$.

Among the early contributions to the literature on the negative translation we should mention especially the treatment in Kleene's textbook *1952*.

Gödel's translation has been extended to stronger mathematical theories, for example, to second-order arithmetic with variables for sets

of natural numbers and with full impredicative comprehension (*Kreisel 1968a*, Section 5; *Kreisel 1968*, page 344), to type theory (*Myhill 1974a*) and to set theory (*Friedman 1973*, *Powell 1975*). In all these cases, as well as for predicate logic, the $'$-translation clause for atomic formulas is $P' := \neg\neg P$; for arithmetic this is redundant since there $\neg\neg P \leftrightarrow P$. An elegant and economical systematization of such extensions is to be found in a paper by Leivant (*198?*).

The extensions just mentioned are in apparent conflict with Gödel's remark in the final paragraph that intuitionism is a *genuine* restriction for analysis and set theory; however, for Gödel intuitionism excluded impredicativity, whereas in more recent publications impredicative versions of intuitionism have also been defended as tenable (see *Kreisel 1968*, §5, *1970*, §5, and *Myhill 1970*). On the narrower view which excludes impredicativity, the systems of intuitionistic second-order arithmetic and intuitionistic set theory referred to above are at best "formally intuitionistic", that is, based on intuitionistic logic.

On the other hand, the extension of Gödel's result to systems formulated in a language having *function* variables and containing versions of the countable axiom of choice is *not* straightforward. For example, the translation of

AC $\forall x \exists y F(x,y) \rightarrow \exists \alpha \forall x F(x, \alpha(x))$

(where α is a function variable) cannot in general be derived from suitable instances of AC itself. This difficulty played an important role in the attempts to extend *Gödel 1958* to analysis and led to the study of "bar recursion" (see our introductory note in Volume II to *Gödel 1958* and *1972*).

The negative translation has been, and still is, an important tool in proof-theoretic work, and we shall not attempt a survey of its applications here. For just one recent example see *Buchholz et alii 1981*, pages 90 and 197.

Semantical counterparts to the negative translation have also been formulated for the various types of truth-value semantics which have been developed for intuitionistic theorems, such as Beth- and Kripke-models, topological models, valuations in Heyting algebras ($=$ pseudo-Boolean algebras), etc. For an illustration in the simplest case, i.e. propositional logic, let H be a Heyting algebra, and let $R(H) := \{\neg\neg a : a \in H\}$ be the set of its regular elements ($\neg$ being the operation of pseudo-complementation). Under the partial order inherited from H, $R(H)$ is a Boolean algebra. To a valuation $\| \ \|$ of formulas in H there corresponds a valuation $\| \ \|^R$ in $R(H)$, given by $\|P\|^R := \neg\neg\|P\|$ for propositional letters P. This implies $\|F\|^R = \|F'\|$ for arbitrary propositional F. Thus we see that here Gödel's translation corresponds to the passage from valuations (models) in H to valuations (models)

in $R(H)$. (See _Rasiowa and Sikorski 1963_, IX, §5, §13, and X, §7. For the more general setting of many-sorted first-order theorems, see e.g. _Fourman and Scott 1979_, 2.14, 2.18(iii), Section 6, in particular 6.15.) It seems that W. Lawvere (_1971_) was the first to give a semantical

Zur intuitionistischen Arithmetik und Zahlentheorie
(_1933e_)

Läßt man den Grundbegriffen des Heytingschen Aussagenkalküls[1] die gleich bezeichneten des klassischen und der "Absurdität" ($\neg$) die Negation ($\sim$) entsprechen, so erscheint der intuitionistische Aussagenkalkül H als echtes Teilsystem des gewöhnlichen A. Bei anderer Zuordnung (Übersetzung) der Begriffe ist jedoch umgekehrt _der klassische ein Teilsystem des intuitionistischen_. Denn es gilt: Jede nur aus Konjunktionen ($\wedge$) und Negationen ($\neg$) aufgebaute in A gültige Formel ist auch in H beweisbar. Jede solche Formel muß nämlich die Gestalt haben:

$$\neg A_1 \wedge \neg A_2 \wedge \cdots \wedge \neg A_n,$$

und ist sie in A gültig, so auch jedes einzelne $\neg A_i$; dann ist nach Glivenko[2] $\neg A_i$ auch in H beweisbar und daher auch die Konjunktion der $\neg A_i$. Daraus folgt: Übersetzt man die klassischen Begriffe:

$$\sim p, \quad p \to q, \quad p \vee q, \quad p \cdot q$$

durch die folgenden intuitionistischen:

$$\neg p, \quad \neg(p \wedge \neg q), \quad \neg(\neg p \wedge \neg q), \quad p \wedge q,$$

so ist jede klassische Formel auch in H gültig.

Das Ziel der vorliegenden Untersuchung ist, zu zeigen, daß etwas Ähnliches auch _für die ganze Arithmetik und Zahlentheorie_ in dem Umfang, wie 35 sie z. B. durch die Axiome von Herbrand[3] | gegeben ist, gilt.[4] Auch hier

---

[1] _Heyting 1930_ (zitiert als H_1).

[2] _Glivenko 1929_.

[3] _Herbrand 1931_.

[4] Das von Glivenko (_1929_) für den Aussagenkalkül abgeleitete Resultat läßt sich auf die Zahlentheorie nicht ausdehnen.

version of Gödel's translation in the very general context of (elementary) toposes.

A. S. Troelstra

The translation is by Stefan Bauer-Mengelberg and Jean van Heijenoort.

On intuitionistic arithmetic and number theory
(*1933e*)

If to the primitive notions of Heyting's propositional calculus[1] we let correspond those notions of the classical propositional calculus that are denoted by the same sign and if to absurdity ($\neg$) we let correspond negation ($\sim$), then the intuitionistic propositional calculus H turns out to be a proper subsystem of the ordinary propositional calculus A. With another correlation (translation) of the notions, however, *the classical propositional calculus is*, conversely, *a subsystem of the intuitionistic one.* For we have: Every formula built up solely by means of conjunctions ($\wedge$) and negations ($\neg$) that holds in A is provable in H as well. Indeed every such formula must have the form:

$$\neg A_1 \wedge \neg A_2 \wedge \cdots \wedge \neg A_n,$$

and, if it holds in A, so does every single $\neg A_i$; then, according to Glivenko,[2] $\neg A_i$ is provable in H also and, therefore, so is the conjunction of the $\neg A_i$. From this it follows that, if we translate the classical notions

$$\sim p, \quad p \to q, \quad p \vee q, \quad p \cdot q$$

by the following intuitionistic notions

$$\neg p, \quad \neg(p \wedge \neg q), \quad \neg(\neg p \wedge \neg q), \quad p \wedge q,$$

respectively, then every ⟦valid⟧ classical formula also holds in H.

The goal of the present investigation is to show that something similar holds[4] also *for all of arithmetic and number theory,* delimited in scope by, say, Herbrand's axioms.[3] Here, too, we can give an interpretation of

---

[1] *Heyting 1930* (cited as H_1).

[2] *Glivenko 1929.*

[3] *Herbrand 1931.*

[4] The result obtained by Glivenko (*1929*) for the propositional calculus cannot be extended to number theory.

kann man eine Interpretation der klassischen Begriffe durch die intuitionistischen geben, *so daß sämtliche aus den klassischen Axiomen beweisbaren Sätze auch für den Intuitionismus gelten.*

Als *Grundzeichen* des Herbrandschen Systems betrachten wir: 1. die Operationen des Aussagenkalküls: $\sim$, $\rightarrow$, $\vee$, . ; 2. Zahlvariable: $x, y, z, \ldots$; 3. die Allzeichen: $(x), (y), \ldots$; 4. $=$; 5. 0 und $+1$; 6. die abzählbare Menge der auf Grund von Axiomgruppe C eingeführten Funktionszeichen f_i, deren jedem eine Zahl n_i (Anzahl der Leerstellen) zugeordnet ist. Um zu präzisieren, wie sich aus diesen Grundzeichen die Formeln aufbauen, definieren wir zunächst den Terminus *"Zahlausdruck"* durch folgende rekursive Vorschrift:

1. 0 und sämtliche Variable $x, y, \ldots$ sind Zahlausdrücke;
2. ist Z ein Zahlausdruck, so auch $Z + 1$;
3. sind $Z_1, Z_2, \ldots, Z_{n_i}$ Zahlausdrücke, so auch $f_i(Z_1, Z_2, \ldots, Z_{n_i})$.

Unter einer *Elementarformel* verstehen wir einen Ausdruck der Gestalt $Z_1 = Z_2$, wobei Z_1, Z_2 Zahlausdrücke sind. "Sinnvolle Formeln der Zahlentheorie" (im folgenden *Z-Formeln* genannt) sind entweder Elementarformeln oder Ausdrücke, die sich aus Elementarformeln mittels der Operationen des Aussagenkalküls und der Allzeichen $(x), (y), \ldots$ aufbauen.

Zu den Herbrandschen Axiomgruppen *A–D* fügen wir folgende dort nicht explizit angeführte *logische Axiomgruppen E–G* hinzu:

E. Jeder Ausdruck, der aus einer richtigen Formel des Aussagenkalküls durch Einsetzung von Z-Formeln für die Variablen entsteht, gilt als Axiom.

F. Alle Formeln der Gestalt: $(x)F(x) \rightarrow F(Z)$, wobei $F(x)$ eine beliebige Z-Formel und Z ein Zahlausdruck ist, sind Axiome (mit der selbstverständlichen Einschränkung, daß die Variablen aus Z in F nicht gebunden sein dürfen).

G. Alle Formeln der Gestalt:

$$x = y . \rightarrow . F(x) \rightarrow F(y)$$

(wobei $F(x)$ eine beliebige Z-Formel ist), sind Axiome.

Als *Schlußregeln* nehmen wir:

I. Aus A und $A \rightarrow B$ folgt B.

II. Aus $A \rightarrow B$ folgt $A \rightarrow (x)B$, wenn x in A nicht frei vorkommt.

In F, G und Regel II bedeuten x, y beliebige Zahlvariable.

Im Gegensatz zum Herbrandschen System gibt es im Heytingschen[5] keine Zahlvariable, sondern nur Variable $x, y, \ldots$ für beliebige Gegenstände. Dadurch ergeben sich für das folgende einige Komplikationen, die man dadurch vermeiden kann, daß man zunächst im Heytingschen System Variable für natürliche Zahlen $x', y', z', \ldots$ einführt in der Weise, daß jeder

---

[5] *Heyting 1930a* (zitiert als H_2).

the classical notions in terms of the intuitionistic ones *so that all proposi-tions provable from the classical axioms hold for intuitionism as well.*

We consider the *primitive signs* of Herbrand's system to be: (1) the operations of the propositional calculus, $\sim$, $\rightarrow$, $\vee$, $.$; (2) number variables, $x, y, z, \ldots$; (3) the universal quantifiers, $(x), (y), \ldots$; (4) $=$; (5) 0 and $x + 1$; (6) the denumerable set of function signs f_i that are introduced on the basis of Axiom Group C and to each of which a number n_i (the number of argument places) is assigned. In order to make precise how formulas are built up from these primitive signs, we first define the term *numerical expression* by means of the following recursive stipulation:

(1) 0 and all the variables $x, y, \ldots$ are numerical expressions;

(2) if Z is a numerical expression, so is $Z + 1$;

(3) if $Z_1, Z_2, \ldots, Z_{n_i}$ are numerical expressions, so is $f_i(Z_1, Z_2, \ldots, Z_{n_i})$.

By an *elementary formula* we understand an expression of the form $Z_1 = Z_2$, where Z_1 and Z_2 are numerical expressions. 'Meaningful formulas of number theory' (called *Z-formulas* in what follows) are either elementary formulas or expressions that are built up from elementary formulas by means of the operations of the propositional calculus and the universal quantifiers, $(x), (y), \ldots$.

To Herbrand's Axiom Groups A–D we adjoin the following Axiom Groups E–G of *logical axioms*, not explicitly given by him:

E. Every expression that results from a correct formula of the propo-sitional calculus when Z-formulas are substituted for the variables is an axiom.

F. All formulas of the form $(x)F(x) \rightarrow F(Z)$, where $F(x)$ is an arbitrary Z-formula and Z is a numerical expression, are axioms (with the obvious restriction that the variables of Z may not be bound in F).

G. All formulas of the form

$$x = y \, . \rightarrow . \, F(x) \rightarrow F(y)$$

(where $F(x)$ is any Z-formula) are axioms.

As *rules of inference* we take:

I. From A and $A \rightarrow B$, B follows.

II. From $A \rightarrow B$, $A \rightarrow (x)B$ follows if x is not free in A.

In F, G and Rule II, x and y stand for arbitrary number variables.

Unlike Herbrand's system, Heyting's[5] has no number variables, but only variables $x, y, \ldots$ for arbitrary objects. This leads to certain complications for what follows, and we can avoid these by first introducing into Hey-ting's system variables $x', y', z', \ldots$ for natural numbers in a way such that every proposition containing variables of the new kind is equivalent to one

---

[5] *Heyting 1930a* (cited as H_2).

die neue Variablenart enthaltende Satz mit einem ohne diese äquivalent ist,
gemäß der Vorschrift: $(x')F(x')$ soll gleichbedeutend sein mit

$$(x) \, . \, x \, \epsilon \, N \supset F(x)$$

und eine Formel $A(x', y', \dots)$, welche $x', y', \dots$ als freie Variable enthält,
soll gleichbedeutend sein mit:

$$x, y, \dots \, \epsilon \, N \supset A(x, y, \dots).^{5a}$$

36 | Es kann aus der eben gegebenen Übersetzungsvorschrift rein formal be-
wiesen werden, daß auch für die neue Variablenart fast alle Sätze aus H_2,
§5, §6, bestehen bleiben. Insbesondere gilt dies von den in der vorliegenden
Arbeit verwendeten: H_2, 5.4, 5.5, 5.8, 6.26, 6.3, 6.4, 6.78, wobei nur in 5.4
$p = p$ durch $p' \, \epsilon \, N$ zu ersetzen ist. In den Sätzen H_2, §10, fallen die
Voraussetzungen: $p \, \epsilon \, N$, $q \, \epsilon \, N$ etc. weg, wenn man sie unter Verwendung
der Zahlenvariablen ausspricht.

Die Definition zahlentheoretischer Funktionen durch Rekursion ist auch
intuitionistisch einwandfrei (vgl. H_2, 10.03, 10.04). Die Funktionen f_i
(Axiomengruppe C) kommen daher sämtlich auch in der intuitionistischen
Mathematik vor und die sie definierenden Formeln denken wir zu den
Heytingschen Axiomen hinzugefügt, ebenso die für den Intuitionismus
selbstverständlich gültigen Formeln der Gruppe D. Das so erweiterte System heiße H'. Nun ordnen wir jeder Z-Formel A eine Formel A' aus H'
(ihre "Übersetzung") zu durch die Festsetzungen: Die Variablen $x, y, \dots$
sind zu übersetzen durch $x', y', \dots$; jedes f_i durch das gleich bezeichnete f_i
aus H'; = durch =; 0 durch 1;[6] +1 durch seq'; die Operationen des Aussagenkalküls in der oben angegebenen Weise. $(x)A$ ist zu übersetzen durch
$(x')A'$, wenn A' die Übersetzung von A ist. Eine Formel, die Übersetzung
einer Z-Formel ist, soll Z'-Formel heißen.

Der zu beweisende Satz I lautet: *Wenn die Formel A im Herbrandschen
System beweisbar ist, so ihre Übersetzung A' in H'.*

Hilfssatz 1. *Für jede Z'-Formel A' ist*

$$\neg\neg A' \supset A' \tag{1}$$

in H' beweisbar.

Dies zeigen wir durch vollständige Induktion.

α) (1) gilt, wenn A eine Elementarformel ist, denn für Zahlausdrücke Z
gilt $Z' \, \epsilon \, N$, was man analog wie H_2, 10.4, beweist. Daher gilt wegen H_2,

---

[5a]Es ist klar, wie (Ex') zu definieren wäre, doch wird dies im Folgenden nicht
verwendet.

[6]Im Heytingschen System beginnt die Zahlenreihe mit 1.

without them; here equivalence is determined by the following stipulation: $(x')F(x')$ is to mean the same as

$$(x) . x \in N \supset F(x)$$

and a formula $A(x', y', \ldots)$ containing $x', y', \ldots$ as free variables is to mean the same as

$$x, y, \ldots \in N \supset A(x, y, \ldots).^{5a}$$

From the rule of translation just given it can be proved in a purely formal way that almost all theorems of H_2, §5 and §6, continue to hold for the new kind of variables as well. In particular, this is the case for the theorems used in the present paper: H_2, 5.4, 5.5, 5.8, 6.26, 6.3, 6.4, 6.78, if only in 5.4 $p = p'$ is replaced by $p' \in N$. In the theorems of H_2, §10, the assumptions $p \in N$, $q \in N$ and so on become superfluous when we use number variables in expressing the theorems.

The definition of number-theoretic functions by recursion is unobjectionable for intuitionism as well (see H_2, 10.03, 10.04). Thus all functions f_i (Axiom Group C), too, occur in intuitionistic mathematics, and we consider the formulas defining them to have been adjoined to Heyting's axioms; the same goes for the formulas of Group D, which are obviously valid for intuitionism. Let H' be this extended system. Now with each Z-formula A we associate a formula A' of H' (its "translation") by the following stipulations: The variables $x, y, \ldots$ are to be translated by $x', y', \ldots$; every f_i by the f_i of H' represented by the same symbol; $=$ by $=$; 0 by 1;[6] $x + 1$ by seq'x; and the operations of the propositional calculus as indicated above. The formula $(x)A$ is to be translated by $(x')A'$ if A' is the translation of A. A formula that is the translation of a Z-formula will be called a Z'-formula.

Theorem I, which is to be proved, reads: *If the formula A is provable in Herbrand's system, then its translation A' is provable in H'.*

Lemma 1. *For every Z'-formula A'*

$$\neg\neg A' \supset A' \tag{1}$$

is provable in H'.

We show this by mathematical induction.

(α) If A is an elementary formula, then (1) holds, because for numerical expressions Z we have $Z' \in N$, as can be proved analogously to 10.4 of H_2. Hence, by 10.25 of H_2, we even have, for elementary formulas, $A' \vee \neg A'$ and, a fortiori, (1) (because of 4.45 of H_1).

[5a]It is clear how (Ex') would have to be defined; but this quantifier will not be used in what follows.

[6]In Heyting's system the number sequence begins with 1.

10.25, für Elementarformeln sogar $A' \lor \neg A'$, um so mehr (1) (wegen H_1, 4.45).

β) Wenn (1) für zwei Z'-Formeln A' und B' gilt, so auch für $A' \land B'$. Denn nach H_1, 4.61, gilt

$$\neg\neg(A' \land B') . \supset . \neg\neg A' \land \neg\neg B'$$

und daher wegen der induktiven Annahme und H_1, 2.23:

$$\neg\neg(A' \land B') . \supset . A' \land B'.$$

γ) Wenn (1) für A' gilt, so auch für $\neg A'$. Denn es gilt ganz allgemein

$$\neg\neg\neg A' \supset \neg A'$$

(nach H_1, 4.32).

δ) Wenn (1) für A' gilt, so auch für $(x')A'$. Beweis: Nach Annahme gilt $\neg\neg A' \supset A'$, daher nach H_2, 5.8, $(x') . \neg\neg A' \supset A'$ und wegen H_2, 6.4,

$$(x')\neg\neg A' . \supset . (x')A'.$$

Nach H_2, 6.78, hat man weiter

$$\neg\neg(x')A' . \supset . (x')\neg\neg A'$$

und die beiden letzten Formeln ergeben

$$\neg\neg(x')A' . \supset . (x')A',$$

w. z. b. w.

Aus α)–δ) folgt Hilfssatz 1, da jede Z'-Formel sich aus Elementarformeln mittels der Operationen $\land$, $\neg$, (x') aufbaut.

37 | Hilfssatz 2. *Für beliebige Z'-Formeln A', B' gilt*

$$A' \supset B' . \supset\subset . \neg(A' \land \neg B')$$

im System H'.

Beweis:

$$A' \supset B' . \supset . \neg(A' \land \neg B') \tag{2}$$

gilt nach H_1, 4.9, ferner

$$\neg(A' \land \neg B') . \supset . A' \supset \neg\neg B'$$

nach H_1, 4.52, und weil B' eine Z'-Formel ist, so folgt nach Hilfssatz 1

$$\neg(A' \land \neg B') . \supset . A' \supset B',$$

(β) If (1) holds for two Z'-formulas A' and B', then it holds also for $A' \wedge B'$. For, by 4.61 of H_1,

$$\neg\neg(A' \wedge B') . \supset . \neg\neg A' \wedge \neg\neg B'$$

holds; hence, by the induction assumption and 2.23 of H_1, so does

$$\neg\neg(A' \wedge B') . \supset . A' \wedge B'.$$

(γ) If (1) holds for A', then it holds also for $\neg A'$. For, quite generally,

$$\neg\neg\neg A' \supset \neg A'$$

holds (by 4.32 of H_1).

(δ) If (1) holds for A', then it also holds for $(x')A'$. Proof: By assumption, $\neg\neg A' \supset A'$ holds; hence, by 5.8 of H_2, (x'). $\neg\neg A' \supset A'$ holds and, by 6.4 of H_2, so does

$$(x')\neg\neg A' . \supset . (x')A'.$$

Furthermore, by 6.78 of H_2, we have

$$\neg\neg(x')A' . \supset . (x')\neg\neg A',$$

and the last two formulas yield

$$\neg\neg(x')A' . \supset . (x')A',$$

q. e. d.

Lemma 1 follows from (α)–(δ), since every Z'-formula is built up from elementary formulas by means of the operations $\wedge$, $\neg$, and (x').

Lemma 2. *For arbitrary Z'-formulas A' and B',*

$$A' \supset B' . \supset\subset . \neg(A' \wedge \neg B')$$

holds in the system H'.

Proof: By 4.9 of H_1 we have

$$A' \supset B' . \supset . \neg(A' \wedge \neg B'); \tag{2}$$

furthermore, by 4.52 of H_1, we have

$$\neg(A' \wedge \neg B') . \supset . A' \supset \neg\neg B'$$

and, since B' is a Z'-formula, it follows by Lemma 1 that

$$\neg(A' \wedge \neg B') . \supset . A' \supset B',$$

which, together with (2), yields what was asserted.

was zusammen mit (2) die Behauptung ergibt.

Wir zeigen jetzt: Für jedes Axiom A des Herbrandschen Systems ist die Übersetzung A' in H' beweisbar.

1. Die Axiome der Gruppe A gehen durch die Übersetzung in Sätze über, die nach Hilfssatz 2 mit H_2, 10.2, 10.22, 10.221, 10.24, 10.26, äquivalent sind.

2. Die Übersetzung eines Axioms der Gruppe B lautet:

$$\neg[\Phi(1) \wedge (x')\neg(\Phi(x') \wedge \neg\Phi(\text{seq}`x')) \wedge \neg(x')\Phi(x')] \tag{3}$$

und dies folgt nach Hilfssatz 2 aus:

$$\Phi(1) \wedge (x') \cdot \Phi(x') \supset \Phi(\text{seq}`x') :\supset (x')\Phi(x'), \tag{4}$$

indem man $p \supset q$ überall durch $\neg(p \wedge \neg q)$ ersetzt. (4) ist aber nichts anderes als H_2, 10.14, in unserer Schreibweise.

3. Die Axiome der Gruppen C und D wurden ausdrücklich zum Heytingschen System adjungiert. Gruppe E folgt aus dem oben über den Heytingschen Aussagenkalkül Bewiesenen und F ergibt sich bei Beachtung von Hilfssatz 2 unmittelbar aus H_2, 6.3 und 5.4; ebenso G aus H_2, 6.26 und 10.01.

Es bleibt noch zu zeigen, daß die Anwendung der Schlußregeln I, II auf Formeln, deren Übersetzung in H' beweisbar ist, wieder solche Formeln erzeugt. Für Regel I bedeutet das: Ist A' und $(A \to B)'$ in H' beweisbar, so auch B', d. h. aus A' und $\neg(A' \wedge \neg B')$ folgt B'. Dies ergibt sich aber unmittelbar aus Hilfssatz 2 und H_1, 1.3. Die entsprechende Tatsache für Regel II folgt ebenso aus H_2, 5.5, und Hilfssatz 2.

Der damit bewiesene Satz I zeigt, *daß die intuitionistische Arithmetik und Zahlentheorie nur scheinbar enger ist als die klassische, in Wahrheit aber die ganze klassische, bloß mit einer etwas abweichenden Interpretation, enthält.* Der Grund dafür liegt darin, daß das intuitionistische Verbot, Allsätze zu negieren und reine Existenzialsätze auszusprechen, in seiner Wirkung dadurch wieder aufgehoben wird, daß das Prädikat der Absurdität auf Allsätze angewendet werden kann, was zu formal genau den gleichen Sätzen führt, wie sie in der klassischen Mathematik behauptet werden. Wirkliche Einschränkungen scheint der Intuitionismus erst für die Analysis und Mengenlehre zu bringen, doch sind diese nicht durch Ablehnung des Tertium non datur, sondern der imprädikativen Begriffsbildungen bedingt. Durch die obigen Betrachtungen ist selbstverständlich ein intuitionistischer | Widerspruchsfreiheitsbeweis für die klassische Arithmetik und Zahlentheorie gegeben. Er ist allerdings nicht "finit" in dem Sinn wie ihn Herbrand[7] im Anschluß an Hilbert angegeben hat.

---

[7] *1930a*, p. 248, und *1930*, p. 3ff.

We now show: For each axiom A of Herbrand's system, the translation A' is provable in H'.

1. By the translation, the axioms of Group A go into propositions that, by Lemma 2, are equivalent to 10.2, 10.22, 10.221, 10.24, and 10.26 of H_2.

2. The translation of an axiom of Group B reads

$$\neg[\Phi(1) \land (x')\neg(\Phi(x') \land \neg\Phi(\text{seq}`x')) \land \neg(x')\Phi(x')] \tag{3}$$

and this, by Lemma 2, follows from

$$\Phi(1) \land (x') \cdot \Phi(x') \supset \Phi(\text{seq}`x') :\supset (x')\Phi(x'), \tag{4}$$

if we replace $p \supset q$ everywhere by $\neg(p \land \neg q)$. But (4) is precisely 10.14 of H_2 recast in our notation.

3. The axioms of Groups C and D were expressly adjoined to Heyting's system. Group E follows from what was proved above about Heyting's propositional calculus and, if we take Lemma 2 into account, we immediately obtain F from 6.3 and 5.4 of H_2; likewise, G follows from 6.26 and 10.01 of H_2.

It remains to be shown that, whenever we apply Rules of inference I and II to formulas whose translations are provable in H', we again obtain such formulas. For Rule I, this means: If A' and $(A \to B)'$ are provable in H', then so is B'; that is, B' follows from A' and $\neg(A' \land \neg B')$. But this is an immediate consequence of Lemma 2 and of 1.3 of H_1. Likewise, the corresponding result for Rule II follows from 5.5 of H_2 and Lemma 2.

Theorem 1, whose proof is thus completed, shows that *the system of intuitionistic arithmetic and number theory is only apparently narrower than the classical one, and in truth contains it, albeit with a somewhat deviant interpretation.* The reason for this is to be found in the fact that the intuitionistic prohibition against restating negated universal propositions as purely existential propositions ceases to have any effect because the predicate of absurdity can be applied to universal propositions, and this leads to propositions that formally are exactly the same as those asserted in classical mathematics. Intuitionism appears to introduce genuine restrictions only for analysis and set theory; these restrictions, however, are due to the rejection, not of the principle of the excluded middle, but of notions introduced by impredicative definitions. The above considerations, of course, provide an intuitionistic consistency proof for classical arithmetic and number theory. This proof, however, is not "finitary" in the sense in which Herbrand,[7] following Hilbert, used the term.

---

[7] *Herbrand 1930a*, p. 248, and *1930*, pp. 3ff.

Introductory note to *1933f*

In this short note Gödel describes an interpretation of intuitionistic propositional logic **IPC** in a system of classical propositional logic enriched with an additional unary operator B. The letter B stands for "beweisbar", i.e., provable, this being the intuitive interpretation of B. Here "provable" should be understood as "provable by any correct means" and not as "provable in a given formal system" (see below). The axioms given for B are now familiar as those for Lewis' system of modal propositional logic **S4** (Gödel's system $\mathfrak{G}$), with B written for the necessity operator N or $\square$. Gödel's result thus takes the following form:

$$(1) \qquad \text{if } \mathbf{IPC} \vdash F, \text{then } \mathbf{S4} \vdash F',$$

where F' is formed from F according to his translation table. In addition, Gödel conjectured that the converse of (1) also holds, that is,

$$(2) \qquad \mathbf{IPC} \vdash F \text{ if and only if } \mathbf{S4} \vdash F'.$$

This conjecture was eventually established by J.C.C. McKinsey and A. Tarski (*1948*), who used algebraic semantics. McKinsey and Tarski also described several alternative interpretations with the same property, for example the interpretation $^-$ given by $p^- := \square\, p$ (for propositional variables p), $(F \vee G)^- := F^- \vee G^-$, $(F \wedge G)^- := F^- \wedge G^-$, $(F \to G)^- := \square\, F^- \to \square\, G^-$, $(\neg F)^- := \square \neg F^-$. (There is a simple relationship between $'$ and $^-$: $\mathbf{S4} \vdash F^- \leftrightarrow \square(F')$.)

In the result (2), **S4** can also be replaced by a stronger system, for example, by addition of the following axiom schema introduced by Grzegorczyk (*1967*), where $F' \Rightarrow F''$ abbreviates $\square(F' \to F'')$:

Grz $\qquad ((F \Rightarrow \square\, G) \Rightarrow \square\, G) \wedge ((\neg F \Rightarrow \square\, G) \Rightarrow \square\, G) \Rightarrow \square\, G,$

which is deductively equivalent to the simpler schema

$$\square(\square(F \to \square\, F) \to F) \to F$$

(see *Boolos 1979*, Chapter 13). That is to say, one has

$$\mathbf{IPC} \vdash F \text{ if and only if } \mathbf{S4} + \text{Grz} \vdash F'.$$

According to Hacking (*1963*) one can also weaken **S4** in (2), namely to **S3**. Hacking's proof is based on cut-elimination.

The result (2) was extended to predicate logic by Rasiowa and Sikorski (*1953*), using algebraic semantics, and, independently, by Maehara (*1954*), using cut-elimination. To be precise, let **IQC** be intuitionistic predicate logic, and let **QS4** be **S4** with quantifiers and the usual axioms and rules for quantification. Extend $'$ by stipulating $(\forall x A)' := \forall x A'$, $(\exists x A)' := \exists x \square\, A'$, or, equivalently, extend $^-$ by $(\forall x A)^- := \square \forall x A^-$, $(\exists x A)^- := \exists x A^-$. Then

(3) $\qquad$ **IQC** $\vdash F$ if and only if **QS4** $\vdash F'$

$\qquad$ if and only if **QS4** $\vdash F^-$.

Prawitz and Malmnäs (*1968*) gave a proof of this result using normalization for suitable natural deduction systems for **IQC** and **QS4**.

Quite recently the result (3) has been extended to systems with mathematical content, namely to intuitionistic arithmetic **HA** on the one hand and to a modal extension of classical first-order arithmetic **PA** on the other (*Mints 1978, Goodman 1984*); the methods used in both cases are proof-theoretical.

As already mentioned above, Gödel's B cannot be interpreted as provability in a given formal system, such as **PA**; for, as Gödel himself observes at the end of his note, this conflicts with the second incompleteness theorem. But it is of interest to see in this connection which laws are preserved by the formal provability interpretation of B. This was accomplished by Solovay (*1976*), who characterized the modal logic (often called **G**) corresponding to formal provability in **PA**. Though the discussion of **G** is really a topic in the history of Gödel's second incompleteness theorem, Solovay's result also leads (as we shall see) to another interpretation of **IPC**.

Let Φ be an interpretation of the formulas in the language of **S4** in **PA** such that: $\Phi(p_i) := \phi_i$ (p_i a proposition letter, ϕ_i a sentence of **PA**); Φ commutes with $\wedge$, $\vee$, $\neg$, $\rightarrow$; and $\Phi(\square\, F) := \mathrm{Prov}(\ulcorner \Phi F \urcorner)$. Here 'Prov' is the canonically defined predicate expressing arithmetized provability in **PA**, and $\ulcorner \Phi F \urcorner$ is the Gödel number of the sentence ΦF. Write $\models^* F$ if we have **PA** $\vdash \Phi(F)$ for all possible interpretations Φ.

Let **G** be the modal system containing all classical tautologies, modus ponens, the necessitation rule (from F infer $\square\, F$), the schema

$$\square(F \rightarrow G) \rightarrow (\square\, F \rightarrow \square\, G)$$

and the following "Löb's axiom":

L $\qquad\qquad\qquad\qquad \square(\square\, F \rightarrow F) \rightarrow \square\, F.$

The reason for so designating this axiom is that Löb (*1955*) proved in

effect that $\models^* F$ holds for all instances F of L.

Solovay's result can now be stated as

$$(4) \qquad\qquad \mathbf{G} \vdash F \text{ if and only if } \models^* F.$$

Let $^\circ$ be the translation of formulas in the language of **S4** such that $p_i^\circ := p_i$; $^\circ$ commutes with $\wedge$, $\vee$, $\rightarrow$, $\neg$, and $(\square\, A)^\circ := A^\circ \wedge \square(A^\circ)$. Goldblatt (*1978*) showed[a]

$$(5) \qquad\qquad \mathbf{IPC} \vdash F \text{ if and only if } \mathbf{G} \vdash (F^-)^\circ,$$

and thus we also obtain a "formal provability" interpretation of **IPC**:

$$(6) \qquad\qquad \mathbf{IPC} \vdash F \text{ if and only if } \models^* (F^-)^\circ.$$

In his note Gödel observes, without proof, that

$$(7) \qquad \text{If } \mathbf{S4} \vdash \square\, F \vee \square\, G, \text{ then } \mathbf{S4} \vdash \square\, F \text{ or } \mathbf{S4} \vdash \square\, G.$$

A proof was given in *McKinsey and Tarski 1948*. By means of (2), this property of **S4** implies the disjunction property for **IPC**:

$$\text{if } \mathbf{IPC} \vdash F \vee G, \text{ then } \mathbf{IPC} \vdash F \text{ or } \mathbf{IPC} \vdash G.$$

---

[a]In a footnote Goldblatt mentions an earlier proof, which we have not seen, due to A. Kuznetsov and A. Muzavitski.

This fact was proved by Gentzen (*1935*) using cut-elimination.

For Gödel, the interest of his result presumably lay in the fact that it gave for **IPC** an interpretation which was meaningful also from a non-intuitionistic point of view. In this connection it is perhaps significant that Gödel mentions in a footnote Kolmogorov's (*1932*) interpretation of **IPC**, which, although different in character, was also put forward as being meaningful independently of intuitionistic bias.

Heyting's interpretation (*1931*) of intuitionistic logic, which was certainly known to Gödel, suggests the identification of "intuitionistically true" with "(intuitionistically) provable". It may well be that this led Gödel to his interpretation. On the other hand, Heyting's interpretation does not quite prepare us for a result like (2), since **S4** is a system based on *classical* logic.

Historically, Gödel's result was instrumental in the development of Kripke's semantics (*1965*) for intuitionistic logic: once the semantics for modal logic, in particular for **QS4**, had been formulated, Gödel's interpretation with its variants showed how one could obtain a semantics for **IQC** (see the introduction to *Kripke 1965*).

Finally, it should be observed that Gödel's axiomatization of **S4** was new and led to a much simpler and more perspicuous axiomatization of systems of modal logic (see *Lemmon 1977*, pages 6–7).

A. S. Troelstra

Eine Interpretation des
intuitionistischen Aussagenkalküls
(*1933f*)

Man kann den Heytingschen Aussagenkalkül mittels der Begriffe des gewöhnlichen Aussagenkalküls und des Begriffes "p ist beweisbar" (bezeichnet mit Bp) interpretieren,[1] wenn man für den letzteren das folgende Axiomensystem $\mathfrak{S}$ annimmt:

1. $Bp \rightarrow p$
2. $Bp \rightarrow . B(p \rightarrow q) \rightarrow Bq$
3. $Bp \rightarrow BBp.$

Außerdem sind für die Begriffe $\rightarrow$, $\sim$, $.$, $\vee$ die Axiome und Schlußregeln des gewöhnlichen Aussagenkalküls anzunehmen, ferner die neue Schlußregel: Aus A darf auf BA geschlossen werden.

Die Heytingschen Grundbegriffe sind folgendermaßen zu übersetzen:

$\neg p$	$\sim Bp$
$p \supset q$	$Bp \rightarrow Bq$
$p \vee q$	$Bp \vee Bq$
$p \wedge q$	$p . q.$

Mit dem selben Erfolg könnte man auch $\neg p$ durch $B \sim Bp$ und $p \wedge q$ durch $Bp . Bq$ übersetzen. Die Übersetzung einer beliebigen im Heytingschen System gültigen Formel folgt aus $\mathfrak{S}$, dagegen folgt aus $\mathfrak{S}$ nicht die Übersetzung von $p \vee \neg p$ und allgemein keine Formel der Gestalt $BP \vee BQ$, für die nicht schon entweder BP oder BQ aus $\mathfrak{S}$ beweisbar ist. Vermutlich gilt eine Formel im Heytingschen Kalkül dann und nur dann, wenn ihre Übersetzung aus $\mathfrak{S}$ beweisbar ist.

Das System $\mathfrak{S}$ ist mit dem Lewisschen System of Strict Implication äquivalent, wenn Bp durch Np übersetzt wird (vgl. *Parry 1933a*), und wenn man das Lewissche System durch das folgende Beckersche Zusatzaxiom $Np < NNp$ ergänzt.[2]

Es ist zu bemerken, daß für den Begriff "beweisbar in einem bestimmten formalen System S" die aus $\mathfrak{S}$ beweisbaren Formeln nicht alle gelten. Es

---

[1] Eine etwas andere Interpretation des intuitionistischen Aussagenkalküls gab Kolmogoroff (*1932*) ohne allerdings einen präzisen Formalismus anzugeben.

[2] *Becker 1930*, S. 497.

An interpretation
of the intuitionistic propositional calculus
(1933f)

One can interpret[1] Heyting's propositional calculus by means of the notions of the ordinary propositional calculus and the notion 'p is provable' (written Bp) if one adopts for that notion the following system $\mathfrak{S}$ of axioms:

  1. $Bp \to p$,
  2. $Bp \to . B(p \to q) \to Bq$,
  3. $Bp \to BBp$.

In addition, for the notions $\to$, $\sim$, ., $\vee$ the axioms and rules of inference of the ordinary propositional calculus are to be adopted, as well as the new rule of inference: From A, BA may be inferred.

Heyting's primitive notions are to be translated as follows:

$\neg p$	$\sim Bp$
$p \supset q$	$Bp \to Bq$
$p \vee q$	$Bp \vee Bq$
$p \wedge q$	$p.q$.

One could also translate $\neg p$ by $B{\sim}Bp$ and $p \wedge q$ by $Bp.Bq$ with equal success. The translation of an arbitrary formula that holds in Heyting's system is derivable in $\mathfrak{S}$; on the other hand, the translation of $p \vee \neg p$ is not derivable in $\mathfrak{S}$, nor in general is any formula of the form $BP \vee BQ$ for which neither BP nor BQ is already provable in $\mathfrak{S}$. Presumably a formula holds in Heyting's calculus if and only if its translation is provable in $\mathfrak{S}$.

The system $\mathfrak{S}$ is equivalent to Lewis' system of strict implication if Bp is translated by Np (see *Parry 1933a*) and one supplements Lewis' system by the following additional axiom[2] of Becker: $Np < NNp$.

It is to be noted that for the notion "provable in a certain formal system S" not all of the formulas provable in $\mathfrak{S}$ hold. For example, $B(Bp \to p)$

---

[1]Kolmogorov (*1932*) has given a somewhat different interpretation of the intuitionistic propositional calculus, without, to be sure, specifying a precise formalism.

[2]*Becker 1930*, p. 497.

gilt z. B. für ihn $B(Bp \rightarrow p)$ niemals, d. h. für kein System S, das die Arith-
40 metik enthält. Denn andernfalls wäre | beispielsweise $B(0 \neq 0) \rightarrow 0 \neq 0$
und daher auch $\sim B(0 \neq 0)$ in S beweisbar, d. h. die Widerspruchsfreiheit
von S wäre in S beweisbar.

Bemerkung über projektive Abbildungen
(*1933g*)

Jede eineindeutige Abbildung ϕ der reellen projektiven Ebene E in sich,
welche Gerade in Gerade überführt, ist eine Kollineation. Bezeichnet man
für jeden Kegelschnitt der Ebene als die zugehörige *Kegelschnittumgebung*
die Menge aller Punkte p von E, für welche jede p enthaltende Gerade genau
zwei Punkte mit dem Kegelschnitt gemein hat, so ist das System $\mathfrak{K}$ aller
Kegelschnittumgebungen ein unbegrenzt feines Überdeckungssystem[1] von
E. Da jeder Kegelschnitt durch zwei Geradenbüschel erzeugt werden kann,
welche projektiv (d. h. durch eine endliche Kette von Perspektivitäten) auf
einander bezogen sind, und da perspektiv auf einander bezogene Geradenbüschel bei jeder eineindeutigen Abbildung, die Gerade in Gerade überführt, in perspektiv auf einander bezogene Geradenbüschel übergehen, so
wird durch die Abbildung ϕ jeder Kegelschnitt auf einen Kegelschnitt
abgebildet. Ferner wird wegen der Eineindeutigkeit von ϕ und der obigen
Definition einer Kegelschnittumgebung jede Kegelschnittumgebung durch
ϕ in eine Kegelschnittumgebung übergeführt. Es wird demnach $\mathfrak{K}$ durch
ϕ in sich übergeführt. Da $\mathfrak{K}$ ein unbegrenzt feines Überdeckungssystem
von E ist, so ist also ϕ eine *stetige* Abbildung und daher nach einem
Fundamentalsatz der projektiven Geometrie eine Kollineation.

[1]So heißt nach Menger ein System von offenen Mengen, in welchem zu jedem Punkt
p des Raumes und zu jeder Umgebung U von p eine p enthaltende offene Menge $\subset U$
existiert.

never holds for that notion, that is, it holds for no system S that contains arithmetic. For otherwise, for example, $B(0 \neq 0) \rightarrow 0 \neq 0$ and therefore also $\sim B(0 \neq 0)$ would be provable in S, that is, the consistency of S would be provable in S.

Remark concerning projective mappings
(*1933g*)

[The introductory note to *Gödel 1933g*, as well as to related items, can be found on page 272, immediately preceding *1933b*.]

Every one-to-one mapping ϕ of the real projective plane E into itself that carries straight lines into straight lines is a collineation. If for each conic section in the plane one designates as the corresponding *conic-section neighborhood* the set of all points p of E for which every straight line containing p has exactly two points in common with the conic section, then the system $\mathfrak{K}$ of all conic-section neighborhoods is an unboundedly fine covering system[1] of E. Since every conic section can be generated by two pencils of straight lines that are projectively correspondent to each other (that is, correspondent by a finite chain of perspectivities) and since perspectively correspondent pencils of straight lines are transformed, by every one-to-one mapping that carries straight lines into straight lines, into perspectively correspondent pencils of straight lines, every conic section is mapped onto a conic section by the mapping ϕ. Furthermore, on account of the one-to-oneness of ϕ and the above definition of a conic-section neighborhood, every conic-section neighborhood is transformed by ϕ into a conic-section neighborhood. Accordingly, $\mathfrak{K}$ is transformed into itself by ϕ. Since $\mathfrak{K}$ is an unboundedly fine covering of E, ϕ is a *continuous* mapping and therefore, by a fundamental theorem of projective geometry, a collineation.

---

[1]Menger so calls a system of open sets in case there exists in it, for each point p of the space and for each neighborhood U of p, an open subset of U that contains p.

Diskussion über koordinatenlose
Differentialgeometrie
(*1933h*)

Um die Aussage, daß die Riemannschen Räume sich im Kleinen wie
euklidische Räume verhalten (daß "für unendlich benachbarte Punkte
Riemannscher Räume die euklidische Geometrie gelte") im Sinne des von
Menger (*1930*, vgl. auch *1932*) entwickelten Programmes einer koordi-
natenlosen Differentialgeometrie zu präzisieren, wären die zur metrischen Kennzeichnung der euklidischen Räume dienenden Determinanten
("Volumsdeterminanten") geeignet. Setzen wir für vier Punkte p_1, p_2, p_3,
p_4 eines metrischen Raumes

$$D(p_1, p_2, p_3, p_4) = \begin{vmatrix} 0 & 1 \\ 1 & (p_i p_j)^2 \end{vmatrix}_{(i,j=1,2,3,4)},$$

wo $p_i p_j$ den Abstand von p_i und p_j bezeichnet, so ist die Ebene unter
den metrischen Räumen nach Menger (*1928*, S. 113 und *1931*) dadurch
gekennzeichnet, daß sie vollständig, konvex und konvex nach außen ist und
daß für je vier ihrer Punkte $D(p_1, p_2, p_3, p_4) = 0$ gilt. Zur Kennzeichnung Gaußscher Flächen müßte statt dessen die Tatsache herangezogen
werden, daß für Punktequadrupel, die gegen einen Punkt konvergieren,
der Wert | von D *stark* (d. h. so wie gewisse Potenzen der Abstände)
gegen 0 konvergiert. Bezeichnen wir mit $S_n(p_1, p_2, p_3, p_4)$ die Summe der
n-ten Potenzen der sechs Abstände $p_i p_j$, so gilt folgender Satz: Ist F eine
Fläche des R_3, die in Parameterdarstellung durch zweimal stetig differen-
zierbare Funktionen $x = x(u, v)$, $y = y(u, v)$, $z = z(u, v)$ gegeben und
entspricht der Punkt p_0 Parameterwerten u_0, v_0, für welche eine der Funk-
tionaldeterminanten, etwa $\dfrac{\partial(x, y)}{\partial(u, v)}$, nicht verschwindet, so besteht, wenn
$\{p_1^n\}$, $\{p_2^n\}$, $\{p_3^n\}$, $\{p_4^n\}$ ($n = 1, 2, \ldots$ ad infinitum) irgend vier gegen p_0
konvergente Folgen von Punkten von F sind, die Beziehung

$$\lim_{n=\infty} \frac{D(p_1^n, p_2^n, p_3^n, p_4^n)}{S_6(p_1^n, p_2^n, p_3^n, p_4^n)} = 0.$$

Discussion concerning coordinate-free
differential geometry
(*1933h*)

[The introductory note to *Gödel 1933h*, as well as to related items, can be found on page 272, immediately preceding *1933b*.]

In order to make precise, in the sense of the program that Menger (*1930*, see also *1932*) has developed for a coordinate-free differential geometry, the assertion that Riemannian spaces behave locally like Euclidean spaces (that "Euclidean geometry holds for infinitesimally adjacent points of Riemannian spaces"), the determinants ("volume determinants") that serve to characterize the metric of Euclidean spaces would be suitable. If for four points p_1, p_2, p_3 and p_4 of a metric space we set

$$D(p_1, p_2, p_3, p_4) = \begin{vmatrix} 0 & 1 \\ 1 & (p_i p_j)^2 \end{vmatrix}_{(i,j=1,2,3,4)},$$

where $p_i p_j$ denotes the distance between p_i and p_j, then according to Menger (*1928* and *1931*) the plane is characterized among metric spaces by the fact that it is complete, convex, and outwardly convex, and that, for any four of its points, $D(p_1, p_2, p_3, p_4) = 0$ holds. For the characterization of Gaussian surfaces, it would be necessary to invoke instead the fact that, for quadruples of points that converge to a point, the value of D *converges strongly* (that is, as certain powers of the distances) *toward 0*. If we denote by $S_n(p_1, p_2, p_3, p_4)$ the sum of the nth powers of the six distances $p_i p_j$, the following theorem holds: If F is a surface in R_3 that has a parametric representation in terms of twice continuously differentiable functions $x = x(u, v)$, $y = y(u, v)$, $z = z(u, v)$ and if the point p_0 corresponds to values u_0, v_0 of the parameters for which one of the Jacobians, say $\dfrac{\partial(x, y)}{\partial(u, v)}$, does not vanish, then, if $\{p_1^n\}$, $\{p_2^n\}$, $\{p_3^n\}$, $\{p_4^n\}$ ($n = 1, 2, \ldots$ ad infinitum) are any four sequences of points of F converging to p_0, the relation

$$\lim_{n=\infty} \frac{D(p_1^n, p_2^n, p_3^n, p_4^n)}{S_6(p_1^n, p_2^n, p_3^n, p_4^n)} = 0$$

holds.

Zum Entscheidungsproblem des logischen Funktionenkalküls
(*1933i*)

In *Ergebnisse eines mathematischen Kolloquiums* (*1932a*) habe ich ein Verfahren kurz skizziert, nach dem man für jede Formel des engeren Funktionenkalküls,[1] die in der Normalform nur zwei (und zwar benachbarte) Allzeichen enthält, entscheiden kann, ob sie erfüllbar[2] ist. Denselben Fall des Entscheidungsproblems hat dann L. Kalmár nach der gleichen Methode ausführlich behandelt.[3] In dem früher von P. Bernays, M. Schönfinkel und W. Ackermann behandelten Fall *eines* Allzeichens hat sich herausgestellt, daß solche Formeln, wenn sie überhaupt erfüllbar sind, schon in einem endlichen Individuenbereich erfüllbar sind. Das Ziel der folgenden Untersuchung ist, dies auch für den Fall zweier Allzeichen nachzuweisen. Zweitens soll gezeigt werden, daß die Lösung des nächst komplizierteren Falles (drei Allzeichen) bereits mit der Lösung des ganzen Entscheidungsproblems äquivalent wäre (vgl. S. 441).

Bei der Behandlung der ersten dieser beiden Fragen kann man sich auf Zählausdrücke der Form:

$$(x_1)(x_2)(Ey_1)(Ey_2)\ldots(Ey_m)\,\mathfrak{A}\,(x_1, x_2, y_1, y_2, \ldots, y_m) \tag{1}$$

beschränken,[3a] da man den Fall, daß irgendwelche *E*-Zeichen den Allzeichen vorangehen, auf den obigen zurückführen kann (vgl. *Kalmár 1933*, S. 478). In (1) mögen die Funktionsvariablen | $F_1, F_2, \ldots, F_s$ vorkommen und die Anzahl der Leerstellen von F_i sei r_i. Ein System von s in einem Individuenbereich $\mathfrak{J}$ definierten Funktionen $\Phi_1, \Phi_2, \ldots, \Phi_s$ mit resp. $r_1, r_2, \ldots, r_s$ Leerstellen soll ein *Modell über* $\mathfrak{J}$ heißen. Die Funktionen Φ_i eines Modells M bezeichnen wir auch mit ϕ_i^M und das Resultat der Einsetzung der Funktionen ϕ_i^M des Modells M für die Funktionsvariablen F_i in $\mathfrak{A}$ bezeichnen wir mit $\mathfrak{A}_M$. Ein Modell über dem Bereich der natürlichen Zahlen $\leq k$ soll eine *Tabelle k-ter Ordnung* heißen. Es gibt

434

---

[1] Vgl. *Hilbert und Ackermann 1928*, S. 43ff.

[2] "Erfüllbar" ohne Zusatz soll bedeuten: es gibt einen Individuenbereich, in dem die Formel erfüllbar ist, was dasselbe bedeutet wie: die Formel ist in einem abzählbaren Individuenbereich erfüllbar.

[3] Vgl. *Kalmár 1933*. Einige der im folgenden verwendeten Bezeichnungen sind dieser Arbeit entnommen.

[3a] Die Buchstaben $\mathfrak{A}, \mathfrak{B}, \ldots$ bedeuten im folgenden immer Formeln, in denen keine All- und Existenzzeichen vorkommen.

On the decision problem for the functional calculus of logic
(1933i)

⟦The introductory note to *1933i*, as well as to related items, can be found on page 226, immediately preceding *1932a*.⟧

In *Ergebnisse eines mathematischen Kolloquiums* (*1932a*) I briefly sketched a procedure by which one can decide, for each formula of the restricted functional calculus[1] that in normal form contains only two (and in fact adjacent) universal quantifiers, whether it is satisfiable.[2] L. Kalmár then dealt with that case of the decision problem in detail by using the same method.[3] In the case of *one* universal quantifier, a case which had been treated earlier by P. Bernays, M. Schönfinkel and W. Ackermann, it turned out that such formulas, if they are satisfiable at all, are satisfiable already in a finite domain of individuals. The goal of the investigation that follows is to prove this also for the case of two universal quantifiers. Secondly, it will be shown that solving the next more complicated case (three universal quantifiers) would already be equivalent to solving the full decision problem (see below, page 322).

In treating the first of these two questions, we can confine ourselves to first-order expressions of the form[3a]

$$(x_1)(x_2)(Ey_1)(Ey_2)\ldots(Ey_m)\mathfrak{A}(x_1, x_2, y_1, y_2, \ldots, y_m), \qquad (1)$$

since the case in which some existential quantifiers precede the universal quantifiers can be reduced to the one above (see *Kalmár 1933*, page 478). Assume that in (1) the functional variables $F_1, F_2, \ldots, F_s$ occur, the number of argument places of F_i being r_i. A system of s functions $\Phi_1, \Phi_2, \ldots, \Phi_s$, defined on a domain $\mathfrak{J}$ of individuals, with $r_1, r_2, \ldots, r_s$ argument places, respectively, will be called a *model over* $\mathfrak{J}$. We also denote by ϕ_i^M the functions Φ_i of a model M, and we denote by $\mathfrak{A}_M$ the result of substituting the functions ϕ_i^M of the model M for the functional variables

---

[1] See *Hilbert and Ackermann 1928*, p. 43 ff.

[2] "Satisfiable" by itself is to mean: there is a domain of individuals in which the formula is satisfiable; this means the same as: the formula is satisfiable in a countable domain of individuals.

[3] See *Kalmár 1933*. Some of the notation used in what follows is taken from that work.

[3a] In what follows, the letters $\mathfrak{A}, \mathfrak{B}, \ldots$ always stand for formulas in which no universal or existential quantifier occurs.

natürlich nur endlich viele Tabellen k-ter Ordnung. Ist M ein Modell über $\mathfrak{J}$ und sind $a_1, a_2, \ldots, a_k$ irgendwelche Elemente aus $\mathfrak{J}$, so bezeichnen wir mit $[M | a_1, a_2, \ldots, a_k]$ diejenige Tabelle T von k-ter Ordnung, für welche gilt:

$$\phi_l^T(i_1, i_2, \ldots, i_{r_l}) \equiv \phi_l^M(a_{i_1}, a_{i_2}, \ldots, a_{i_{r_l}})$$

für $l = 1, 2, \ldots, s$ und für alle r_l-tupel natürlicher Zahlen $i_1, i_2, \ldots, i_{r_l}$ aus dem Intervall $[1, \ldots, k]$.

Der zu beweisende Satz I lautet:

Wenn (1) *erfüllbar ist, so auch in einem endlichen Individuenbereich.*

Nun impliziert (1) offenbar die folgende Formel:

$$(x_1)(Ez_1)(Ez_2)\ldots(Ez_m)\mathfrak{A}(x_1, x_1, z_1, z_2, \ldots, z_m). \tag{2}$$

Daher ist (1) äquivalent mit der Konjunktion von (1) und (2), welche folgende Normalform hat:

$$(x_1)(x_2)(Ey_1)\ldots(Ey_m)(Ez_1)\ldots(Ez_m)[\mathfrak{A}(x_1, x_2, y_1, \ldots, y_m) \ \&$$
$$\mathfrak{A}(x_1, x_1, z_1, \ldots, z_m)]. \tag{3}$$

Indem wir setzen $2m = n$ und

$$\mathfrak{A}(x_1, x_2, y_1, \ldots, y_m) \ \& \ \mathfrak{A}(x_1, x_1, z_1, \ldots, z_m) =$$
$$\mathfrak{B}(x_1, x_2, y_1, \ldots, y_m, z_1, \ldots, z_m),$$

können wir statt (3) schreiben:

$$(x_1)(x_2)(Ey_1)(Ey_2)\ldots(Ey_n)\mathfrak{B}(x_1, x_2, y_1, y_2, \ldots, y_n) \tag{4}$$

und es genügt, Satz I für (4) zu beweisen. Um aber von einem Modell M über $\mathfrak{J}$ zu zeigen, daß es (4) erfüllt, genügt es zu beweisen, daß es für je zwei *verschiedene* Elemente a_1, a_2 aus $\mathfrak{J}$ Elemente $b_1, b_2, \ldots, b_n \ \epsilon \ \mathfrak{J}$ gibt, so daß

$$\mathfrak{B}_M(a_1, a_2, b_1, \ldots, b_n)$$

gilt, was für das folgende einige Vereinfachungen bringt.[4]

Angenommen die Formel (4) sei erfüllbar durch das Modell M über dem Individuenbereich $\mathfrak{J}$. Dann bilde man die Menge $\mathfrak{P}$ aller Tabellen erster Ordnung $[M | a]$, wobei a sämtliche Individuen aus $\mathfrak{J}$ durchläuft, und die
435 Menge $\mathfrak{Q}$ aller Tabellen zweiter Ordnung $[M | a, b]$, | wobei a, b unabhängig

---

[4]Der Übergang von Formel (1) zu Formel (4) wurde in meiner oben zitierten Note *1932a* versehentlich nicht angeführt, ist aber nötig, damit die dort angegebenen Bedingungen für die Erfüllbarkeit wirklich hinreichen.

F_i in $\mathfrak{A}$. A model over the domain of natural numbers $\leq k$ will be called a *kth-order table*. There are, of course, only finitely many kth-order tables. If M is a model over $\mathfrak{J}$ and $a_1, a_2, \ldots, a_k$ are any elements of $\mathfrak{J}$, we denote by $[M|a_1, a_2, \ldots, a_k]$ that kth-order table T for which

$$\phi_l^T(i_1, i_2, \ldots, i_{r_l}) \equiv \phi_l^M(a_{i_1}, a_{i_2}, \ldots, a_{i_{r_l}})$$

holds for $l = 1, 2, \ldots, s$ and for all r_l-tuples of natural numbers $i_1, i_2, \ldots, i_{r_l}$ from the interval $[1, \ldots, k]$.

Theorem I, which is to be proved, reads:

If (1) *is satisfiable, it is satisfiable also in a finite domain of individuals.*

Now (1) evidently implies the following formula:

$$(x_1)(Ez_1)(Ez_2)\ldots(Ez_m)\,\mathfrak{A}(x_1, x_1, z_1, z_2, \ldots, z_m). \tag{2}$$

Hence (1) is equivalent to the conjunction of (1) and (2), which has the following normal form:

$$(x_1)(x_2)(Ey_1)\ldots(Ey_m)(Ez_1)\ldots(Ez_m)[\mathfrak{A}(x_1, x_2, y_1, \ldots, y_m)\,\&$$
$$\mathfrak{A}(x_1, x_1, z_1, \ldots, z_m)]. \tag{3}$$

By setting $2m = n$ and

$$\mathfrak{A}(x_1, x_2, y_1, \ldots, y_m)\,\&\,\mathfrak{A}(x_1, x_1, z_1, \ldots, z_m) =$$
$$\mathfrak{B}(x_1, x_2, y_1, \ldots, y_m, z_1, \ldots, z_m),$$

we can write, in place of (3),

$$(x_1)(x_2)(Ey_1)(Ey_2)\ldots(Ey_n)\,\mathfrak{B}(x_1, x_2, y_1, y_2, \ldots, y_n), \tag{4}$$

and it will suffice to prove Theorem I for (4). But, in order to show of a model M over $\mathfrak{J}$ that it satisfies (4), it suffices to prove that for any two *distinct* elements a_1 and a_2 of $\mathfrak{J}$ there are elements $b_1, b_2, \ldots, b_n \in \mathfrak{J}$ such that

$$\mathfrak{B}_M(a_1, a_2, b_1, \ldots, b_n)$$

holds, which brings about a few simplifications in what follows.[4]

Assume that the formula (4) is satisfiable in the model M over the domain $\mathfrak{J}$ of individuals. Then form the set $\mathfrak{P}$ of all 1st-order tables $[M|a]$, where a runs over all individuals of $\mathfrak{J}$, and the set $\mathfrak{Q}$ of all 2nd-order tables $[M|a, b]$, where a and b, independently of each other, run over all elements

---

[4] The passage from formula (1) to formula (4) was inadvertently not presented in my note *1932a*, cited above, but is necessary if the conditions for satisfiability given there are actually to suffice.

voneinander alle Elemente aus $\mathfrak{J}$ durchlaufen. Die nicht leeren Mengen $\mathfrak{P}, \mathfrak{Q}$ genügen folgenden Bedingungen (I und II):

I. Zu je zwei Tabellen $R, S \in \mathfrak{P}$ gibt es eine Tabelle $T \in \mathfrak{Q}$, so daß $[T|1] = R$, $[T|2] = S$.

Denn wenn $R, S \in \mathfrak{P}$, so gibt es $a, b \in \mathfrak{J}$, so daß $R = [M|a]$, $S = [M|b]$, und man braucht nur zu setzen: $T = [M|a, b]$.

II. Zu jeder Tabelle $R \in \mathfrak{Q}$ gibt es eine Tabelle T der Ordnung $n + 2$, derart daß:

1.) $R = [T|1, 2]$,

2.) $\mathfrak{B}_T(1, 2, \ldots, n + 2)$ wahr ist,

3.) für jede natürliche Zahl $i \leq n + 2$ gilt: $[T|i] \in \mathfrak{P}$ und für je zwei natürliche Zahlen $i, k \leq n + 2$ gilt: $[T|i, k] \in \mathfrak{Q}$.

Denn wenn $R \in \mathfrak{Q}$, so gibt es $a, b \in \mathfrak{J}$, so daß $R = [M|a, b]$, und weil M die Formel (4) erfüllt, so gibt es Individuen $c_1, c_2, \ldots, c_n$ aus $\mathfrak{J}$, so daß $\mathfrak{B}_M(a, b, c_1, c_2, \ldots, c_n)$ wahr ist, und man braucht nur zu setzen:

$$T = [M|a, b, c_1, c_2, \ldots, c_n].$$

Um Satz I zu beweisen, bleibt also zu zeigen:

Satz II. *Wenn es zwei nicht leere Mengen $\mathfrak{P}, \mathfrak{Q}$ von Tabellen erster bzw. zweiter Ordnung gibt, welche den Bedingungen I und II genügen, so ist die Formel (4) in einem endlichen Individuenbereich erfüllbar.*

Wir beweisen zunächst zwei Hilfssätze.

Hilfssatz 1. *Sei $\mathfrak{M}$ eine endliche Menge mit k Elementen und $\mathfrak{M}^*$ die Menge der geordneten Paare $[a, b]$,[5] für welche $a, b \in \mathfrak{M}, a \neq b$; sei ferner $\chi(k)$ die kleinste natürliche Zahl von der Art, daß man $\mathfrak{M}^*$ in $\chi(k)$ paarweise fremde Klassen $K_1, K_2, \ldots, K_{\chi(k)}$ einteilen kann, so daß, wenn $[a, b]$ und $[c, d]$ zur selben Klasse K_i gehören, immer $a \neq d$ und $b \neq c$ ist, dann ist $\chi(k) \leq 2\dfrac{\log k}{\log 2} + 2$.*

Beweis: Die Funktion $\chi(k)$ genügt der Ungleichung

$$\chi(2k) \leq \chi(k) + 2, \tag{5}$$

denn seien $\mathfrak{M}_1, \mathfrak{M}_2$ zwei fremde Mengen mit je k Elementen und $\mathfrak{M} = \mathfrak{M}_1 + \mathfrak{M}_2$, seien ferner $K_1', K_2', \ldots, K_{\chi(k)}'$ bzw. $K_1'', K_2'', \ldots, K_{\chi(k)}''$ die durch den Hilfssatz geforderten Teilklassen von $\mathfrak{M}_1^*$ bzw. $\mathfrak{M}_2^*$. Teilklassen $K_1, K_2, \ldots, K_{\chi(k)+2}$ von $\mathfrak{M}^*$, welche der Bedingung des Hilfssatzes genügen, kann man folgendermaßen bilden: Es sei $K_i = K_i' + K_i''$ für $i = 1, 2, \ldots, \chi(k)$, ferner bestehe $K_{\chi(k)+1}$ aus allen Paaren $[a, b]$, für welche

---

[5]Das k-tupel aus irgendwelchen Elemente $a_1, a_2, \ldots, a_k$ (in dieser Reihenfolge) bezeichnen wir mit $[a_1, a_2, \ldots, a_k]$, die nicht geordnete Menge dieser Elemente mit $\{a_1, a_2, \ldots, a_k\}$.

of $\mathfrak{J}$. The non-empty sets $\mathfrak{P}$ and $\mathfrak{Q}$ satisfy the following conditions (I and II):

I. For any two tables $R, S \, \epsilon \, \mathfrak{P}$ there exists a table $T \, \epsilon \, \mathfrak{Q}$ such that $[T|1] = R$ and $[T|2] = S$.

For, if $R, S \, \epsilon \, \mathfrak{P}$, there exist $a, b \, \epsilon \, \mathfrak{J}$ such that $R = [M|a]$ and $S = [M|b]$, and we need only set $T = [M|a, b]$.

II. For every table $R \, \epsilon \, \mathfrak{Q}$ there exists a table T of order $n + 2$ such that:
(1) $R = [T|1, 2]$,
(2) $\mathfrak{B}_T(1, 2, \ldots, n + 2)$ is true,
(3) for every natural number $i \leq n + 2$, $[T|i] \, \epsilon \, \mathfrak{P}$ holds, and, for any two natural numbers $i, k \leq n + 2$, $[T|i, k] \, \epsilon \, \mathfrak{Q}$ holds.

For, if $R \, \epsilon \, \mathfrak{Q}$, there exist $a, b \, \epsilon \, \mathfrak{J}$ such that $R = M[a, b]$, and, since M satisfies formula (4), there exist individuals $c_1, c_2, \ldots, c_n$ of $\mathfrak{J}$ such that $\mathfrak{B}_M(a, b, c_1, c_2, \ldots, c_n)$ is true, and we need only set

$$T = [M|a, b, c_1, c_2, \ldots, c_n].$$

In order to prove Theorem I we thus still must show

Theorem II. *If there exist two non-empty sets, $\mathfrak{P}$ and $\mathfrak{Q}$, of tables, of the first and second order respectively, that satisfy conditions* I *and* II, *then formula* (4) *is satisfiable in a finite domain of individuals.*

We first prove two lemmas.

Lemma I. *Let $\mathfrak{M}$ be a finite set of k elements and let $\mathfrak{M}^*$ be the set of ordered pairs[5] $[a, b]$ for which $a, b \, \epsilon \, \mathfrak{M}$, $a \neq b$; further, let $\chi(k)$ be the least natural number such that one can partition $\mathfrak{M}^*$ into $\chi(k)$ pairwise disjoint classes $K_1, K_2, \ldots, K_{\chi(k)}$ so that, whenever $[a, b]$ and $[c, d]$ belong to the same class K_i, $a \neq d$ and $b \neq c$, then $\chi(k) \leq 2\dfrac{\log k}{\log 2} + 2.$*

Proof: The function $\chi(k)$ satisfies the inequality

$$\chi(2k) \leq \chi(k) + 2. \tag{5}$$

For let $\mathfrak{M}_1$ and $\mathfrak{M}_2$ be two disjoint sets of k elements each and let $\mathfrak{M} = \mathfrak{M}_1 + \mathfrak{M}_2$; further, let $K'_1, K'_2, \ldots, K'_{\chi(k)}$, or $K''_1, K''_2, \ldots, K''_{\chi(k)}$ respectively, be the subclasses of $\mathfrak{M}_1^*$, or $\mathfrak{M}_2^*$, required by the lemma. Subclasses $K_1, K_2, \ldots, K_{\chi(k)+2}$ of $\mathfrak{M}^*$ satisfying the conditions of the lemma can be formed in the following manner: let $K_i = K'_i + K''_i$ for $i = 1, 2, \ldots, \chi(k)$; further, let $K_{\chi(k)+1}$ consist of all pairs $[a, b]$ for which $a \, \epsilon \, \mathfrak{M}_1$ and $b \, \epsilon \, \mathfrak{M}_2$,

---

[5] We denote the k-tuple of the elements $a_1, a_2, \ldots, a_k$ (in that order) by $[a_1, a_2, \ldots, a_k]$ and the unordered set of these elements by $\{a_1, a_2, \ldots, a_k\}$.

436   $a \in \mathfrak{M}_1$, $b \in \mathfrak{M}_2$, und $K_{\chi(k)+2}$ aus allen Paaren $[a, b]$, für | welche $a \in \mathfrak{M}_2$, $b \in \mathfrak{M}_1$. Da $\chi(2k)$ die kleinstmögliche Anzahl von Klassen K_i ist, so ist damit (5) bewiesen. Da ferner $\chi(2) = 2$, so folgt durch iterierte Anwendung von (5) $\chi(2^k) \leq 2k$. Sei nun n eine beliebige natürliche Zahl. Man setze $n = 2^r$, $r = \dfrac{\log n}{\log 2}$. Da $\chi(k)$ monoton ist, so hat man

$$\chi(n) = \chi(2^r) \leq \chi(2^{[r]+1}) \leq 2[r] + 2 \leq 2\frac{\log n}{\log 2} + 2.$$

Hilfssatz 2. *Sei $\rho(x, y)$ die im Bereich der natürlichen Zahlen ≤ 7 definierte Relation, welche dann und nur dann besteht, wenn $x \neq y$ und wenn $x - y$ quadratischer Rest* mod 7 *ist;*[6] *dann gilt:* 1.) *$\rho(x, y)$ und $\rho(y, x)$ gelten niemals beide gleichzeitig (Asymmetrie).* 2.) *Für je zwei natürliche Zahlen $x, y \leq 7$ gibt es ein $z \leq 7$, so daß $\rho(z, x)$ und $\rho(z, y)$.*[7]

Das letztere besagt, daß die Kongruenz $x + u^2 \equiv y + v^2 \pmod{7}$ immer durch solche u, v, die beide $\neq 0 \pmod{7}$ sind, lösbar ist, was man sofort einsieht, wenn man sie in der Form $(u+v)(u-v) \equiv y-x \pmod{7}$ schreibt. Die kleinste Zahl z, für welche $\rho(z, x)$ und $\rho(z, y)$, bezeichne man mit $\alpha(x, y)$, so daß gilt:

3.) *Aus $z = \alpha(x, y)$ folgt $\rho(z, x)$ und $\rho(z, y)$.*

Ich gehe jetzt zum Beweise des Satzes II über und nehme also an, es seien zwei Mengen $\mathfrak{P}, \mathfrak{Q}$ gegeben, welche den Bedingungen I, II (vgl. S. 435) genügen. Unter den gemäß I existierenden Tabellen T greifen wir für jedes Paar $R, S \in \mathfrak{P}$ eine heraus und bezeichnen sie mit $\delta(R, S)$; ebenso unter den gemäß II existierenden Tabellen T $(n + 2)$-ter Ordnung je eine für jedes $R \in \mathfrak{Q}$, die wir mit $\eta(R)$ bezeichnen. Für alle $R, S \in \mathfrak{P}$ gilt also:

$$\delta(R, S) \in \mathfrak{Q} \quad (6.1) \qquad [\delta(R,S)|1] = R \quad (6.2) \qquad [\delta(R,S)|2] = S \quad (6.3)$$

und für alle $R \in \mathfrak{Q}$ gilt:

$$[\eta(R)|1, 2] = R \quad (7.1) \qquad \mathfrak{B}_{\eta(R)}(1, 2, \ldots, n + 2) \quad (7.2)$$

und

$$[\eta(R)|i] \in \mathfrak{P} \quad (7.3) \qquad [\eta(R)|i, j] \in \mathfrak{Q} \quad (7.4)$$

für alle $i, j \leq n + 2$.

---

[6] Die Relation ρ besteht also zwischen x und y, wenn $x - y$ einen der folgenden Werte hat: $1, 2, 4, -3, -5, -6$.

[7] In einem Bereich von weniger als 7 Elementen existiert eine den Bedingungen 1.) 2.) genügende Relation nicht.

and let $K_{\chi(k)+2}$ consist of all pairs $[a, b]$ for which $a \in \mathfrak{M}_2$ and $b \in \mathfrak{M}_1$. Since $\chi(2k)$ is the least possible number of classes K_i, (5) is thereby proved. Since furthermore $\chi(2) = 2$, then $\chi(2^k) \leq 2k$ follows by iterated application of (5). Now let n be an arbitrary natural number. Set $r = \dfrac{\log n}{\log 2}$ or $n = 2^r$.

Since $\chi(k)$ is monotonic, we have

$$\chi(n) = \chi(2^r) \leq \chi(2^{[r]+1}) \leq 2[r] + 2 \leq 2\frac{\log n}{\log 2} + 2.$$

Lemma 2. *Let $\rho(x, y)$ be the relation, defined on the domain of natural numbers ≤ 7, that obtains if and only if $x \neq y$ and $x - y$ is a quadratic residue mod 7;[6] then the following hold:*

(1) *$\rho(x, y)$ and $\rho(y, x)$ never both hold at the same time (asymmetry);*

(2) *for any two natural numbers $x, y \leq 7$ there exists a $z \leq 7$ such that $\rho(z, x)$ and $\rho(z, y)$.[7]*

The latter means that the congruence $x + u^2 \equiv y + v^2 \pmod{7}$ can always be solved by numbers u and v neither of which is congruent to 0 $\pmod 7$, as one sees immediately if one writes it in the form $(u + v)(u - v) \equiv y - x \pmod 7$. Denote by $\alpha(x, y)$ the least number z for which $\rho(z, x)$ and $\rho(z, y)$, so that the following holds:

(3) *from $z = \alpha(x, y)$, $\rho(z, x)$ and $\rho(z, y)$ follow.*

I now turn to the proof of Theorem II and therefore assume that two sets $\mathfrak{P}$ and $\mathfrak{Q}$ are given that satisfy conditions I and II (see above, page 311). Among the tables T that exist according to I, we select one for each pair $R, S \in \mathfrak{P}$ and denote it by $\delta(R, S)$; likewise, among the tables T of $(n+2)$th order that exist according to II, we select one for each $R \in \mathfrak{Q}$ and denote it by $\eta(R)$. Thus, for all $R, S \in \mathfrak{P}$,

$$\delta(R, S) \in \mathfrak{Q} \ \ (6.1), \qquad [\delta(R, S)|1] = R \ \ (6.2), \qquad [\delta(R, S)|2] = S \ \ (6.3)$$

hold and, for all $R \in \mathfrak{Q}$,

$$[\eta(R)|1, 2] = R \quad (7.1), \qquad \mathfrak{B}_{\eta(R)}(1, 2, \ldots, n + 2) \quad (7.2),$$

$$[\eta(R)|i] \in \mathfrak{P} \quad (7.3), \qquad [\eta(R)|i, j] \in \mathfrak{Q} \quad (7.4)$$

hold for all $i, j \leq n + 2$.

---

[6] The relation ρ thus obtains between x and y if $x - y$ has one of the following values: $1, 2, 4, -3, -5, -6$.

[7] In a domain of fewer than 7 elements no relation satisfying conditions (1) and (2) exists.

Die Anzahl der Elemente von $\mathfrak{Q}$ sei q, die Anzahl der Es-gibt-Zeichen in Formel (4) ist n. Man bestimme eine natürliche Zahl L, für welche

$$qn\chi(7L) \le L, \tag{8}$$

437 | was wegen Hilfssatz 1 möglich ist. Der Individuenbereich $\mathfrak{J}$ des zu definierenden Modells, welches die Formel (4) erfüllen wird, sei eine Menge von Quadrupeln, und zwar soll ein Quadrupel $[x, y, z, u]$ dann zu $\mathfrak{J}$ gehören, wenn $u \,\epsilon\, \mathfrak{Q}$ und wenn x, y, z natürliche Zahlen sind und $x \le 7$, $y \le \chi(7L)$, $z \le n$ ist. $\mathfrak{J}$ ist also eine Menge mit $7nq\chi(7L)$, d. h. nach (8) mit höchstens $7L$ Elementen. Die gemäß Hilfssatz 1 bestimmten Teilklassen von $\mathfrak{J}^*$ seien $K_1, K_2, \ldots, K_\nu$ $(\nu \le \chi(7L))$; den Index i derjenigen Klasse K_i, zu welcher ein Paar $[a, b]$ aus $\mathfrak{J}^*$ gehört, bezeichne ich mit $Z(a, b)$. Hinsichtlich der Bezeichnungen bemerke ich, daß ich unter einem *Komplex* eine endliche Folge von Dingen verstehe, also ein k-tupel, wobei k eine beliebige natürliche Zahl ist. Ist a ein Komplex, so bezeichne ich mit a_i das i-te Glied von a, wobei diese Bezeichnungsweise auch iteriert werden kann, wenn die Glieder von a selbst wieder Komplexe sind ($a_{i,k}$ bedeutet also: Das k-te Glied des i-ten Gliedes von a). $\bar{a}$ bezeichne die Menge, die aus den Gliedern des Komplexes a besteht, und ist a ein Komplex, in dem keine zwei Glieder identisch sind und $x \,\epsilon\, \bar{a}$, so bedeute $N(x, a)$ die Zahl, welche angibt, das wievielte Glied x in a ist.

Ein $(n+2)$-tupel a von Elementen aus $\mathfrak{J}$ möge ein *E-Komplex* heißen, wenn $a_1 \ne a_2$ und wenn es ein $T \,\epsilon\, \mathfrak{Q}$ gibt, so daß

$$a_{i+2} = [\alpha(a_{1,1}, a_{2,1}), Z(a_1, a_2), i, T] \tag{9}$$

für $i = 1, 2, \ldots, n$, wobei α die Bedeutung aus Hilfssatz 2 hat. Für jeden *E-Komplex* a bezeichne ich mit $A(a)$ das Paar $[a_1, a_2]$ und mit $B(a)$ das n-tupel $[a_3, a_4, \ldots, a_{n+2}]$, ferner mit $\mathfrak{Q}(a)$ das (durch a offenbar eindeutig bestimmte) Element T aus $\mathfrak{Q}$, für welches (9) gilt (also $\mathfrak{Q}(a) = a_{i,4}$ für $i = 3, 4, \ldots, n+2$). *Zu jedem Paar von Individuen x, y aus $\mathfrak{J}$ $(x \ne y)$ und jeder Tabelle T aus $\mathfrak{Q}$ gibt es offenbar einen E-Komplex a derart, daß $x = a_1$, $y = a_2$, $T = \mathfrak{Q}(a)$*, denn um einen solchen zu finden, braucht man nur a_i (für $i > 2$) gemäß (9) zu bestimmen.

Man bestätigt leicht folgende zum Beweise von Hilfssatz 3 erforderlichen Tatsachen:

(10) Für zwei beliebige *E*-Komplexe a, b ist

$$\text{entweder} \quad B(a) = B(b) \quad \text{oder} \quad \overline{B(a)} \cdot \overline{B(b)} = \Lambda.$$

Denn wenn

$$\alpha(a_{1,1}, a_{2,1}) = \alpha(b_{1,1}, b_{2,1}), \quad Z(a_1, a_2) = Z(b_1, b_2) \quad \text{und} \quad \mathfrak{Q}(a) = \mathfrak{Q}(b),$$

Let q be the number of elements of $\mathfrak{Q}$. The number of existential quantifiers in formula (4) is n. Determine a natural number L such that

$$q n \chi(7L) \leq L, \qquad (8)$$

as is possible on account of Lemma 1. Let the domain $\mathfrak{J}$ of individuals of the model to be defined, which will satisfy formula (4), be a set of quadruples; here a quadruple $[x, y, z, u]$ is to belong to $\mathfrak{J}$ if $u \in \mathfrak{Q}$ and if x, y, z are natural numbers with $x \leq 7$, $y \leq \chi(7L)$, and $z \leq n$. $\mathfrak{J}$ is thus a set with $7 n q \chi(7L)$ elements, that is, according to (8), with at most $7L$ elements. Let the subclasses of $\mathfrak{J}^*$ that are determined in conformity with Lemma 1 be $K_1, K_2, \ldots, K_\nu$ $(\nu \leq \chi(7L))$; by $Z(a, b)$ I denote the subscript i of that class K_i to which a pair $[a, b]$ of $\mathfrak{J}^*$ belongs. With regard to the notation, I remark that by a *complex* I understand a finite sequence of objects, that is, a k-tuple, where k is an arbitrary natural number. If a is a complex, I denote by a_i the ith term of a, and the use of this notation can in fact be iterated if the terms of a are themselves complexes (thus $a_{i,k}$ is the kth term of the ith term of a). Let $\bar{a}$ be the set that consists of the terms of the complex a and, if a is a complex in which no two terms are identical and $x \in \bar{a}$, let $N(x, a)$ be the number that gives the rank of x in a.

Let an $(n+2)$-tuple a of elements of $\mathfrak{J}$ be called an *E-complex* if $a_1 \neq a_2$ and there exists a $T \in \mathfrak{Q}$ such that

$$a_{i+2} = [\alpha(a_{1,1}, a_{2,1}), Z(a_1, a_2), i, T] \qquad (9)$$

for $i = 1, 2, \ldots, n$, where α has the meaning it has in ⟦the paragraph following⟧ Lemma 2. For every E-complex a I denote by $A(a)$ the pair $[a_1, a_2]$, and by $B(a)$ the n-tuple $[a_3, a_4, \ldots, a_{n+2}]$ and, further, by $\mathfrak{Q}(a)$ the element (obviously uniquely determined by a) T of $\mathfrak{Q}$ for which (9) holds (thus $\mathfrak{Q}(a) = a_{i,4}$ for $i = 3, 4, \ldots, n+2$). *For each pair of individuals x, y of $\mathfrak{J}$ $(x \neq y)$ and each table T of $\mathfrak{Q}$ there obviously exists an E-complex a such that $x = a_1$, $y = a_2$, and $T = \mathfrak{Q}(a)$*; for, in order to find such an a, one need only determine a_i (for $i > 2$) according to (9).

The following facts, required for the proof of Lemma 3, are easily ascertained:

(10) For two arbitrary E-complexes a and b,

$$\text{either} \quad B(a) = B(b) \quad \text{or} \quad \overline{B(a)} \cdot \overline{B(b)} = \Lambda.$$

For, if

$$\alpha(a_{1,1}, a_{2,1}) = \alpha(b_{1,1}, b_{2,1}), \quad Z(a_1, a_2) = Z(b_1, b_2) \quad \text{and} \quad \mathfrak{Q}(a) = \mathfrak{Q}(b),$$

so ist wegen (9) $B(a) = B(b)$; andernfalls ist $\overline{B(a)} \cdot \overline{B(b)} = \Lambda$.

(11) Sind a, b E-Komplexe und $x, y \in \mathfrak{I}$, so können die folgenden beiden Fälle α), β) niemals beide zugleich eintreten:

$$\alpha) \quad x \in \overline{A(a)}, \quad y \in \overline{B(a)}, \qquad \beta) \quad y \in \overline{A(b)}, \quad x \in \overline{B(b)}.$$

438 | Denn wegen (9) folgt aus α) $y_1 = \alpha(a_{1,1}, a_{2,1})$, daher wegen Hilfssatz 2 $\rho(y_1, a_{1,1})$, $\rho(y_1, a_{2,1})$. Da aber im Falle α) $x = a_1$ oder $x = a_2$, so folgt $\rho(y_1, x_1)$. Ebenso folgt aus β) $\rho(x_1, y_1)$. Nach Hilfssatz 2 können aber $\rho(x_1, y_1)$ und $\rho(y_1, x_1)$ nicht beide gelten.

Aus (11) folgt sofort, daß $\overline{A(a)}$ und $\overline{B(a)}$ immer fremd sind. Denn wäre x ein gemeinsames Element, so würden α) und β) für $x = y$, $a = b$ beide gelten. Jeder E-Komplex besteht also aus $n + 2$ verschiedenen Elementen und daher hat $N(x, a)$ für jeden E-Komplex a und jedes $x \in \bar{a}$ einen Sinn.

Hilfssatz 3. *Sind a, b zwei E-Komplexe und $\mathfrak{m}$ eine Teilmenge von $\mathfrak{I}$, für welche gilt:*

$$\mathfrak{m} \subset \bar{a}, \quad \mathfrak{m} \subset \bar{b}; \quad \mathfrak{m}\overline{B(a)} \neq \Lambda \quad (*) \qquad \mathfrak{m}\overline{B(b)} \neq \Lambda \quad (**),$$

so ist $\mathfrak{Q}(a) = \mathfrak{Q}(b)$ und für jedes $z \in \mathfrak{m}$ ist $N(z, a) = N(z, b)$. (Falls $\mathfrak{m}$ mindestens drei Elemente hat, sind $(*), (**)$ von selbst erfüllt.)

Beweis: Wegen $(*)$ gibt es ein $y \in \mathfrak{m}\overline{B(a)}$ und wegen $(**)$ ein $x \in \mathfrak{m}\overline{B(b)}$. Wegen (11) kann nicht gleichzeitig $x \in \overline{A(a)}$ und $y \in \overline{A(b)}$ sein. Daher muß entweder $x \in \overline{B(a)}$ oder $y \in \overline{B(b)}$. In jedem der beiden Fälle ist aber $\overline{B(a)} \cdot \overline{B(b)} \neq \Lambda$, daher wegen (10) $B(a) = B(b)$. Daraus folgt $\mathfrak{Q}(a) = \mathfrak{Q}(b)$ ($\mathfrak{Q}(a)$ hängt ja nur von den letzten n Elementen von a ab) und $N(z, a) = N(z, b)$ für $z \in \mathfrak{m}\overline{B(a)}$. Falls aber $z \in \mathfrak{m}\overline{A(a)}$ (und daher auch $z \in \overline{A(b)}$), so bleibt nur zu zeigen, daß dann weder $z = a_1, z = b_2$ noch $z = a_2, z = b_1$ sein kann. Träte aber einer dieser beiden Fälle ein, so würden $[a_1, a_2]$ und $[b_1, b_2]$ zu zwei verschiedenen Klassen K_i gehören (nach Definition der K_i), d. h. $Z(a_1, a_2) \neq Z(b_1, b_2)$ und daher wegen (9) $a_{i+2} \neq b_{i+2}$, was wegen $B(a) = B(b)$ nicht sein kann.

Um das Modell M über $\mathfrak{I}$, welches die Formel (4) erfüllen wird, bequemer definieren zu können, schicke ich folgendes voraus. Falls R und S zwei Tabellen k-ter Ordnung sind, so schreibe ich

$$R \sim S,$$

wenn $\phi_l^R(u) \equiv \phi_l^S(u)$ für $l = 1, 2, \ldots, s$ und für alle solchen r_l-tupel u, für welche $\bar{u} = \{1, 2, \ldots, k\}$ (d. h. in denen sämtliche Zahlen von 1 bis k wirklich auftreten). Aus $R = S$ folgt $R \sim S$, das Umgekehrte gilt aber nur für Tabellen erster Ordnung. Sei ferner $\mathfrak{I}^k$ ($k = 1, 2, \ldots$) eine Menge von k-tupeln von Elementen aus $\mathfrak{I}$ derart, daß jedes k-tupel aus $\mathfrak{I}^k$ lauter untereinander verschiedene Glieder hat und daß es für jede genau k Elemente enthaltende Teilmenge $\mathfrak{n}$ von $\mathfrak{I}$ genau ein $u \in \mathfrak{I}^k$ gibt, für welches

then, on account of (9), $B(a) = B(b)$; otherwise $\overline{B(a)} . \overline{B(b)} = \Lambda$.

(11) If a and b are E-complexes and $x, y \in \mathfrak{I}$, the following two cases, (α) and (β), can never both occur at once:

$$(\alpha) \quad x \in \overline{A(a)}, \quad y \in \overline{B(a)}, \qquad (\beta) \quad y \in \overline{A(b)}, \quad x \in \overline{B(b)}.$$

For, on account of (9), $y_1 = \alpha(a_{1,1}, a_{2,1})$ follows from (α); hence, by virtue of Lemma 2, $\rho(y_1, a_{1,1})$ and $\rho(y_1, a_{2,1})$. But, since in case (α) $x = a_1$ or $x = a_2$, $\rho(y_1, x_1)$ follows. Likewise, $\rho(x_1, y_1)$ follows from (β). But, according to Lemma 2, $\rho(x_1, y_1)$ and $\rho(y_1, x_1)$ cannot both hold.

From (11) it follows immediately that $A(a)$ and $\overline{B(a)}$ are always disjoint. For, if x were a common element, both (α) and (β) would hold when $x = y$ and $a = b$. Every E-complex thus consists of $n + 2$ distinct elements, and therefore $N(x, a)$ has a meaning for every E-complex a and every $x \in \bar{a}$.

Lemma 3. *If a and b are two E-complexes and $\mathfrak{m}$ is a subset of $\mathfrak{I}$ for which*

$$\mathfrak{m} \subset \bar{a}, \quad \mathfrak{m} \subset \bar{b}, \quad \mathfrak{m}\overline{B(a)} \neq \Lambda \quad (*), \qquad \mathfrak{m}\overline{B(b)} \neq \Lambda \quad (**),$$

hold, then $\mathfrak{Q}(a) = \mathfrak{Q}(b)$ and, for each $z \in \mathfrak{m}$, $N(z, a) = N(z, b)$. (In case $\mathfrak{m}$ has at least three elements, $(*)$ and $(**)$ are automatically satisfied.)

Proof: By virtue of $(*)$ there is a $y \in \mathfrak{m}\overline{B(a)}$ and by virtue of $(**)$ there is an $x \in \mathfrak{m}\overline{B(b)}$. By virtue of (11), we cannot have both $x \in \overline{A(a)}$ and $y \in \overline{A(b)}$ at once. Therefore, we must have either $x \in \overline{B(a)}$ or $y \in \overline{B(b)}$. But in each of the two cases $\overline{B(a)} . \overline{B(b)} \neq \Lambda$, hence, by virtue of (10), $B(a) = B(b)$. From that, $\mathfrak{Q}(a) = \mathfrak{Q}(b)$ follows (indeed, $\mathfrak{Q}(a)$ depends only on the last n elements of a) and $N(z, a) = N(z, b)$ for $z \in \mathfrak{m}\overline{B(a)}$. But in case $z \in \mathfrak{m}\overline{A(a)}$ (and therefore also $z \in \overline{A(b)}$), it only remains to show that then we can have neither $z = a_1$ and $z = b_2$ nor $z = a_2$ and $z = b_1$. But, if one of these two cases were to occur, $[a_1, a_2]$ and $[b_1, b_2]$ would belong to two different classes K_i (by definition of the K_i); that is, $Z(a_1, a_2) \neq Z(b_1, b_2)$, hence, by virtue of (9), $a_{i+2} \neq b_{i+2}$, which cannot be since $B(a) = B(b)$.

In order to be able to define more conveniently the model M over $\mathfrak{I}$ that will satisfy formula (4), I make the following prefatory remarks. In case R and S are two kth-order tables, I write

$$R \sim S$$

if $\phi_l^R(u) \equiv \phi_l^S(u)$ for $l = 1, 2, \ldots, s$ and for all r_l-tuples u for which $\bar{u} = \{1, 2, \ldots, k\}$ (that is, in which all of the numbers from 1 to k actually appear). $R \sim S$ follows from $R = S$, but the converse holds only for first-order tables. Further, let $\mathfrak{I}^k$ $(k = 1, 2, \ldots)$ be a set of k-tuples of elements of $\mathfrak{I}$ such that every k-tuple of $\mathfrak{I}^k$ has only terms distinct from one another and, for every subset $\mathfrak{n}$ of $\mathfrak{I}$ containing exactly k elements, there is exactly one $u \in \mathfrak{I}^k$ for which $\mathfrak{n} = \bar{u}$. Thus $\mathfrak{I}^k$ is essentially the set of k-element

$\mathfrak{n} = \overline{u}$. $\mathfrak{J}^k$ ist also im wesentlichen die Menge der k-elementigen Teilmengen von $\mathfrak{J}$, wobei für jede solche Teilmenge eine Anordnung ihrer Elemente festgesetzt ist. Man bestätigt leicht die folgenden beiden Sätze:

439 | Hilfssatz 4. *Wenn jedem $u \, \epsilon \, \mathfrak{J}^k$ eine Tabelle k-ter Ordnung T_u zuge- ordnet ist, und zwar simultan für $k = 1, 2, \ldots, r$ (wo r eine beliebige natür- liche Zahl ist), so gibt es ein Modell M über $\mathfrak{J}$, für welches gilt:*

$$[M|u] \sim T_u \tag{12}$$

für alle $u \, \epsilon \, (\mathfrak{J}^1 + \mathfrak{J}^2 + \cdots + \mathfrak{J}^r)$.

Denn durch die Beziehung (12) (für irgend ein bestimmtes u) werden die Wahrheitswerte der ϕ_i^M für alle solchen Argument-r_i-tupel v festgelegt, für welche $\overline{v} = \overline{u}$, und für zwei verschiedene u sind das immer auch verschiedene Argument-r_i-tupel.

Hilfssatz 5. *Ist T eine Tabelle k-ter Ordnung und ist a ein k-tupel von Elementen aus $\mathfrak{J}$ und M ein Modell über $\mathfrak{J}$, so folgt die Gleichung $[M|a_1, a_2, \ldots, a_k] = T$ daraus, daß $[M|a_{i_1}, a_{i_2}, \ldots, a_{i_p}] \sim [T|i_1, i_2, \ldots, i_p]$ für alle solchen $i_1, i_2, \ldots, i_p$, für welche*

$$[a_{i_1}, a_{i_2}, \ldots, a_{i_p}] \, \epsilon \, \mathfrak{J}^p \quad und \quad \{a_{i_1}, a_{i_2}, \ldots, a_{i_p}\} \subset \overline{a}$$

($p = 1, 2, \ldots, k$).

Nun spezialisieren wir die Beziehungen (12) aus Hilfssatz 4 folgender- maßen:

A) Sei $x \, \epsilon \, \mathfrak{J}$:

I. Falls es E-Komplexe a gibt, so daß $x \, \epsilon \, \overline{B(a)}$, so greife man irgend- einen solchen E-Komplex a heraus und setze

$$[M|x] = [\eta(\mathfrak{Q}(a))|N(x, a)]. \tag{13}$$

II. Falls es keine derartigen E-Komplexe gibt, setze man $[M|x] = V$, wo V ein beliebiges Element aus $\mathfrak{P}$ ist.

Wegen (7.3) ist auch im Fall I und daher ausnahmslos $[M|x] \, \epsilon \, \mathfrak{P}$.

B) Sei $[x, y] \, \epsilon \, \mathfrak{J}^2$:

I. Falls es E-Komplexe a gibt von der Art, daß

$$\{x, y\} \subset \overline{a} \quad und \quad \{x, y\} \cdot \overline{B(a)} \neq \Lambda$$

und so, daß außerdem für die (durch die Festsetzung A) bereits festgelegten) Tabellen $[M|x], [M|y]$ folgendes gilt:

$$[M|x] = [\eta(\mathfrak{Q}(a))|N(x, a)], \quad [M|y] = [\eta(\mathfrak{Q}(a))|N(y, a)], \tag{14}$$

so greife man irgend einen derartigen E-Komplex a heraus und setze

$$[M|x, y] \sim [\eta(\mathfrak{Q}(a))|N(x, a), N(y, a)]. \tag{15}$$

subsets of $\mathfrak{I}$, an ordering being specified for the elements of each such subset. The following two theorems are easily seen to hold:

Lemma 4. If to each $u \in \mathfrak{I}^k$ there is assigned a kth-order table T_u, and indeed simultaneously for $k = 1, 2, \ldots, r$ (where r is an arbitrary natural number), then there exists a model M over $\mathfrak{I}$ for which

$$[M|u] \sim T_u \tag{12}$$

holds for all $u \in (\mathfrak{I}^1 + \mathfrak{I}^2 + \cdots + \mathfrak{I}^r)$.

Thus by the relation (12) (for some one fixed u) the truth values of the ϕ_i^M, for all argument r_i-tuples v for which $\bar{v} = \bar{u}$, are determined, and for two different values of u these argument r_i-tuples are always different also.

Lemma 5. If T is a kth-order table, a is a k-tuple of elements of $\mathfrak{I}$, and M is a model over $\mathfrak{I}$, then the equation $[M|a_1, a_2, \ldots, a_k] = T$ follows from the fact that $[M|a_{i_1}, a_{i_2}, \ldots, a_{i_p}] \sim [T|i_1, i_2, \ldots, i_p]$ holds for all $i_1, i_2, \ldots, i_p$ for which

$$[a_{i_1}, a_{i_2}, \ldots, a_{i_p}] \in \mathfrak{I}^p \quad and \quad \{a_{i_1}, a_{i_2}, \ldots, a_{i_p}\} \subset \bar{a}$$

$(p = 1, 2, \ldots, k)$.

We now specialize the relations (12) occurring in Lemma 4 as follows:

(A) Let $x \in \mathfrak{I}$:

    I. In case there exist E-complexes a such that $x \in \overline{B(a)}$, select some such E-complex a and set

$$[M|x] = [\eta(\mathfrak{Q}(a))|N(x, a)]. \tag{13}$$

    II. In case there exists no such E-complex, set $[M|x] = V$, where V is an arbitrary element of $\mathfrak{P}$.

By virtue of (7.3), $[M|x] \in \mathfrak{P}$ holds also in case I, and therefore without exception.

(B) Let $[x, y] \in \mathfrak{I}^2$:

    I. In case there exist E-complexes a such that

$$\{x, y\} \subset \bar{a} \quad and \quad \{x, y\} \cdot \overline{B(a)} \neq \Lambda$$

and such that, moreover, the following holds for the tables $[M|x]$ and $[M|y]$ (already determined by the stipulation (A)),

$$[M|x] = [\eta(\mathfrak{Q}(a))|N(x, a)], \quad [M|y] = [\eta(\mathfrak{Q}(a))|N(y, a)], \tag{14}$$

select one such E-complex a and set

$$[M|x, y] \sim [\eta(\mathfrak{Q}(a))|N(x, a), N(y, a)]. \tag{15}$$

II. Falls es keinen E-Komplex a mit den in B) I geforderten Eigenschaften gibt, so sei

$$[M|x, y] \sim \delta([M|x], [M|y]). \tag{16}$$

C) Sei $[x_1, x_2, \ldots, x_k] \, \epsilon \, \mathfrak{I}^k, k > 2$:

    I. Falls es E-Komplexe a gibt, so daß $\{x_1, x_2, \ldots, x_k\} \subset \bar{a}$, so greife man einen solchen heraus und setze

$$[M|x_1, x_2, \ldots, x_k] \sim [\eta(\mathfrak{Q}(a))|N(x_1, a), N(x_2, a), \ldots, N(x_k, a)]. \tag{17}$$

440 | II. Falls es keinen derartigen E-Komplex gibt, sei $[M|x_1, x_2, \ldots, x_k]$ $\sim W$, wobei W eine ganz beliebige Tabelle k-ter Ordnung ist.

Für das so definierte Modell M gelten nun folgende Sätze:

(18) Die Beziehungen (13) bzw. (15) bzw. (17) bestehen für *alle* E-Komplexe a, welche die in A) I bzw. B) I bzw. C) I geforderten Eigenschaften haben.

Denn indem man in Hilfssatz 3 $\mathfrak{m}$ gleichsetzt $\{x\}$ bzw. $\{x, y\}$ bzw. $\{x_1, x_2, \ldots, x_k\}$, so erkennt man, daß die rechten Seiten von (13) bzw. (15) bzw. (17) für alle derartigen a gleich sind.

Hilfssatz 6. *Ist* $x, y \, \epsilon \, \mathfrak{I}, x \neq y$, *so ist* $[M|x, y] \, \epsilon \, \mathfrak{Q}$.

Beweis: Ist $[x, y] \, \epsilon \, \mathfrak{I}^2$ und tritt für $[x, y]$ Fall B) I ein, so ist wegen (14) und (15) $[M|x, y] = [\eta(\mathfrak{Q}(a))|N(x, a), N(y, a)]$ und die Behauptung folgt wegen (7.4). Ist $[x, y] \, \epsilon \, \mathfrak{I}^2$ und tritt Fall B) II ein, so ist wegen (16) und (6.2), (6.3) $[M|x, y] = \delta([M|x], [M|y])$ und die Behauptung folgt aus (6.1). Ist $[x, y]$ nicht Element von $\mathfrak{I}^2$, so ist $[y, x] \, \epsilon \, \mathfrak{I}^2$, daher nach dem eben Bewiesenen $[M|y, x] \, \epsilon \, \mathfrak{Q}$. Nun ist wegen (7.1) $[M|y, x] = [\eta([M|y, x])|1, 2]$ und daher $[M|x, y] = [\eta([M|y, x])|2, 1]$, woraus die Behauptung nach (7.4) folgt.

Hilfssatz 7. *Ist* a *ein* E-*Komplex und* $[M|a_1, a_2] = \mathfrak{Q}(a)$, *so ist* $\mathfrak{B}_M(a_1, a_2, \ldots, a_{n+2})$ *wahr*.

Ich zeige $[M|a_1, a_2, \ldots, a_{n+2}] = \eta(\mathfrak{Q}(a))$, woraus die Behauptung wegen (7.2) folgt. Zum Beweis verwende ich Hilfssatz 5.

    α) Für jedes a_i $(i = 1, 2, \ldots, n + 2)$ gilt $[M|a_i] = [\eta(\mathfrak{Q}(a))|i]$.

Beweis: Ist $i \geq 3$, d. h. $a_i \, \epsilon \, B(a)$, so tritt für a_i der Fall A) I ein und die Behauptung folgt aus (13) und (18).

Ist $i = 1$ so hat man nur zu beachten, daß wegen der Voraussetzung $[M|a_1, a_2] = \mathfrak{Q}(a)$ die Gleichung $[M|a_1] = [\mathfrak{Q}(a)|1]$ besteht und wegen (7.1) $[\mathfrak{Q}(a)|1] = [\eta(\mathfrak{Q}(a))|1]$. Analog ist für $i = 2$

$$[M|a_2] = [\mathfrak{Q}(a)|2] = [\eta(\mathfrak{Q}(a))|2].$$

    β) Ist $\{a_i, a_k\} \subset \bar{a}$ und $[a_i, a_k] \, \epsilon \, \mathfrak{I}^2$, so ist $[M|a_i, a_k] \sim [\eta(\mathfrak{Q}(a))|i, k]$.

Beweis: Ist $\{a_i, a_k\} \cdot \overline{B(a)} \neq \Lambda$, so tritt für das Paar $[a_i, a_k]$ der Fall B) I ein, da ja nach α) $[M|a_i] = [\eta(\mathfrak{Q}(a))|i]$, $[M|a_k] = [\eta(\mathfrak{Q}(a))|k]$. Daher ist

II. In case there exists no E-complex a with the properties required by (B) I, let

$$[M|x, y] \sim \delta([M|x], [M|y]).\qquad (16)$$

(C) Let $[x_1, x_2, \ldots, x_k] \, \epsilon \, \mathfrak{J}^k, k > 2$:

I. In case there exist E-complexes a such that $\{x_1, x_2, \ldots, x_k\} \subset \bar{a}$, select one and set

$$[M|x_1, x_2, \ldots, x_k] \sim [\eta(\mathfrak{Q}(a))|N(x_1, a), N(x_2, a), \ldots, N(x_k, a)].\qquad (17)$$

II. In case there exists no such E-complex, let $[M|x_1, x_2, \ldots, x_k] \sim W$, where W is a completely arbitrary kth-order table.

Now, for the model M so defined, the following theorems hold:

(18) The relations (13) (or (15) or (17)) hold for *all* E-complexes a that have the properties required by (A) I (or (B) I or (C) I, respectively).

For, by setting $\mathfrak{m}$ equal to $\{x\}$ (or $\{x, y\}$ or $\{x_1, x_2, \ldots, x_k\}$, respectively) in Lemma 3, one sees that the right sides of (13) (or (15) or (17), respectively) are the same for all such a.

Lemma 6. *If $x, y \, \epsilon \, \mathfrak{J}$ and $x \neq y$, then $[M|x, y] \, \epsilon \, \mathfrak{Q}$.*

Proof: If $[x, y] \, \epsilon \, \mathfrak{J}^2$ and $[x, y]$ falls under case (B) I, then, by virtue of (14) and (15), $[M|x, y] = [\eta(\mathfrak{Q}(a))|N(x, a), N(y, a)]$, and the assertion follows by (7.4). If $[x, y] \, \epsilon \, \mathfrak{J}^2$ and case (B) II obtains, then, by virtue of (16), (6.2) and (6.3), $[M|x, y] = \delta([M|x], [M|y])$, and the assertion follows from (6.1). If $[x, y]$ is not an element of $\mathfrak{J}^2$, then $[y, x] \, \epsilon \, \mathfrak{J}^2$, hence, by what was just proved, $[M|y, x] \, \epsilon \, \mathfrak{Q}$. Now, by virtue of (7.1), $[M|y, x] = [\eta([M|y, x])|1, 2]$, and therefore $[M|x, y] = [\eta([M|y, x]|2, 1]$, from which the assertion follows by (7.4).

Lemma 7. *If a is an E-complex and $[M|a_1, a_2] = \mathfrak{Q}(a)$, then $\mathfrak{B}_M(a_1, a_2, \ldots, a_{n+2})$ is true.*

I show that $[M|a_1, a_2, \ldots, a_{n+2}] = \eta(\mathfrak{Q}(a))$, from which the assertion follows by (7.2). For the proof I make use of Lemma 5.

(α) For every a_i $(i = 1, 2, \ldots, n + 2)$, $[M|a_i] = [\eta(\mathfrak{Q}(a))|i]$ holds.

Proof: If $i \geq 3$, that is, $a_i \, \epsilon \, B(a)$, then a_i falls under case (A) I and the assertion follows from (13) and (18).

If $i = 1$, one has only to observe that, on account of the assumption $[M|a_1, a_2] = \mathfrak{Q}(a)$, the equation $[M|a_1] = [\mathfrak{Q}(a)|1]$ obtains and, on account of (7.1), $[\mathfrak{Q}(a)|1] = [\eta(\mathfrak{Q}(a))|1]$. Analogously for $i = 2$,

$$[M|a_2] = [\mathfrak{Q}(a)|2] = [\eta(\mathfrak{Q}(a))|2].$$

(β) If $\{a_i, a_k\} \subset \bar{a}$ and $[a_i, a_k] \, \epsilon \, \mathfrak{J}^2$, then $[M|a_i, a_k] \sim [\eta(\mathfrak{Q}(a))|i, k]$.

Proof: If $\{a_1, a_k\} . \overline{B(a)} \neq \Lambda$, then the pair $[a_i, a_k]$ falls under case (B) I since, after all, according to (α), $[M|a_i] = [\eta(\mathfrak{Q}(a))|i]$ and $[M|a_k] = [\eta(\mathfrak{Q}(a))|k]$. Therefore, by (15) and (18), $[M|a_i, a_k] \sim [\eta(\mathfrak{Q}(a))|i, k]$.

nach (15) und (18) $[M|a_i, a_k] \sim [\eta(\mathfrak{Q}(a))|i, k]$.

Ist $\{a_i, a_k\} . B(a) = \Lambda$, also $\{a_i, a_k\} = \{a_1, a_2\}$, so braucht man nur zu beachten, daß nach Voraussetzung $[M|a_1, a_2] = \mathfrak{Q}(a)$ und wegen (7.1) $\mathfrak{Q}(a) = [\eta(\mathfrak{Q}(a))|1, 2]$, woraus $[M|a_1, a_2] = [\eta(\mathfrak{Q}(a))|1, 2]$ und daher auch $[M|a_2, a_1] = [\eta(\mathfrak{Q}(a))|2, 1]$ folgt.

441 $\quad\gamma)$ Ist $\{a_{i_1}, \ldots, a_{i_k}\} \subset \bar{a}, [a_{i_1}, \ldots, a_{i_k}] \epsilon \mathfrak{J}^k, k > 2,$
| so gilt $[M|a_{i_1}, a_{i_2}, \ldots, a_{i_k}] \sim [\eta(\mathfrak{Q}(a))|i_1, i_2, \ldots, i_k]$, was sich sofort aus C) I und (18) ergibt.

Aus $\alpha), \beta), \gamma)$ und Hilfssatz 5 folgt die Behauptung.

Aus Hilfssatz 6 und 7 folgt unmittelbar, daß M die Formel (4) erfüllt. Dazu braucht man nur zu zeigen, daß es für je zwei Elemente $x, y \epsilon \mathfrak{J}$, $x \neq y$, Elemente $u_1, u_2, \ldots, u_n \epsilon \mathfrak{J}$ gibt, so daß $\mathfrak{B}_M(x, y, u_1, \ldots, u_n)$. Sei also $x, y \epsilon \mathfrak{J}$, $x \neq y$, nach Hilfssatz 6 ist $[M|x, y] \epsilon \mathfrak{Q}$. Nach einer oben (S. 437) gemachten Bemerkung gibt es einen E-Komplex a, so daß $x = a_1$, $y = a_2$ und $\mathfrak{Q}(a) = [M|x, y]$. Dieser E-Komplex erfüllt die Voraussetzung von Hilfssatz 7 und daher ist $\mathfrak{B}_M(x, y, a_3, \ldots, a_{n+2})$ wahr, womit Satz II bewiesen ist.

Zwei Zählausdrücke mögen gleichwertig heißen, wenn sie entweder beide erfüllbar oder beide nicht erfüllbar sind, und zur Erledigung des zweiten eingangs angekündigten Punktes ist zu zeigen:

Satz III: *Zu jedem Zählausdruck kann man einen gleichwertigen der Gestalt*

$$(x_1)(x_2)(x_3)(Ey_1)(Ey_2)\ldots(Ey_n)\mathfrak{A}(x_1, x_2, x_3, y_1, y_2, \ldots, y_n) \qquad (19)$$

angeben, wobei man übrigens auch erreichen kann, daß in (19) nur zweistellige Funktionsvariable vorkommen.

Ein Zählausdruck möge *ausgezeichnete Normalform* heißen, wenn er die Gestalt hat

$$(x_1)(x_2)\ldots(x_p)(Ey_1)(Ey_2)\ldots(Ey_q)\mathfrak{A}(x_1, x_2, \ldots, x_p, y_1, y_2, \ldots, y_q).$$

Jeder Zählausdruck ist mit einer ausgezeichneten binären[8] Normalform gleichwertig. Um dies einzusehen stelle man zunächst nach Löwenheim[9] einen gleichwertigen binären Ausdruck her und wende auf diesen das Skolemsche Verfahren[10] zur Herstellung einer gleichwertigen ausgezeichneten Normalform an, in der Form, wie ich es in *1930* zum Beweise von Satz IV (S. 352 dieser Arbeit) verwendet habe, nur mit der Modifikation,

---

[8]Ein Zählausdruck heißt binär, wenn er nur zweistellige Funktionsvariable enthält.

[9]Vgl. *Löwenheim 1915.*

[10]Vgl. *Skolem 1920.*

If $\{a_i, a_k\} \cdot \overline{B(a)} = \Lambda$, hence $\{a_i, a_k\} = \{a_1, a_2\}$, one need only observe that, by assumption, $[M|a_1, a_2] = \mathfrak{Q}(a)$ and, on account of (7.1), $\mathfrak{Q}(a) = [\eta(\mathfrak{Q}(a))|1, 2]$, from which $[M|a_1, a_2] = [\eta(\mathfrak{Q}(a))|1, 2]$, and therefore $[M|a_2, a_1] = [\eta(\mathfrak{Q}(a))|1, 2]$, too, follows.

(γ) If $\{a_{i_1}, \ldots, a_{i_k}\} \subset \overline{a}$, $[a_{i_1}, \ldots, a_{i_k}] \, \epsilon \, \mathfrak{J}^k$, $k > 2$, then

$$[M|a_{i_1}, a_{i_2}, \ldots, a_{i_k}] \sim [\eta(\mathfrak{Q}(a))|i_1, i_2, \ldots, i_k]$$

holds, which results at once from (C) I and (18).

The assertion follows from (α), (β), (γ) and Lemma 5.

From Lemmas 6 and 7 it follows immediately that M satisfies formula (4). To that end, one need only show that for any two elements $x, y \, \epsilon \, \mathfrak{J}$, $x \neq y$, there are elements $u_1, u_2, \ldots, u_n \, \epsilon \, \mathfrak{J}$ such that $\mathfrak{B}_M(x, y, u_1, \ldots, u_n)$. Thus, if $x, y \, \epsilon \, \mathfrak{J}$, $x \neq y$, then, by Lemma 6, $[M|x, y] \, \epsilon \, \mathfrak{Q}$. By a remark made above (page 315) there is an E-complex a such that $x = a_1$, $y = a_2$ and $\mathfrak{Q}(a) = [M|x, y]$. This E-complex satisfies the hypothesis of Lemma 7, and therefore $\mathfrak{B}_M(x, y, a_3, \ldots, a_{n+2})$ is true, whereby Theorem II is proved.

Two first-order expressions will be said to be equivalent if they are either both satisfiable or both unsatisfiable, and, to take care of the second of the points announced at the outset, we must show:

Theorem III. *For every first-order expression one can determine an equivalent one of the form*

$$(x_1)(x_2)(x_3)(Ey_1)(Ey_2)\ldots(Ey_n)\,\mathfrak{A}\,(x_1, x_2, x_3, y_1, y_2, \ldots, y_n), \qquad (19)$$

and moreover we can arrange matters so that no functional variable with more than two arguments occurs in (19).

A first-order expression will be said to be in *distinguished normal form* if it has the form

$$(x_1)(x_2)\ldots(x_p)(Ey_1)(Ey_2)\ldots(Ey_q)\,\mathfrak{A}\,(x_1, x_2, \ldots, x_p, y_1, y_2, \ldots, y_q).$$

Every first-order expression is equivalent to a binary[8] *one in distinguished normal form.* To see this, first obtain, in the manner of Löwenheim,[9] an equivalent binary expression and then apply to it Skolem's procedure[10] for producing an equivalent distinguished normal form, in the way in which this procedure is used in my *1930* for the proof of Theorem IV (page 352 of that paper [above, page 109]), but with the modification that in formula

---

[8] A first-order expression is said to be binary if it contains no functional variables with more than two arguments.

[9] See *Löwenheim 1915.*

[10] See *Skolem 1920.*

daß man in Formel B (S. 353 oben) statt $F(\mathfrak{x}, \mathfrak{y})$ (dies ist eine Abkürzung für $F(x_1, \ldots, x_n, y_1, \ldots, y_m)$) den folgenden Ausdruck setzt

$$(Eu)[R_1(u, x_1)\ \&\ \cdots\ \&\ R_n(u, x_n)\ \&$$
$$R_{n+1}(u, y_1)\ \&\ \cdots\ \&\ R_{n+m}(u, y_m)], \qquad (20)$$

442 | wobei R_i irgendwelche in A nicht vorkommende Funktionsvariable sind; ebenso statt $F(\mathfrak{x}', \mathfrak{y}')$ einen entsprechenden Ausdruck mit denselben R_i. Auch die so modifizierte Formel B (die binär ist, wenn $P(A)$ binär ist) kann auf eine Normalform vom Grad $k - 1$ gebracht werden und ist mit $(P)A$ gleichwertig. Das letztere ergibt sich sofort daraus, daß jede $(n+m)$-stellige Funktion F in einem abzählbaren Individuenbereich sich durch geeignete R_i in der Form (20) darstellen läßt.

Es bleibt also nur noch zu zeigen, *daß man zu jeder binären ausgezeichneten Normalform A mit n Allzeichen eine gleichwertige B mit $n - 1$ Allzeichen angeben kann, wenn $n > 3$,* woraus durch iterierte Anwendung Satz III folgt.

Beweis: Ist A der Ausdruck

$$(x_1) \ldots (x_n)(Ey_1) \ldots (Ey_m) \mathfrak{A}(x_1, \ldots, x_n, y_1, \ldots, y_m),$$

so bezeichne ich mit A' die Konjunktion der folgenden drei Formeln (21), (22), (23):

$$(x_3) \ldots (x_n)(u)(Ev)(Ew)(Ey_1) \ldots (Ey_m)[R_1(u, v)\ \&\ R_2(u, w)\ \&$$
$$\mathfrak{A}(v, w, x_3, \ldots, x_n, y_1, \ldots, y_m)] \qquad (21)$$

$$(z_1)(z_2)(Eu)[R_1(u, z_1)\ \&\ R_2(u, z_2)] \qquad (22)$$

$$(x)(y)(z)[(R_1(x, y)\ \&\ R_1(x, z) \to y = z)\ \&$$
$$(R_2(x, y)\ \&\ R_2(x, z) \to y = z)], \qquad (23)$$

wobei R_1, R_2 zwei in $\mathfrak{A}$ nicht vorkommende Variable sind. Die Funktionsvariablen aus $\mathfrak{A}$ seien $S_1, S_2, \ldots, S_k$. Ist $\overline{S}_1, \overline{S}_2, \ldots, \overline{S}_k$ ein A erfüllendes System von Funktionen in einem abzählbaren Bereich $\mathfrak{J}$ und sind $\overline{R}_1, \overline{R}_2$ zwei so beschaffene Relationen in $\mathfrak{J}$, daß durch die Relation $x\,\overline{R}_1\,y\ \&\ x\,\overline{R}_2\,z$ eine eineindeutige Abbildung der Elemente x aus $\mathfrak{J}$ auf die Paare von Elementen $[y, z]$ aus $\mathfrak{J}$ geliefert wird, so erfüllt $\overline{R}_1, \overline{R}_2, \overline{S}_1, \overline{S}_2, \ldots, \overline{S}_k$ offenbar die Formel A'; also: Wenn A erfüllbar ist, so auch A'. Ist umgekehrt $\overline{R}_1, \overline{R}_2, \overline{S}_1, \overline{S}_2, \ldots, \overline{S}_k$ ein A' erfüllendes System, so ist $\overline{S}_1, \overline{S}_2, \ldots, \overline{S}_k$ ein A erfüllendes System. Denn sind $x_1, x_2, \ldots, x_n$ Elemente aus $\mathfrak{J}$, so gibt es wegen (22) ein Element $u \,\epsilon\, \mathfrak{J}$, so daß $\overline{R}_1(u, x_1), \overline{R}_2(u, x_2)$. Wegen (21) gibt es weiter Elemente v, w und $y_1, \ldots, y_m$ aus $\mathfrak{J}$, so daß $\overline{R}_1(u, v), \overline{R}_2(u, w)$, $\overline{\mathfrak{A}}(v, w, x_3, \ldots, x_n, y_1, \ldots, y_m)$, wobei $\overline{\mathfrak{A}}$ den durch Einsetzung der $\overline{R}_i, \overline{S}_i$

B (*1930*, page 353, top [above, page 109]), replace $F(\mathfrak{x}, \mathfrak{y})$ (which is an abbreviation for $F(x_1, \ldots, x_n, y_1, \ldots, y_m)$) by the following expression,

$$(Eu)[R_1(u, x_1) \,\&\, \cdots \,\&\, R_n(u, x_n) \,\& \\ R_{n+1}(u, y_1) \,\&\, \cdots \,\&\, R_{n+m}(u, y_m)], \qquad (20)$$

where the R_i are any functional variables not occurring in A; likewise, replace $F(\mathfrak{x}', \mathfrak{y}')$ by a corresponding expression with the same R_i. Formula B so modified (which is binary if $P(A)$ is) can be brought into normal form of degree $k - 1$ and is equivalent to $P(A)$. The latter results at once from the fact that every $(n + m)$-place function F in a countable domain of individuals can be represented in the form (20) by means of suitable R_i.

Thus it only remains to show that, *if $n > 3$, one can assign to every binary distinguished normal form A with n universal quantifiers an equivalent one B with $n - 1$ universal quantifiers*, from which Theorem III follows by iterated application.

Proof: If A is the expression

$$(x_1) \ldots (x_n)(Ey_1) \ldots (Ey_m) \,\mathfrak{A}\,(x_1, \ldots, x_n, y_1, \ldots, y_m),$$

I denote by A' the conjunction of the following three formulas, (21), (22) and (23):

$$(x_3) \ldots (x_n)(u)(Ev)(Ew)(Ey_1) \ldots (Ey_m)[R_1(u, v) \,\&\, R_2(u, w) \,\& \\ \mathfrak{A}\,(v, w, x_3, \ldots, x_n, y_1, \ldots, y_m)], \quad (21)$$

$$(z_1)(z_2)(Eu)[R_1(u, z_1) \,\&\, R_2(u, z_2)], \qquad (22)$$

$$(x)(y)(z)[(R_1(x, y) \,\&\, R_1(x, z) \to y = z) \,\& \\ (R_2(x, y) \,\&\, R_2(x, z) \to y = z)], \qquad (23)$$

where R_1 and R_2 are two variables not occurring in $\mathfrak{A}$. Let the functional variables of $\mathfrak{A}$ be $S_1, S_2, \ldots, S_k$. If $\overline{S}_1, \overline{S}_2, \ldots, \overline{S}_k$ is a system of functions, in a countable domain $\mathfrak{J}$, satisfying A and if $\overline{R}_1, \overline{R}_2$ are two relations in $\mathfrak{J}$ such that the relation $x \overline{R}_1 y \,\&\, x \overline{R}_2 z$ yields a one-to-one mapping of the elements x of $\mathfrak{J}$ into the pairs $[y, z]$ of elements of $\mathfrak{J}$, then $\overline{R}_1, \overline{R}_2, \overline{S}_1, \overline{S}_2, \ldots, \overline{S}_k$ obviously satisfy the formula A'. Thus, if A is satisfiable, so is A'. If, conversely, $\overline{R}_1, \overline{R}_2, \overline{S}_1, \overline{S}_2, \ldots, \overline{S}_k$ is a system satisfying A', then $\overline{S}_1, \overline{S}_2, \ldots, \overline{S}_k$ is a system satisfying A. For, if $x_1, x_2, \ldots, x_n$ are elements of $\mathfrak{J}$, there is on account of (22) an element $u \,\epsilon\, \mathfrak{J}$ such that $\overline{R}_1(u, x_1)$ and $\overline{R}_2(u, x_2)$. On account of (21) there are, further, elements v, w and $y_1, \ldots, y_m$ of $\mathfrak{J}$ such that $\overline{R}_1(u, v), \overline{R}_2(u, w)$ and $\overline{\mathfrak{A}}(v, w, x_3, \ldots, x_n, y_1, \ldots, y_m)$, where $\overline{\mathfrak{A}}$ stands for the expression that results from $\mathfrak{A}$ when $\overline{R}_i$ and $\overline{S}_i$ are put in place of

an Stelle der R_i, S_i aus $\mathfrak{A}$ entstehenden Ausdruck bedeutet. Wegen (23) ist $v = x_1$ und $w = x_2$. Also gilt $\overline{\mathfrak{A}}(x_1, \ldots, x_n, y_1, \ldots, y_m)$, womit die Behauptung bewiesen ist und damit auch, daß A und A' gleichwertige Ausdrücke sind.

443   Die Konjunktion zweier ausgezeichneter Normalformen, von denen jede höchstens n Allzeichen enthält, ist wieder mit einer ausgezeich|neten Normalform mit höchstens n Allzeichen äquivalent (weil ja

$$(x)F(x) \;\&\; (y)G(y) \equiv (x)[F(x) \;\&\; G(x)]).$$

Auf A' angewendet ergibt dies eine ausgezeichnete Normalform mit $n - 1$ Allzeichen, die aber noch das =-Zeichen enthält. Eliminiert man dieses, nach dem von L. Kalmár in *1929*, und von mir in *1930*, p. 356, entwickelten Verfahren und beachtet die obige Bemerkung über die Konjunktion ausgezeichneter Normalformen, so erhält man (weil in A' nur zweistellige Funktionsvariable vorkommen) wieder eine ausgezeichnete Normalform mit $n-1$ Allzeichen und diese liefert das B in dem zu beweisenden Satze.

Zum Schluß möchte ich noch bemerken, daß sich Satz I auch für Formeln, welche das =-Zeichen enthalten, nach demselben Verfahren beweisen läßt.

Besprechung von *Kaczmarz 1932:* Axioms for arithmetic (*1933j*)

Es wird ein aus vier Axiomen bestehendes Axiomensystem für die Arithmetik der natürlichen Zahlen mit den Grundbegriffen "Zahl" und "kleiner als" angegeben und die Unabhängigkeit der Axiome sowie ihre Äquivalenz mit den von L. Neder (*1931*) aufgestellten nachgewiesen.

R_i and S_i, respectively. On account of (23), $v = x_1$ and $w = x_2$. Hence $\overline{\mathfrak{A}}(x_1, \ldots, x_n, y_1, \ldots, y_m)$ holds, whereby the assertion is proved and, with it, the claim that A and A' are equivalent expressions.

The conjunction of two distinguished normal forms each of which contains at most n universal quantifiers is again equivalent to a distinguished normal form with at most n universal quantifiers (because

$$(x)F(x) \ \& \ (y)G(y) \equiv (x)[F(x) \ \& \ G(x)]).$$

Applied to A', this yields a distinguished normal form with $n - 1$ universal quantifiers, which, however, still contains the identity sign. If one eliminates it according to the procedure developed by L. Kalmár in *1929* and by me on page 356 of *1930* [above, page 117] and if one takes into consideration the remark above concerning the conjunction of distinguished normal forms, one again obtains (since no functional variable with more than two arguments occurs in A') a distinguished normal form with $n - 1$ universal quantifiers, and this yields the B in the theorem to be proved.

In conclusion, I would still like to remark that Theorem I can also be proved, by the same method, for formulas that contain the identity sign.

Review of *Kaczmarz 1932*: Axioms for arithmetic (*1933j*)

A system of axioms for the arithmetic of natural numbers, consisting of four axioms with the primitive notions "number" and "less than", is specified, and the independence of the axioms, as well as their equivalence with those put forward by L. Neder (*1931*), is demonstrated.

Besprechung von *Lewis 1932*:
Alternative systems of logic
(*1933k*)

Der Verfasser behauptet die Existenz verschiedener Systeme der Logik (und zwar des Aussagenkalküls) nur in dem Sinn, daß man verschiedene logische Begriffe zugrunde legen | kann, von denen dann selbstverständlich auch verschiedene Sätze gelten. Das System der Grundbegriffe ist immer eine endliche Klasse K von Eigenschaften (Wahrheitswerten), so daß jedem Satz eine dieser Eigenschaften zukommt und daß es darunter eine (E) von der Art gibt, daß jeder Satz mit der Eigenschaft E behauptet werden kann. Die Sätze der Logik handeln von Aussagefunktionen $f(p_1, \ldots, p_n)$, deren Wahrheitswert nur von den Wahrheitswerten der Argumente p_i abhängt. Besteht z. B. K: 1. aus den beiden Eigenschaften "wahr", "falsch", so erhält man das System der *Principia mathematica*, 2. aus den Eigenschaften "sicher", "sicher falsch", "zweifelhaft", so erhält man das dreiwertige System von Lukasiewicz, 3. aus den Eigenschaften "notwendig", "unmöglich", "wahr aber nicht notwendig", "falsch aber nicht unmöglich", so erhält man das Lewissche System of strict implication. Für die Aussagefunktionen der Systeme 2. und 3. werden Interpretationen gegeben, z. B. für $p \supset q$ im dreiwertigen System die folgende: Bei Umwandlung einer Wette für p in eine Wette für q, erleidet man (abgesehen von Graden der Ungewißheit) keinen Nachteil. Die wahren Sätze eines Systems sind durch die Bedeutung der Grundbegriffe bestimmt und daher analytisch. Verschiedene Systeme (von denen es unendlich viele gibt) widersprechen einander nicht, und wenn man sich für eines entschließt (der Verfasser glaubt, daß eine solche Auswahl nötig ist, um Logik zu betreiben), so geschieht dies nach psychologischen und pragmatischen Gesichtspunkten. Gründe werden angeführt, warum vielleicht eine dreiwertige Logik der zweiwertigen vorzuziehen ist, unter anderem die Tatsache, daß Aussagefunktionen für gewisse Argumente weder wahr noch falsch, sondern sinnlos werden.

Review of *Lewis 1932*:
Alternative systems of logic
(*1933k*)

The author asserts the existence of different systems of logic (and indeed of the propositional calculus) only in the sense that we can take as primitive different logical notions, whence the propositions that hold in them obviously differ also. The system of primitive notions is always a finite class K of properties (truth values) such that to each statement one of these properties is assigned and that, among them, there is one property, E, such that every statement with property E can be asserted. The theorems of logic concern propositional functions $f(p_1, \ldots, p_n)$ whose truth values depend only on the truth values of the arguments p_i. For example: (1) if K consists of the two properties "true" and "false", one obtains the system of *Principia mathematica*; (2) if K consists of the properties "certain", "certainly false", and "doubtful", one obtains the three-valued system of Lukasiewicz; (3) if K consists of the properties "necessary", "impossible", "true but not necessary", and "false but not impossible", one obtains Lewis' system of strict implication. Interpretations are given for the propositional functions of systems (2) and (3); for example, the following interpretation is given for $p \supset q$ in the three-valued system: no disadvantage is sustained in converting a wager on p into a wager on q (apart from the degree of uncertainty). The true statements of a system are determined by the meaning of the primitive notions and are therefore analytic. Different systems (of which there are infinitely many) do not contradict one another, and, if one opts for one of them, this is done in accordance with psychological and pragmatic viewpoints. (The author believes such a choice to be necessary if one wants to do logic.) Reasons are adduced why a three-valued logic is perhaps to be preferred to the two-valued one—among others, the fact that for certain arguments propositional functions become neither true nor false, but meaningless.

Besprechung von *Kalmár 1933*:
Über die Erfüllbarkeit derjenigen Zählausdrücke, welche in der Normalform zwei benachbarte Allzeichen enthalten
(*1933l*)

Die Arbeit gibt für den im Titel angeführten Fall eine Lösung des Entscheidungsproblems, die sich nicht wesentlich von der in *1932a* vom Referent skizzierten unterscheidet. Nach einem für den Fall *eines* Allzeichens von W. Ackermann in *1928a*, S. 647 angegebenem Verfahren wird das Problem zunächst auf das entsprechende für Ausdrücke der Form

$$(x_1)(x_2)(Ex_3)(Ex_4)\ldots(Ex_n)\,\mathfrak{K}(x_1,\ldots,x_n) \tag{$*$}$$

zurückgeführt. Kommen in $\mathfrak{K}$ die Funktionsvariablen $F_1, F_2, \ldots, F_l$ vor und ist I eine beliebige Menge, so versteht der Verfasser unter einem "Modell über I" ein System von im Individuenbereich I definierten Funktionen $\Phi_1, \Phi_2, \ldots, \Phi_l$, welche resp. dieselben Leerstellenanzahlen haben wie $F_1, F_2, \ldots, F_l$; ferner unter einer "Tabelle k-ter Ordnung" ein Modell über der Menge der natürlichen Zahlen $\leq k$. Ist M ein Modell über I und sind $a_1, a_2, \ldots, a_k$ beliebige Elemente aus I, so bezeichnet der Verfasser mit $[M|a_1, a_2, \ldots, a_k]$ diejenige Tabelle k-ter Ordnung, deren Funktionen für irgendwelche Zahlen $i_1, i_2, \ldots, i_s \mid (\leq k)$ als Argumente dieselben Wahrheitswerte ergeben, wie die entsprechenden Funktionen aus M für $a_{i_1}, a_{i_2}, \ldots, a_{i_s}$. Ferner bedeute $\mathfrak{K}(M, x_1, \ldots, x_n)$ den Ausdruck, welcher aus $\mathfrak{K}$ entsteht, wenn man die F_i durch die Funktionen des Modells M ersetzt. Ist nun M ein die Formel ($*$) erfüllendes Modell über I, und bildet man die Menge $\mathfrak{T}$ derjenigen Tabellen n-ter Ordnung, welche man erhält, indem man in $[M|a_1, a_2, \ldots, a_n]$ für $a_1, a_2, \ldots, a_n$ alle solchen n-Tupel aus I setzt, für welche $\mathfrak{K}(M, a_1, a_2, \ldots, a_n)$ gilt, so hat $\mathfrak{T}$ folgende vier Eigenschaften:

I. Für jedes $T \in \mathfrak{T}$ gilt $\mathfrak{K}(T, 1, 2, \ldots, n)$.

II. Ist $T_0 \in \mathfrak{T}$ und $\nu, \nu' = 1, 2, \ldots, n$, $\nu \neq \nu'$, so gibt es eine Tabelle $T \in \mathfrak{T}$ derart, daß $[T|1, 2] = [T_0|\nu, \nu']$.

III. Ist $T_0 \in \mathfrak{T}$, so gibt es eine Tabelle $T' \in \mathfrak{T}$, so daß $[T'|1] = [T_0|1]$ und $T' = [T'|1, 1, 3, 4, \ldots, n]$.

IV. Ist $T_1 \in \mathfrak{T}$, $T_2 \in \mathfrak{T}$, so gibt [[es]] eine Tabelle $T \in \mathfrak{T}$, so daß $[T|1] = [T_1|1]$ und $[T|2] = [T_2|1]$.

Die Existenz einer nicht leeren Menge $\mathfrak{T}$ von Tabellen n-ter Ordnung mit diesen Eigenschaften erweist sich aber für die Erfüllbarkeit von ($*$) auch

Review of *Kalmàr 1933*:
On the satisfiability of those
first-order expressions that in normal form
contain two adjacent universal quantifiers
(*1933l*)

⟦The introductory note to *1933l*, as well as to related items, can be found on page 226, immediately preceding *1932a*.⟧

This work presents a solution of the decision problem for the case cited in the title, one that does not differ essentially from that sketched by the reviewer in his *1932a*. By a procedure given by W. Ackermann in his *1928a*, page 647, for the case of *one* universal quantifier, the problem is first reduced to the corresponding one for expressions of the form

$$(x_1)(x_2)(Ex_3)(Ex_4)\ldots(Ex_n)\,\Re(x_1,\ldots,x_n). \tag{*}$$

If the functional variables $F_1, F_2, \ldots, F_l$ occur in $\Re$ and if I is an arbitrary set, then by a "model over I" the author understands a system of functions $\Phi_1, \Phi_2, \ldots, \Phi_l$, defined on the domain I of individuals, that have the same number of argument places as $F_1, F_2, \ldots, F_l$, respectively. Furthermore, by a "kth-order table" he understands a model over the set of natural numbers $\leq k$. If M is a model over I and $a_1, a_2, \ldots, a_k$ are arbitrary elements of I, the author denotes by $[M|a_1, a_2, \ldots, a_k]$ that kth-order table whose functions, for arbitrary numerical arguments $i_1, i_2, \ldots, i_s$ $(\leq k)$, yield the same truth values as the corresponding functions from M do for $a_{i_1}, a_{i_2}, \ldots, a_{i_s}$. Moreover, let $\Re(M, x_1, \ldots, x_n)$ denote the expression that results from $\Re$ if the F_i are replaced by the functions of the model M. If now M is a model over I satisfying the formula (*) and if one forms the set $\mathfrak{T}$ of those nth-order tables obtained by taking for $a_1, a_2, \ldots, a_n$ in $[M|a_1, a_2, \ldots, a_n]$ all n-tuples from I for which $\Re(M, a_1, a_2, \ldots, a_n)$ holds, then $\mathfrak{T}$ has the following four properties:

I. For every $T \in \mathfrak{T}$, $\Re(T, 1, 2, \ldots, n)$ holds.

II. If $T_0 \in \mathfrak{T}$ and $\nu, \nu' = 1, 2, \ldots, n$, with $\nu \neq \nu'$, there is a table $T \in \mathfrak{T}$ such that $[T|1, 2] = [T_0|\nu, \nu']$.

III. If $T_0 \in \mathfrak{T}$, there is a table $T' \in \mathfrak{T}$ such that $[T'|1] = [T_0|1]$ and $T' = [T'|1, 1, 3, 4, \ldots, n]$.

IV. If $T_1 \in \mathfrak{T}$ and $T_2 \in \mathfrak{T}$, there is a table $T \in \mathfrak{T}$ such that $[T|1] = [T_1|1]$ and $[T|2] = [T_2|1]$.

But the existence of a non-empty set $\mathfrak{T}$ of nth-order tables with these prop-

als hinreichend. Da es nur endlich viele Tabellen n-ter Ordnung gibt, ist damit ein Entscheidungsverfahren gegeben.

Besprechung von *Hahn 1932*:
Reelle Funktionen
(*1933m*)

Die seit langem vergriffene *Theorie der reellen Funktionen* [[*1921*]] desselben Autors erscheint hier in einer sowohl inhaltlich als in der Darstellungsform völlig neuen Bearbeitung. Die im vorliegenden ersten Band enthaltenen Kapitel (die Theorie der Mengenfunktionen ist einem 21   zweiten Band vorbehalten) sind ent|sprechend den seither erzielten neuen Ergebnissen weit eingehender behandelt als in dem früheren Buche und ein neues Kapitel über analytische Mengen ist dazugekommen. Die Beweise sind mit einer Exaktheit und Ausführlichkeit gegeben, die in der mathematischen Lehrbuchliteratur wohl kaum ihresgleichen hat und die nicht mehr weit von der völligen "Formalisierung" (im Sinne etwa der *Principia mathematica*) entfernt ist. Dabei ist die Anzahl der explizit formulierten Lehrsätze so groß, daß jeder einzelne unter Berufung auf die vorhergehenden verhältnismäßig kurz bewiesen werden kann (ein Beweis überschreitet nur selten eine halbe Seite und ist meistens noch wesentlich kürzer). Ein weiterer charakteristischer Zug des Buches ist das überall durchgeführte Prinzip, mit der geringsten Zahl von Voraussetzungen auszukommen und daher auch alles möglichst abstrakt zu formulieren.

Trotz der detaillierten Angabe der Beweise ist das Werk auch inhaltlich außerordentlich reichhaltig. Viele Resultate aus der modernsten Zeit, die in den verschiedensten Zeitschriften und Büchern verstreut sind, wurden hier zum ersten Male systematisch dargestellt, mit zahlreichen Verbesserungen, Ergänzungen und neuen Beweisen, die vielfach (insbesondere in dem Kapitel über Bairesche Funktionen) auch für den Kenner Neues bringen werden. Jedem Abschnitt ist ein ausführliches Verzeichnis der Originalarbeiten beigegeben.

Vorkenntnisse werden außer den elementarsten Dingen keine vorausgesetzt und im I. Kapitel werden daher die wichtigsten Sätze aus der abstrakten Mengenlehre über Kardinalzahlen, Ordinalzahlen und Mengensysteme abgeleitet.

Das II. Kapitel behandelt die Theorie der Punktmengen (topologische und metrische Räume) und beschränkt sich dabei durchaus nicht auf das

erties also turns out to be sufficient for the satisfiability of (∗). Since there are only finitely many nth-order tables, this gives a decision procedure.

Review of *Hahn 1932*:
Real functions
(*1933m*)

The same author's *Theorie der reellen Funktionen* [*1921*], long out of print, appears here in a revised version that is completely new, with respect to both content and form of presentation. The chapters of the present first volume (the theory of set functions being reserved for a second volume) contain a far more detailed treatment than those in the earlier book, in accordance with the new results obtained in the interim, and a new chapter on analytic sets has been added. The proofs are given with a rigor and attention to detail that probably have hardly an equal in the textbooks of mathematics and are not far removed from complete "formalization" (in the sense of, say, *Principia mathematica*). At the same time, the number of explicitly formulated theorems is so large that the proof of each one of them can be kept comparatively brief by appeal to the theorems preceding it. (Proofs only rarely exceed half a page and are usually substantially shorter.) A further characteristic feature of the book is the principle, followed throughout, of making do with the fewest assumptions and therefore also of formulating everything as abstractly as possible.

In spite of the detailed statement of the proofs, the work is also extraordinarily rich in content. Many of the most recent results, which are scattered about in the most diverse journals and books, are presented here systematically for the first time, with numerous improvements, additions, and new proofs that in many cases (especially in the chapter on Baire functions) will also impart something new to the specialist. A detailed listing of original sources is appended to each section.

Aside from the most elementary matters, no background knowledge is presumed, and therefore the most important theorems from abstract set theory concerning cardinal and ordinal numbers as well as systems of sets are derived in the first chapter.

The second chapter deals with the theory of point sets (in topological and metric spaces) and does not at all confine itself to what is used later. For example, the metrizability of regular separable spaces and their embeddability in Hilbert spaces are proved. The notions "being of first or second

im folgenden Verwendete; z. B. wird die Metrisierbarkeit der regulären separablen Räume und ihre Einbettbarkeit in den Hilbertschen Raum bewiesen. Besonders ausführlich sind die Begriffe "von I. und von II. Kategorie", "residual" etc. sowie die entsprechenden lokalen Begriffe "von I. Kategorie in einem Punkt" etc. behandelt.

Es folgt ein Kapitel (III) über den Begriff der Stetigkeit. Zunächst werden beliebige (auch mehr-mehrdeutige) Abbildungen zweier metrischer Räume aufeinander betrachtet, wobei der Begriff der Stetigkeit aufgespalten wird in "oberhalb stetig" (d. h. die Urbilder der abgeschlossenen Mengen sind abgeschlossen) und "unterhalb stetig" (d. h. die Urbilder der offenen Mengen sind offen), wodurch auch die Lehrsätze entsprechend in Paare zerfallen. Auch die Sätze über die Erweiterbarkeit von in Teilräumen definierten stetigen Funktionen auf den ganzen Raum werden auf mehr-mehrdeutige Abbildungen verallgemeinert. Im Zusammenhang mit den stetigen Abbildungen werden die oberhalb stetigen Zerlegungssysteme behandelt und wird ferner die Homöomorphie jedes absoluten G_δ mit einem vollständigen Raum (nach Hausdorff), sowie die stetige Durchlaufbarkeit der im kleinen zusammenhängenden Kontinua nach einem neuen Verfahren von Whyburn bewiesen. Es folgt die Theorie der Schrankenfunktionen und der Schwankung, weiter ein Abschnitt über die verschiedenen Arten der gleichmäßigen Konvergenz, über Ungleichmäßigkeitsgrad, gleichgradige Stetigkeit und ähnliches. Der (aus dem Jahre 1922 stammende) Satz, daß es zu jeder in einem separablen vollständigen Raum R definierten Funktion eine in R dichte Menge gibt, auf der sie stetig ist, wird bewiesen, samt der Ergänzung, daß es nicht immer eine abzählbare Menge dieser Art gibt; ferner auch die Hurewiczsche Verallgemeinerung des Satzes von der Beschränktheit jeder auf einer kompakten Menge stetigen Funktion auf Folgen stetiger Funktionen.

Kapitel IV behandelt die Borelschen Mengen und Baireschen Funktionen. Die dieser Theorie zugrunde liegenden Sätze werden ganz abstrakt für beliebige, gewissen einfachen Bedingungen genügende Funktionssysteme mit einer beliebigen Menge (nicht notwendig metrischem Raum) als Definitionsbereich ausgesprochen. Zwischen den Begriffen und Sätzen über Borelsche Mengen und Baireschen Funktionen wird ein vollkommener Parallelismus hergestellt. Zum Beispiel erscheinen die Lusinschen Trennungssätze (deren einfachster Fall ist die Trennbarkeit zweier fremder G_δ durch Mengen, die sowohl F_σ als G_δ sind) als Analoga zu den Einschiebungssätzen (deren einfachster Fall die Existenz einer stetigen Funktion zwischen einer oberhalb stetigen und einer größeren unterhalb stetigen Funktion ist). Neu 22 gegenüber der früheren *Theorie der reellen Funk|tionen* [*1921*] sind ferner unter anderem Bedingungen dafür, daß der Limes einer Folge von Funktionen α-ter Klasse zur selben Klasse gehört und dafür, daß zwei Funktionen f und g $\underline{\lim}$ und $\overline{\lim}$ derselben Funktionsfolge eines Funktionssystems $\mathfrak{S}$ sind. Das Kapitel schließt mit Untersuchungen über partiell stetige Funktionen,

category", "residual", and so on, as well as the corresponding local notions, "being of first category at a point", and so on, are treated in particular detail.

There follows a chapter (III) on the notion of continuity. First, arbitrary (even many-to-many) mappings of two metric spaces into one another are considered, the notion of continuity being split into those of being "upper semi-continuous" (that is, inverse images of closed sets are closed) and of being "lower semi-continuous" (that is, inverse images of open sets are open); as a result, the theorems, too, are correspondingly divided into pairs. Also, the theorems concerning the possibility of extending continuous functions defined on subspaces to the whole space are generalized to many-to-many mappings. Upper semi-continuous decomposition systems are treated in connection with continuous mappings, and in addition it is proved that every absolute G_δ is homeomorphic to a complete [metric] space (in Hausdorff's sense). The path-connectedness of locally connected continua is proved as well, by a new method due to Whyburn. This is followed by the theory of limit functions [lim sup and lim inf] and variation and, further, by a section on the different kinds of uniform convergence, the degree of non-uniformity, uniform continuity, and the like. A proof is given of the theorem (which goes back to the year 1922) that for every function defined on a separable complete [metric] space R there is a set dense in R on which the function is continuous, together with the supplementary fact that there does not always exist a countable set of this kind. The author also proves Hurewicz's generalization, to sequences of continuous functions, of the theorem that every function continuous on a compact set is bounded.

Chapter IV deals with Borel sets and Baire functions. The basic theorems of this theory are stated quite abstractly for arbitrary systems of functions satisfying certain simple conditions, with an arbitrary set (not necessarily a metric space) as the domain of definition. A complete parallelism is established between the notions and theorems concerning Borel sets and those concerning Baire functions. For example, Luzin's separation theorems (whose simplest case is the separability of two disjoint G_δ sets by sets that are both F_σ and G_δ) appear as analogues to the interpolation theorems (whose simplest case is the existence of a continuous function lying between an upper semi-continuous function and a [pointwise] greater lower semi-continuous function). Furthermore, new in relation to the earlier *1921* are, among others, conditions for the limit of a sequence of functions of the αth class to belong to that same class, and for two functions f and g to be $\underline{\lim}$ and $\overline{\lim}$, respectively, of the same sequence of functions of a system $\mathfrak{S}$ of functions. The chapter concludes with investigations of partially continuous functions. There, among other things, it is proved that, if $f(x_1, x_2, \ldots, x_n)$ is a function that is partially continuous (that is, continuous in each coordinate) in R_n, then $f(x, x, \ldots, x)$ is a function of

wobei unter anderem bewiesen wird, daß, wenn $f(x_1, x_2, \ldots, x_n)$ eine partiell (d. h. nach jeder Koordinate) stetige Funktion im R_n ist, $f(x, x, \ldots, x)$ eine Funktion höchstens $(n-1)$-ter Klasse im R_1 ist, und daß umgekehrt jede Funktion $(n-1)$-ter Klasse im R_1 sich in dieser Form durch eine partiell stetige Funktion darstellen läßt.

Das letzte (V.) Kapitel bringt eine ausführliche Theorie der analytischen Mengen, unter anderem ihre Darstellbarkeit als Wertmengen stetiger und halbstetiger Funktionen und als Menge der Werte k-facher (abzählbarer, unabzählbarer) Vielfachheit stetiger Funktionen im Baireschen Nullraum, ferner eine Reihe von Sätzen über das Verhältnis der analytischen zu den Borelschen Mengen. Es folgt eine Anwendung der analytischen Mengen auf die durch Bairesche Funktionen implizit definierten Funktionen und den Schluß bildet die Theorie der Lusinschen Siebe.

Im Rahmen einer Besprechung ist es leider völlig unmöglich, den behandelten Stoff auch nur annähernd zu erschöpfen, doch dürfte auch schon aus dem Gesagten zur Genüge die außerordentliche Reichhaltigkeit des Werkes insbesondere an modernen Ergebnissen hervorgehen, welche es vor allem auch als Nachschlagewerk für selbständig Arbeitende vorzüglich geeignet macht.

at most the $(n-1)$th class in R_1 and that, conversely, every function of the $(n-1)$th class in R_1 can be represented in this form by a partially continuous function.

The fifth and final chapter presents a detailed theory of analytic sets: among other results, their representability as the ranges of continuous and of semi-continuous functions and as the set of values, of k-fold (or countable or uncountable) multiplicity, of continuous functions on the Baire null space; and, furthermore, a sequence of theorems about the relation of analytic sets to Borel sets. This is followed by an application of analytic sets to functions implicitly defined by Baire functions. The theory of Luzin sieves forms the conclusion.

It is, unfortunately, quite impossible within the framework of a review to cover exhaustively, or even approximately so, the material treated by the author. Yet even what has been said may already suffice to make apparent the extraordinary richness of the work, especially in terms of modern results, a richness that, above all, also makes it eminently suitable as a reference tool for those working independently.

Introductory note to *1934*

Gödel lectured on his *1931* results at the Institute for Advanced Study from February to May 1934. Notes were taken by S. C. Kleene and J. B. Rosser. These notes were put out in installments to the persons who had subscribed to them. After the lectures were finished, the whole set of the notes, supplemented by two pages of "Notes and errata", was approved by Gödel. At the time of the *Davis 1965* printing of the lecture notes, those "Notes and errata" (and some minor typographical corrections) were incorporated into the text, either directly or as footnotes, and what had been on page 28 became the last paragraph of §8. For that printing, Gödel made available a substantial number of corrections and emendations (enclosed here in square brackets), and supplied the "Postscriptum" which follows the text.

Gödel's 1934 lectures and their notes cover the territory treated in *Gödel 1931* and, as we shall see, more.

As a specific example of a formal system, the existence in which of formally undecidable propositions is demonstrated, Gödel uses one with variables for propositions, for natural numbers, and for one-place number-theoretic functions. This makes possible a considerable simplification of the proof that every primitive recursive relation is numeralwise expressible in the system (what was Theorem V in *1931*). Gödel mentions the possibility of working in a system with higher-type variables; but in his specific formal system (introduced in Section 3) he does not use variables of type 2 or higher.

It would be usual now to call his "expressions of the I-st kind" *terms*, and his "expressions of the II-nd kind" *formulas*. The class of the terms includes 0 and the individual variables, and is closed under the application of successor N and of function variables f, g, $h, \ldots$; if $A(x)$ is a formula, $\epsilon x A(x)$ is a term (Hilbert's ϵ-operator or selection operator for individuals). The class of the formulas includes the propositional variables p, q, $r, \ldots$ and equations between terms; it is closed under the propositional operations and quantification with respect to each of the three kinds of variables (individual, function and propositional).

Three kinds of substitution are described in Section 3: $\mathrm{Subst}[A_B^x]$ where B is a term, $\mathrm{Subst}'[A_{G(x),x}^f]$ where $G(x)$ is a term, and $\mathrm{Subst}[A_P^p]$ where P is a formula. In each case there are some evident syntactic restrictions to be made. The informal intention in the second case is that we are substituting $\lambda x G(x)$ for f. This is complicated when there are nestings $f(\ldots f(\ldots) \ldots)$, and that is the reason for the more complicated definition of substitution $\mathrm{Subst}'[A_{G(x),x}^f]$.

Rule 4b is $A/\mathrm{Subst}'[A^f_{G(x),x}]$ for any formula A. This allows us to obtain a form of full second-order comprehension, as follows. Let $B(x,y)$ be any formula such that $\Pi x \Sigma! y[B(x,y)]$ has been proved, and let $G(x)$ be $\epsilon y[B(x,y)]$; then by Rule 4b $\Pi x \Pi z[B(x,z) \equiv z = G(x)]$. Let A be $\Sigma g \Pi x[g(x) = f(x)]$; then $\Sigma g \Pi x[g(x) = G(x)]$, i.e., there is a function defined by $G(x)$. If we take $B(x,y)$ to be

$$(C(x) \ \& \ y = 0) \vee (\sim C(x) \ \& \ y = 1)$$

for any given $C(x)$, we obtain the existence of the representing function of C.

Quantified propositional variables are eliminable in favor of function quantifiers. Thus the whole system is a form of full second-order arithmetic (now frequently called the system of "analysis").

Gödel reaches the main results of *1931* for this new specific system at the end of Section 5. As was explained above, every primitive recursive function is numeralwise representable in it. The crucial thing is that the finite sequences needed to effect primitive recursions can be represented by functions f, provably in the system. He might have noted, but apparently didn't, that much weaker axioms for function existence suffice for this proof—most simply, arithmetical comprehension without function parameters; the resulting system is conservative over first-order arithmetic.

In Section 6 Gödel states general conditions on a formal system for his arguments to apply. This moves to a much more general formulation of the incompleteness results.

In Section 7 he discusses the relation of his arguments to the paradoxes. This contains two points of special significance: the general self-referential lemma and the indefinability of truth in the formal system. The first is credited to *Carnap 1934a*, and the second to *Tarski 1933a* and *Carnap 1934*, by footnotes added in the *Davis 1965* printing.

In Section 8 Gödel obtains a version of his undecidable proposition in a form concerning solutions to Diophantine problems (extending Theorem VIII of *1931*). Suppose $F(x_1, \ldots, x_n)$ is a polynomial with integral coefficients. By applying a logical quantifier, (x) ("for all x") or (Ex) ("there exists an x such that ") to each of the variables $x_1, \ldots, x_n$, thus building a sequence of n quantifiers (write it "(P)", and write "F" for the polynomial), we can make a statement $(P)(F = 0)$ about the solutions in natural numbers of the Diophantine equation $F(x_1, \ldots, x_n) = 0$. Gödel proves that there is a Diophantine equation $F = 0$ and a sequence of quantifiers (P) such that his undecidable proposition is equivalent to $(P)(F = 0)$.

This result was improved considerably, taking (P) to consist only of existential quantifiers, by Matiyasevich in *1970* in connection with

results on Hilbert's tenth problem (cf. *Davis, Matiyasevich and Robinson 1976*, page 349). Then one can take as the true but unprovable proposition the assertion that a solution in natural numbers of $F(x_1, \ldots, x_n) = 0$ does not exist: $\overline{(Ex_1, \ldots, x_n)}(F(x_1, \ldots, x_n) = 0)$.

In Section 9 Gödel introduces a new idea into mathematics, that of "general recursive function". He is led to this by considering various sorts of definitions of functions by inductions (often called recursions). This raises "the question what one would mean by 'every recursive function'". He obtained his notion by modifying the suggestion (which had been communicated to him privately in 1931 by Herbrand[a]) that, if a set of functional equations involving known functions $\psi_1, \ldots, \psi_k$ (already accepted as being recursive) and an unknown function ϕ has a unique solution for ϕ, then ϕ is a recursive function. Gödel's modification consisted in requiring the left side of each equation to have the form

$$\phi(\psi_{i1}(x_1, \ldots, x_n), \psi_{i2}(x_1, \ldots, x_n), \ldots, \psi_{il}(x_1, \ldots, x_n)),$$

and, much more importantly, in requiring that, for each l-tuple of numerals $n_1, \ldots, n_l$, a unique equation $\phi(n_1, \ldots, n_l) = m$ (m a numeral) be formally derivable from the set of functional equations by use of a replacement rule and a substitution rule. This makes the definition of ϕ from $\psi_{i1}, \ldots, \psi_{il}$ constructive.

This notion of what a *general recursive function* is, or of the *general recursiveness* of a number-theoretic function, was the second of a number of equivalent exact characterizations of a class of number-theoretic functions that came to the attention of mathematicians in the period 1932–1961 (the history is sketched in *Kleene 1981* and *Davis 1982*). It and the equivalent notion of λ-*definability* were taken by Church in *1936* as exact characterizations of the number-theoretic functions that are "effectively calculable" (in the somewhat vague intuitive sense), or for which, in terminology that has come down through the centuries, there are "algorithms". The proposition that every "effectively calculable" function is general recursive (or the equivalent proposition with one of the other notions) has been called (since *Kleene 1943* and *1952*) "Church's thesis". The notions extend to predicates $R(x_1, \ldots, x_n)$ by applying them to the representing functions $\phi(x_1, \ldots, x_n)$, which take 0 or 1 as values according as $R(x_1, \ldots, x_n)$ is true or false.[b] Church's thesis had already come to Church's mind by his using the notion of λ-*definability*, which had arisen from work of Church and Kleene in

---

[a]The history is elaborated in *van Heijenoort 1982*, pp. 72–75.

[b]Here we are saying "predicate" (after *Hilbert and Bernays 1934* and *Kleene 1943*) where Gödel usually says "relation" and sometimes "class or relation".

1932 and 1933, before Gödel in his lectures of May 1934 came out with general recursiveness as a modification of Herbrand's suggestion.

Gödel in 1934 did not subscribe to Church's thesis, which as formulated in terms of λ-definability Church had proposed to him in a personal communication around February or March 1934. Thus Gödel wrote on 15 February 1965 to Davis, "it is *not true* that footnote 3 [of the present paper *1934*] is a statement of Church's Thesis. The conjecture stated there only refers to the *equivalence* of 'finite (computation) procedure' and 'recursive procedure'. However, I was, at the time of these lectures, not at all convinced that my concept of recursion comprises all possible recursions; and in fact the equivalence between my definition and Kleene's in *Math. Ann. 112* [*1936*] is not quite trivial."

The appearance of Gödel's *general recursiveness* made the evidence for Church's thesis stronger, by there being then two notions, arising independently but proved equivalent (in *Kleene 1936a*), as candidates for being precise characterizations of the class of all the effectively calculable number-theoretic functions. Thus Church came to the point of publishing his thesis in *1936* in terms of either of these equivalent notions.

Church's thesis, and the theorem he based on it ("Church's theorem") in *1936*, gave a new kind of undecidability result. Gödel had shown the existence in certain formal systems S of *formally undecidable propositions* A, i.e., propositions for which neither the formula A expressing A in S nor its negation ~A is provable in S. Church now showed the existence of *informally undecidable predicates* $P(a)$, i.e., ones for which no algorithm exists for deciding as to the truth or falsity of each of the propositions $P(a)$ taken (for particular a's) as the values of the predicate $P(a)$. Thus logicians began to talk about "undecidable predicates" (and presently also "undecidable theories") as well as "undecidable propositions".

In his celebrated paper *1937*, Turing (who only learned of Church's *1936* work when his own paper was ready for publication) gave a direct analysis of effective calculability in terms of the possibility of computation by what we now call a "Turing machine", abstracting from an analysis of the operations performed by a human computer. Post in *1936* (received for publication later than Turing's paper, but written with Church's *1936* at hand) gave very briefly a similar analysis.

Gödel in *1931* had introduced the idea of *numeralwise expressibility* (*Entscheidungsdefinitheit*) of a relation in a formal system S. In a remark added in proof to his paper *1936a* he notes that it is absolute, i.e., the same for all formal systems S including elementary number theory. He repeated this in the first paragraph of *1946*, where the equivalence of that notion to *Turing computability* and *general recursiveness* is mentioned.

All these exact formulations which the Church–Turing thesis asserts to represent the intuitive notion of "effective calculability" (or that of there being an "algorithm" for the function, or for the representing function of the predicate, in question), plus several more which arose later (due to Post, Markov and Smullyan; cf. *Kleene 1981*), were proved by various combined efforts to be equivalent.

Church in *1936* as already mentioned, Kleene in *1936* (who already knew of Church's thesis and theorem), and Turing in *1937* independently, gave examples of informally undecidable predicates arising in the respective theories of λ-definability, general recursiveness, and Turing computability. All rest obviously on the famous Cantor diagonalization method of *1891* (as do Gödel's *1931* results). Church and Kleene in their details used Gödel's technique of numbering formulas, while Turing employed a different technique of his own.

Other informal undecidability results are for the provability predicate for a formal system S,[c] the system S then being said to be *undecidable*. The problem of finding a decision procedure (or algorithm) for this predicate (for a formal system S) has gone under the name of the *Entscheidungsproblem (for S)*.

Specifically, at the end of his paper *1936*, Church gives the unsolvability of the Entscheidungsproblem for the same ω-consistent systems S as are considered in *Gödel 1931*. The (informal) undecidability of such systems S implies their incompleteness, i.e., that they have formally undecidable propositions. This was noticed in passing in *Turing 1937*, page 259, and is implicit in *Kleene 1936*, pages 740–741.[d] A little care is needed to make this result constructive. (The converse is not true; e.g., the theory of dense linear order is incomplete but decidable.)

Then Church in *1936a* and Turing in *1937* established the undecidabil-

---

[c]This is the metamathematical predicate $P^*(A)$: "A is provable in S", where A ranges over the formulas (or more generally over the finite sequences of symbols) of the system S in question. After picking an obviously constructive Gödel numbering for S, to have an algorithm for $P^*(A)$ is equivalent to having one for the number-theoretic predicate $P(a)$: "a is the Gödel number of a formula provable in S". By Church's thesis, no algorithm exists for $P(a)$ if $P(a)$ is not general recursive. So the "undecidability" of a system S (or the "unsolvability" of its Entscheidungsproblem) is established by showing the predicate $P(a)$ to be not general recursive.

[d]This can be regarded as an application of the theorem of *Kleene 1943*, *Post 1944* and *Mostowski 1947* that each Δ_1^0 set is general recursive. Here $\Delta_1^0 = \Pi_1^0 \cap \Sigma_1^0$, where Σ_1^0 is the sets expressible in the form $\hat{x}(Ey)R(x,y)$ with a general (or equivalently, by *Kleene 1943*, a primitive) recursive R, and the Π_1^0 are those whose complements are so expressible. The Σ_1^0 sets are called *recursively enumerable*, being either empty or enumerable allowing repetitions as the sequence of the values $\phi(0)$, $\phi(1)$, $\phi(2), \ldots$ of a general (or equivalently, by *Rosser 1936*, a primitive) recursive function ϕ. (Cf. footnote f in the introductory note to *Gödel 1930b, 1931* and *1932b* (p. 133 above).)

ity of the first-order predicate calculus. As has already been remarked in the introductory note to *Gödel 1930b, 1931* and *1932b* (page 136 above), Gödel's proof of Theorem X of *1931* could have been combined with *Kleene 1936* to obtain this result quickly in two versions.

Rosser in *1936* (drawing on *Church 1936* and *Kleene 1936*), as we have already noted in discussing *Gödel 1931*, replaced Gödel's hypothesis of ω-consistency by simple consistency; and he accomplished the same for Church's undecidability of the systems S.

The formal systems which were shown in *Rosser 1936* to have unsolvable decision problems, if they are simply consistent, retain that feature when one extends them by adding further axioms, provided the consistency is preserved. Such systems Tarski in *1949* called "essentially undecidable theories". In Tarski, Mostowski and Robinson's *1953* monograph *Undecidable theories*, it is shown that: (1) If a theory T is essentially undecidable and T is (relatively) interpretable in S, then S is essentially undecidable. (2) If, further, T is finitely axiomatizable (i.e., has only finitely many non-logical axioms), then any subtheory S' of S with the same language as S is undecidable (not necessarily essentially). Using these general results, they establish the essential undecidability, or just the undecidability, of a wide variety of theories of mathematical interest.

Rosser's argument in *1936* for the undecidability of arithmetic uses an internal self-referential construction similar to that for his incompleteness theorem. The point of this construction, which is more complicated than is needed for the immediate applications, appears in terms of later work: The essential undecidability can be improved to the statement (using modern terminology) that the sets of the provable and of the refutable sentences of Peano arithmetic are *effectively inseparable*.[e]

By use of Church's thesis to render "effectively" as "general recursively", this means the following: Let P (respectively Q) be the set of the Gödel numbers of sentences provable (refutable) in Peano arithmetic. To any pair B_0 and B_1 of disjoint recursively enumerable sets with $B_0 \supset P$ and $B_1 \supset Q$, a number g can be found such that $g \notin B_0 \cup B_1$ (whence it follows that there is no general recursive predicate $R(x)$ such that $P \subset \hat{x}R(x)$ and $Q \subset \hat{x}\overline{R}(x)$).[f] Effective inseparability thus gives a

---

[e]We recall that "sentence" means closed formula, "refutable" means that the negation of the sentence is provable, and "Peano arithmetic" means the first-order predicate calculus with equality and, as non-logical axioms, Peano's axioms for the natural numbers (with the first-order induction schema replacing Peano's second-order induction axiom) together with the recursion equations for addition $+$ and multiplication $\cdot$ (cf. footnote k in the introductory note to *Gödel 1930b, 1931* and *1932b*, p. 138 above).

[f]The recursively enumerable sets (cf. footnote d) include the general recursive sets.

constructive witness to the undecidability, and thence to the incompleteness of extensions; so undecidability-implies-incompleteness is given constructive force. A proof of this inseparability result for a rather general class of systems can be found in *R. Smullyan 1958*. But it is remarked there that the result is also immediate from the construction in *Kleene 1950* (repeated in *1952*, §61) of a pair of disjoint recursively enumerable sets $C_0 = \hat{x}(Ey)W_0(x,y)$ and $C_1 = \hat{x}(Ey)W_1(x,y)$ which are not general recursively separable (being such that, to any pair of disjoint recursively enumerable sets D_0 and D_1 with $D_0 \supset C_0$ and $D_1 \supset C_1$, a number f can be found such that $f \notin D_0 \cup D_1$).[g]

The appearance of Church's thesis and theorem in his *1936* presented the opportunity to reconsider Gödel's undecidable propositions from the point of view of the new theory dealing with all effectively calculable functions and predicates. Kleene took advantage of this opportunity in *1936*, and more simply in *1943*, to formulate some generalized versions of Gödel's first incompleteness theorem. The result of *Kleene 1950* just stated is another such version.

In Kleene's *1943* version, his argument is that to formalize the theory of a number-theoretic predicate $P(a)$ in a given formal system S, correct and *complete* for that theory, entails expressing $P(a)$ in the form $(Ex)R(a,x)$ where $R(a,x)$ is a general recursive predicate. Indeed, $R(a,x)$ can simply be the predicate saying that x is the Gödel number of a proof in S of the formula A_a expressing in S the proposition $P(a)$ for a given a (Thesis II of *Kleene 1943* and *1952*, Church's thesis being Thesis I). For, our fundamental aim in constructing a formal system S is to make completely effective the recognition of when we have before us a proof of a given formula. And S could not serve us as a formalization of the theory of $P(a)$ if we could not effectively find, for any given a, the formula A_a which we are choosing to express the proposition $P(a)$ in S. So $R(a,x)$ should be effectively decidable, and hence by Church's thesis general recursive. It was a result of Kleene's

---

[g]Kreisel wrote Kleene on 4 October 1983 that Gödel had told him "on several occasions that he regarded this improvement [of his incompleteness results] as very significant." To infer the effective inseparability of the provable and the refutable sentences of arithmetic, we need simply add the detail that (under Kleene's definition of W_0 and W_1) the propositions $(Ey)W_0(x,y)$ for $x = 0, 1, 2, \ldots$ are expressible in Peano arithmetic by sentences $B(x)$ such that, for each x,
$$x \in C_0 \rightarrow \{B(x) \text{ is provable}\},$$
$$x \in C_1 \rightarrow \{B(x) \text{ is refutable}\}$$
(*Kleene 1952*, p. 310, Example 1). Now, for any B_0 and B_1 as described, taking
$$D_0 = \hat{x}[(\text{the Gödel number of } B(x)) \in B_0],$$
$$D_1 = \hat{x}[(\text{the Gödel number of } B(x)) \in B_1],$$
we note that D_0 and D_1 are disjoint, recursively enumerable, and contain C_0 and C_1 respectively; so by *Kleene 1950* we can find a number f such that $f \notin D_0 \cup D_1$, and thence a number g (= the Gödel number of $B(f)$) such that $g \notin B_0 \cup B_1$.

study in *1936* of Gödel's *general recursiveness* that there is a predicate $(x)\overline{T}_1(a,a,x)$, with $\overline{T}_1$ primitive recursive, which is not expressible in the form $(Ex)R(a,x)$ with R general recursive.[h] Thus (in *1943*) Kleene gave a generalized version of Gödel's first incompleteness theorem in which the undecidable propositions singled out by it in all conceivable formal systems correctly formalizing a certain modicum of elementary number theory are values of the preassigned predicate $(x)\overline{T}_1(a,a,x)$.

As Kleene said in a lecture at Wisconsin while his *1936* was in press, it has been discovered that mathematics, indeed just elementary number theory, offers "inexhaustible scope for ingenuity". We shall never succeed in writing down in completely usable form all the principles required for determining the truth or falsity of its propositions, so that thereafter all that will be needed is sufficient patience (or insight into the best searches for investing one's time) in applying those principles.

Gödel eventually came to accept Church's thesis in the intrinsically plausible version given by Turing's *1937* characterization of what we now call the *Turing computable functions* (and in Church's two versions using λ-*definability* and *general recursiveness*). Gödel's pronouncement to this effect, and his acceptance of resulting generalized versions of his two incompleteness theorems, is in the *"Note added 28 August 1963"* to the translation in *van Heijenoort 1967* of his *1931*, and in the "Postscriptum", dated 3 June 1964, to the *Davis 1965* printing of his *1934* lecture notes (where he also cites *Post 1936*), both reproduced in the present volume. In that Postscriptum, he mentions the effective unsolvability of the (semi-)Diophantine problem $(P)[F = 0]$, again improving his earlier results.

Stephen C. Kleene

[h]In modern terminology, $(Ex)T_1(z,a,x)$ is a universal Σ_1^0 predicate, so $(x)\overline{T}_1(a,a,x)$ is not Σ_1^0. See footnote d above.

On undecidable propositions of formal mathematical systems

(*1934*)

1. Introduction

A *formal mathematical system* is a system of symbols together with rules for employing them. The individual symbols are called *undefined terms*. *Formulas* are finite sequences of the undefined terms. There shall be defined a class of formulas called *meaningful formulas*, and a class of meaningful formulas called *axioms*. There may be a finite or infinite number of axioms. Further, there shall be specified a list of rules, called *rules of inference;* if such a rule be called R, it defines the relation of *immediate consequence by R* between a set of meaningful formulas $M_1, \ldots, M_k$, called the *premises*, and a meaningful formula N, called the *conclusion* (ordinarily $k = 1$ or 2). We require that the rules of inference, and the definitions of meaningful formulas and axioms, be constructive; that is, for each rule of inference there shall be a finite procedure for determining whether a given formula B is an immediate consequence (by that rule) of given formulas $A_1, \ldots, A_n$, and there shall be a finite procedure for determining whether a given formula A is a meaningful formula or an axiom.

A formula N shall be called an *immediate consequence* of $M_1, \ldots, M_n$ if N is an immediate consequence of $M_1, \ldots, M_n$ by any one of the rules of inference. A finite sequence of formulas shall be a *proof* (specifically, a proof of the last formula of the sequence) if each formula of the sequence is either an axiom, or an immediate consequence of one or more of the preceding formulas. A formula is *provable* if a proof of it exists. Let the symbol $\sim$ be one of the undefined terms, and suppose it to express negation. Then the formal system shall be said to be *complete* if for every [closed] meaningful formula A either A or $\sim A$ is provable. We shall prove later that (under conditions to be stated) a system in which all propositions of arithmetic can be expressed as meaningful formulas is not complete.

2 |
2. Recursive functions and relations[1]

Now we turn to some considerations which for the present have nothing to do with a formal system.

---

[1][What is called "recursive" in these lectures (except for §9) is now called "primitive recursive".]

Lower-case letters $x, y, z, \ldots$ will denote arbitrary natural numbers (i.e., non-negative integers); and German letters will be used in abbreviation for finite sequences of the former, i.e., $\mathfrak{x}$ for $x_1, \ldots, x_n$; $\mathfrak{y}$ for $y_1, \ldots, y_m$. Greek letters $\phi, \psi, \chi, \ldots$ will represent functions, of one or more natural numbers, whose values are natural numbers. Capitals R, S, $T, \ldots$ will stand for classes of, or relations among, natural numbers. $R(x)$ shall stand for the proposition that x is in the class R, and $S(x_1, \ldots, x_n)$ for the proposition that $x_1, \ldots, x_n$ stand in the relation S. Classes may be considered as relations with only one term, and relations as classes of ordered n-tuples. There shall correspond to each class or relation R a *representing function* ϕ such that $\phi(x_1, \ldots, x_n) = 0$ if $R(x_1, \ldots, x_n)$ and $\phi(x_1, \ldots, x_n) = 1$ if $\sim R(x_1, \ldots, x_n)$.

We use the following notations as abbreviations (p, q are to be replaced by any propositions): $(x)[A(x)]$ (for every natural number $x, A(x)$), $(Ex)[A(x)]$ (there exists a natural number x such that $A(x)$), $\epsilon x[A(x)]$ (the least natural number x such that $A(x)$ if $(Ex)[A(x)]$; otherwise 0), $\sim p$ (not p), $p \vee q$ (p or q), $p \mathbin{\&} q$ (p and q), $p \to q$ (p implies q, i.e., $(\sim p) \vee q$), $p \equiv q$ (p is equivalent to q, i.e., $(p \to q) \mathbin{\&} (q \to p)$).

The function $\phi(x_1, \ldots, x_n)$ shall be *compound* with respect to $\psi(x_1, \ldots, x_m)$ and $\chi_i(x_1, \ldots, x_n)$ $(i = 1, \ldots, m)$ if, for all natural numbers $x_1, \ldots, x_n$,

(1) $$\phi(x_1, \ldots, x_n) = \psi(\chi_1(x_1, \ldots, x_n), \ldots, \chi_m(x_1, \ldots, x_n)).$$

$\phi(x_1, \ldots, x_n)$ shall be said to be *recursive* with respect to $\psi(x_1, \ldots, x_{n-1})$ and $\chi(x_1, \ldots, x_{n+1})$ if, for all natural numbers $k, x_2, \ldots, x_n$,

(2) $$\phi(0, x_2, \ldots, x_n) = \psi(x_2, \ldots, x_n)$$
$$\phi(k + 1, x_2, \ldots, x_n) = \chi(k, \phi(k, x_2, \ldots, x_n), x_2, \ldots, x_n).$$

| In both (1) and (2), we allow the omission of each of the variables in any (or all) of its occurrences on the right side (e.g., $\phi(x, y) = \psi(\chi_1(x), \chi_2(x, y))$ is permitted under (1)).[2] We define the class of *recursive* functions to be the totality of functions which can be generated by substitution, according to the scheme (1), and recursion, according to the scheme (2), from the successor function $x + 1$, constant functions $f(x_1, \ldots, x_n) = c$, and identity functions $U_j^n(x_1, \ldots, x_n) = x_j$ $(1 \le j \le n)$. In other words, a function ϕ shall be recursive if there exists a finite sequence of functions $\phi_1, \ldots, \phi_n$ which terminates with ϕ such that each function of the sequence is either the successor function $x + 1$, or a constant function $f(x_1, \ldots, x_n) = c$, or

---

[2]The first sentence may be omitted, since the removal of any of the occurrences of variables on the right may be effected by means of the function U_j^n.

an identity function $U_j^n(x_1, \ldots, x_n) = x_j$, or is compound with respect to preceding functions, or is recursive with respect to preceding functions. A relation R shall be *recursive* if the representing function is recursive.

Recursive functions have the important property that, for each given set of values of the arguments, the value of the function can be computed by a finite procedure.[3] Similarly, recursive relations (classes) are decidable in the sense that, for each given n-tuple of natural numbers, it can be determined by a finite procedure whether the relation holds or does not hold (the number belongs to the class or not), since the representing function is computable.

The functions $x+y$, xy, x^y and $x!$ are clearly recursive. Hence $\phi(\mathfrak{x})+\psi(\mathfrak{y})$, $\phi(\mathfrak{x})\psi(\mathfrak{y})$, $\phi(\mathfrak{x})^{\psi(\mathfrak{y})}$, and $\phi(\mathfrak{x})!$ are recursive, if $\phi(\mathfrak{x})$ and $\psi(\mathfrak{y})$ are.

I. *If the relations* $R(\mathfrak{x})$ *and* $S(\mathfrak{y})$ *are recursive, then* $\sim R(\mathfrak{x})$, $R(\mathfrak{x}) \vee S(\mathfrak{y})$, $R(\mathfrak{x})$ & $S(\mathfrak{y})$, $R(\mathfrak{x}) \to S(\mathfrak{y})$, $R(\mathfrak{x}) \equiv S(\mathfrak{y})$ *are recursive.*

By hypothesis, the representing functions $\rho(\mathfrak{x})$ and $\sigma(\mathfrak{y})$ of R and S,

4 | respectively, are recursive. If

$$\alpha(0) = 1, \quad \alpha(k + 1) = 0,$$

then $\alpha(x)$ and hence $\alpha(\rho(\mathfrak{x}))$ are recursive. But, since $\alpha(\rho(\mathfrak{x}))$ is 1 or 0 according as $\rho(\mathfrak{x})$ is 0 or 1, $\alpha(\rho(\mathfrak{x}))$ is the representing function of $\sim R(\mathfrak{x})$. Thus $\sim R(\mathfrak{x})$ is recursive. If $\beta(0, x) = 0$, $\beta(k + 1, x) = \alpha(\alpha(x))$, then

$$\beta(0, x) = \beta(x, 0) = 0 \quad \text{and} \quad \beta(x, y) = 1 \quad \text{when } x, y > 0.$$

Hence $\beta(\rho(\mathfrak{x}), \sigma(\mathfrak{y}))$, which is recursive, represents $R(\mathfrak{x}) \vee S(\mathfrak{y})$; that is, $R(\mathfrak{x}) \vee S(\mathfrak{y})$ is recursive. Since $R(\mathfrak{x})$ & $S(\mathfrak{y}) \equiv \sim(\sim R(\mathfrak{x}) \vee \sim S(\mathfrak{y}))$, it follows that $R(\mathfrak{x})$ & $S(\mathfrak{y})$ is recursive. Similarly, $R(\mathfrak{x}) \to S(\mathfrak{y})$, $R(\mathfrak{x}) \equiv S(\mathfrak{y})$ and all other relations definable from $R(\mathfrak{x})$ and $S(\mathfrak{y})$ by use of $\sim$ and $\vee$ are recursive.

II. *If the functions* $\phi(\mathfrak{x})$, $\psi(\mathfrak{y})$ *are recursive, then the relations* $\phi(\mathfrak{x}) = \psi(\mathfrak{y})$, $\phi(\mathfrak{x}) < \psi(\mathfrak{y})$, $\phi(\mathfrak{x}) \leq \psi(\mathfrak{y})$ *are recursive.*

Let

$$\delta(0) = 0, \quad \delta(k + 1) = k,$$

and $x \mathbin{\dot{-}} 0 = x$, $x \mathbin{\dot{-}} (k + 1) = \delta(x \mathbin{\dot{-}} k)$. Then

$$x \mathbin{\dot{-}} y = x - y \quad \text{if } x \geq y \quad \text{and} \quad x \mathbin{\dot{-}} y = 0 \quad \text{if } x \leq y.$$

---

[3] The converse seems to be true if, besides recursions according to the scheme (2), recursions of other forms (e.g., with respect to two variables simultaneously) are admitted. This cannot be proved, since the notion of finite computation is not defined, but it serves as a heuristic principle. [This statement is now outdated; see the Postscriptum, pp. 369–371.]

Hence $\alpha(y \dotminus x)$ is a representing function for $x < y$, and $\alpha(\psi(\mathfrak{n}) \dotminus \phi(\mathfrak{x}))$ for $\phi(\mathfrak{x}) < \psi(\mathfrak{n})$. Thus $\phi(\mathfrak{x}) < \psi(\mathfrak{n})$ is recursive. $\phi(\mathfrak{x}) = \psi(\mathfrak{n})$ and $\phi(\mathfrak{x}) \leq \psi(\mathfrak{n})$ are likewise recursive, as may be seen directly, or inferred from the theorem for $\phi(\mathfrak{x}) < \psi(\mathfrak{n})$ by use of I.

III. *If the function $\phi(\mathfrak{x})$ and the relation $R(x, \mathfrak{n})$ are recursive, then the relations S, T, where*

$$S(\mathfrak{x}, \mathfrak{n}) \equiv (Ex)[x \leq \phi(\mathfrak{x}) \ \& \ R(x, \mathfrak{n})],$$
$$T(\mathfrak{x}, \mathfrak{n}) \equiv (x)[x \leq \phi(\mathfrak{x}) \rightarrow R(x, \mathfrak{n})],$$

and the function ψ, where

$$\psi(\mathfrak{x}, \mathfrak{n}) = \epsilon x[x \leq \phi(\mathfrak{x}) \ \& \ R(x, \mathfrak{n})],$$

are recursive.

Let the representing function of $R(x, \mathfrak{n})$ be $\rho(x, \mathfrak{n})$. Let $\mid \pi(0, \mathfrak{n}) = \rho(0, \mathfrak{n})$ 5 and $\pi(k + 1, \mathfrak{n}) = \pi(k, \mathfrak{n})\rho(k + 1, \mathfrak{n})$. Then $\pi(x, \mathfrak{n}) = \rho(0, \mathfrak{n})\rho(1, \mathfrak{n}) \ldots \rho(x, \mathfrak{n})$. Hence $\pi(x, \mathfrak{n})$ is 0 or 1 according as some or none of $\rho(0, \mathfrak{n}), \ldots, \rho(x, \mathfrak{n})$ are 0; that is, according as there do or do not exist natural numbers $n \leq x$ for which $R(n, \mathfrak{n})$ holds. Hence $\pi(\phi(\mathfrak{x}), \mathfrak{n})$, which is recursive, represents $(Ex)[x \leq \phi(\mathfrak{x}) \ \& \ R(x, \mathfrak{n})]$. Thus $(Ex)[x \leq \phi(\mathfrak{x}) \ \& \ R(x, \mathfrak{n})]$ is a recursive relation. It follows from this result and I that $(x)[x \leq \phi(\mathfrak{x}) \rightarrow R(x, \mathfrak{n})]$ is recursive, since $(x)[x \leq \phi(\mathfrak{x}) \rightarrow R(x, \mathfrak{n})] \equiv {\sim}(Ex)[x \leq \phi(\mathfrak{x}) \ \& \ {\sim}R(x, \mathfrak{n})]$. Let $\mu(0, \mathfrak{n}) = 0$ and

$$\mu(k+1, \mathfrak{n}) = (k+1)[\pi(k, \mathfrak{n}) \dotminus \pi(k+1, \mathfrak{n})] + \mu(k, \mathfrak{n})[\alpha(\pi(k, \mathfrak{n}) \dotminus \pi(k+1, \mathfrak{n}))].$$

Since $1 \geq \pi(k, \mathfrak{n}) \geq \pi(k + 1, \mathfrak{n}) \geq 0$, $\mu(k + 1, \mathfrak{n}) = k + 1$ if $\pi(k, \mathfrak{n}) = 1$ and $\pi(k + 1, \mathfrak{n}) = 0$, and otherwise $\mu(k + 1, \mathfrak{n}) = \mu(k, \mathfrak{n})$. Both $\pi(k, \mathfrak{n}) = 1$ and $\pi(k + 1, \mathfrak{n}) = 0$ hold only when ${\sim}R(1, \mathfrak{n}), \ldots, {\sim}R(k, \mathfrak{n})$ and $R(k + 1, \mathfrak{n})$; that is, when $k + 1$ is the least value x' of x such that $R(x, \mathfrak{n})$. Hence, if such an x' exists and is > 1, $\mu(0, \mathfrak{n}) = \cdots = \mu(x' - 1, \mathfrak{n}) = 0$ and $\mu(x, \mathfrak{n}) = x'$ for all $x \geq x'$. If $x' = 0$, or x' does not exist, all $\mu(x, \mathfrak{n})$ are 0. Hence $\mu(\phi(\mathfrak{x}), \mathfrak{n})$ is the least $x \leq \phi(\mathfrak{x})$ such that $R(x, \mathfrak{n})$, if such exists, otherwise 0; i.e., $\mu(\phi(\mathfrak{x}), \mathfrak{n}) = \epsilon x[x \leq \phi(\mathfrak{x}) \ \& \ R(x, \mathfrak{n})]$.

3. A formal system

We now describe in some detail a formal system which will serve as an example for what follows. While a formal system consists only of symbols and mechanical rules relating to them, the meaning which we attach to the symbols is a leading principle in the setting up of the system.

We shall depend on the theory of types as our means for avoiding paradox. Accordingly, we exclude the use of variables running over all

objects, and use different kinds of variables for different domains. Specifically, $p, q, r, \ldots$ shall be variables for propositions. Then there shall be variables of successive types as follows:

6 | $x, y, z, \ldots$ for natural numbers,

$f, g, h, \ldots$ for functions (of one variable) whose domain and values are natural numbers,

$F, G, H, \ldots$ for functions (of one variable) whose domain and values are functions $f, g, h, \ldots$, and so on.[4]

Different formal systems are determined according to how many of these types of variables are used. We shall restrict ourselves to the first two types; that is, we shall use variables of the three sorts $p, q, r, \ldots$; $x, y, z, \ldots$; $f, g, h, \ldots$. We assume that a denumerably infinite number of each are included among the undefined terms (as may be secured, for example, by the use of letters with numerical subscripts).

The undefined terms, in addition to variables, shall be 0 (the number 0), N ($N(x)$ denotes the next greater number than x, i.e., the successor of x), $\sim, \vee, \&, \rightarrow, \equiv, \Pi$ ($\Pi x(F(x))$ means "$F(x)$ is true for all natural numbers x", and may be regarded as the logical product of $F(x)$ over all x), Σ ($\Sigma x(F(x))$ means "there is at least one natural number x such that $F(x)$ is true", and may be regarded as the logical sum of $F(x)$ over all x), ϵ, = (equals), (,) ($f(x)$ is the value of f for the argument x, (and) being then interpreted as symbols for the operation of *application* of a function to an argument). Parentheses are also used as signs of inclusion, as in $\Pi x(A)$, $(A) \rightarrow (B)$, etc.).[5]

7 | Next, the class of meaningful formulas must be defined. To do this we describe two classes of formulas which have significance: formulas which denote numbers, and formulas which denote propositions. The first comprises *numerical symbols* or expressions representing numbers (as 0, $N(0)$, ...) together with *functional expressions* or expressions which become numerical expressions when numerical expressions are substituted in a suitable manner for variables which occur in them (as $\epsilon x[y = N(x)]$).[6] The second comprises *propositions* (e.g., $\Pi x[\sim(0 = N(x))]$), together with *propositional functions* or expressions which become propositions when numerical expressions are substituted in a suitable manner for variables which occur in them

---

[4]Functions for several variables need not be provided separately, since n-tuples of objects of each of these types can be mapped one-to-one on single objects of the same type. [Moreover inhomogeneous functions can be represented by homogeneous ones.]

    Variables for classes and relations are unnecessary, since we can use, instead of the classes and relations, their representing functions.

[5]$\sim, \vee, \&, \rightarrow, \equiv$, and ϵ have the significances assigned to them in §2. $\Pi x(A)$ and $\Sigma x(A)$, when A does not involve x, mean the same as A. The fact that the logical notions among our undefined terms are not independent does not matter for our purpose.

[6]As indicated on page 8, the substitutions which are meant in the case of $\epsilon x[y = N(x)]$ and $\Sigma x[y = N(x)]$ are substitutions for y.

(e.g., $\Sigma x[y = N(x)]$).[6] The exact definitions we give by complete induction, thus:

1. 0 and $x, y, z, \ldots$ (variables for numbers) are expressions of the I-st kind [exp. I], and $p, q, r, \ldots$ (variables for propositions) are expressions of the II-nd kind [exp. II].

2. If A and B are expressions of the I-st kind, then $A = B$ is an expression of the II-nd kind.

3. If A exp. I, then $N(A)$ exp. I.

4. If A exp. I, and f is a variable for a function, then $f(A)$ exp. I.

5. If A and B exp. II, then $\sim(A)$, $(A) \vee (B)$, $(A)\ \&\ (B)$, $(A) \to (B)$ and $(A) \equiv (B)$ exp. II.

6. If A exp. II, and x is a variable for a number, then $\Pi x(A)$ and $\Sigma x(A)$ exp. II, and $\epsilon x(A)$ exp. I.

7. If A exp. II, and f is a variable for a function, then $\Pi f(A)$ and $\Sigma f(A)$ exp. II.

8. If A exp. II, and p is a variable for a proposition, then $\Pi p(A)$ and $\Sigma p(A)$ exp. II.

9. The classes of expressions of the I-st and of the II-nd kind shall be the pair of classes with the least union satisfying 1–8.

A formula shall be *meaningful* if it is either an expression of the I-st kind or an expression of the II-nd kind.

The occurrences[7] of variables in a meaningful expression can be classified as free and bound in the following manner: There corresponds to each occurrence of Π in a meaningful expression A a unique part of A, beginning with the occurrence of $|\ \Pi$, of the form $\Pi t(B)$, where t is a variable and B is meaningful. This part of A will be called the *scope* of the given occurrence of Π in A. Similarly we define the *scope* of an occurence of Σ or ϵ in A. A given occurrence of the variable t in A shall be *bound* or *free* according as it is or is not in the scope of a Π, Σ or ϵ followed by t.

In the above definitions of functional expressions and propositional functions, the substitutions which are meant are substitutions for the free occurrences of variables. (y is free and x is bound in $\epsilon x[y = N(x)]$ and $\Sigma x[y = N(x)]$.)

We use $\mathrm{Subst}[A_G^t]$ to denote the expression obtained from A by substituting G for each occurrence of t in A as a free variable.[8]

We may use $F(t)$ to represent a meaningful formula in which t occurs as a free variable,[9] and $F(A)$ to denote $\mathrm{Subst}[F(t)_A^t]$.

---

[7] By an *occurrence* of a symbol (or formula) K in an expression A, we shall mean a particular part of A of the form K.

[8] "Subst" by itself is not a formula of our system. ["Subst", of course, is not a symbol of the formal system, but rather of metamathematics.]

[9] Then F by itself does not represent a formula.

If A is a meaningful formula, and f a variable for a function, then the occurrences of f in A as a free symbol are as the first symbol of parts of A of the form $f(U)$. We may list these parts as $f(U_1), \ldots, f(U_n)$ in such an order that, if U_j contains $f(U_i)$, then $i < j$. Let $G(x)$ be a meaningful formula in which x occurs as a free variable. Let $A', f(U_2'), \ldots, f(U_n')$ be obtained from $A, f(U_2), \ldots, f(U_n)$ be substituting $G(U_1)$ for $f(U_1)$;[10] then let $A'', f(U_3''), \ldots, f(U_n'')$ be obtained from $A', f(U_3'), \ldots, f(U_n')$ by substituting $G(U_2')$ for $f(U_2')$; and so on. We shall denote by $\mathrm{Subst}'[A_{G(x),x}^f]$ the expression $A^{(n)}$.[11]

9 | An expression A in which the distinct free variables $t_1, \ldots, t_n$ occur shall mean the same as $\Pi t_1(\ldots (\Pi t_n(A)) \ldots)$.

The axioms shall be the following formulas (A1–C2).[12]

A. Axioms concerning the notions of the calculus of propositions.

1. $(p \to q) \to ((q \to r) \to (p \to r))$.
2. $((\sim p) \to p) \to p$.
3. $p \to ((\sim p) \to q)$.
4. $p \,\&\, q . \equiv . \sim[(\sim p) \vee (\sim q)]$.
5. $p \vee q . \equiv . (\sim p) \to q$.
6. $p \equiv q . \equiv . (p \to q) \,\&\, (q \to p)$.
7. $p \equiv q . \to . p \to q$.
8. $p \equiv q . \to . q \to p$.

To give the theory of this group of axioms would require a study of the theory of the calculus of propositions.

B. Axioms concerning the notion of identity.

1. $x = x$.
2. $x = y . \to . f(x) = f(y)$.

---

[10]More explicitly, let A' be the expression obtained from A by substituting $G(U_1)$ for the part $f(U_1)$, and let $f(U_2'), \ldots, f(U_n')$ be the parts of A' into which the parts $f(U_2), \ldots, f(U_n)$ of A are transformed by the substitution.

[11]Here we describe the proper method of substituting an expression $G(x)$ for a functional variable f in an expression A, and denote the result of the substitution by $\mathrm{Subst}'[A_{G(x),x}^f]$. If $G(x)$ does not contain f (which can always be made the case in the course of formal proofs by a change in the notation), $\mathrm{Subst}'[A_{G(x),x}^f]$ may also be defined as follows: Replace $f(U)$ by $G(U)$ for one of the free occurrences of f in A, do the same thing with the resulting expression, and so on, until an expression is obtained in which f no longer occurs as a free variable.

[x must be mentioned explicitly as the variable of $G(x)$ to be used in the process of substitution, since $G(x)$ may contain other free variables too.]

[In the original mimeographed version the operation was written $\mathrm{Subst}'(A_{G(x)}^f)$.]

[12]In writing down these axioms and other meaningful formulas we employ the usual conventions concerning the omission of parentheses.

All abbreviation of formulas is to be regarded as extraneous to the formal system; and each statement about a formula of the system refers to its unabbreviated form.

3. $(x = y)$ & $(y = z) . \rightarrow . z = x.$

C. Axioms which correspond to certain of Peano's axioms for the natural numbers.

1. $\sim(0 = N(x))$.

2. $N(x) = N(y) . \rightarrow . x = y.$

To complete the definition of the formal system under consideration, it remains to list the rules of inference. Each rule is to be interpreted as a statement of the conditions under which a formula N shall be an immediate consequence, by | that rule, of the meaningful formula(s) $M (M_1, M_2)$.[13]   10

1. *If* $(A) \rightarrow (B)$ *and* A, *then* B.

2. Suppose that A is meaningful,[14] that t is a variable, and that t does not occur in A.

> a. *If* $(A) \rightarrow (B)$, *then* $(A) \rightarrow (\Pi t(B))$.
> b. *If* $(A) \rightarrow (\Pi t(B))$, *then* $(A) \rightarrow (B)$.

3. Suppose that A is meaningful, that t is a variable, and that t does not occur in B.

> a. *If* $(A) \rightarrow (B)$, *then* $(\Sigma t(A)) \rightarrow (B)$.
> b. *If* $(\Sigma t(A)) \rightarrow (B)$, *then* $(A) \rightarrow (B)$.

4a. Suppose that x is a variable for a number, that A contains x as a free variable, that G is an expression of the I-st kind, and that no free variable of G is bound in A.

> *If* A, *then* $\mathrm{Subst}[A_G^x]$.

4b. Suppose that f is a variable for a function, that A contains f as a free variable, that x is a variable for a number, that $G(x)$ is an expression of the I-st kind in which x occurs as a free variable, and that no free variable of $G(x)$ [or A is bound in either $G(x)$ or A, and that, moreover, $G(x)$ and A have no common bound variables].

---

[13] N also will be meaningful whenever the conditions are realized.
   Rule 1, for example, can be written more explicitly thus: N shall be an *immediate consequence by Rule 1* of M_1 and M_2 if and only if [M_1 and M_2 are meaningful formulas and] there exist formulas A and B such that M_1 is $(A) \rightarrow (B)$, M_2 is A, and N is B.

[14] This condition ensures that, when $(A) \rightarrow (B)$ is a meaningful formula, the occurrence of $\rightarrow$ which separates (A) from (B) in $(A) \rightarrow (B)$ should be the last occurrence of $\rightarrow$ introduced in the construction of $(A) \rightarrow (B)$ according to the definition of meaningful formula. (We may say then that the *main operation* of $(A) \rightarrow (B)$ is an implication, whose first and second *terms* are A and B, respectively.) It excludes such possibilities as that A be $p \rightarrow q) \rightarrow ((q \rightarrow r$, when $(A) \rightarrow (B)$ is Axiom A1.

$$\text{If } A, \text{ then } \mathrm{Subst}'[A^f_{G(x),x}].$$

4c. Suppose that p is a variable for a proposition, that A contains p as a free variable, that P is an expression of the II-nd kind, and that no free variable of P is bound in A.

$$\text{If } A, \text{ then } \mathrm{Subst}[A^p_P].$$

4d. Suppose that x is a variable for a number, and that $F(x)$ is meaningful and contains x as a free variable [and no free variable of F occurs as a bound variable in F].

$$\text{If } (A) \rightarrow (F(x)), \text{ then } (A) \rightarrow (F(\epsilon x[F(x)])).^{15}$$

5. Suppose that x is a variable for a number, and that $F(x)$ is a meaningful formula in which x occurs as a free variable.

$$\text{If } F(0) \text{ and } (F(x)) \rightarrow (F(N(x))) <, \text{ then } F(x).$$

6. Suppose that s and t are variables of the same kind, that s does not occur in A as a free variable, and that t does not occur in A. Let A' denote the result of substituting t for s throughout A. Suppose that A is meaningful, and let B' denote the expression obtained from B by the substitution of A' for a given occurrence of A in B.

$$\text{If } B, \text{ then } B'.^{16}$$

---

[15]If $F(x)$ is an expression of the II-nd kind containing the variable x for a number as a free variable, then, with the aid of this rule, $\Sigma x F(x) \rightarrow F(\epsilon x[F(x)])$ is provable in our formal system. For Rule 4c allows us to infer $(p \rightarrow [(\sim p) \rightarrow p]) \rightarrow (([[(\sim p) \rightarrow p] \rightarrow p) \rightarrow (p \rightarrow p))$ from Axiom A1 (i.e., by substituting $(\sim p) \rightarrow p$ for q and p for r), and $p \rightarrow [(\sim p) \rightarrow p]$ from Axiom A3 (by substituting p for q). Then Rule 1 allows us from these two results to infer $([(\sim p) \rightarrow p] \rightarrow p) \rightarrow (p \rightarrow q)$, and then from the latter and Axiom A2 to infer $p \rightarrow p$. Thence we can successively infer $F(x) \rightarrow F(x)$ by Rule 4c (by substituting $F(x)$ for p), $F(x) \rightarrow F(\epsilon x[F(x)])$ by Rule 4d, $F(x) \rightarrow F(\epsilon y[F(y)])$ by one or more applications of Rule 6, $\Sigma x[F(x)] \rightarrow F(\epsilon y[F(y)])$ by Rule 3a, and $\Sigma x[F(x)] \rightarrow F(\epsilon x[F(x)])$ by Rule 6. Thus the last formula, which expresses the essential property of ϵ, is proved in our formal system. If the system admitted the use of ϵ with variables for functions (i.e., if a rule of inference 4d', obtained from 4d by replacing "x" by "f" and "number" by "function", were added), then similarly, for any expression $G(f)$ containing the variable f for a function as a free variable, $\Sigma f[G(f)] \rightarrow G(\epsilon f[G(f)])$ would be provable. The latter formula expresses the axiom of choice for classes of functions of integers.

Note that by our formal rule for ϵ we cannot prove that $\epsilon x F(x)$ is the *smallest* integer x for which $F(x)$, nor that $\epsilon x F(x) = 0$ if there is no such integer, but we can prove only that, if there are integers x satisfying $F(x)$, then $\epsilon x F(x)$ is one of them (i.e., $\Sigma x[F(x)] \rightarrow F(\epsilon x[F(x)])$). This however suffices for all applications.

[16]Note that B may be A itself (then B' is A').

One process used in mathematical proof is not represented in this system, namely the definition and introduction of new symbols. However, this process is not essential, but merely a matter of abbreviation.

| ## 4. A representation of the system by a system of positive integers 11

For the considerations which follow, the meaning of the symbols is immaterial, and it is desirable that it be forgotten. Notions which relate to the system considered purely formally may be called *metamathematical*.

For undefined terms (hence the formulas and proofs) are countable, and hence a representation of the system by a system of positive integers can be constructed, as we shall now do.

We order the numbers 1–13 to symbols thus:

$$\begin{array}{ccccccccccccc} 0 & N & = & \sim & \vee & \& & \rightarrow & \equiv & \Pi & \Sigma & \epsilon & (&) \\ 1 & 2 & 3 & 4 & 5 & 6 & 7 & 8 & 9 & 10 & 11 & 12 & 13, \end{array}$$

the integers > 13 and $\equiv 0 \pmod 3$ to the variables for propositions, the integers > 13, $\equiv 1 \pmod 3$ to the variables for numbers, and the integers > 13, $\equiv 2 \pmod 3$ to the variables for functions. Thus a one-to-one correspondence is established between the undefined terms and the positive integers.

We order single integers to finite sequences of positive integers by means of the scheme

$$k_1, \ldots, k_n \quad \text{corresponds to} \quad 2^{k_1} \cdot 3^{k_2} \cdot 5^{k_3} \cdot \ldots \cdot p_n^{k_n},$$

where p_i is the ith prime number (in order of magnitude). A formula is a finite sequence of undefined terms, and a proof a finite sequence of formulas. To each formula we order the integer which corresponds to the sequence of the integers ordered to its symbols; and to each proof we order the integer which corresponds to the sequence of the integers which are then ordered to its member formulas. Then a one-to-one correspondence is determined between formulas (proofs) and a subset of the positive integers.

We may now define various metamathematical classes and relations of positive integers, including one corresponding to each class and relation of formulas. x shall be an f number ($\mathfrak{B}$ number) if there is a formula (proof) to which x corre|sponds.[17] The relation $x, y \, U z$ between numbers shall 12

---

[17] $\mathfrak{B}$ for "Beweis". Below occur U for "unmittelbar Folge", Gl for "Glied", $\mathfrak{E}$ for "einklammern".

mean that x, y and z are f numbers, and the formula which z represents is an immediate consequence of the formulas which x and y represent. $x\mathfrak{B}y$ shall mean that x is a $\mathfrak{B}$ number and y an f number, and the proof which x represents is a proof of the formula which y represents. Also there are metamathematical functions of integers such as the following: $\text{Neg}(x) =$ the number representing $\sim(X)$ if x represents the formula X; and $= 0$ if x is not an f number. $Sb[x^y_z] =$ the number which represents the result $\text{Subst}[F^t_G]$ of substituting G for the free occurrences of t in F, if x, z represent the formulas F, G, respectively, and y the variable t; and $= 0$ otherwise. These relations and functions, which we have defined indirectly by using the correspondence between formulas and numbers, are constructive. Hence it is not surprising to find that they are recursive. We shall show this for some of the more important of them, by defining them directly, from relations and functions previously known to be recursive, by methods shown in §2 to generate recursive relations and functions out of recursive relations and functions.[18]

1. $x \mid y \equiv (Ez)[z \leq x \ \& \ x = yz]$.

"$x \mid y$" means "x is divisible by y". (yz is recursive. Hence by II of §2, $x = yz$ is recursive. It follows by III that $x \mid y$ is recursive. $z \leq x$ is inserted in the definition to make it clear that III applies and could be omitted without changing the meaning.)

2. $\text{Prime}(x) \equiv x > 1 \ \& \sim(Ez)[z \leq x \ \& \sim(z = 1) \ \& \sim(z = x) \ \& \ x \mid z]$.

"x is a prime number".

13 | 3. $\text{Pr}(0) = 0$,

$\text{Pr}(n+1) = \epsilon y[y \leq \{\text{Pr}(n)\}! + 1 \ \& \ \text{Prime}(y) \ \& \ y > \text{Pr}(n)]$.

$\text{Pr}(n)$ is the nth prime number (in order of magnitude).

4. $n\text{Gl}x = \epsilon y[y \leq x \ \& \ x \mid \{\text{Pr}(n)\}^y \ \& \sim(x \mid \{\text{Pr}(n)\}^{y+1})]$.

$n\text{Gl}x$ is the nth member of the sequence of positive integers which x represents (i.e., $n\text{Gl}x$ is k_n if $x = 2^{k_1} \cdot 3^{k_2} \cdot \ldots \cdot p_n^{k_n} \cdot \ldots \cdot p_l^{k_l}$).

5. $L(x) = \epsilon y[y \leq x \ \& \ (y+1)\text{Gl}x = 0]$.

$L(x)$ is the number of members in the sequence represented by x (if x represents a sequence of positive integers).

6. $x * y = \epsilon z\{z \leq [\text{Pr}(L(x) + L(y))]^{x+y} \ \&$
$\quad (n)[n \leq L(x) \to n\text{Gl}z = n\text{Gl}x] \ \&$
$\quad (n)[0 < n \leq L(y) \to (n + L(x))\text{Gl}z = n\text{Gl}y]\}$.

$*$ represents the operation of joining one finite sequence to another (i.e., if $x = 2^{k_1} \cdot \ldots \cdot p_r^{k_r}$ and $y = 2^{l_1} \cdot \ldots \cdot p_s^{l_s}$, then $x*y = 2^{k_1} \cdot \ldots \cdot p_r^{k_r} \cdot p_{r+1}^{l_1} \cdot \ldots \cdot p_{r+s}^{l_s}$).

Note that the number of the sequence consisting of the single number x is 2^x.

---

[18] We use formal notations (including those explained in §2) in the following definitions for the purpose of abbreviating the discussion. These formal notations must not be confused with the formulas of the formal mathematical system under consideration.

7. $\mathfrak{E}(x) = 2^{12} * x * 2^{13}$.

If x represents the formula A, $\mathfrak{E}(x)$ represents (A) (for then the sequence of the numbers ordered to the symbols of (A) is 12, $k_1, \ldots, k_n$, 13, where $k_1, \ldots, k_n$ is the sequence of the numbers ordered to the symbols of A).

8. $\mathrm{Neg}(x) = 2^4 * \mathfrak{E}(x)$.

If x represents the formula A, $\mathrm{Neg}(x)$ represents $\sim(A)$.

9. $\mathrm{Imp}(x, y) = \mathfrak{E}(x) * 2^7 * \mathfrak{E}(y)$.

If x, y represent the formulas A, B respectively, then $\mathrm{Imp}(x, y)$ represents $(A) \to (B)$.

10. $u\mathrm{Gen}x = 2^9 * 2^u * \mathfrak{E}(x)$.

If u represents the variable t, and x the formula A, then $u\mathrm{Gen}x$ represents $\Pi t(A)$.

Similarly for $\Sigma t(A)$ and $\epsilon x(A)$.

| 11. $x \equiv y \,(\mathrm{mod}\, n) . \equiv . (Ez)[z \leq x + y \;\&\; (x = y + zn \vee y = x + zn)]$.   14

$x \equiv y \,(\mathrm{mod}\, n)$ has the usual significance.

$t > 13$ expresses "t represents a variable". Also, using 11, recursive classes $\mathrm{Var}_p(t)$, $\mathrm{Var}_x(t)$, $\mathrm{Var}_f(t)$ can be defined to express "t represents a variable for a proposition", "t represents a variable for a number", "t represents a variable for a function", respectively.

Recursive classes $M_I(x)$, $M_{II}(x)$, $M(x)$ expressing "x represents an exp. I", "x represents an exp. II", "x represents a meaningful formula", respectively, recursive relations corresponding to the relations "t occurs in A as a free (bound) variable", and recursive functions corresponding to the operations of substitution used in the rules of inference, can be defined.[19]

12. $x, yU_1z \equiv x = \mathrm{Imp}(y, z) \;\&\; M(y) \;\&\; M(z)$.

"z represents a formula which is an immediate consequence by Rule 1 of the formulas represented by x, y".

13. $xU_{2a}z \equiv (Et, v, w)[t, v, w \leq z \;\&\; M(v) \;\&\; M(w) \;\&\; x = \mathrm{Imp}(v, w) \;\& $
$z = \mathrm{Imp}(v, t\mathrm{Gen}w) \;\&\; t > 13 \;\&\; \sim(Ek)[k \leq L(v) \;\&\; k\mathrm{Gl}v = t]]$.[20]

"z represents a formula which is an immediate consequence by Rule 2a of the formula represented by x".

Similarly for each of the other rules of inference.

| 14. $x, y\, Uz \equiv x, yU_1z \vee xU_{2a}z \vee \cdots \vee xU_6z$.   15

"z represents a formula which is an immediate consequence of the formula(s) represented by x (x and y)".

Each axiom is represented by a number. Let the numbers corresponding to the axioms be $\alpha_1, \ldots, \alpha_{13}$.

---

[19]For the details of the definition of classes, relations and functions of these sorts, relating to a formal system similar to the one under consideration, see *Gödel 1931*; specifically, see the definitions 1–31, pp. 182–184 [above, pp. 163ff.]

[20]$(Et, v, w)[t, v, w \leq z \;\&\; \ldots]$ stands for
$$(Et)[(Ev)[(Ew)[t \leq z \;\&\; v \leq z \;\&\; w \leq z \;\&\; \ldots]]].$$
Similarly, $(x, y, z)[A]$ stands for $(x)[(y)[(z)[A]]]$.

15. $\mathrm{Ax}(x) \equiv x = \alpha_1 \lor x = \alpha_2 \lor \cdots \lor x = \alpha_{13}$.
"x represents an axiom".

16. $\mathfrak{B}(x) \equiv (n)[0 < n \le L(x) \to \{\mathrm{Ax}(n\mathrm{Gl}x)\lor$
$\qquad (Ep, q)[0 < p, q < n \mathrel{\&} p\mathrm{Gl}x, q\mathrm{Gl}x\, Un\mathrm{Gl}x]\}] \mathrel{\&} L(x) > 0$.
"x represents a proof".

17. $x\mathfrak{B}y \equiv \mathfrak{B}(x) \mathrel{\&} L(x)\mathrm{Gl}x = y$.
"x represents a proof and y a formula, and the proof which x represents is a proof of the formula which y represents".

The assertion that the system is free from contradiction can be written as a proposition of arithmetic thus:

$$(x, y, z)[\sim(x\mathfrak{B}z \mathrel{\&} y\mathfrak{B}\mathrm{Neg}(z))]$$

(i.e., for all natural numbers x, y and z, x does not represent a proof of the formula A, and y of $\sim(A)$, where z represents A).

5. Representation of recursive functions by formulas of our formal system

We abbreviate certain formal expressions as follows: z_0 for 0, z_1 for $N(0)$, z_2 for $N(N(0))$, etc. The z's then *represent* the natural numbers in the formal logic. Again, if $\phi(x_1, x_2, \ldots)$ is a function of positive integers, we shall say that the formal functional expression $G(u_1, u_2, \ldots)$ *represents* $\phi(x_1, x_2, \ldots)$ if $G(z_m, z_n, \ldots) = z_{\phi(m,n\ldots)}$ is provable formally for each given set of natural numbers $m, n, \ldots$; in other words, if $G(z_m, z_n, \ldots) = z_k$ is provable formally whenever $\phi(m, n, \ldots) = k$ holds. If the value of $\phi(x_1, x_2, \ldots)$ is independent of some variable x_p, then $G(u_1, u_2, \ldots)$ need not contain the corresponding variable u_p. Similarly, if $R(x_1, x_2, \ldots)$ is a class or relation of natural numbers, we shall say that the formal propositional function $H(u_1, u_2, \ldots)$ *represents* $R(x_1, x_2, \ldots)$ if we can prove formally $H(z_m, z_n, \ldots)$ whenever $R(m, n, \ldots)$ holds, and $\sim H(z_m, z_n, \ldots)$ whenever $R(m, n, \ldots)$ does not hold.

We now sketch a proof that every recursive function, class, and relation is represented by some formula of our formal system.

The recursive function $x + 1$ is represented by $N(w)$, because $N(z_n) = z_{n+1}$ can be proved formally for each natural number n. The proof is immediate, since under our abbreviations z_{n+1} *is* $N(z_n)$. The constant function $f(x_1, x_2, \ldots, x_n) = c$ is represented by z_c, and the identity function $U_j^n(x_1, \ldots, x_n) = x_j$ is represented by u_j.

If $\phi(x_1, \ldots, x_n)$ is compound with respect to $\psi(x_1, \ldots, x_m)$ and $\chi_i(x_1, \ldots, x_n)$, where $i = 1, \ldots, m$, and if $G(w_1, \ldots, w_m)$ represents

$\psi(x_1, \ldots, x_m)$ and $H_i(w_1, \ldots, w_n)$ represents $\chi_i(x_1, \ldots, x_n)$, then

$$G(H_1(w_1, \ldots, w_n), \ldots, H_m(w_1, \ldots, w_n))$$

represents $\phi(x_1, \ldots, x_n)$.

If $\phi(x_1, \ldots, x_n)$ is recursive with respect to $\psi(x_1, \ldots, x_{n-1})$ and $\chi(x_1, \ldots, x_{n+1})$, and if $G(w_1, \ldots, w_{n-1})$ represents $\psi(x_1, \ldots, x_{n-1})$ and $H(w_1, \ldots, w_{n+1})$ represents $\chi(x_1, \ldots, x_{n+1})$, then

$$\epsilon z[\Sigma f\{f(0) = G(w_2, \ldots, w_n) \,\&$$
$$\Pi u[f(N(u)) = H(u, f(u), w_2, \ldots, w_n)] \,\& f(w_1) = z\}]$$

| represents $\phi(x_1, \ldots, x_n)$. This formula (call it $K(w_1, \ldots, w_n)$) intuitively   17
has the desired significance. For each set of natural numbers $w_1, \ldots, w_n$, there is one and only one function f satisfying the conditions

$$f(0) = G(w_2, \ldots, w_n), \quad f(k+1) = H(k, f(k), w_2, \ldots, w_n),$$

and therefore $K(w_1, \ldots, w_n)$ means "The value which the function f satisfying the above conditions takes on for the argument w_1". This value obviously is $\phi(w_1, \ldots, w_n)$. The proof that $K(w_1, \ldots, w_n)$ actually represents $\phi(x_1, \ldots, x_n)$, if G represents ψ and H represents χ, is too long to give here.[21]

If $R(x, y, \ldots)$ is a recursive class or relation, there is a recursive function $\phi(x, y, \ldots)$ such that $\phi(x, y, \ldots) = 0$ if $R(x, y, \ldots)$ and $\phi(x, y, \ldots) = 1$ if $\sim R(x, y, \ldots)$. Then there is a $G(u, v, \ldots)$ which represents $\phi(x, y, \ldots)$. $G(u, v, \ldots) = 0$ represents $R(x, y, \ldots)$. For, if $R(m, n, \ldots)$, then $G(z_m, z_n, \ldots) = z_0 = 0$ is provable formally; and, if $\sim R(m, n, \ldots)$, then $G(z_m, z_n, \ldots) = z_1$, and therefore $\sim[G(z_m, z_n, \ldots) = 0]$, is provable formally.

Because certain metamathematical relations and propositions about our formal system can be expressed by recursive relations and statements [about them], these relations and propositions can be expressed in the formal system. Thus parts of the theory whose object is our formal system can be expressed in the same formal system. This leads to interesting results.

We have noted that $x\mathfrak{B}y$ is a recursive relation; and we can also prove that $\mathfrak{S}(x, y)$ is recursive, where $\mathfrak{S}(x, y)$ is the number of the formula which results when we replace all free ocurrences of w by z_y in the formula whose

---

[21][Note that this proof is not necessary for the demonstration of the existence of undecidable arithmetic propositions in the system considered. For, if some recursive function were not "represented" by the corresponding formula constructed on pp. 16–17 [above, pp. 358–359], this would trivially imply the existence of undecidable propositions unless some wrong proposition on integers were demonstrable.]

number is x. (In fact, if a is the number of w and $\chi(n)$ the number of z_n, $\mathfrak{S}(x,y)$ is $Sb[x^a_{\chi(y)}]$). Hence there is a formula $B(u,v)$ which represents $x\mathfrak{B}y$, and a formula $S(u,v)$ which represents $\mathfrak{S}(x,y)$.

Let $U(w)$ be the formula $\Pi v[{\sim}B(v,S(w,w))]$ and let p be the number of $U(w)$. Now $U(z_p)$ is the formula which results when we replace all free occurrences of w by z_p in the formula whose number is p, and hence has the number $\mathfrak{S}(p,p)$. | Hence, if $U(z_p)$ is provable, there is a k such that $k\mathfrak{B}\mathfrak{S}(p,p)$. But, since $S(u,v)$ represents $\mathfrak{S}(x,y)$ and $B(u,v)$ represents $x\mathfrak{B}y$, it follows that $B(z_k,S(z_p,z_p))$ is provable. Also, it is a property of our system that, if $\Pi vF(v)$ is provable, then $F(z_l)$ is provable for all l; consequently, if $U(z_p)$ is provable, ${\sim}B(z_k,S(z_p,z_p))$, as well as $B(z_k,S(z_p,z_p))$, is provable, and the system contains a contradiction. Thus we conclude that $U(z_p)$ cannot be proved unless the system contains a contradiction.[22]

Next we raise the question of whether ${\sim}U(z_p)$ can be proved if the system is not contradictory. If the system is not contradictory, $U(z_p)$ cannot be proved, as just seen. But $U(z_p)$ is the formula with the number $\mathfrak{S}(p,p)$, so that, for all k, ${\sim}k\mathfrak{B}\mathfrak{S}(p,p)$. Therefore ${\sim}B(z_k,S(z_p,z_p))$ is provable for all k. If furthermore ${\sim}U(z_p)$, i.e., ${\sim}\Pi v[{\sim}B(v,S(z_p,z_p))]$, is provable, then we have that a formula is provable which asserts that ${\sim}B(z_k,S(z_p,z_p))$ is not true for all k, and this, together with the fact that ${\sim}B(z_k,S(z_p,z_p))$ is provable for all k, makes the system intuitively contradictory. In other words, if we consider the system to be contradictory not merely if there is an A such that both A and ${\sim}A$ are provable, but also if there is an F such that all of the formulas ${\sim}\Pi vF(v),F(z_0),F(z_1),\ldots$ are provable, then, if ${\sim}U(z_p)$ is provable, the system is contradictory in this weaker sense. Hence neither $U(z_p)$ nor ${\sim}U(z_p)$ is provable, unless the system is contradictory.

If our system is free from contradiction in the strong sense (i.e., if A and ${\sim}A$ are not both provable for any A), then $U(z_p)$ is not provable. But

$$(x,y,z)[{\sim}\{x\mathfrak{B}y \ \& \ z\mathfrak{B}\,\mathrm{Neg}\,y\}]$$

is a statement that our system is free from contradiction in the strong sense. Hence we have shown that

$$(x,y,z)[{\sim}\{x\mathfrak{B}y \ \& \ z\mathfrak{B}\mathrm{Neg}\,y\}] \to (x){\sim}x\mathfrak{B}\mathfrak{S}(p,p).$$

---

[22][This version of the argument is along the lines of one used by Herbrand in an informal exposition (*1931*, p. 7). It is a little shorter than my original proof in my *1931* [Theorem VI]. However, in order to make it completely precise, a few words would have to be added about the properties of the symbolism used to denote the formulas of the system. Note that "$\Pi v[{\sim}B(v,S(z_p,z_p))]$" *is not* the undecidable sentence, but only *denotes* it.]

The fairly simple arguments of this proof can be paralleled in the formal logic to give a formal proof of

$$\text{Contrad} \to \Pi v[\sim B(v, S(z_p, z_p))],$$

| where Contrad is a formula of the system which expresses the proposition 19 $(x, y, z)[\sim\{x\mathfrak{B}y \ \& \ z\mathfrak{B}\text{Neg} \, y\}]$. Then, if Contrad could also be proved formally, we could use Rule I to infer $\Pi v[\sim B(v, S(z_p, z_p))]$ or $U(z_p)$, in which case, as we have seen, the system would contain a contradiction. Hence Contrad cannot be proved in the system itself, unless the system contains a contradiction.

6. Conditions that a formal system must satisfy in order that the foregoing arguments apply

Now consider any formal system (in the sense of §1) satisfying the following five conditions:

(1) Supposing the symbols and formulas to be numbered in a manner similar to that used for the particular system considered above, then the class of axioms and the relation of immediate consequence shall be recursive.

This is a precise [condition which in practice suffices as a substitute for the unprecise] requirement of §1 that the class of axioms and relation of immediate consequence be constructive.

(2) There shall be a certain sequence of meaningful formulas z_n (standing for the natural numbers n) such that the relation between n and the number representing z_n is recursive.

(3) There shall be a symbol $\sim$ (negation) and two symbols v and w (variables) such that, to every recursive relation of two variables, there corresponds a formula $R(v, w)$ of the system such that $R(z_p, z_q)$ is provable if the relation holds of p and q, and $\sim R(z_p, z_q)$ is provable if the relation does not hold of p and q; or, instead of a single symbol $\sim$, there may be a formula $F(x)$ not containing v or w such that the foregoing holds when $\sim(A)$ stands for the formula $F(A)$.

The formulas $R(v, w)$ which represent recursive relations, and their negations $\sim R(v, w)$, shall be called *recursive propositional functions of two variables*; and $R(v, z_n)$ and $\sim R(v, z_n)$ *recursive propositional functions of one variable*.

(4) There shall be a symbol Π such that, if $\Pi v F(v)$ is provable for a recursive propositional function $F(v)$ of one variable, then $F(z_k)$ shall be provable for all k; | or, instead of a single symbol Π, there may be a formula 20 $G(x)$ not containing w such that the foregoing holds when $\Pi v F(v)$ stands for $G(F(v))$.

(5) The system shall be free from contradiction in the two following senses:

    (a) If $R(v, w)$ is a recursive propositional function of two variables, then $R(z_p, z_q)$ and $\sim R(z_p, z_q)$ shall not both be provable.

    (b) If $F(v)$ is a recursive propositional function of one variable, then the formulas $\sim\Pi v F(v), F(z_0), F(z_1), F(z_2), \ldots$ shall not all be provable.

Now, using condition (1), $x\mathfrak{B}y$, $\mathfrak{S}(x, y)$ (defined as before) and $k\mathfrak{B}\mathfrak{S}(l, l)$ are recursive. Then, by (3), there is an $R(v, w)$ such that $R(z_k, z_l)$ is provable if $k\mathfrak{B}\mathfrak{S}(l, l)$, and $\sim R(z_k, z_l)$ is provable if $\sim k\mathfrak{B}\mathfrak{S}(l, l)$. Noting that $R(v, w)$ plays the same role as $B(v, S(w, w))$ in our special system, we can prove by reasoning similar to that of §5 that, if p is the number of $\Pi v \sim R(v, w)$, (5a) implies that $\Pi v \sim R(v, z_p)$ is not provable, and (5b) [in conjunction with (5a)] implies that $\sim\Pi v \sim R(v, z_p)$ is not provable. Also, as before,

$$(x, y, z)[\sim\{x\mathfrak{B}y \ \& \ z\mathfrak{B}\mathrm{Neg}\, y\}] \to (x)\sim x\mathfrak{B}\mathfrak{S}(p, p)$$

can be established. We shall not list the further conditions under which it is possible to convert the intuitive proof of this into a formal proof of Contrad $\to \Pi v \sim R(v, z_p)$ (Contrad defined as before). However, they are conditions satisfied by all systems of the type under consideration which contain a certain amount of ordinary arithmetic, and these systems therefore cannot contain a proof of their own freedom from contradiction.

7. Relation of the foregoing arguments to the paradoxes

We have seen that in a formal system we can construct statements about the formal system, of which some can be proved and some cannot, according to what they say about the system. We shall compare this fact with the famous Epimenides paradox ("Der Lügner"). Suppose that on 4 May 1934, A makes the single statement, "Every statement which A makes on 4 May 1934 is false." This statement clearly cannot be true. Also it cannot be false, since the only way for it to be false is for A to | have made a true statement in the time specified and in that time he made only the single statement.

The solution suggested by Whitehead and Russell, that a proposition cannot say something about itself, is too drastic. We saw that we can construct propositions which make statements about themselves, and, in fact, these are arithmetic propositions which involve only recursively defined functions, and therefore are undoubtedly meaningful statements. It is even possible, for any metamathematical property f which can be expressed in the system, to construct a proposition which says of itself that it

has this property.[23] For suppose that $F(z_n)$ means that n is the number of a formula that has the property f. Then, if $F(S(w, w))$ has the number p, $F(S(z_p, z_p))$ says that it itself has the property f.[24] This construction can only be carried out if the property f can be expressed in the system, and the solution of the Epimenides paradox lies in the fact that the latter is not possible for every metamathematical property. For consider the above statement made by A. A must specify a language B and say that every statement that he made in the given time was a false statement in B. But "false statement in B" cannot be expressed in B, and so his statement was in some other language, and the paradox disappears.

The paradox can be considered as a proof that "false statement in B" cannot be expressed in B.[25] We shall now establish this fact in a more formal manner, and in doing so obtain a heuristic argument for the existence of undecidable propositions. Suppose that $T(z_n)$ means that the formula whose number is n is true. That is, if n is the number of N, $T(z_n)$ shall be equivalent to N. Then we could apply our procedure to $\sim T(S(w, w))$, obtaining $\sim T(S(z_p, z_p))$, which says that it is itself false, and this leads to a contradiction similar to the "Epimenides". But, on the other hand, $\Sigma v B(v, z_k)$ is a statement in the system of the fact that the formula with number k is provable. So we see that the class α of numbers of true | formulas cannot be expressed by a propositional function of our system, whereas the class β of provable formulas can. Hence $\alpha \neq \beta$ and if we assume $\beta \subseteq \alpha$ (i.e., every provable formula is true) we have $\beta \subset \alpha$, i.e., there is a proposition A which is true but not provable. $\sim A$ then is not true and therefore not provable either, i.e., A is undecidable.[26]

22

8. Diophantine equivalents of undecidable propositions

Suppose $F(x_1, \ldots, x_n)$ a polynomial with integral coefficients. By use of logical quantifiers (x) and (Ex),[27] we can make certain statements about

---

[23][This was first noted by Carnap in *1934a*, p. 91.]

[24]Of course we can find properties f such that $F(S(z_p, z_p))$ is provable, just as we found ones for which it was not provable.

[25][For a closer examination of this fact see A. Tarski's papers *1933a* (in particular p. 247 ff. of the translation in *1956*) and *1944*. In these two papers the concept of truth relating to sentences of a language is discussed systematically. See also *Carnap 1934*.]

[26][Note that this argument can be carried through with full precision for any system whose formulas have a well-defined meaning, provided the axioms and rules of inference are correct for this meaning, and arithmetic is contained in the system. One thus obtains a proof for the *existence* of undecidable propositions in that system, but no individual instance of an undecidable proposition.]

[27]Where x is any variable for a natural number.

the solutions in natural numbers of the Diophantine equation $F(x_1, \ldots, x_n)$ $= 0$. Thus $(Ex_1)(Ex_2)\ldots(Ex_n)(F(x_1,\ldots,x_n) = 0)$ says that there is a solution; $(x_3)(Ex_1)(Ex_2)(Ex_4)\ldots(Ex_n)(F(x_1,\ldots,x_n) = 0)$ says that, for any assigned value of x_3, the resulting equation has a solution; etc. We wish to prove that there is a sequence of logical quantifiers, say (P), and a Diophantine equation, $F = 0$, such that our undecidable proposition is equivalent to $(P)(F = 0)$.

To prove this we find it convenient to make use of the intermediary concept of an *arithmetical* expression, that is, an expression built up out of $\sim$, $\vee$, &, $\rightarrow$, $\equiv$, $+$, $\times$, $=$, natural numbers, variables running over natural numbers, and the quantifiers (x) and (Ex),[27] according to the following induction:

1. If f and g are built up out of variables, natural numbers, $+$, and $\times$,[28] then $f = g$ is an arithmetical expression.

2. If A and B are arithmetical expressions, then $\sim A$, $A \vee B$, A & B, $A \rightarrow B$, and $A \equiv B$ are arithmetical expressions.

3. If A is an arithmetical expression which contains x as a free variable, then $(x)A$ and $(Ex)A$ are arithmetical expressions.

23   We shall prove first that, if $\phi(x_1, \ldots, x_n)$ is recursive, then there is | an arithmetical expression $A(x_1, \ldots, x_n, y)$ such that

$$A(x_1, \ldots, x_n, y) . \equiv . \phi(x_1, \ldots, x_n) = y;$$

and second that, if $B(x_1, \ldots, x_n)$ is an arithmetical expression, then there are polynomials $Q(x_1, \ldots, x_n, y_1, \ldots, y_m)$ and $R(x_1, \ldots, x_n, y_1, \ldots, y_m)$ with natural number coefficients and a sequence (P) of quantifiers such that

$$B(x_1, \ldots, x_n) . \equiv . (P)[Q(x_1, \ldots, x_n, y_1, \ldots, y_m)$$
$$= R(x_1, \ldots, x_n, y_1, \ldots, y_m)],$$

where the x's and y's range over the natural numbers. Since our undecidable proposition has the form $(x)F(x)$ where F is recursive, there is a recursive function $\phi(x)$ such that our undecidable proposition is equivalent to $(x)[\phi(x) = 0]$. Then there is an arithmetical expression $A(x, y)$ such that $\phi(x) = y . \equiv . A(x, y)$, and there are polynomials $Q(x, y, z_1, \ldots, z_m)$ and $R(x, y, z_1, \ldots, z_m)$ with natural number coefficients, and a sequence of quantifiers (P), such that

$$A(x, y) . \equiv . (P)[Q(x, y, z_1, \ldots, z_m) = R(x, y, z_1, \ldots, z_m)].$$

---

[28]That is, if f and g are polynomials with natural number coefficients.

Then our undecidable proposition is equivalent to

$$(x)(P)[Q(x, 0, z_1, \ldots, z_m) = R(x, 0, z_1, \ldots, z_m)].$$

We prove first that recursive functions are expressible arithmetically.
If $f(x) = x + 1$, then $f(x) = y . \equiv . x + 1 = y$.
If $f(x_1, \ldots, x_n) = c$, then $f(x_1, \ldots, x_n) = w . \equiv . w = c$.
Similarly for the identity functions $U_j^n(x_1, \ldots, x_n)$.
If

$$\psi(x_1, \ldots, x_m) = y . \equiv . A(x_1, \ldots, x_m, y),$$
$$\chi_i(x_1, \ldots, x_n) = y . \equiv . B_i(x_1, \ldots, x_n, y),$$

and

$$\phi(x_1, \ldots, x_n) = \psi(\chi_1(x_1, \ldots, x_n), \ldots, \chi_m(x_1, \ldots, x_n)),$$

then

$$\phi(x_1, \ldots, x_n) = y . \equiv . (Et_1) \ldots (Et_m)[B_1(x_1, \ldots, x_n, t_1) \And \cdots \And$$
$$B_m(x_1, \ldots, x_n, t_m) \And A(t_1, \ldots, t_m, y)].$$

To handle the case where ϕ is recursive with respect to ψ and χ, we require an arithmetical expression for $\beta(c, d, i) = y$, where $\beta(c, d, i)$ is a certain function which has the property that, if a function $f(i)$ of natural numbers and a natural number l are given, then natural numbers c and d such that $\beta(c, d, i) = f(i)$ $(i = 0, \ldots, l)$ can be found. We may define $x \equiv y \,(\text{mod } z)$ as $\mid (Et)[x = y + tz \lor y = x + tz]$, and $x \geq y$ as $(Et)[x = y + t]$.[29] 24
Then we define $\beta(c, d, i)$ to be the least non-negative residue of c modulo $1 + (i + 1)d$, i.e., $\beta(c, d, i) = z . \equiv . z \equiv c \,(\text{mod}\,[1 + (i + 1)d]) \And z \leq (i + 1)d$.
To prove that $\beta(c, d, i)$ has the aforesaid property, suppose $f(i)$ and l given. Choose s greater than all of the numbers $l, f(0), f(1), \ldots, f(l)$. Then the numbers $1 + s!, 1 + 2s!, \ldots, 1 + (l + 1)s!$ are relatively prime. For, if a prime number divides two of them, it divides their difference $(i - j)s!$; but it cannot divide $s!$, since it divides $1 + is!$; then also it cannot divide $i - j$, since $i - j \leq l < s$ and hence $i - j$ is a factor of $s!$. Then, if we let $d = s!$, we can find a c such that $c \equiv f(i) \,(\text{mod}\,[1 + (i + 1)d])$ $(i = 0, \ldots, l)$ since $1 + s!, \ldots, 1 + (l + 1)s!$ are relatively prime. Since $s > f(i)$ and therefore $1 + (i + 1)s! > f(i)$, we have $f(i) = \beta(c, d, i)$, as was to be shown.
If

$$\psi(x_1, \ldots, x_{n-1}) = y . \equiv . A(x_1, \ldots, x_{n-1}, y),$$
$$\chi(x_1, \ldots, x_{n+1}) = y . \equiv . B(x_1, \ldots, x_{n+1}, y),$$
$$\phi(0, x_2, \ldots, x_n) = \psi(x_2, \ldots, x_n),$$

[29]If we were allowing the variables to run over the integers instead of just the natural numbers, we could define $x \geq y$ as $(Et_1) \ldots (Et_4)[x = y + t_1^2 + t_2^2 + t_3^2 + t_4^2]$, since every positive integer is the sum of four squares.

and

$$\phi(k+1, x_2, \ldots, x_n) = \chi(k, \phi(k, x_2, \ldots, x_n), x_2, \ldots, x_n),$$

then

$$\phi(x_1, \ldots, x_n) = y . \equiv . (Ef)[A(x_2, \ldots, x_n, f(0)) \& (t)\{t+1 \le x_1 \rightarrow$$
$$B(t, f(t), x_2, \ldots, x_n, f(t+1))\} \& f(x_1) = y].$$

But, if there is an f satisfying the condition in square brackets, then there is a c and a d such that $\beta(c, d, i) = f(i)$ $(i = 0, \ldots, x_1)$ and therefore

$$(Ec)(Ed)[A(x_2, \ldots, x_n, \beta(c, d, 0)) \& (t)\{t+1 \le x_1 \rightarrow$$
$$B(t, \beta(c, d, t), x_2, \ldots, x_n, \beta(c, d, t+1))\} \& \beta(c, d, x_1) = y].$$

Conversely, this obviously implies the original expression. The latter formula can be transformed into the arithmetical one

$$(Ec)(Ed)[(Ev)\{A(x_2, \ldots, x_n, v) \& v = \beta(c, d, 0)\} \& (t)\{t+1 \le x_1 \rightarrow$$
$$(Ev)(Ew)[B(t, v, x_2, \ldots, x_n, w) \& v = \beta(c, d, t) \& w = \beta(c, d, t+1)]\}$$
$$\& y = \beta(c, d, x_1)]$$

25 |

by substituting $(Ev)[A(x_2, \ldots, x_n, v) \& v = \beta(c, d, 0)]$ for $A(x_2, \ldots, x_n, \beta(c, d, 0))$ and $(Ev)(Ew)[B(t, v, x_2, \ldots, x_n, w) \& v = \beta(c, d, t) \& w = \beta(c, d, t+1)]$ for $B(t, \beta(c, d, t), x_2, \ldots, x_n, \beta(c, d, t+1))$. This completes the proof that all recursive functions are arithmetical.

We next show that all arithmetical expressions can be given the equivalent normal form $(P)[Q = R]$ where Q and R are polynomials with natural number coefficients.

If $\sim$, $\lor$, $\&$, $\rightarrow$, $\equiv$, and quantifiers do not occur in an arithmetical expression, then it has the required normal form by definition (page 22 [above, page 364]).

Suppose that $A \equiv (P)[Q = R]$, where x does not occur in the quantifiers denoted by (P). Then $(x)A . \equiv . (x)(P)[Q = R]$ and $(Ex)A . \equiv . (Ex)(P)[Q = R]$.

Suppose also that $B \equiv (P')[Q' = R']$, where the variables of (P') are distinct from those of (P). Then, owing to the fact that

$$p \lor (Ex)F(x) . \equiv . (Ex)[p \lor F(x)]$$

and

$$p \lor (x)F(x) . \equiv . (x)[p \lor F(x)],$$

we have

$$A \lor B . \equiv . (P)(P')[Q = R \lor Q' = R']$$
$$\equiv . (P)(P')[Q - R = 0 \lor Q' - R' = 0]$$
$$\equiv . (P)(P')[(Q - R)(Q' - R') = 0]$$
$$\equiv . (P)(P')[QQ' + RR' = Q'R + QR'].$$

Moreover $\sim A . \equiv . \sim(P)[Q = R]$. Then, since $\sim(x)p . \equiv . (Ex)\sim p$ and $\sim(Ex)p . \equiv . (x)\sim p$, we can shift the negation sign through the prefix (P) and find a P'' such that

$$\sim A . \equiv . (P'')[\sim Q = R]$$
$$\equiv . (P'')[(Q - R)^2 > 0]$$
$$\equiv . (P'')[Q^2 + R^2 \geq 2QR + 1]$$
$$\equiv . (P'')(Et)[Q^2 + R^2 = 2QR + t + 1].$$

$\&$, $\rightarrow$, and $\equiv$ are expressible by means of $\sim$ and $\vee$.

If the argument is modified slightly, the variables can be allowed to run over the integers instead of just the natural numbers.

| Thus there exists a statement about the solutions of a Diophantine equa- 26 tion which is not decidable in our formal system. It can be shown that it is decidable in the next higher type, but there is another such statement which is not decidable even in that type, but which is decidable by going into the next higher type; and so on [including transfinite iterations describable in set theory, such as occur, e.g., in the higher axioms of infinity]. In other words, [on the basis of the principles of proof used in mathematics today[29a]] there can be no complete theory of Diophantine analysis[, not even of the problems of the form $(P)[F = 0]$].[30]

Presburger has given a set of axioms for the relations built up out of $+$, $=$, and logical symbols, together with a method of deciding such relations.[31] Skolem has sketched a method of deciding relations constructed similarly using $\times$ instead of $+$.[32] [However, on the basis of the principles of proof used in mathematics today, no general method of deciding relations in which both $+$ and $\times$ occur can be established, since (as shown above) there can be, on this basis, no complete theory of the Diophantine problems of the form $(P)[F = 0]$.][30]

[29a][Note that the axioms about *all* sets or about classes of sets that are assumed today do not carry any farther, because they are assumed to hold also for the sets of some definite type (or "rank", according to current terminology). See *Levy 1960a*. The principles of proof of intuitionistic mathematics are not taken into account here, because, at any rate up to now, they have proved weaker than those of classical mathematics.]

[30][By a complete theory of some class of problems on the basis of certain principles of proof we here mean a theorem, demonstrable on this basis, which states that (and how) the solution of any problem of the class can be obtained on this basis. For a different and more definitive version of this incompleteness result see the postscript.]

[31]*Presburger 1930.*

[32]*Skolem 1931.*

9. General recursive functions

If $\psi(y)$ and $\chi(x)$ are given recursive functions, then the function $\phi(x, y)$, defined inductively by the relations

$$\phi(0, y) = \psi(y),$$
$$\phi(x + 1, 0) = \chi(x),$$
$$\phi(x + 1, y + 1) = \phi(x, \phi(x + 1, y)),$$

is not in general recursive in the limited sense of §2. This is an example of a definition by induction with respect to two variables simultaneously.[33]

To get arithmetical definitions of such functions, we have to generalize our β function. The consideration of various sorts of functions defined by inductions leads to the question what one would mean by "every recursive function".

One may attempt to define this notion as follows: If ϕ denotes an unknown function, and $\psi_1, \ldots, \psi_k$ are known functions, and if the ψ's and ϕ are substituted in one another in the most general fashions and certain pairs of the resulting expressions are equated, then, if the resulting set of functional equations has one and only one solution for ϕ, ϕ is a recursive function.[34]

Thus we might have

$$\phi(x, 0) = \psi_1(x),$$
$$\phi(0, y + 1) = \psi_2(y),$$
$$\phi(1, y + 1) = \psi_3(y),$$
$$\phi(x + 2, y + 1) = \psi_4(\phi(x, y + 2), \phi(x, \phi(x, y + 2))).$$

27 | We shall make two restrictions on Herbrand's definition. The first is that the left-hand side of each of the given functional equations defining ϕ shall be of the form

$$\phi(\psi_{i1}(x_1, \ldots, x_n), \psi_{i2}(x_1, \ldots, x_n), \ldots, \psi_{il}(x_1, \ldots, x_n)).$$

---

[33][For a very similar function W. Ackermann in *1928* proved that it cannot be defined by recursion with respect to one variable.]

[34]This was suggested by Herbrand in a private communication. [A slightly different definition was given by him in his *1931*, p. 5, where he postulated "computability". However, also in this definition, he did not require computability by any definite formal rules (note the phrase "considérées intuitionnistiquement" and footnote 5). In intuitionistic mathematics the two Herbrand definitions are trivially equivalent. In classical mathematics the non-equivalence of general recursiveness with the first mentioned concept of Herbrand was proved by L. Kalmár in *1955*. Whether Herbrand's second concept is equivalent with general recursiveness is a largely epistemological question which has not yet been answered. See the postscript.]

The second (as stated below) is equivalent to the condition that all possible sets of arguments $(n_1, \ldots, n_l)$ of ϕ can be so arranged that the computation of the value of ϕ for any given set of arguments $(n_1, \ldots, n_l)$ by means of the given equation requires a knowledge of the values of ϕ only for sets of arguments which precede $(n_1, \ldots, n_l)$.

From the given set of functional equations, we define by induction a set of derived equations, thus:

(1a) Any expression obtained by replacing all the variables of one of the given equations by natural numbers shall be a derived equation.

(1b) $\psi_{ij}(k_i, \ldots, k_n) = m$ shall be a derived equation if $k_1, \ldots, k_n$ are natural numbers and $\psi_{ij}(k_1, \ldots, k_n) = m$ is a true equality.

(2a) If $\psi_{ij}(k_1, \ldots, k_n) = m$ is a derived equation, the equality obtained by substituting m for an occurrence of $\psi_{ij}(k_1, \ldots, k_n)$ in a derived equation shall be a derived equation.

(2b) If $\phi(k_1, \ldots, k_l) = m$ is a derived equation, where $k_1, \ldots, k_l$, m are natural numbers, the expression obtained by substituting m for an occurrence of $\phi(k_1, \ldots, k_l)$ on the right-hand side of a derived equation shall be a derived equation.

Now our second restriction on Herbrand's definition of recursive function is that for each set of natural numbers $k_1, \ldots, k_l$ there shall be one and only one m such that $\phi(k_1, \ldots, k_l) = m$ is a derived equation.

Using this definition of the notion of a recursive function, we can prove that, if $\phi(x_1, \ldots, x_l)$ is recursive, there is an arithmetical expression $A(x_1, \ldots, x_l, y)$ such that $\phi(x_1, \ldots, x_l) = y . \equiv . A(x_1, \ldots, x_l, y)$.

Postscriptum
(3 June 1964)

In consequence of later advances, in particular of the fact that, due to A. M. Turing's work, a precise and unquestionably adequate definition of the general concept of formal system can now be given, the existence of undecidable arithmetical propositions and the non-demonstrability of the consistency of a system in the same system can now be proved rigorously for *every* consistent formal system containing a certain amount of finitary number theory.

Turing's work gives an analysis of the concept of "mechanical procedure" (alias "algorithm" or "computation procedure" or "finite combinatorial procedure"). This concept is shown to be equivalent with that of a

"Turing machine".[35] A formal system can simply be defined to be any mechanical procedure for producing formulas, called provable formulas. For any formal system in this sense there exists one in the sense of page 346 above that has the same provable formulas (and likewise vice versa), provided the term "finite procedure" occurring on page 346 is understood to mean "mechanical procedure". This meaning, however, is required by the concept of formal system, whose essence it is that reasoning is completely replaced by mechanical operations on formulas. (Note that the question of whether there exist finite *non-mechanical* procedures,[36] not equivalent with any algorithm, has nothing whatsoever to do with the adequacy of the definition of "formal system" and of "mechanical procedure".)

On the basis of the definitions just mentioned, condition (1) in §6 becomes superfluous, because for any formal system provability is a predicate of the form $(Ex)x\mathfrak{B}y$, where $\mathfrak{B}$ is primitive recursive. Moreover, the two incompleteness results mentioned in the end of §8 can now be proved in the definitive form: "There exists no *formalized theory* that answers all Diophantine questions of the form $(P)[F = 0]$", and "There is no *algorithm* for deciding relations in which both $+$ and $\times$ occur". (For theories and procedures in the more general sense indicated in footnote 36, the situation may be different. Note that the results mentioned in this postscript do not establish any bounds for the powers of human reason, but rather for the potentialities of pure formalism in mathematics.) Thirdly, if "finite procedure" is understood to mean "mechanical procedure", the question raised in footnote 3 can be answered affirmatively for recursiveness as defined in §9, which is equivalent with general recursiveness as defined today (see *Kleene 1936*, page 730, and *1952*, pages 220ff, 232ff).

As for the elimination of ω-consistency (first accomplished by J. B. Rosser (*1936*)), see *Tarski, Mostowski and Robinson 1953*, page 49, Corollary 2. The proof of the unprovability in the same system of the consistency of a system was carried out for number theory in *Hilbert and Bernays 1939*, pages 297–324. The proof carries over almost literally to any system containing, among its axioms and rules of inference, the axioms and rules of inference of number theory. As to the consequence for Hilbert's program, see my *1958* and the material cited there. See also *Kreisel 1958*.

By slightly strengthening the methods used above in §8, it can easily be accomplished that the prefix of the undecidable proposition consists of only one block of universal quantifiers followed by one block of existential

---

[35]See *Turing 1937*, p. 249, and the almost simultaneous paper by E. L. Post (*1936*). As for previous equivalent definitions of computability, which, however, are much less suitable for our purpose, see *Church 1936*, pp. 356–358. One of those definitions is given in §9 of these lectures.

[36]I.e., such as involve the use of abstract terms on the basis of their meaning. See my *1958*.

quantifiers, and that, moreover, the degree of the polynomial is 4 (⟦hitherto⟧ unpublished result).

A number of misprints and oversights in the original mimeographed lecture notes were corrected when they were reprinted in *Davis 1965*. I am indebted to Professor Martin Davis for calling my attention to some of them.

Besprechung von *Skolem 1933*:
Ein kombinatorischer Satz mit Anwendung
auf ein logisches Entscheidungsproblem
(*1934a*)

Es handelt sich um folgenden von F. P. Ramsey in *1929* bewiesenen Satz: Sind n, μ, ν drei beliebige natürliche Zahlen, so kann man eine natürliche Zahl N bestimmen derart, daß, wenn man die sämtlichen ν-Tupel aus den Elementen einer mindestens N-elementigen Menge M irgendwie in μ fremde Systeme $C_1, C_2, \ldots, C_\mu$ verteilt, immer eine n-elementige Teilmenge von M existiert, so daß die aus den Elementen dieser Teilmenge gebildeten ν-Tupel alle zu demselben C_i gehören. Der Beweis des Verfassers ist einfacher als der Ramseysche und liefert auch kleinere Werte für N. F. P. Ramsey hatte auf Grund dieses Satzes das Entscheidungs|problem für Formeln des engeren Funktionenkalküls, die in der Normalform keine Existenzzeichen haben, aber das Gleichheitszeichen enthalten dürfen, gelöst. Diese Anwendung wird vom Verfasser in etwas anderer und verkürzter Form gebracht, indem bewiesen wird, daß man zu jeder Formel F der genannten Art (sie möge n Allzeichen enthalten) eine Zahl N so bestimmen kann, daß F in einem Individuenbereich mit N oder mehr Elementen dann und nur dann erfüllbar ist, wenn sie im Individuenbereich von $2n - 1$ Elementen $a_1, a_2, \ldots, a_{2n-1}$ erfüllbar ist durch ein Modell mit einer gewissen Symmetrieeigenschaft. Die letztere besteht darin, daß für jede Funktion ϕ des Modells $\phi(a_{i_1}, a_{i_2}, \ldots, a_{i_k}) \equiv \phi(a_{i'_1}, a_{i'_2}, \ldots, a_{i'_k})$ gelten soll, wenn zugleich mit $i_p > i_q$, bzw. $i_p = i_q$, bzw. $i_p < i_q$, stets auch $i'_p > i'_q$, bzw. $i'_p = i'_q$, bzw. $i'_p < i'_q$ ist (für $p, q = 1, 2, \ldots, k$).

98

Review of *Skolem 1933*:
A combinatorial theorem with an application
to a decision problem in logic
(*1934a*)

This article concerns the following theorem, which was proved by F. P. Ramsey in his *1929*: If n, μ and ν are three arbitrary natural numbers, a natural number N can be determined such that, if all ν-tuples of the elements of a set M containing at least N elements are partitioned in any way into μ disjoint systems $C_1, C_2, \ldots, C_\mu$, there always exists an n-element subset of M such that all of the ν-tuples formed from elements of this subset belong to the same C_i. The author's proof is simpler than Ramsey's and also furnishes smaller values for N. On the basis of this theorem, Ramsey had solved the decision problem for formulas of the restricted functional calculus that in normal form have no existential quantifier but may contain the identity sign. This application is presented by the author in a somewhat different and abbreviated form, in that it is proved that, for each formula F of the kind mentioned (assumed to contain n universal quantifiers), a number N can be determined so that F is satisfiable in a domain of individuals with N or more elements if and only if it is satisfiable in the domain containing as its $2n - 1$ elements the individuals $a_1, a_2, \ldots, a_{2n-1}$ by a model with a certain symmetry property. The latter consists in that, for every function ϕ of the model, $\phi(a_{i_1}, a_{i_2}, \ldots, a_{i_k}) \equiv \phi(a_{i'_1}, a_{i'_2}, \ldots, a_{i'_k})$ is to hold if, according as $i_p > i_q$, $i_p = i_q$ or $i_p < i_q$ holds, then $i'_p > i'_q$, $i'_p = i'_q$ or $i'_p < i'_q$ holds, respectively (for $p, q = 1, 2, \ldots, k$).

Introductory note to *1934b*

In this review of my earliest article but one, Gödel's last sentence can be puzzling. Given just μ classes, and $\nu > \mu$, how can Gödel speak of a product of ν of them? The answer lies in the formalism of *Principia*

Besprechung von *Quine 1933*:
A theorem in the calculus of classes
(*1934b*)

Die folgende Formel des Klassenkalküls

$$ab + ac + bc = (a + b)(a + c)(b + c)$$

wird verallgemeinert zu dem Satz: Ist μ eine endliche Zahl und $\nu \leq \mu$, so ist die Summe aller möglichen Produkte aus je ν von μ Klassen gleich dem Produkt aller möglichen Summen aus je $\mu - \nu + 1$ dieser μ Klassen. Der Beweis wird in der Symbolik der *Principia mathematica* rein formal durchgeführt unter alleiniger Verwendung von Sätzen, die in den *Principia* formal bewiesen sind. Ebenso wird gezeigt, daß für $\nu > \mu$ das Theorem falsch ist.

mathematica. Definitions are couched in terms of variables whose values are not restricted to the ranges for which the definitions were wanted, and meanings thus accrue to the waste cases that made no intuitive sense. They are the "don't-cares", but evidently I cared a little, and Gödel indulged me.

W. V. Quine

Review of *Quine 1933*: A theorem in the calculus of classes (*1934b*)

The following formula of the calculus of classes,

$$ab + ac + bc = (a + b)(a + c)(b + c),$$

is generalized to the theorem: If μ is a finite number and $\nu \leq \mu$, then the sum of all possible products from any ν of μ classes is equal to the product of all possible sums from any $\mu - \nu + 1$ of these μ classes. The proof is carried out purely formally in the symbolism of *Principia mathematica*, using only theorems that are formally proved in *Principia*. It is also shown that the theorem is false for $\nu > \mu$.

Introductory note to *1934c* and *1935*

These are reviews by Gödel of Skolem's papers, *1933a* and *1934*, on non-standard models of arithmetic. As Gödel remarks, Skolem's two papers are essentially one. This work of Skolem's was one of six or seven fundamental and wide-ranging contributions which he made to logic in the years 1919–1934. Skolem's contributions were honored then, but they look still better as time goes on.

The result in *Skolem 1933a* and *1934* can be stated as follows. Put $N = \{1, 2, 3 \ldots\}$ and let $\mathfrak{N}$ be the structure $(N, +, \cdot, <)$ or, more generally (in the second paper), let $\mathfrak{N} = (N, +, \cdot, <, o_1, o_2, \ldots, o_n, \ldots)$, where each o_i is a finitary operation over N. Then

(1)   *there exists a structure $\mathfrak{N}^*$ not isomorphic to $\mathfrak{N}$ which has the same true (first-order) sentences as $\mathfrak{N}$.*

(We do not say that $\mathfrak{N}^*$ can be taken to be countable, since that followed from (1) by the submodel-form of the Löwenheim–Skolem theorem, even in 1933.) In fact, a particular model $\mathfrak{N}^*$ is described as follows: a certain infinite subset G of N is specified (our G is the range of Skolem's g) in a way we describe later. Now, if f and f' on N to N are any definable functions (that is, first-order definable in $\mathfrak{N}$), put fEf' if and only if $f(n) = f'(n)$ for almost all (that is, all but finitely many) $n \, \epsilon \, G$. The set N^* consists of all f/E such that f is definable. The $+$, $\cdot$, $<$, and o_i for $\mathfrak{N}^*$ are defined in the same 'almost all' way. Then (2) this $\mathfrak{N}^*$ is as in (1). The set G is obtained by applying to a certain list $(f_1, \ldots, f_k, \ldots)$ of functions Skolem's Theorem 1 (not hard to prove), which says about any such list: There is an infinite set $G \subseteq N$ such that whenever $i < j$ either $f_i(n) < f_j(n)$ for almost all $n \, \epsilon \, G$, or the same holds for $=$, or the same holds for $>$.

The summary just given of Skolem's work is almost identical with that in Gödel's review (*1934c*). Gödel's precise and simple writing style is in contrast here with Skolem's. For example, in these papers Skolem does not use the word 'definable' and replaces it in each paper with a different discussion over a page long!

Today we can also state Skolem's results as follows: Take D to be any non-principal ultrafilter on N. (To get exactly the same $\mathfrak{N}^*$ as above, take a D which contains all subsets U of N such that G is almost included in U.) Form the ultrapower $\mathfrak{N}^N/D$ and take for $\mathfrak{N}^*$ its elementary substructure consisting of all f/D such that f is definable in $\mathfrak{N}$ (or, what is the same, containing all elements definable in $\mathfrak{N}^N/D$ from the single parameter I/D, where I is the identity function on N).

Skolem infers from (1) the corollary that no finite axiom system can characterize $\mathfrak{N}$ up to isomorphism. Gödel states a similar consequence of (1) but says simply "no axiom system", which presumably means no recursive (or possibly primitive recursive) set of axioms. Then Gödel remarks that this consequence of (1) follows from his *1931 incompleteness* theorem. Indeed, if Σ is any, say, recursive, set of sentences true in $\mathfrak{N}$, the incompleteness theorem tells us there is even a model $\mathfrak{N}'$ of Σ in which some sentence true in $\mathfrak{N}$ is false. (As Kleene (*1952*, page 430) notes, this argument in fact also uses Gödel's completeness theorem from *1930*.)

However, the main result of Skolem's paper (as Skolem and Gödel both say) is certainly (1), that the set of *all* true sentences in $\mathfrak{N}$ does not characterize $\mathfrak{N}$. And the strange fact is that nowadays (1) is proved in a few lines by a "compactness argument", that is, from Gödel's compactness theorem (*1930*). (To do so, take a new individual constant c and form the set W containing all sentences true in $\mathfrak{N}$ plus all the sentences $c \neq 1$, $c \neq 1 + 1, \ldots$. Every finite subset of W has a model; so W has a model, which is clearly not isomorphic to $\mathfrak{N}$.) Thus it seems extraordinary that Skolem and especially Gödel himself did not observe that (1) is a simple consequence of the compactness theorem. Nevertheless, it appears that the idea of such simple (but important) applications of the compactness theorem was probably unknown before 1936—in particular to Gödel and Skolem, and also to Tarski. (Indeed, in *1936*, page 300, Tarski showed that the class of all well-orderings is not the class of models of any set of sentences—by using Langford's (*1927*) elimination of quantifiers. This result can be proved easily by a compactness argument.) It appears that A. Maltsev, in *1936* and *1941*, was the first person to publish such applications to algebra via "compactness arguments". (These may well have motivated his well-known extension of Gödel's compactness theorem to uncountable languages in the same two papers.) By 1941, Maltsev was making some quite sophisticated compactness arguments. But it was only after the Second World War that other logicians began to exploit the compactness theorem.

Beyond yielding the existence statement (1), Skolem's specially constructed model $\mathfrak{N}^*$ seems likely to be interesting in itself. Apparently Gödel thought so, as more than a third of his main review (*1934c*) is devoted to describing $\mathfrak{N}^*$ fully. Nevertheless, no further use was made of Skolem's construction for about twenty-five years. Then two uses were made of it. First, J. Łoś (*1955*) introduced the notion of ultraproduct and implicitly gave the basic result about when a sentence holds in an ultraproduct. Three years later, a number of people began to study ultraproducts. In just 1958 and 1959, important contributions to the subject were made by Chang, Frayne, Keisler, Kochen, Morel, Rabin, Scott and Tarski. (For references, see *Frayne, Morel and Scott 1962*.)

At about the same time it was "discovered" that ultraproducts had been
partly anticipated in two places, namely in Skolem (*1933a*, *1934*)—as
seen above—and also in a paper of E. Hewitt (*1948*).

Now we turn to the second use of Skolem's ideas, which was made by
MacDowell and Specker (*1961*). They showed that any model $\mathfrak{M}$ of the
usual first-order Peano axioms *PA* has a proper elementary extension $\mathfrak{M}^*$
in which *all new elements follow all old*. They use an argument related
to Skolem's; but now Skolem's use of *definable* things is critical, and
not his use of (something like) ultraproducts. Actually, ultraproducts
do appear in the paper by MacDowell and Specker, but can easily be
replaced by a compactness argument. However, in the model $\mathfrak{M}'$ so

Besprechung von *Skolem 1933a*:
Über die Unmöglichkeit einer vollständigen Charakterisierung der Zahlenreihe mittels eines endlichen Axiomensystems
(*1934c*)

Man betrachte diejenigen Ausdrücke, die sich aufbauen aus: 1. Variablen $x, y, \ldots$, deren Wertbereich die natürlichen Zahlen sind, 2. + (Ad-
194   dition) und · (Multiplikation), | 3. > und =, 4. den Operationen des Aus-
sagenkalküls, 5. den Quantifikatoren, bezogen auf Zahlvariable. (Kompliziertere Funktionen, wie z. B. x^y, $x!$, lassen sich durch +, · und die
angeführten logischen Begriffe definieren.) Der Verfasser beweist, daß es
ein System N^* von Dingen mit zwei darin definierten Operationen +, · und
mit zwei Relationen >, = gibt, welches mit dem System N der natürlichen
Zahlen *nicht* isomorph ist, für welches aber trotzdem alle mittels der eingangs erwähnten Symbole ausdrückbaren Sätze gelten, die für das System
N gelten. Daraus folgt, daß es kein, nur die eingangs erwähnten Begriffe
verwendendes (und daher überhaupt kein, bloß zahlentheoretische
Begriffe verwendendes) Axiomensystem gibt, welches die Struktur
der Zahlenreihe eindeutig festlegt, ein Resultat, das sich auch unschwer
aus den Untersuchungen des Referents in *1931* ergibt. Das vom Verfasser
konstruierte System N^* besteht aus den mittels der eingangs erwähnten
Begriffe definierbaren Funktionen $f_i(x)$, zwischen denen eine >-Relation
dadurch festgelegt wird, daß zunächst eine Funktion $g(x)$ bestimmt wird,

constructed, extending $\mathfrak{M}$, the desired model $\mathfrak{M}^*$ must be taken as the set of all elements *definable in $\mathfrak{M}'$* using as parameters elements of M plus a certain single element of $M' - M$ (compare the use of I/D above). A key point in the proof of MacDowell and Specker is a result on page 262 which is more or less a formalization in *PA* of Skolem's Theorem 1. Its not-so-simple proof, which they omit, is discussed and generalized in *Gaifman 1976*.

Since 1960, both ultraproducts and non-standard models of arithmetic have been intensively studied and a number of deep results have been obtained.

Robert Vaught

Review of *Skolem 1933a*:
On the impossibility of a complete characterization of the number sequence by means of a finite axiom system
(*1934c*)

Consider the expressions that are built up from: (1) variables $x, y, \ldots$, ranging over the natural numbers; (2) $+$ (addition) and $\cdot$ (multiplication); (3) $>$ and $=$; (4) the operations of the propositional calculus; (5) the quantifiers, applied to number variables. (More complicated functions, as for example x^y and $x!$, can be defined in terms of $+$, $\cdot$ and the logical concepts cited.) The author proves that there is a system N^* of entities, with two operations, $+$ and $\cdot$, defined on it and with two relations, $>$ and $=$, that is *not* isomorphic to the system N of natural numbers, but for which nevertheless all statements hold that are expressible by means of the symbols mentioned at the outset and hold for the system N. From this it follows that there is no axiom system employing only the notions mentioned at the outset (and therefore none at all employing only number-theoretic notions) that uniquely determines the structure of the sequence of natural numbers, a result that also follows without difficulty from the investigations of the reviewer in his *1931*. The system N^* constructed by the author consists of the functions $f_i(x)$ definable by means of the notions mentioned at the outset, among which a $>$-relation is determined as follows: first, a function $g(x)$ is specified such that, for every pair f_i, f_k, either $f_i[g(x)] > f_k[g(x)]$

derart daß für jedes Paar f_i, f_k entweder $f_i[g(x)] > f_k[g(x)]$ oder $f_i[g(x)] = f_k[g(x)]$ oder $f_i[g(x)] < f_k[g(x)]$ für fast alle x gilt. Die Operationen $+, \cdot$ für die f_i werden in der gewöhnlichen Weise definiert. Mit Hilfe anderer als der eingangs erwähnten Begriffe kann man natürlich Sätze bilden, die für N^*, aber nicht für N gelten, wofür einige Beispiele angeführt werden.

Besprechung von *Chen 1933*:
Axioms for real numbers
(*1934d*)

Es handelt sich bloß um ein Axiomensystem für den Begriff $<$ (Addition und Multiplikation kommen nicht vor), d. h. für den Ordnungstypus der reellen Zahlen. Von der üblichen Charakterisierung dieses Ordnungstypus unterscheidet es sich nur dadurch, daß die Existenz einer abzählbaren dichten Teilmenge nicht gefordert wird, wodurch natürlich die Vollständigkeit (Kategorizität) verlorengeht, welche aber doch (es ist nicht klar, in welchem Sinne) behauptet wird.

Besprechung von *Church 1933*:
A set of postulates for the foundation of logic
(second paper)
(*1934e*)

Das vom Verfasser in *1932* angegebene Axiomensystem hat sich als widerspruchsvoll herausgestellt. Um den Widerspruch (eine modifizierte Form der Russellschen Paradoxie) zu vermeiden, werden in der vorliegenden Arbeit einige Modifikationen an den Axiomen vorgenommen, welche die Wirkung haben sollen, den Beweis "leerer" Implikationen (d. h. solcher, deren Vorderglied immer falsch ist) unmöglich zu machen. Der Verfasser entwickelt dann (streng formal) eine Reihe von Folgerungen aus den Axiomen, welche insbesondere die Begriffe der Identität, der Klasse und der

or $f_i[g(x)] = f_k[g(x)]$ or $f_i[g(x)] < f_k[g(x)]$ holds for almost all x. The operations $+$ and $\cdot$ for the f_i are defined in the usual way. Of course, by means of notions other than those mentioned at the outset one can form statements that hold for N^* but not for N, some examples of which are given.

Review of *Chen 1933*:
Axioms for real numbers
(*1934d*)

What is presented is merely an axiom system for the notion $<$ (addition and multiplication do not occur), that is, for the order type of the real numbers. It differs from the usual characterization of this order type only in that the existence of a countable dense subset is not demanded, which of course causes the loss of completeness (categoricity). It is claimed nonetheless that the system has this property (in what sense is not clear).

Review of *Church 1933*:
A set of postulates for the foundation of logic
(second paper)
(*1934e*)

[The introductory note to *1934e*, as well as to related items, can be found on page 256, immediately preceding *1932k*.]

The axiom system given by the author in his *1932* has turned out to be contradictory. To avoid the contradiction (a modified form of Russell's paradox), the axioms are subjected to a few modifications in the work at hand; these are supposed to have the effect of making it impossible to prove "empty" implications (that is, those in which the antecedent is always false). The author then develops (strictly formally) a sequence

sogenannten "Vervollständigung einer Aussagefunktion" betreffen. Das letztere bedeutet die Konstruktion einer neuen Aussagefunktion, deren Definitionsbereich möglichst umfassend ist und welche im alten Bereich mit der früheren Aussagefunktion übereinstimmt. Seit dem Erscheinen der Arbeit hat sich herausgestellt, daß die vom Verfasser angegebenen Modifikationen seiner ursprünglichen Axiome zur Vermeidung von Widersprüchen nicht hinreichen, d. h. daß auch das neue System Antinomien enthält. Davon wird nicht betroffen eine im letzten Abschnitt gegebene Definition der natürlichen Zahlen, nach welcher die Zahl n diejenige Operation ist, welche aus jeder Funktion $f(x)$ ihre n-te Potenz $\underbrace{f(\cdots f(f(x))\cdots)}_{n}$ erzeugt.

Dies ermöglicht eine besonders einfache Darstellung der rekursiven Definitionen. Zum Beispiel ist $m+n = \{n(S)\}(m)$, wenn $S(x)$ die Funktion $x+1$ bedeutet.

Besprechung von *Notcutt 1934*:
A set of axioms for the theory of deduction
(*1934f*)

Der Verfasser ist der Meinung, daß in den üblichen Darstellungen formaler Systeme gewisse zu ihrer Handhabung notwendige Regeln stillschweigend vorausgesetzt werden (z. B. daß die Formeln von links nach rechts zu lesen sind) und stellt ein Axiomensystem für den Aussagenkalkül auf, das von diesem Mangel frei sein soll, d. h. in dem "nichts der Intelligenz des Lesers überlassen ist, sondern alle Annahmen explizit formuliert werden". Ein zweiter Teil der Arbeit beschäftigt sich mit der (verzweigten) Typentheorie. Es werden sogenannten "intertypical variables" eingeführt, d. h. Variable, die gleichzeitig über alle möglichen Typen laufen. Um Widersprüche zu vermeiden, dürfen diese Variablen niemals gebunden auftreten. Die "intertypical variables" sollen das Reduzibilitätsaxiom überflüssig machen.

of consequences of the axioms that concern, in particular, the notions of
identity, class, and what is called the "completion of a propositional function". The latter signifies the construction of a new propositional function
whose domain of definition is as comprehensive as possible and which coincides on the old domain with the original propositional function. Since
the appearance of the work, it has turned out that the author's proposed
modifications of his original axioms do not suffice for the avoidance of contradictions, that is, that the new system, too, contains antinomies. But
this does not affect a definition of the natural numbers, given in the last
section, according to which the number n is the operation that generates
from each function $f(x)$ its nth power $\underbrace{f(\ldots f(f(x))\ldots)}_{n\ \text{times}}$. This makes possible an especially simple presentation of [[primitive]] recursive definitions.
For example, $m + n = \{n(S)\}(m)$, if $S(x)$ stands for the function $x + 1$.

Review of *Notcutt 1934*:
A set of axioms for the theory of deduction
(*1934f*)

The author is of the opinion that in the usual presentations of formal
systems certain rules necessary for their manipulation are tacitly assumed
(for example, that the formulas are to be read from left to right), and he
sets up an axiom system for the propositional calculus that is supposed
to be free from this flaw, that is, one in which "nothing is left to the
intelligence of the reader, where everything assumed is explicitly stated".
A second part of the work is concerned with the (ramified) theory of types.
What are called "inter-typical variables" are introduced, that is, variables
that simultaneously range over all possible types. If contradictions are to
be avoided, these variables may never occur bound. The "inter-typical
variables" are supposed to make the axiom of reducibility superfluous.

Besprechung von *Skolem 1934*:
Über die Nicht-charakterisierbarkeit der Zahlenreihe mittels endlich oder abzählbar unendlich vieler Aussagen mit ausschließlich Zahlenvariablen
(*1935*)

Die Arbeit stimmt hinsichtlich des Resultates mit der als *1933a* vom gleichen Verfasser erschienenen überein, gibt bloß einen etwas vereinfachten Beweis und die folgende Formulierung des Ergebnisses: Endlich oder abzählbar viele Aussagen mit lauter Individuenvariablen, die für eine Reihe vom Typus ω gelten, können diese Reihe nicht von gewissen Reihen von höherem Ordnungstypus unterscheiden. Daß dies auch richtig bleibt, wenn höhere Variable (für Mengen oder Satzfunktionen) eingeführt und im Sinne der axiomatischen Mengenlehre interpretiert werden, wird ohne Beweis behauptet.

Review of *Skolem 1934*:
On the non-characterizability of the
number sequence by means of finitely
or denumerably many propositions
containing variables only for numbers
(*1935*)

[The introductory note to *1935*, as well as to related items, can be found on page 376, immediately preceding *1934c*.]

With respect to its result, this work is identical to the same author's *1933a*; it merely gives a somewhat simplified proof and the following formulation of the result: Finitely or denumerably many propositions that contain no variables other than those for individuals and hold for a sequence of type ω cannot distinguish that sequence from certain sequences of higher order type. It is asserted without proof that this remains true if variables of higher type (for sets or propositional functions) are introduced and interpreted as they are in axiomatic set theory.

Introductory note to *1935a*

Huntington's system is not a modal logic, for he uses predicates rather than iterable functors to express necessity and impossibility. Like

Besprechung von *Huntington 1934*:
Independent postulates related to
C. I. Lewis' theory of strict implication
(*1935a*)

Der Verfasser stellt ein aus 13 unabhängigen Postulaten bestehendes Axiomensystem mit folgenden Grundbegriffen auf: K (Klasse der Sätze), Q (Klasse der falschen Sätze), D (Klasse der unmöglichen Sätze), $a \times b$ (Konjunktion), a' (Negation), $a = b$ (Gleichheit), wobei $\times$ und $'$ als Operationen aufzufassen sind, die aus Elementen von K wieder Elemente von K erzeugen. Sätze der Form $a \, \epsilon \, D$ sind inhaltliche Aussagen (nicht Elemente von K). Daher kann der axiomatische Unmöglichkeitsbegriff D nicht iteriert angewendet werden, worin der Hauptunterschied gegenüber dem Lewisschen System liegt. Für Aussagen der Form $a \, \epsilon \, Q$, $a \, \epsilon \, D$ und die aus solchen mittels Konjunktion und Negation aufgebauten definiert der Verfasser einen inhaltlichen Unmöglichkeitsbegriff (unmöglich ist, was auf eine Widerspruch führt) und zeigt, daß die Lewisschen Axiome aus seinen folgen, wenn man in ihnen die Unmöglichkeitszeichen erster Stufe durch seinen axiomatischen und die zweiter Stufe durch seinen inhaltlichen Unmöglichkeitsbegriff ersetzt. Im selben Sinn folgen die Axiome des Verfassers aus den Lewisschen, woraus aber noch nicht geschlossen werden kann, daß die beiden Systeme auch in ihren Folgerungen übereinstimmen, da ja für die Unmöglichkeitszeichen dritter und höherer Stufe gar keine Übersetzung vorliegt.

Lewis, he stops short of quantification. But, if he were to introduce it, he would still be unable, on this approach, to quantify into modal contexts and thus precipitate the referential opacity and related perplexities that beset modal logic.

W. V. Quine

Review of *Huntington 1934*: Independent postulates related to C. I. Lewis' theory of strict implication (*1935a*)

The author sets up an axiom system, consisting of thirteen independent postulates, with the following primitive notions: K (class of propositions), Q (class of false propositions), D (class of impossible propositions), $a \times b$ (conjunction), a' (negation), $a = b$ (identity), where $\times$ and $'$ are interpreted as operations that, when applied to elements of K, again generate elements of K. Propositions of the form $a \, \epsilon \, D$ are contentual statements (not elements of K). Therefore the axiomatic notion of impossibility, D, cannot be iteratively applied, and therein lies the principal distinction between this system and Lewis'. For statements of the form $a \, \epsilon \, Q$ or $a \, \epsilon \, D$, and those built up from such statements by means of conjunction and negation, the author defines a notion of contentual impossibility (impossible is that which leads to a contradiction) and shows that Lewis' axioms follow from his if in them we replace the impossibility signs of the first degree by his axiomatic notion of impossibility and those of the second degree by his contentual notion of impossibility. Correspondingly, the author's axioms follow from those of Lewis, from which, however, we cannot yet conclude that the two systems agree also in what they entail, since no translation at all is at hand for the impossibility signs of third and higher degrees.

Besprechung von *Carnap 1934*:
Die Antinomien und
die Unvollständigkeit der Mathematik
(*1935b*)

In dieser Arbeit zieht der Verfasser die Konsequenzen, welche sich aus der Konstruktion formal unentscheidbarer Sätze für das Problem der Antinomien zweiter Art (z. B. Epimenides) ergeben, nämlich die folgenden: Der logische Fehler dieser Antinomien liegt *nicht* in der Selbstbezogenheit gewisser in ihnen auftretender Begriffe und Sätze (diese Selbstbezogenheit kommt ja z. B. auch den erwähnten unentscheidbaren Sätzen zu), sondern in der Verwendung des Begriffes "wahr". D. h. genauer, es wird fälschlich angenommen, man habe einen Begriff $\mathfrak{W}$ (wahr) von der Art, daß für jeden Satz A die Formel

$$\mathfrak{W}(\text{``}A\text{''}) \equiv A \tag{1}$$

beweisbar ist. Aus dem Zustandekommen der Antinomien kann man schließen, daß es einen solchen begriff $\mathfrak{W}$ in keiner widerspruchsfreien Sprache geben kann. Man kann zwar jede Sprache so erweitern, daß sie einen Begriff $\mathfrak{W}$ enthält, der (1) für alle Sätze A der *ursprünglichen* Sprache befriedigt, nicht aber so, daß (1) auch für die Sätze der *erweiterten* Sprache gelten würde. Jedes formale System ist also in zweifacher Hinsicht unvollständig: 1. insofern, als es darin unentscheidbare Sätze gibt, 2. insofern als es Begriffe gibt, die sich darin nicht definieren lassen (z. B. $\mathfrak{W}$ oder die nach dem Diagonalverfahren konstruierten Zahlenfolgen). So kommt man zu dem Schluß, daß, obwohl alles Mathematische formalisierbar ist, doch nicht die ganze Mathematik in *einem* formalen System formalisiert werden kann, eine seit jeher vom Intuitionismus behauptete Tatsache.—Ein zweiter Teil der Arbeit beschäftigt sich mit der Paradoxie der abzählbaren Modelle der Mengenlehre und präzisiert die übliche Auflösung dieser scheinbaren Paradoxie in dem Sinne, daß je zwei Mengen der axiomatischen Mengenlehre syntaktisch (d. h. in einer geeigneten Metasprache) gleichmächtig sind, nicht aber innerhalb des ursprünglichen Systems.—Bezüglich der Antinomien zweiter Art und des Wahrheitsbegriffs wurde die gleiche Auffassung von A. Tarski in *1932*, ferner in *1933a* und in *1935* vertreten.

Review of *Carnap 1934*:
The antinomies and
the incompleteness of mathematics
(*1935b*)

In this work the author derives the consequences that result from the construction of formally undecidable propositions for the problem of antinomies of the second kind (for instance that of Epimenides). Namely, the logical flaw in these antinomies lies *not* in the self-referential character of certain notions and propositions occurring in them (this self-referential character, after all, also attaches, for example, to the undecidable propositions mentioned), but rather in the use of the notion "true". That is, more precisely, it is falsely assumed that there is a notion $\mathfrak{W}$ (true) such that for each sentence A the formula

$$\mathfrak{W}(\text{``}A\text{''}) \equiv A \tag{1}$$

is provable. From the occurrence of the antinomies it can be concluded that such a notion $\mathfrak{W}$ cannot be expressed in a consistent language. To be sure, every language can be extended so as to contain a notion $\mathfrak{W}$ that satisfies (1) for all sentences A of the *original* language, but not so that (1) holds for the sentences of the *extended* language as well. Every formal system is thus incomplete in two respects: $1°$ insofar as there are propositions undecidable within it, and $2°$ insofar as there are notions that cannot be defined within it (such as $\mathfrak{W}$ or the number sequences constructed by the diagonal method). Thus we are led to conclude that, although everything mathematical is formalizable, it is nonetheless impossible to formalize all of mathematics in a *single* formal system, a fact that intuitionism has asserted all along.

A second part of the work deals with the paradox of countable models of set theory and makes the usual resolution of this apparent paradox more precise, in the sense that any two sets in axiomatic set theory are of equal cardinality syntactically (that is, in a suitable metalanguage), but not within the original system.

With respect to the antinomies of the second kind and the notion of truth, the same view was advocated by A. Tarski in his *1932* and further in his *1933a* and *1935*.

Besprechung von *Kalmár 1934*:
Über einen Löwenheimschen Satz
(*1935c*)

Für den Löwenheimschen Satz, daß man zu jedem Zählausdruck $\mathfrak{A}$ einen binären $\mathfrak{B}$ angeben kann, so daß $\mathfrak{A}$ dann und nur dann allgemeingültig ist, wenn $\mathfrak{B}$ es ist, wird | ein einfacherer Beweis gegeben ($\mathfrak{B}$ heißt binär, wenn es nur zweistellige Funktionsvariable enthält). $\mathfrak{B}$ wird aus $\mathfrak{A}$ konstruiert, indem die Funktionszeichen $F_\lambda(x_1, x_2, \ldots, x_{r_\lambda})$ aus $\mathfrak{A}$ ($\lambda = 1, 2, \ldots, l$) durch die Formeln

$$(u)[H_1(x_1, u) \ \& \ H_2(x_2, u) \ \& \ \ldots \ \& \ H_{r_\lambda}(x_{r_\lambda}, u) \rightarrow G_\lambda(u)]$$

ersetzt werden. Es wird gezeigt, daß, falls für $\mathfrak{B}$ bekannt ist, für Individuenbereiche welcher Mächtigkeiten es allgemeingültig ist, dieselbe Frage auch für $\mathfrak{A}$ entschieden werden kann, ferner daß man, falls ein Beweis für $\mathfrak{B}$ aus den Axiomen des engeren Funktionenkalküls vorgelegt ist, einen solchen Beweis auch für $\mathfrak{A}$ konstruieren kann.

Review of *Kalmár 1934*:
On a theorem of Löwenheim
(*1935c*)

A simpler proof is given for Löwenheim's theorem that for each first-order expression $\mathfrak{A}$ a binary one $\mathfrak{B}$ can be specified such that $\mathfrak{A}$ is valid if and only if $\mathfrak{B}$ is ($\mathfrak{B}$ is said to be binary if it does not contain functional variables with more than two argument places). $\mathfrak{B}$ is constructed from $\mathfrak{A}$ by replacement of the functional symbols $F_\lambda(x_1, x_2, \ldots, x_{r_\lambda})$ in $\mathfrak{A}$ ($\lambda = 1, 2, \ldots, l$) with formulas [of the form]

$$(u)[H_1(x_1, u) \mathbin{\&} H_2(x_2, u) \mathbin{\&} \ldots \mathbin{\&} H_{r_\lambda}(x_{r_\lambda}, u) \to G_\lambda(u)].$$

It is shown that, if it is known what the cardinalities are of the domains of individuals in which $\mathfrak{B}$ is valid, then the same question can also be decided for $\mathfrak{A}$, and further that, if a proof of $\mathfrak{B}$ from the axioms of the restricted functional calculus is exhibited, then such a proof can also be constructed for $\mathfrak{A}$.

Introductory note to *1936*

That mathematical economics figured prominently among the subjects discussed in Karl Menger's colloquium is not surprising, since Menger's father Carl was a distinguished economist; but that Kurt Gödel should have contributed a published remark on such a topic is surely somewhat unexpected. Nevertheless, Gödel reportedly discussed the foundations of economics with Oskar Morgenstern, then a fellow participant in the Schlick circle (see *Henn and Moeschlin 1977*, page 8), and in an unpublished interview with Axel Leijonhufvud in 1977, shortly before Morgenstern's death, Morgenstern named Gödel as one of the colleagues who had most influenced his work.

Gödel's brief published comment came during the 80th session of the colloquium, on 6 November 1934, following Abraham Wald's presentation "Über die Produktionsgleichungen der ökonomischen Wertlehre, II".[a] Wald's work was based on a paper of the same title by Karl

---

[a] "On the equations of production in the economic theory of value, II", published as *Wald 1936*.

Diskussionsbemerkung
(*1936*)

In Wirklichkeit hängt für jeden einzelnen Wirtschafter die Nachfrage auch von seinem Einkommen ab und dieses wieder vom Preis der Produktionsmittel. Man kann ein dementsprechendes Gleichungssystem formulieren und auf seine Lösbarkeit hin untersuchen.

Schlesinger (*1935*), which was presented at the 74th meeting of the colloquium, on 19 March 1934, along with Wald's own *1935* (to which his *1936* is a sequel). A technical overview of all three papers, as well as a detailed discussion of the role the Menger colloquium played in the development of economic theory, is given in *Weintraub 1983*, pages 5–12. Briefly, Weintraub notes that Gustav Cassel, following Walras, had formulated non-linear systems of equations in which demand quantities of goods were expressed as functions of their prices. Schlesinger had then inverted this relationship, formulating a system of equations in which demand *prices* were expressed as functions of the quantities of goods produced. Wald, in turn, investigated conditions under which Schlesinger's equations would have unique solutions. In this context, Gödel's remark may be viewed not as chiding Wald for having neglected the role of income, but as suggesting a generalization of Wald's investigations to systems in which factor prices were included. Weintraub, however, quoting Lionel McKenzie, notes (footnote 11, pages 10–11) that, although Walras' original equations *had* included factor prices as arguments, there is, contrary to Gödel's suggestions, no obvious way to formulate a meaningful system of inverse equations involving factor prices if there is more than one consumer.

John W. Dawson, Jr.

Discussion remark
(*1936*)

Actually, for each individual entrepreneur the demand also depends on his income, and that in turn depends on the price of the factors of production. One can formulate an appropriate system of equations and investigate whether it is solvable.

Introductory note to *1936a*

In this abstract, published in *Ergebnisse eines mathematischen Kolloquiums*, Gödel announced a "speed-up" theorem that holds when one switches from a weaker logic to a stronger one. (The term "logic" may be somewhat misleading since apparatus for proving facts about numbers is, of course, present.) Gödel had already shown that a logic S_{n+1} of a higher order could prove formulas that a logic S_n of a lower order could not prove. Now he considered the question of formulas that can be proved in both the weaker and the stronger logics, and his theorem stated that, if the length of a proof is defined to be the number of lines in it, there are formulas that can be proved in both S_n and S_{n+1} but that have a proof in S_{n+1} much shorter than their shortest proof in S_n. Thus S_{n+1} achieves a speed-up over S_n. This speed-up can be by an arbitrary function computable in S_n, in the sense that, for any function ϕ computable in S_n, there would be a formula A and a number k such that A can be proved in S_{n+1} in k lines but, although provable in S_n also, A cannot be proved in S_n in $\phi(k)$ lines.

The notion of computability in a formal system that is used here is what Gödel had previously called *Entscheidungsdefinitheit* in *1931*. Interestingly, Gödel did not remark at first that the notion of computable function in S_n is independent of n, but this crucial observation was added in the page proof before publication.[a] Gödel later remarked, in the first paragraph of his paper *1946*, that his notion of computability in a logic is equivalent to Turing computability, and he stressed that this is an absolute notion independent of the formalism chosen. Incidentally, it is worth mentioning here that Gödel does not bother to make a distinction between a function ϕ and the syntactic object corresponding to its representation in S_n. It is true of course, by virtue of Gödel's own incompleteness results, that two different representations of the same function may fail to be provably equivalent.

Gödel did not give a proof of his result, but an analogous result taking the length of a proof to be its Gödel number, rather than the number of lines in it, was given by Mostowski in *1952*. Similar results were also proved by Ehrenfeucht and Mycielski (*1971*) and by Parikh (*1971*). The case where $n > 1$ is not explicitly stated by these authors, but *Solomon 1981* gives an argument based on some complexity-theoretic results of *Hartmanis 1978*.

---

[a]See *Davis 1982*, pp. 15–16, for a discussion, as well as pp. 340–342 of the introductory note to *1934* in this volume.

The case where the length of proof is measured by the number of lines appears to be open for $n > 1$. Thus, in the following, we shall consider only the case where S_n is replaced by some version PA of Peano arithmetic, and S_{n+1} by some system S in which the consistency of Peano arithmetic is provable.

A crucial difference between taking the length of a proof to be the number of lines and taking it to be its Gödel number (equivalently, the number of symbols) is that there are infinitely many proofs with a given number of lines, but only finitely many with a given number of symbols. This affects both the character of the demonstration and the impact of the result. For example, a result formulated in terms of Gödel numbers tends to apply also to arbitrary recursive axiomatizations of PA and S. However, it is easy to find recursive axiomatizations of *any* recursively enumerable theory (PA in particular) so that all theorems have proofs of only three lines at most. Thus a result proved in terms of the length of a proof as the number of lines must make essential use of the fact that the usual formalizations of familiar systems are schematic in nature; all axioms and rules fall under a *finite* number of schemata.

Returning to Gödel's work, we note that there is some ambiguity about the precise result being claimed. Namely, it is not obvious from the abstract whether Gödel intended $+$ and $\times$ to be function symbols or predicate symbols in his logics. A remark at the beginning of Section 2 of his *1931* seems to indicate that it might be the latter.

This point is of some importance. While systems formalized using function or predicate symbols respectively for the same objects have the same proof-theoretic strength, the proofs tend to be shorter with the former. For instance, the formula $t = t$, where t is a complicated term, can be an axiom if function symbols are available, but must be stated (and proved) in a roundabout way in terms of predicate symbols if that is all we have. (Gödel does not remark on this last fact.) It was shown in *Parikh 1973* that, if PA^* is the version of PA where $+$ and $\times$ are ternary predicate symbols, then there are sequences A_i of formulas whose proofs in S are uniformly bounded in length but whose proofs in PA^* must be unbounded in length. Gödel's theorem for $n = 1$ follows as a corollary, and we can eliminate the requirement that the speed-up be by a computable function.

The argument of *Parikh 1973* depended on showing that the question "Does a given formula have a proof in PA^* of k lines or less?" is decidable. While the corresponding question for the usual formalization of PA is open, a recent result by Goldfarb (*1981*) seems to indicate that a positive result is unlikely. However, Statman (*1981*) showed that when a theory T, axiomatized with a finite number of schemata, is extended by adding an undecidable formula A, and $T \cup \{\neg A\}$ has an infinite model,

then even tautologies become easier to prove. The Gödel theorem for
$n = 1$ follows immediately.

A general study of the speed-up phenomenon in the context of proofs
appears in *Statman 1978*. The term "speed-up" is not due to Gödel
but was introduced by Blum (*1967*) in the context of complexity theory. Subsequently, the phenomenon of speed-up became a major topic

Über die Länge von Beweisen
(*1936a*)

Unter der *Länge* eines Beweises in einem formalen System S möge die
Anzahl der Formeln, aus denen er besteht, verstanden werden. Ist ferner
jede natürliche Zahl in bestimmter Weise durch ein Symbol (Zahlzeichen)
in S darstellt (z. B. durch ein Symbol der Form $1 + 1 + \cdots + 1$) so möge
eine Funktion $\phi(x)$ *berechenbar in S* heißen, falls es zu jedem Zahlzeichen m
ein Zahlzeichen n gibt, so daß $\phi(m) = n$ in S beweisbar ist. Insbesondere
sind z. B. alle rekursiv definierten Funktionen schon in der klassischen
Arithmetik (d. h. im System S_1 der unten definierten Folge) berechenbar.

Sei nun S_i das System der Logik i-ter Stufe, wobei die natürlichen Zahlen
als Individuen betrachtet werden. D. h. genauer: S_i soll Variable und
Quantifikatoren für natürliche Zahlen, für Klassen natürlicher Zahlen, für
Klassen von Klassen natürlicher Zahlen u. s. w. bis Klassen von i-tem Typus
samt den zugehörigen logischen Axiomen enthalten, aber keine Variable von
höherem Typus. Dann gibt es bekanntlich Sätze aus S_i, die in S_{i+1}, aber
nicht in S_i, beweisbar sind. Betrachten wir hingegen diejenigen Formeln
f, die sowohl in S_i als in S_{i+1} beweisbar sind, so gilt folgendes: Zu jeder
in S_i berechenbaren Funktion ϕ gibt es unendlich viele Formeln f von der
Art, daß, wenn k die Länge eines kürzesten Beweises für f in S_i und l die
Länge eines kürzesten Beweises für f in S_{i+1} ist, $k > \phi(l)$. Setzt man z.
B. $\phi(n) = 10^6 n$, so folgt: Es gibt unendlich viele Formeln, deren kürzester
Beweis in S_i mehr als 10^6 mal länger ist als ihr kürzester Beweis in S_{i+1}.
*Der Übergang zur Logik der nächst höheren Stufe bewirkt also nicht bloß,
daß gewisse früher unbeweisbare Sätze beweisbar werden, sondern auch,
daß unendlich viele der schon vorhandenen Beweise außerordentlich stark
abgekürzt werden können.*

Die Formeln f, für welche die obige Ungleichung $k > \phi(l)$ gilt, sind
arithmetische Sätze von demselben Charakter wie die von mir konstruierten

in theoretical computer science and, indeed, the celebrated $P = NP$? question can itself be thought of as a speed-up question. To discuss that here, however, would take us too far afield.

Rohit Parikh

The translation is by Stefan Bauer-Mengelberg and Jean van Heijenoort.

On the length of proofs
(*1936a*)

By the *length* of a proof in a formal system S we shall understand the number of formulas of which the proof consists. Further, if every natural number is represented in S in a definite way by a symbol (a numeral) (for example, by a symbol of the form $1 + 1 + \cdots + 1$), then a function $\phi(x)$ will be said to be *computable in S* in case for each numeral m there exists a numeral n such that $\phi(m) = n$ is provable in S. In particular, all recursively defined functions, for example, are already computable in classical arithmetic (that is, in the system S_1 of the sequence defined below).

Now let S_i be the system of logic of the ith order, the natural numbers being taken as individuals. That is, more precisely: S_i shall contain, together with the appropriate logical axioms, variables and quantifiers for natural numbers, for classes of natural numbers, for classes of classes of natural numbers, and so on, up to classes of the ith type, but no variable of a higher type. Then, as is known, there are propositions of S_i that are provable in S_{i+1} but not in S_i. If, on the other hand, we consider those formulas f that are provable in S_i as well as in S_{i+1}, then the following holds: For each function ϕ that is computable in S_i there exist infinitely many formulas f such that, if k is the length of a shortest proof of f in S_i and l is the length of a shortest proof of f in S_{i+1}, then $k > \phi(l)$. If, for example, we set $\phi(n) = 10^6 n$, it follows that there are infinitely many formulas whose shortest proof in S_i is more than 10^6 times as long as their shortest proof in S_{i+1}. *Thus, passing to the logic of the next higher order has the effect, not only of making provable certain propositions that were not provable before, but also of making it possible to shorten, by an extraordinary amount, infinitely many of the proofs already available.*

The formulas f for which the above inequality $k > \phi(l)$ holds are arithmetical propositions of the same kind as the propositions undecidable in S_i that I have constructed; that is, they can be brought into the following

in S_i unentscheidbaren Sätze, d. h. sie können auf folgende Normalform gebracht werden:

$$(\mathfrak{P})[Q(x_1, x_2, \ldots, x_n) = 0],$$

wobei $x_1, x_2, \ldots, x_n$ Variable für ganze Zahlen sind, Q ein bestimmtes Polynom mit ganzzahligen Koeffizienten in den n Variablen $x_1, x_2, \ldots, x_n$ ist und $\mathfrak{P}$ ein Präfix bezeichnet, d. h. eine bestimmte Folge von All- und Existenzzeichen für die Variabeln $x_1, x_2, \ldots, x_n$. Ein solcher Satz spricht also eine Eigenschaft der Diophantischen Gleichung $Q = 0$ aus. Z. B. besagt der Satz

$$(x_1)(x_2)(Ex_3)(Ex_4)[Q(x_1, x_2, x_3, x_4) = 0],$$

daß die Diophantische Gleichung $Q(a, b, x, y) = 0$ für alle Werte der Parameter a und b ganzzahlige Lösungen x, y besitzt. Enthält | das Präfix die Allzeichen für $x_{i_1}, x_{i_2}, \ldots, x_{i_k}$ und die Existenzzeichen für $x_{j_1}, x_{j_2}, \ldots, x_{j_{n-k}}$, so hat man $x_{i_1}, \ldots, x_{i_k}$ als Parameter aufzufassen und die obige Formel besagt, daß für beliebige Werte der Parameter Lösungen $x_{j_1}, x_{j_2}, \ldots, x_{j_{n-k}}$ existieren, wobei der Wert von x_{j_r} nur abhängt von den Werten derjenigen Parameter, die x_{j_r} im Präfix vorangehen.

Bemerkung während der Drucklegung:

Es läßt sich übrigens zeigen, daß eine Funktion, die in einem der Systeme S_i oder auch in einem System transfiniter Stufe berechenbar ist, schon in S_1 berechenbar ist, so daß also der Begriff "berechenbar" in gewissem Sinn "absolut" ist, während fast alle sonst bekannten metamathematischen Begriffe (z. B. beweisbar, definierbar etc.) sehr wesentlich vom zu Grunde gelegten System abhängen.

Besprechung von *Church 1935*:
A proof of freedom from contradiction
(*1936b*)

Durch weitgehende Einschränkungen an seinem früheren System erhält der Verfasser ein neues mit nur drei Grundbegriffen, nämlich: 1. der Operation der "Anwendung" einer Funktion auf ein Argument, 2. der "Identität", 3. der Variablen bindenden Operation "λx", welche aus einem Ausdruck mit der freien Variablen x die durch ihn definierte Funktion erzeugt. Dieses System kann einerseits noch finit als widerspruchsfrei er-

normal form:

$$(\mathfrak{P})[Q(x_1, x_2, \ldots, x_n) = 0],$$

where $x_1, x_2, \ldots, x_n$ are variables for natural numbers, Q is a certain poly-nomial, with integral coefficients, in the n variables $x_1, x_2, \ldots, x_n$, and $(\mathfrak{P})$ stands for a prefix, that is, a certain sequence of universal and existen-tial quantifiers for the variables $x_1, x_2, \ldots, x_n$. Such a proposition thus expresses a property of the Diophantine equation $Q = 0$. For example, the proposition

$$(x_1)(x_2)(Ex_3)(Ex_4)[Q(x_1, x_2, x_3, x_4) = 0]$$

states that the Diophantine equation $Q(a, b, x, y) = 0$ possesses integral solutions x and y for all values of the parameters a and b. If the prefix contains the universal quantifiers for $x_{i_1}, x_{i_2}, \ldots, x_{i_k}$ and the existential quantifiers for $x_{j_1}, x_{j_2}, \ldots, x_{j_{n-k}}$, then one must consider $x_{i_1}, \ldots, x_{i_k}$ as parameters, and the formula above asserts that, for arbitrary values of the parameters, there exist solutions $x_{j_1}, x_{j_2}, \ldots, x_{j_{n-k}}$, the value of x_{j_r} depending solely upon the values of those parameters that precede x_{j_r} in the prefix.

Remark added in proof:

It can, moreover, be shown that a function computable in one of the sys-tems S_i, or even in a system of transfinite order, is computable already in S_1. Thus the notion 'computable' is in a certain sense 'ab-solute', while almost all metamathematical notions otherwise known (for example, provable, definable, and so on) quite essentially depend upon the system adopted.

Review of *Church 1935*:
A proof of freedom from contradiction
(*1936b*)

[The introductory note to *1936b*, as well as to related items, can be found on page 256, immediately preceding *1932k*.]

By placing far-reaching restrictions on his earlier system, the author obtains a new one with only three primitive notions, namely: (1) the operation of "application" of a function to an argument, (2) "identity",

wiesen werden und ist andererseits so umfassend, daß jede durch Rekursion (beliebig hoher Ordnung) definierte zahlentheoretische Funktion in ihm enthalten ist, d. h. zu jeder solchen Funktion ϕ gibt es eine Formel F des Systems, so daß, wenn n eine beliebige natürliche Zahl und $m = \phi(n)$ ist, die Formel $F(n) = m$ beweisbar ist, wobei die natürlichen Zahlen $1, 2, 3, \ldots$ usw. durch folgende Formeln ausgedrückt werden: $\lambda fx \cdot f(x)$, $\lambda fx \cdot f(f(x))$, $\lambda fx \cdot f(f(f(x))) \ldots$ usw. (vgl. *Gödel 1934e*). Die Zahlen 1, 2 werden gleichzeitig als Wahrheitswerte "falsch", "wahr" verwendet und die Operationen des Aussagenkalküls als zahlentheoretische Funktionen 242 | durch Definition eingeführt. Die Formel 2 ist das einzige Axiom, und die Schlußregeln (die der Verfasser "Konversionen" nennt) sind sämtlich umkehrbar, so daß eine Formel dann als bewiesen gilt, wenn sie in 2 konvertiert werden kann. Der Alloperator tritt nicht als Grundbegriff auf, sondern wird vertreten durch gewisse mittels der drei Grundbegriffe aufgebaute Formeln Π mit der Eigenschaft, daß $\Pi(F, G)$ dann und nur dann beweisbar ist, wenn $G(x)$ aus $F(x)$ vermittels gewisser von den Konversionen im allgemeinen verschiedenen Schlußregeln abgeleitet werden kann. Zufolge der Tatsache, daß jedes formale System unvollständig ist, kann eine transfinite Folge immer schärferer Schlußregeln und dementsprechend eine transfinite Folge verschiedener Alloperatoren definiert werden. Es wird vermutet, daß diese den verschiedenen Ordnungen des Beweises durch vollständige Induktion entsprechen.

(3) the variable-binding operation "λx", which generates from an expression with the free variable x the function defined by it. This system, on the one hand, can still be proved consistent by finitary means and, on the other, is so comprehensive that every number-theoretic function defined by recursion (of arbitrarily high order) is contained in it; that is, for every such function ϕ there is a formula F of the system such that, if n is an arbitrary natural number and $m = \phi(n)$, the formula $F(\boldsymbol{n}) = \boldsymbol{m}$ is provable, where the natural numbers $1, 2, 3, \ldots$ and so on are represented by the following formulas: $\lambda f x \cdot f(x)$, $\lambda f x \cdot f(f(x))$, $\lambda f x \cdot f(f(f(x)))$, $\ldots$ and so on (see *Gödel 1934e*). The numbers 1 and 2 are at the same time employed as the truth values "false" and "true", and the operations of the propositional calculus are introduced by definition as number-theoretic functions. The formula 2 is the sole axiom, and all the rules of inference (which the author calls "conversions") are reversible, so that a formula is regarded as proved if it can be converted into 2. The universal quantifier does not appear as a primitive notion but is represented by certain formulas Π, built up by means of the three primitive notions, having the property that $\Pi(F, G)$ is provable if and only if $G(x)$ can be derived from $F(x)$ by means of certain rules of inference differing in general from the conversions. In consequence of the fact that every formal system is incomplete, a transfinite sequence of ever stronger rules of inference, and correspondingly a transfinite sequence of different universal quantifiers, can be defined. It is conjectured that these correspond to the different orders of proof by mathematical induction.

Textual notes

The copy-text for each work is the corresponding published text, except in the case of *Gödel 1929*, for which the copy-text is the copy of the dissertation on file at the University of Vienna.

Pagination, given for all texts except *1929*, follows the copy-text and is indicated by a page number in the margin, a vertical bar in the text indicating where the page begins. The first page number is always omitted.

In these textual notes, the pairs of numbers on the left indicate page and line in this volume.

Additions enclosed within square brackets [ ] are Gödel's own. Editorial additions are indicated by double square brackets ⟦ ⟧ in the text or by textual notes (see below).

Abbreviations in German (such as Anm., Ax., betr., ev., i.a., inkl., nat., Ref., sog., u.a., u.z., v., z.t.), in English (such as cor., def., p., prop. funct., resp., Th.) and in Latin (such as ad inf.) have been expanded throughout, except for p. and pp. in footnotes. On occasion, when the German abbreviation Verf. has been expanded to Verfasser, it has been necessary to change its ending or to add an article. Within the text of *1931*, *1931b*, *1932d, k, 1933b, i, k* and *l*, some numerals have been spelled out.

Concerning our translations, a few points should be mentioned. In *1933i* Gödel uses "Zählausdruck", borrowed from Löwenheim, and in *1929* he uses "Zählaxiomsystem". For him, the first expression refers to a closed well-formed formula of first-order logic, and the second to an applied (in the sense of Church) first-order axiom system. Various names are used by Gödel and others for first-order logic: "restricted functional calculus" (or, in German, "engerer Funktionenkalkül", used by Hilbert and Ackermann), "first-order predicate calculus" (or simply "predicate calculus"), "quantification theory" (used by Quine), and perhaps others. The editors have not imposed uniformity here, and the reader should be aware of this diversity. Gödel's "Funktionenvariable" has been translated as "functional variable" and refers to what is now commonly known as a predicate letter. At times Stefan Bauer-Mengelberg's neologism "contentual" has been used to translate "inhaltlich".

Gödel 1929

When used with function names that have superscripts, like f_l^m, the prime notation introduced on page 90 becomes ambiguous, and we have inserted parentheses to mark the scope: $(f_l^m)'$. Moreover, in Gödel's text up to the end of Section 6, the superscript m was, at some occurrences, mistakenly interchanged with the subscript l.

	Original	Replaced by
86, 12	S_k von k-ter Stufe	S_i von i-ter Stufe
86, 24	Stelle X_i	Stelle der X_i

Gödel 1931

Changes marked by an asterisk were introduced by Gödel in an offprint of *1931* found in his *Nachlass*. Two changes that he made there have not been incorporated: "jede" to "jede[s] Glied" (166, 4) and roman to italics (our small capitals) for a number of instances of metamathematical notions. On the use of italics (in the original) or small capitals (in our version) for metamathematical notions, see Gödel's own remarks on page 595 of *van Heijenoort 1967*.

	Original	Replaced by
*148, 22	$\overline{n \; \epsilon \; K}$	$\overline{q \; \epsilon \; K}$
152, 18	18a	19a
152, 38	18a	19a
158, 19	auch R	auch $\overline{R}$
166, 19	$u * R(n \, Gl \, x) \; v$	$u * R(n \, Gl \, x) * v$
172, 4	rekursiv	*rekursiv*
*176, 31	Existenz	Existenz von aus κ

Gödel 1931e

	Original	Replaced by
216, 1	C. J.	C. I.

Gödel 1932

The first sentence of *1932* reads: "Für das von A. Heyting[1] aufgestellte System H des intuitionistischen Aussagenkalküls gelten folgende Sätze:". This was modified in the *1933n* reprinting to the following: "In Beantwortung einer von Hahn aufgeworfenen Frage gelten für das von A. Heyting[1] aufgestellte System H folgende Sätze:".

Gödel 1932b

The remark on line 3 of footnote 1, page 235 was sent by Gödel to van Heijenoort on 18 May 1966, with the following note: "I am sorry I am so

late in returning the revised draft translation of my *1932b*. The reason mainly was that I found an inaccuracy in footnote 1. This gave rise to certain considerations which I am exposing in a remark enclosed herewith. This matter seems important to me."

Gödel 1932c

	Original	Replaced by
240, 13	$(a \supset)x \in \mathfrak{T}_\alpha$	$(a \supset x) \in \mathfrak{T}_\alpha$

Gödel 1932g

	Original	Replaced by
248, 25	Widerspruchfreiheitsbeweise	Widerspruchsfreiheitsbeweise
248, 27	Widerspruchfreiheitsbeweis	Widerspruchsfreiheitsbeweis
248, 31	Widerspruchfreiheitsbeweisen	Widerspruchsfreiheitsbeweisen

Gödel 1932n

	Original	Replaced by
262, 15	in welche	in welches

Gödel 1933e

Here Gödel did not follow Heyting's *1930a* notation $\equiv$ for intuitionistic equality but used $=$ for both intuitionistic and classical equality.

Gödel 1933i

	Original	Replaced by
306, 20	Lehrstellen	Leerstellen
320, 22	$\mathfrak{J}_2$, daher nach	$\mathfrak{J}^2$, daher nach

Gödel 1933m

	Original	Replaced by
334, 3	und II.	und von II.

Gödel 1934

When this set of lectures, mimeographed as *1934*, was printed in *Davis 1965*, various textual modifications were made. In particular, the "Notes and errata" that had appeared at the end of *1934* were incorporated either directly into the text or into the footnotes. As a result, the original footnotes 1–12, 14–18, 20 and 21 became footnotes 3–5, 8–10, 12–14, 17–20, 24, 27–29, 31 and 32, respectively, of *Gödel 1965* as well as of the text printed in this volume. The original footnote 13 was suppressed in *1965*, as now, along with "(if x, y represent meaningful formulas)" in Definition 12 of Section 4, and a similar clause was omitted in Definitions 13 and 14. The first sentence of the present footnote 34 was the original footnote 19. As for the "Notes and errata", the original items 3–5 and 8 became the new footnotes 6, 7, 11 and 15, respectively. Footnote 16 now consists of the last part of item 9, the first part having been incorporated into the *1965* text. Footnote 2, incorporating item 1 of the "Notes and errata", was modified in the *1965* text to begin "This sentence could have been"; but we have retained the original wording of item 1: "This sentence may be". Finally, in *1965* the text and footnotes originally on page 28 became what is now the last paragraph of Section 8, and the text itself was amplified.

	Original	Replaced by
347, 1	Small Roman	Lower-case
347, 5	Roman capitals	Capitals
351, 18	class of expressions of the I-st (II-nd) kind shall be the least class	classes of expressions of the I-st and of the II-nd kind shall be the pair of classes with the least union
353, 7	rules of procedure	rules of inference
357, 25	$\mathrm{Imp}(y,z)$ & $M(z)$	$\mathrm{Imp}(y,z)$ & $M(y)$ & $M(z)$
358, 25	functional expression	propositional function
358, 37	$(i = 1,\ldots,m)$,	, where $i = 1,\ldots,m$,
361, 18	precise formulation of the requirement	precise [condition ... unprecise] requirement
363, 12	We now will	We shall now
364, 9	arithmetical	*arithmetical*
365, 4	$x + 1$,	$x + 1$, then
365, 5	c,	c, then
365, 36	the positive and negative integers	the integers
366, 33	$A \vee B$	we have $A \vee B$
371, 5	notes have been corrected in this volume.	notes were corrected when they were reprinted in *Davis 1965*.

Gödel 1934e

	Original	Replaced by
382, 10	$f\{f \dots f[f(x)]\}$	$f(\dots f(f(x))\dots)$

Gödel 1935

	Original	Replaced by
384, 1	mit der in Norsk Mat. Forenings Skr., [[...]] (vgl. dies Zbl. **7**, 193) vom	mit der als *1933a* vom

Gödel 1935a

In this review of *Huntington 1934* Gödel used the term "unmöglich" ("impossible") where Huntington in his article had used "necessarily false".

Gödel 1936a

This paper was delivered at the ninety-second meeting of Menger's colloquium on 19 June 1935. In the original printing of *1936a* this date is mis-cited as 19 June 1934. The error can be recognized through comparison with the dates of other sessions of the colloquium.

	Original	Replaced by
398, 13	x_{i_l}	x_{i_1}
398, 16	x_{i_1}	x_{j_1}

Gödel 1936b

	Original	Replaced by
398, 31	variablen bindenden	Variablen bindenden

References

Aanderaa, Stål
    See Dreben, Burton, Peter Andrews and Stål Aanderaa.

Aanderaa, Stål and Warren D. Goldfarb
    1974   The finite controllability of the Maslov case, *The journal of
            symbolic logic 39*, 509–518.

Ackermann, Wilhelm
    1924   Begründung des "tertium non datur" mittels der Hilbertschen
            Theorie der Widerspruchsfreiheit, *Mathematische Annalen 93*,
            1–36.
    1928   Zum Hilbertschen Aufbau der reellen Zahlen, *ibid. 99*, 118–
            133; English translation by Stefan Bauer-Mengelberg in *van
            Heijenoort 1967*, 493–507.
    1928a   Über die Erfüllbarkeit gewisser Zählausdrücke, *Mathematische
            Annalen 100*, 638–649.
    1940   Zur Widerspruchsfreiheit der Zahlentheorie, *ibid. 117*, 162–
            194.
    1951   Konstruktiver Aufbau eines Abschnitts der zweiten Cantor-
            schen Zahlenklasse, *Mathematische Zeitschrift 53*, 403–413.
    1954   *Solvable cases of the decision problem* (Amsterdam: North-
            Holland).
    See also Hilbert, David, and Wilhelm Ackermann.

Addison, John W.
    1958   Separation principles in the hierarchies of classical and effective
            descriptive set theory, *Fundamenta mathematicae 46*, 123–135.
    See also *Henkin et alii 1974*.

Albeverio, Sergio, Jens E. Fenstad, Raphael Hoegh-Krøhn, and Tom
Lindstrøm
    1985   (eds.) *Nonstandard methods in stochastic analysis and mathe-
            matical physics* (New York: Academic Press).

Aleksandrov, Pavel Sergeyevich (Alexandroff, Paul; Александров, Павел
Сергеевич)
    1916   Sur la puissance des ensembles mesurables B, *Comptes rendus
            hebdomadaires des séances de l'Académie des Sciences, Paris
            162*, 323–325.

Alt, Franz
1933   Zur Theorie der Krümmung, *Ergebnisse eines mathematischen Kolloquiums 4*, 4.
See also *Menger 1936*.

Andrews, Peter
See Dreben, Burton, Peter Andrews and Stål Aanderaa.

Bachmann, Heinz
1955   *Transfinite Zahlen*, Ergebnisse der Mathematik und ihrer Grenzgebiete, vol. 1 (Berlin: Springer).

Barendregt, Hendrick P.
1981   *The lambda calculus. Its syntax and semantics* (Amsterdam: North-Holland).
1984   Second edition of *Barendregt 1981*, with revisions (mainly in Part V) and addenda.

Bar-Hillel, Yehoshua
1965   (ed.) *Logic, methodology, and philosophy of science. Proceedings of the 1964 International Congress* (Amsterdam: North-Holland).
See also Fraenkel, Abraham A., and Yehoshua Bar-Hillel.

Bar-Hillel, Yehoshua, E. I. J. Poznanski, Michael O. Rabin and Abraham Robinson
1961   (eds.) *Essays on the foundations of mathematics, dedicated to A. A. Fraenkel on his seventieth anniversary* (Jerusalem: Magnes Press; Amsterdam: North-Holland).

Barwise, Jon
1977   (ed.) *Handbook of mathematical logic* (Amsterdam: North-Holland).

Barzin, Marcel
1940   Sur la portée du théorème de M. Gödel, *Académie royale de Belgique, Bulletin de la classe des sciences (5) 26*, 230–239.

Becker, Oskar
1930   Zur Logik der Modalitäten, *Jahrbuch für Philosophie und phänomenologische Forschung 11*, 497–548.

Beeson, Michael
1978   A type-free Gödel interpretation, *The journal of symbolic logic 43*, 213–227.

Benacerraf, Paul, and Hilary Putnam
1964   (eds.) *Philosophy of mathematics: selected readings* (Engle-wood Cliffs, N. J.: Prentice-Hall; Oxford: Blackwell).

Bergmann, Gustav
1931   Zur Axiomatik der Elementargeometrie, *Ergebnisse eines mathematischen Kolloquiums 1*, 28–30.

Bernays, Paul
1923   Erwiderung auf die Note von Herrn Aloys Müller: "Über Zahlen als Zeichen", *Mathematische Annalen 90*, 159–163; reprinted in *Annalen der Philosophie und philosophischen Kritik 4* (1924), 492–497.

1926   Axiomatische Untersuchung des Aussagen-Kalkuls der *Principia mathematica*, *Mathematische Zeitschrift 25*, 305–320.

1927   Probleme der theoretischen Logik, *Unterrichtsblätter für Mathematik und Naturwissenschaften 33*, 369–377; reprinted in *Bernays 1976*, 1–16.

1935   Sur le platonisme dans les mathématiques, *L'enseignement mathématique 34*, 52–69; English translation by Charles D. Parsons in *Benacerraf and Putnam 1964*, 274–286.

1935a   Hilberts Untersuchungen über die Grundlagen der Arithmetik, in *Hilbert 1935*, 196–216.

1937   A system of axiomatic set theory. Part I, *The journal of symbolic logic 2*, 65–77; reprinted in *Bernays 1976a*, 1–13.

1941   A system of axiomatic set theory. Part II, *The journal of symbolic logic 6*, 1–17; reprinted in *Bernays 1976a*, 14–30.

1941a   Sur les questions méthodologiques actuelles de la théorie hilbertienne de la démonstration, in *Gonseth 1941*, 144–152.

1942   A system of axiomatic set theory. Part III. Infinity and enumerability. Analysis, *The journal of symbolic logic 7*, 65–89; reprinted in *Bernays 1976a*, 31–55.

1943   A system of axiomatic set theory. Part V. General set theory (continued), *The journal of symbolic logic 8*, 89–106; reprinted in *Bernays 1976a*, 69–86.

1954   Zur Beurteilung der Situation in der beweistheoretischen Forschung, *Revue internationale de philosophie 8*, 9–13.

1961   Zur Frage der Unendlichkeitsschemata in der axiomatischen Mengenlehre, in *Bar-Hillel et alii 1961*, 3–49; English translation by J. Bell and M. Plänitz in *Bernays 1976a*, 121–172.

1967   Hilbert, David, in *Edwards 1967*, vol. 3, 496–504.

1976   *Abhandlungen zur Philosophie der Mathematik* (Darmstadt: Wissenschaftliche Buchgesellschaft).

1976a   *Sets and classes: on the work by Paul Bernays*, edited by Gert
        H. Müller (Amsterdam: North-Holland).
1976b   Kurze Biographie, in *Bernays 1976a*, xiv–xvi; English transla-
        tion in *Bernays 1976a*, xi–xiii.
See also Hilbert, David, and Paul Bernays.

Bernays, Paul, and Abraham A. Fraenkel
1958   *Axiomatic set theory* (Amsterdam: North-Holland).

Bernays, Paul, and Moses Schönfinkel
1928   Zum   Entscheidungsproblem   der   mathematischen   Logik,
        Mathematische Annalen 99, 342–372.

Bernstein, Benjamin A.
1931   Whitehead and Russell's theory of deduction as a mathemati-
        cal science, *Bulletin of the American Mathematical Society 37*,
        480–488.

Bernstein, Felix
1901   *Untersuchungen aus der Mengenlehre* (doctoral dissertation,
        Göttingen; printed at Halle); reprinted with several alterations
        in *Mathematische Annalen 61* (1905), 117–155.

Betsch, Christian
1926   *Fiktionen in der Mathematik* (Stuttgart: Frommanns).

Bianchi, Luigi
1918   *Lezioni sulla teoria dei gruppi continui finiti di transformazioni*
        (Pisa: E. Spoerri).

Birkhoff, Garrett
1933   On the combination of subalgebras, *Proceedings of the Cam-
        bridge Philosophical Society 29*, 441–464.
1935   Combinatorial relations in projective geometries, *Annals of
        mathematics (2) 36*, 743–748.
1938   Lattices and their applications, *Bulletin of the American
        Mathematical Society 44*, 793–800.
1940   *Lattice theory*, Colloquium publications, vol. 25 (New York:
        American Mathematical Society).

Blackwell, Kenneth
1976   A non-existent revision of *Introduction to mathematical phi-
        losophy*, *Russell: the journal of the Bertrand Russell Archives*
        no. 20, 16–18.

Blum, Manuel
1967   A machine-independent theory of the complexity of recursive functions, *Journal of the Association for Computing Machinery 14*, 322–336.

Blumenthal, Leonard M.
1940   "A paradox, a paradox, a most ingenious paradox", *American mathematical monthly 47*, 346–353.
See also Menger, Karl, and Leonard M. Blumenthal.

Boolos, George
1979   *The unprovability of consistency. An essay in modal logic* (Cambridge, U.K.: Cambridge University Press).

Braun, Stefania, and Wacław Sierpiński
1932   Sur quelques propositions équivalentes à l'hypothèse du continu, *Fundamenta mathematicae 19*, 1–7.

Brouwer, Luitzen E. J.
1907   *Over de grondslagen der wiskunde* (Amsterdam: Maas & van Suchtelen); English translation by Arend Heyting and Dr. Gibson in *Brouwer 1975*, 11–101.
1909   Die möglichen Mächtigkeiten, *Atti dei IV Congresso Internazionale dei Matematici, Roma, 6–11 Aprile 1908* (Rome: Accademia dei Lincei), III, 569–571.
1929   Mathematik, Wissenschaft und Sprache, *Monatshefte für Mathematik und Physik 36*, 153–164; reprinted in *Brouwer 1975*, 417–428.
1930   *Die Struktur des Kontinuums* (Vienna: Gistel); reprinted in *Brouwer 1975*, 429–440.
1975   *Collected works*, edited by Arend Heyting, vol. 1 (Amsterdam: North Holland).

Browder, Felix E.
1976   (ed.) *Mathematical developments arising from Hilbert problems*, Proceedings of symposia in pure mathematics, vol. 28 (Providence, R.I.: American Mathematical Society).

Buchholz, Wilfried, Solomon Feferman, Wolfram Pohlers and Wilfried Sieg
1981   *Iterated inductive definitions and subsystems of analysis: recent proof-theoretical studies*, Springer lecture notes in mathematics, no. 897 (Berlin: Springer).

Cantor, Georg
1874 Über eine Eigenschaft des Inbegriffes aller reellen algebraischen Zahlen, *Journal für die reine und angewandte Mathematik 77*, 258–262; reprinted in *Cantor 1932*, 115–118.

1878 Ein Beitrag zur Mannigfaltigkeitslehre, *Journal für die reine und angewandte Mathematik 84*, 242–258; reprinted in *Cantor 1932*, 119–133.

1883 Ueber unendliche, lineare Punktmannichfaltigkeiten. V, *Mathematische Annalen 21*, 545–591; reprinted in *Cantor 1932*, 165–209.

1884 De la puissance des ensembles parfaits de points, *Acta mathematica 4*, 381–392; reprinted in *Cantor 1932*, 252–260.

1891 Über eine elementare Frage der Mannigfaltigkeitslehre, *Jahresbericht der Deutschen Mathematiker-Vereinigung 1*, 75–78; reprinted in *Cantor 1932*, 278–281.

1932 *Gesammelte Abhandlungen mathematischen und philosphischen Inhalts. Mit erläuternden Anmerkungen sowie mit Ergänzungen aus dem Briefwechsel Cantor–Dedekind*, edited by Ernst Zermelo (Berlin: Springer); reprinted in 1962 (Hildesheim: Olms).

Capelli, Alfredo
1897 Saggio sulla introduzione dei numeri irrazionali col metodo delle classi contigue, *Giornale di matematiche di Battaglini 35*, 209–234.

Carnap, Rudolf
1931 Die logizistische Grundlegung der Mathematik, *Erkenntnis 2*, 91–105; English translation by Erna Putnam and Gerald J. Massey in *Benacerraf and Putnam 1964*, 31–41.

1934 Die Antinomien und die Unvollständigkeit der Mathematik, *Monatshefte für Mathematik und Physik 41*, 263–284.

1934a *Logische Syntax der Sprache* (Vienna: Springer); translated into English as *Carnap 1937*.

1935 Ein Gültigkeitskriterium für die Sätze der klassischen Mathematik, *Monatshefte für Mathematik und Physik 42*, 163–190.

1937 *The logical syntax of language* (London: Paul, Trench, Trubner; New York: Harcourt, Brace, and Co.); English translation by Amethe Smeaton of *Carnap 1934a*, with revisions.
See also *Hahn et alii 1931*.

Chen, Kien-Kwong
1933 Axioms for real numbers, *Tôhoku mathematical journal 37*, 94–99.

Chihara, Charles S.
1973 *Ontology and the vicious-circle principle* (Ithaca, N.Y.: Cornell University Press).
1982 A Gödelian thesis regarding mathematical objects: Do they exist? And can we perceive them?, *The philosophical review 91*, 211–227.

Christian, Curt
1980 Leben und Wirken Kurt Gödels, *Monatshefte für Mathematik 89*, 261–273.

Church, Alonzo
1932 A set of postulates for the foundation of logic, *Annals of mathematics (2) 33*, 346–366.
1933 A set of postulates for the foundation of logic (second paper), *ibid. (2) 34*, 839–864.
1935 A proof of freedom from contradiction, *Proceedings of the National Academy of Sciences, U.S.A. 21*, 275–281.
1936 An unsolvable problem of elementary number theory, *American journal of mathematics 58*, 345–363; reprinted in *Davis 1965*, 88–107.
1936a A note on the Entscheidungsproblem, *The journal of symbolic logic 1*, 40–41; correction, *ibid.* 101–102; reprinted in *Davis 1965*, 108–115, with the correction incorporated.
1941 *The calculi of lambda-conversion*, Annals of mathematics studies, vol. 6 (Princeton: Princeton University Press); second printing, 1951.
1942 Review of *Quine 1941*, *The journal of symbolic logic 7*, 100–101.
1943 Carnap's introduction to semantics, *Philosophical review 52*, 298–304.
1976 Comparison of Russell's resolution of the semantical antinomies with that of Tarski, *The journal of symbolic logic 41*, 747–760.

Church, Alonzo, and J. Barkley Rosser
1936 Some properties of conversion, *Transactions of the American Mathematical Society 39*, 472–482.

Chwistek, Leon
1933 Die nominalistische Grundlegung der Mathematik, *Erkenntnis 3*, 367–388.

Cohen, Paul J.
1963   The independence of the continuum hypothesis. I, *Proceedings of the National Academy of Sciences, U.S.A. 50*, 1143–1148.
1964   The independence of the continuum hypothesis. II, *ibid. 51*, 105–110.
1966   *Set theory and the continuum hypothesis* (New York: Benjamin).

Crossley, John N., and Michael A. E. Dummett
1965   (eds.) *Formal systems and recursive functions* (Amsterdam: North-Holland).

Dauben, Joseph W.
1982   Peirce's place in mathematics, *Historia mathematica 9*, 311–325.

Davies, Roy O.
1963   Covering the plane with denumerably many curves, *The journal of the London Mathematical Society 38*, 433–438.

Davis, Martin
1965   (ed.) *The undecidable: basic papers on undecidable propositions, unsolvable problems, and computable functions* (Hewlett, N.Y.: Raven Press).
1982   Why Gödel didn't have Church's thesis, *Information and control 54*, 3–24.

Davis, Martin, Yuri Matiyasevich and Julia Robinson
1976   Hilbert's tenth problem. Diophantine equations: positive aspects of a negative solution, in *Browder 1976*, 323–378.

Dawson, John W., Jr.
1983   The published work of Kurt Gödel: an annotated bibliography, *Notre Dame journal of formal logic 24*, 255–284; addenda and corrigenda, *ibid. 25*, 283–287.
1984   Discussion on the foundation of mathematics, *History and philosophy of logic 5*, 111–129.
1984a   Kurt Gödel in sharper focus, *The mathematical intelligencer 6*, no. 4, 9–17.
1985   The reception of Gödel's incompleteness theorems, *PSA 1984: proceedings of the 1984 biennial meeting of the Philosophy of Science Association 2*, to appear.
1985a   Completing the Gödel–Zermelo correspondence, *Historia mathematica 12*, 66–70.

Dedekind, Richard

1872   *Stetigkeit und irrationale Zahlen* (Braunschweig: Vieweg);
       English translation in *Dedekind 1901*, 1–27.

1888   *Was sind und was sollen die Zahlen?* (Braunschweig: Vieweg);
       English translation in *Dedekind 1901*, 31–115.

1901   *Essays on the theory of numbers: Continuity and irrational
       numbers. The nature and meaning of numbers*, English trans-
       lation of *1872* and *1888* by Wooster W. Beman (Chicago: Open
       Court); reprinted in 1963 (New York: Dover).

Dehn, Max

1926   Die Grundlagen der Geometrie in historischen Entwicklung, in
       Pasch 1926, 185–271.

Dekker, J. C. E.

1962   (ed.) *Recursive function theory*, Proceedings of symposia in
       pure mathematics, vol. 5 (Providence R.I.: American Mathe-
       matical Society).

Denton, John

See Dreben, Burton, and John Denton.

Diller, Justus

1968   Zur Berechenbarkeit primitiv-rekursiver Funktionale endlicher
       Typen, in *Schmidt et alii 1968*, 109–120.

1979   Functional interpretations of Heyting's arithmetic in all finite
       types, *Nieuw archief voor wiskunde (3) 27*, 70–97.

Diller, Justus, and Gert H. Müller

1975   (eds.) $\models$ *ISLIC Proof theory symposion. Dedicated to Kurt
       Schütte on the occasion of his 65th birthday. Proceedings of
       the International Summer Institute and Logic Colloquium, Kiel
       1974*, Springer lecture notes in mathematics, no. 500 (Berlin:
       Springer).

Diller, Justus, and Werner Nahm

1974   Eine Variante zur Dialectica-Interpretation der Heyting-Arith-
       metik endlicher Typen, *Archiv für mathematische Logik und
       Grundlagenforschung 16*, 49–66.

Diller, Justus, and Kurt Schütte

1971   Simultane Rekursionen in der Theorie der Funktionale end-
       licher Typen, *Archiv für mathematische Logik und Grundla-
       genforschung 14*, 69–74.

Diller, Justus, and Helmut Vogel
1975   Intensionale Funktionalinterpretation der Analysis, in *Diller and Müller 1975*, 56–72.

Dingler, Hugo
1931   *Philosophie der Logik und Arithmetik* (Munich: Reinhardt).

Dragalin, Albert G. (Драгалин, Альберт Г.)
1968   The computability of primitive recursive terms of finite type and primitive recursive realizability (Russian), *Zapiski nauchnyk seminarov Leningradskogo otdeleniya Matematicheskogo Instituta im. V. A. Steklova, Akademii nauk S.S.S.R. (Leningrad) 8*, 32–45; English translation in *Slisenko 1970*, 13–18.
1980   New forms of realizability and Markov's rule, *Soviet mathematics doklady 21*, 461–464.

Dreben, Burton
1952   On the completeness of quantification theory, *Proceedings of the National Academy of Sciences, U.S.A. 38*, 1047–1052.
1962   Solvable Surányi subclasses: an introduction to the Herbrand theory, *Proceedings of a Harvard symposium on digital computers and their applications, 3–6 April 1961 (The annals of the Computation Laboratory of Harvard University 31)* (Cambridge, Mass.: Harvard University Press), 32–47.

Dreben, Burton, Peter Andrews and Stål Aanderaa
1963   False lemmas in Herbrand, *Bulletin of the American Mathematical Society 69*, 699–706.

Dreben, Burton, and John Denton
1970   Herbrand-style consistency proofs, in *Myhill et alii 1970*, 419–433.

Dreben, Burton, and Warren D. Goldfarb
1979   *The decision problem: solvable classes of quantificational formulas* (Reading, Mass.: Addison–Wesley).

Dummett, Michael A. E.
1959   A propositional calculus with denumerable matrix, *The journal of symbolic logic 24*, 97–106.
1978   *Truth and other enigmas* (London: Duckworth).
See also Crossley, John N., and Michael A. E. Dummett.

Dyson, Freeman
1983   Unfashionable pursuits, *The mathematical intelligencer 5*, no. 3, 47–54.

Easton, William B.
1964   *Powers of regular cardinals* (doctoral dissertation, Princeton University); reprinted in part as *Easton 1970*.

1964a   Proper classes of generic sets, *Notices of the American Mathematical Society 11*, 205.

1970   Powers of regular cardinals, *Annals of mathematical logic 1*, 139–178.

Edwards, Paul
1967   (ed.) *The encyclopedia of philosophy* (New York: Macmillan and the Free Press).

Ehrenfeucht, Andrzej, and Jan Mycielski
1971   Abbreviating proofs by adding new axioms, *Bulletin of the American Mathematical Society 77*, 366–367.

Ellis, G. F. R.
See Hawking, Stephen W. and G. F. R. Ellis.

Erdös, Paul
1963   On a problem in graph theory, *Mathematical gazette 47*, 220–223.

Errera, Alfred
1952   Le problème du continu, *Atti della Accademia Ligure di Scienze e Lettere (Roma) 9*, 176–183.

Feferman, Solomon
1955   Review of *Wang 1951*, *The journal of symbolic logic 20*, 76–77.

1960   Arithmetization of metamathematics in a general setting, *Fundamenta mathematicae 49*, 35–92.

1962   Transfinite recursive progressions of axiomatic theories, *The journal of symbolic logic 27*, 259–316.

1964   Systems of predicative analysis, *ibid. 29*, 1–30.

1966   Predicative provability in set theory, *Bulletin of the American Mathematical Society 72*, 486–489.

1971   Ordinals and functionals in proof theory, *Actes du Congrès international des mathématiciens, 1–10 septembre 1970, Nice, France* (Paris: Gauthier-Villars), vol. I, 229–233.

1974   Predicatively reducible systems of set theory, in *Jech 1974*, 11–32.

1977   Theories of finite type related to mathematical practice, in *Barwise 1977*, 913–971.

1984   Toward useful type-free theories. I., *The journal of symbolic logic 49*, 75–111.

1984a   Kurt Gödel: conviction and caution, *Philosophia naturalis 21*, 546–562.

See also Buchholz, Wilfried, et alii.

Feferman, Solomon, and Clifford Spector
1962   Incompleteness along paths in progressions of theories, *The journal of symbolic logic 27*, 383–390.

Feferman, Solomon, and Alfred Tarski
1953   Review of *Rasiowa and Sikorski 1951*, *The journal of symbolic logic 18*, 339–340.

Feigl, Herbert
1969   The Wiener Kreis in America, in *Fleming and Bailyn 1969*, 630–673.

Fenstad, Jens E.
1971   (ed.) *Proceedings of the second Scandinavian logic symposium* (Amsterdam: North-Holland).
1985   Is nonstandard analysis relevant for the philosophy of mathematics?, *Synthèse 62*, 289–301.

Fleming, Donald, and Bernard Bailyn
1969   (eds.) *The intellectual migration: Europe and America, 1930–1960* (Cambridge, Mass.: Harvard University Press).

Fourman, Michael P., C. J. Mulvey and Dana S. Scott
1979   (eds.) *Applications of sheaves*, Springer lecture notes in mathematics, no. 753 (Berlin: Springer).

Fourman, Michael P., and Dana S. Scott
1979   Sheaves and logic, in *Fourman, Mulvey and Scott 1979*, 302–401.

Fraenkel, Abraham A.
1919   *Einleitung in die Mengenlehre* (Berlin: Springer).
1922   Der Begriff 'definit' und die Unabhängigkeit des Auswahlaxioms, *Sitzungsberichte der Preussischen Akademie der Wissenschaften, Physikalisch-mathematische Klasse*, 253–257; English translation by Beverly Woodward in *van Heijenoort 1967*, 284–289.
1925   Untersuchungen über die Grundlagen der Mengenlehre, *Mathematische Zeitschrift 22*, 250–273.
1927   *Zehn Vorlesungen über die Grundlegung der Mengenlehre* (Leipzig: Teubner).

1928   Third, revised edition of *Fraenkel 1919* (Berlin: Springer).
See also Bernays, Paul, and Abraham A. Fraenkel.

Fraenkel, Abraham A., and Yehoshua Bar-Hillel
1958   *Foundations of set theory* (Amsterdam: North-Holland).

Frayne, Thomas, Anne Morel and Dana S. Scott
1962   Reduced direct products, *Fundamenta mathematicae 51*, 195–228.

Frege, Gottlob
1879   *Begriffsschrift, eine der arithmetischen nachgebildete Formel-sprache des reinen Denkens* (Halle: Nebert); reprinted in *Frege 1964*; English translation by Stefan Bauer-Mengelberg in *van Heijenoort 1967*, 1—82, and by Terrell W. Bynum in *Frege 1972*, 101–203.
1892   Über Sinn und Bedeutung, *Zeitschrift für Philosophie und phi-losophische Kritik* (n. s.) *100*, 25–50; English translation by Max Black in *Frege 1952*, 56–78.
1903   *Grundgesetze der Arithmetik, begriffsschriftlich abgeleitet* (Jena: Pohle), vol. 2.
1952   *Translations from the philosophical writings of Gottlob Frege*, edited by Peter Geach and Max Black (Oxford: Blackwell); third edition, 1980.
1964   *Begriffsschrift und andere Aufsätze*, edited by Ignacio Angelelli (Hildesheim: Olms).
1972   *Conceptual notation and related articles*, translated and edited by Terrell Ward Bynum (Oxford: Clarendon Press).

Friedman, Harvey
1973   The consistency of classical set theory relative to a set theory with intuitionistic logic, *The journal of symbolic logic 38*, 315–319.
1978   Classically and intuitionistically provably recursive functions, in *Müller and Scott 1978*, 21–27.

Friedrich, Wolfgang
1984   Spielquantorinterpretation unstetiger Funktionale der höheren Analysis, *Archiv für mathematische Logik und Grundlagen-forschung 24*, 73–99.
1985   Gödelsche Funktionalinterpretation für eine Erweiterung der klassischen Analysis um einen Spielquantor, Extensionalität von Typ 2 und Stetigkeit, *Zeitschrift für mathematische Logik und Grundlagen der Mathematik*, to appear.

Gaifman, Haim
1976   Models and types of Peano's arithmetic, *Annals of mathematical logic 9*, 223–306.

Gentzen, Gerhard
1935   Untersuchungen über das logische Schließen, *Mathematische Zeitschrift 39*, 176–210, 405–431; English translation by M. E. Szabo in *Gentzen 1969*, 68–131.
1936   Die Widerspruchsfreiheit der reinen Zahlentheorie, *Mathematische Annalen 112*, 493–565; English translation by M. E. Szabo in *Gentzen 1969*, 132–213.
1969   *The collected papers of Gerhard Gentzen*, edited and translated into English by M. E. Szabo (Amsterdam: North-Holland).

Girard, Jean-Yves
1971   Une extension de l'interprétation de Gödel à l'analyse, et son application à l'élimination des coupures dans l'analyse et la théorie des types, in *Fenstad 1971*, 63–92.
1972   *Interprétation fonctionelle et élimination des coupures de l'arithmétique d'ordre supérieur* (doctoral dissertation, Université de Paris VII).
1982   Herbrand's theorem and proof theory, in *Stern 1982*, 29–38.

Glivenko, Valerii Ivanovich (Гливенко, Валерий Иванович)
1929   Sur quelques points de la logique de M. Brouwer, *Académie royale de Belgique, Bulletin de la classe des sciences (5) 15*, 183–188.

Gödel, Kurt
1929   *Über die Vollständigkeit des Logikkalküls* (doctoral dissertation, University of Vienna).
1930   Die Vollständigkeit der Axiome des logischen Funktionenkalküls, *Monatshefte für Mathematik und Physik 37*, 349–360.
1930a   Über die Vollständigkeit des Logikkalküls, *Die Naturwissenschaften 18*, 1068.
1930b   Einige metamathematische Resultate über Entscheidungsdefinitheit und Widerspruchsfreiheit, *Anzeiger der Akademie der Wissenschaften in Wien 67*, 214–215.
1931   Über formal unentscheidbare Sätze der *Principia mathematica* und verwandter Systeme I, *Monatshefte für Mathematik und Physik 38*, 173–198.
1931a   Diskussion zur Grundlegung der Mathematik (Gödel's remarks in *Hahn et alii 1931*), *Erkenntnis 2*, 147–151.
1931b   Review of *Neder 1931*, *Zentralblatt für Mathematik und ihre Grenzgebiete 1*, 5–6.

1931c Review of *Hilbert 1931*, *ibid. 1*, 260.

1931d Review of *Betsch 1926*, *Monatshefte für Mathematik und Physik (Literaturberichte) 38*, 5.

1931e Review of *Becker 1930*, *ibid. 38*, 5–6.

1931f Review of *Hasse and Scholz 1928*, *ibid. 38*, 37.

1931g Review of *von Juhos 1930*, *ibid. 38*, 39.

1932 Zum intuitionistischen Aussagenkalkül, *Anzeiger der Akademie der Wissenschaften in Wien 69*, 65–66; reprinted, with additional comment, as *1933n*.

1932a Ein Spezialfall des Entscheidungsproblems der theoretischen Logik, *Ergebnisse eines mathematischen Kolloquiums 2*, 27–28.

1932b Über Vollständigkeit und Widerspruchsfreiheit, *ibid. 3*, 12–13.

1932c Eine Eigenschaft der Realisierungen des Aussagenkalküls, *ibid. 3*, 20–21.

1932d Review of *Skolem 1931*, *Zentralblatt für Mathematik und ihre Grenzgebiete 2*, 3.

1932e Review of *Carnap 1931*, *ibid. 2*, 321.

1932f Review of *Heyting 1931*, *ibid. 2*, 321–322.

1932g Review of *von Neumann 1931*, *ibid. 2*, 322.

1932h Review of *Klein 1931*, *ibid. 2*, 323.

1932i Review of *Hoensbroech 1931*, *ibid. 3*, 289.

1932j Review of *Klein 1932*, *ibid. 3*, 291.

1932k Review of *Church 1932*, *ibid. 4*, 145–146.

1932l Review of *Kalmár 1932*, *ibid. 4*, 146.

1932m Review of *Huntington 1932*, *ibid. 4*, 146.

1932n Review of *Skolem 1932*, *ibid. 4*, 385.

1932o Review of *Dingler 1931*, *Monatshefte für Mathematik und Physik (Literaturberichte) 39*, 3.

1933 Untitled remark following *Parry 1933*, *Ergebnisse eines mathematischen Kolloquiums 4*, 6.

1933a Über Unabhängigkeitsbeweise im Aussagenkalkül, *ibid. 4*, 9–10.

1933b Über die metrische Einbettbarkeit der Quadrupel des R_3 in Kugelflächen, *ibid. 4*, 16–17.

1933c Über die Waldsche Axiomatik des Zwischenbegriffes, *ibid. 4*, 17–18.

1933d Zur Axiomatik der elementargeometrischen Verknüpfungsrelationen, *ibid. 4*, 34.

1933e Zur intuitionistischen Arithmetik und Zahlentheorie, *ibid. 4*, 34–38.

1933f Eine Interpretation des intuitionistischen Aussagenkalküls, *ibid. 4*, 39–40.

1933g Bemerkung über projektive Abbildungen, *ibid. 5*, 1.

1933h (with K. Menger and A. Wald) Diskussion über koordinaten-lose Differentialgeometrie, *ibid.* 5, 25–26.

1933i Zum Entscheidungsproblem des logischen Funktionenkalküls, *Monatshefte für Mathematik und Physik 40*, 433–443.

1933j Review of *Kaczmarz 1932*, Zentralblatt für Mathematik und ihre Grenzgebiete 5, 146.

1933k Review of *Lewis 1932*, *ibid.* 5, 337–338.

1933l Review of *Kalmár 1933*, *ibid.* 6, 385–386.

1933m Review of *Hahn 1932*, Monatshefte für Mathematik und Physik (*Literaturberichte*) 40, 20–22.

1933n Reprint of *Gödel 1932*, with additional comment, *Ergebnisse eines mathematischen Kolloquiums 4*, 40.

1934 *On undecidable propositions of formal mathematical systems* (mimeographed lecture notes, taken by Stephen C. Kleene and J. Barkley Rosser); reprinted with revisions in *Davis 1965*, 39–74.

1934a Review of *Skolem 1933*, Zentralblatt für Mathematik und ihre Grenzgebiete 7, 97–98.

1934b Review of *Quine 1933*, *ibid.* 7, 98.

1934c Review of *Skolem 1933a*, *ibid.* 7, 193–194.

1934d Review of *Chen 1933*, *ibid.* 7, 385.

1934e Review of *Church 1933*, *ibid.* 8, 289.

1934f Review of *Notcutt 1934*, *ibid.* 9, 3.

1935 Review of *Skolem 1934*, *ibid.* 10, 49.

1935a Review of *Huntington 1934*, *ibid.* 10, 49.

1935b Review of *Carnap 1934*, *ibid.* 11, 1.

1935c Review of *Kalmár 1934*, *ibid.* 11, 3–4.

1936 Untitled remark following *Wald 1936*, *Ergebnisse eines mathematischen Kolloquiums 7*, 6.

1936a Über die Länge von Beweisen, *ibid.* 7, 23–24.

1936b Review of *Church 1935*, Zentralblatt für Mathematik und ihre Grenzgebiete 12, 241–242.

1938 The consistency of the axiom of choice and of the generalized continuum hypothesis, *Proceedings of the National Academy of Sciences, U.S.A. 24*, 556–557.

1939 The consistency of the generalized continuum hypothesis, *Bulletin of the American Mathematical Society 45*, 93.

1939a Consistency proof for the generalized continuum hypothesis, *Proceedings of the National Academy of Sciences, U.S.A. 25*, 220–224; errata, *1947*, footnote 23.

1940   *The consistency of the axiom of choice and of the generalized continuum hypothesis with the axioms of set theory*, Annals of mathematics studies, vol. 3 (Princeton: Princeton University Press), lecture notes taken by George W. Brown; reprinted with additional notes in 1951 and with further notes in 1966.

1944   Russell's mathematical logic, in *Schilpp 1944*, 123–153.

1946   Remarks before the Princeton bicentennial conference on problems in mathematics, 1–4; first published in *Davis 1965*, 84–88.

1947   What is Cantor's continuum problem?, *American mathematical monthly 54*, 515–525; errata, *55*, 151.

1949   An example of a new type of cosmological solutions of Einstein's field equations of gravitation, *Reviews of modern physics 21*, 447–450.

1949a   A remark about the relationship between relativity theory and idealistic philosophy, in *Schilpp 1949*, 555–562.

1952   Rotating universes in general relativity theory, *Proceedings of the International Congress of Mathematicians; Cambridge, Massachusetts, U.S.A. August 30-September 6, 1950* (Providence, R.I.: American Mathematical Society, 1952), I, 175–181.

1955   Eine Bemerkung über die Beziehungen zwischen der Relativitätstheorie und der idealistischen Philosophie (German translation of *Gödel 1949a* by Hans Hartmann), in *Schilpp 1955*, 406–412.

1958   Über eine bisher noch nicht benützte Erweiterung des finiten Standpunktes, *Dialectica 12*, 280–287.

1962   Postscript to *Spector 1962*, 27.

1964   Revised and expanded version of *Gödel 1947*, in *Benacerraf and Putnam 1964*, 258–273.

1964a   Reprint, with some alterations, of *Gödel 1944*, in *Benacerraf and Putnam 1964*, 211–232.

1965   Expanded version of *Gödel 1934*, in *Davis 1965*, 39–74.

1967   English translation of *Gödel 1931* by Jean van Heijenoort, in *van Heijenoort 1967*, 596–616.

1968   Reprint, with some alterations, of *Gödel 1946*, in *Klibansky 1968*, 250–253.

1972   On an extension of finitary mathematics which has not yet been used (to have appeared in *Dialectica*; to be published in volume II of Gödel's *Collected works*), revised and expanded English translation of *Gödel 1958*.

1972a   Some remarks on the undecidability results (to have appeared in *Dialectica*; to be published in volume II of Gödel's *Collected works*).

1972b   Reprint, with some alterations, of *Gödel 1944*, in *Pears 1972*, 192–226.
1974   Untitled remarks, in *Robinson 1974*, x.

Goldblatt, Robert
1978   Arithmetical necessity, provability and intuitionistic logic, *Theoria 44*, 38–46.

Goldfarb, Warren D.
1971   Review of *Skolem 1970*, *The journal of philosophy 68*, 520–530.
1979   Logic in the twenties: the nature of the quantifier, *The journal of symbolic logic 44*, 351–368.
1981   The undecidability of the second-order unification problem, *Theoretical computer science 13*, 225–230.
1984   The Gödel class with identity is unsolvable, *Bulletin of the American Mathematical Society 10*, 113–115.
1984a   The unsolvability of the Gödel class with identity, *The journal of symbolic logic 49*, 1237–1252.
See also Aanderaa, Stål, and Warren D. Goldfarb.
See also Dreben, Burton, and Warren D. Goldfarb.

Goldstine, Herman H.
1972   *The computer from Pascal to von Neumann* (Princeton: Princeton University Press).

Gonseth, Ferdinand
1941   (ed.) *Les entretiens de Zurich, 6-9 décembre 1938* (Zurich: Leemann).

Goodman, Nicolas D.
1984   Epistemic arithmetic is a conservative extension of intuitionistic arithmetic, *The journal of symbolic logic 49*, 192–203.

Goodstein, Reuben L.
1945   Function theory in an axiom-free equation calculus, *Proceedings of the London Mathematical Society (2) 48*, 401–434.
1957   *Recursive number theory* (Amsterdam: North-Holland).

Grassl, Wolfgang
1982   (ed.) *Friedrich Waismann, lectures on the philosophy of mathematics*, Studien zur österreichischen Philosophie, vol. 4 (Amsterdam: Rodopi).

Grattan-Guinness, Ivor
1979   In memoriam Kurt Gödel: his 1931 correspondence with Zermelo on his incompletability theorem, *Historia mathematica 6*, 294–304.

Greenberg, Marvin J.
1974   *Euclidean and non-Euclidean geometries: development and history* (San Francisco: Freeman).
1980   Second edition of *Greenberg 1974*.

Grzegorczyk, Andrzej
1964   Recursive objects in all finite types, *Fundamenta mathematicae 54*, 73–93.
1967   Some relational systems and the associated topological spaces, *Fundamenta mathematicae 60*, 223-231.

Gurevich, Yuri, and Saharon Shelah
1983   Random models and the Gödel case of the decision problem, *The journal of symbolic logic 48*, 1120–1124.

Hacking, Ian
1963   What is strict implication?, *The journal of symbolic logic 28*, 51–71.

Hahn, Hans
1921   *Theorie der reellen Funktionen* (Berlin: Springer).
1932   *Reelle Funktionen* (Leipzig: Akademische Verlagsgesellschaft).
1980   *Empiricism, logic and mathematics: philosophical papers*, edited by Brian McGuinness (Dordrecht: Reidel).

Hahn, Hans, Rudolf Carnap, Kurt Gödel, Arend Heyting, Kurt Reidemeister, Arnold Scholz and John von Neumann
1931   Diskussion zur Grundlegung der Mathematik, *Erkenntnis 2*, 135–151; English translation by John W. Dawson, Jr., in *Dawson 1984*.

Hajnal, András
1956   On a consistency theorem connected with the generalized continuum problem, *Zeitschrift für mathematische Logik und Grundlagen der Mathematik 2*, 131–136.

Hanatani, Yoshito
1975   Calculability of the primitive recursive functionals of finite type over the natural numbers, in *Diller and Müller 1975*, 152–163.

Hanf, William P., and Dana S. Scott
1961 Classifying inaccessible cardinals, *Notices of the American Mathematical Society 8*, 445.

Harrington, Leo
See Paris, Jeff, and Leo Harrington.

Hartmanis, Juris
1978 *Feasible computations and provable complexity properties*, CBMS–NSF regional conference series in applied mathematics (Philadelphia: Society for Industrial and Applied Mathematics).

Hasse, Helmut and Heinrich Scholz
1928 *Die Grundlagenkrisis der griechischen Mathematik* (Charlottenburg: Metzner).

Hausdorff, Felix
1914 *Grundzüge der Mengenlehre* (Leipzig: Veit); reprinted in 1949 (New York: Chelsea).
1916 Die Mächtigkeit der Borelschen Mengen, *Mathematische Annalen 77*, 430–437.
1935 Third, revised edition of *Hausdorff 1914* (Berlin and Leipzig: W. de Gruyter).
See also Paul Mongré.

Hawking, Stephen W., and G. F. R. Ellis
1973 *The large scale structure of space-time* (Cambridge, U.K.: Cambridge University Press).

Heims, Steve
1980 *John von Neumann and Norbert Wiener: from mathematics to the technologies of life and death* (Cambridge, Mass.: M.I.T. Press).

Henkin, Leon
1949 The completeness of the first-order functional calculus, *The journal of symbolic logic 14*, 159–166.

Henkin, Leon, John Addison, Chen Chung Chang, William Craig, Dana S. Scott and Robert Vaught
1974 (eds.) *Proceedings of the Tarski symposium*, Proceedings of symposia in pure mathematics, vol. 25 (Providence, R.I.: American Mathematical Society).

Henn, Rudolph, and Otto Moeschlin
1977   The scientific work of Oskar Morgenstern, in *Mathematical economics and game theory: Essays in honor of Oskar Morgenstern*, Springer lecture notes in economics and mathematical systems, no. 141 (New York: Springer), 1–9.

Herbrand, Jacques
1930   *Recherches sur la théorie de la démonstration* (doctoral dissertation, University of Paris); English translation by Warren D. Goldfarb in *Herbrand 1971*, 44–202.
1930a   Les bases de la logique hilbertienne, *Revue de métaphysique et de morale 37*, 243–255; English translation by Warren D. Goldfarb in *Herbrand 1971*, 203–214.
1931   Sur la non-contradiction de l'arithmétique, *Journal für die reine und angewandte Mathematik 166*, 1–8; English translation by Jean van Heijenoort in *van Heijenoort 1967*, 618–628, and in *Herbrand 1971*, 282–298.
1931a   Sur le problème fondamental de la logique mathématique, *Sprawozdania z posiedzeń Towarzystwa Naukowego Warszawskiego wydział III, 24*, 12–56; English translation by Warren D. Goldfarb in *Herbrand 1971*, 215–271.
1968   *Ecrits logiques*, edited by Jean van Heijenoort (Paris: Presses Universitaires de France).
1971   *Logical writings*, English translation of *Herbrand 1968* by Warren D. Goldfarb (Dordrecht: Reidel).

Hewitt, Edwin
1948   Rings of real-valued continuous functions I., *Transactions of the American Mathematical Society 64*, 45–99, 596.

Heyting, Arend
1930   Die formalen Regeln der intuitionistischen Logik, *Sitzungsberichte der Preussischen Akademie der Wissenschaften, physikalisch-mathematische Klasse*, 42–56.
1930a   Die formalen Regeln der intuitionistischen Mathematik, *ibid.*, 57–71, 158–169.
1931   Die intuitionistische Grundlegung der Mathematik, *Erkenntnis 2*, 106–115; English translation by Erna Putnam and Gerald J. Massey in *Benacerraf and Putnam 1964*, 42–49.
1956   *Intuitionism: an introduction* (Amsterdam: North-Holland).
1959   (ed.) *Constructivity in mathematics. Proceedings of the colloquium held at Amsterdam, 1957* (Amsterdam: North-Holland).
See also *Hahn et alii 1931*.

Hilbert, David

1899   *Grundlagen der Geometrie. Festschrift zur Feier der Enthüllung des Gauss–Weber Denkmals in Göttingen* (Leipzig: Teubner).

1900   Mathematische Probleme. Vortrag, gehalten auf dem internationalen Mathematiker-Kongress zu Paris 1900, *Nachrichten von der Königlichen Gesellschaft der Wissenschaften zu Göttingen*, 253–297; English translation by Mary W. Newson in *Bulletin of the American Mathematical Society 8* (1902), 437–479, reprinted in *Browder 1976*, 1–34.

1918   Axiomatisches Denken, *Mathematische Annalen 78*, 405–415; reprinted in *Hilbert 1935*, 146–156.

1922   Neubegründung der Mathematik (Erste Mitteilung), *Abhandlungen aus dem mathematischen Seminar der Hamburgischen Universität 1*, 157–177; reprinted in *Hilbert 1935*, 157–177.

1923   Die logischen Grundlagen der Mathematik, *Mathematische Annalen 88*, 151–165; reprinted in *Hilbert 1935*, 178–191.

1926   Über das Unendliche, *Mathematische Annalen 95*, 161–190; English translation by Stefan Bauer-Mengelberg in *van Heijenoort 1967*, 367–392.

1928   Die Grundlagen der Mathematik, *Abhandlungen aus dem mathematischen Seminar der Hamburgischen Universität 6*, 65–85; English translation by Stefan Bauer-Mengelberg and Dagfinn Føllesdal in *van Heijenoort 1967*, 464–479.

1929   Probleme der Grundlegung der Mathematik, *Atti del Congresso internazionale dei matematici, Bologna 3–10 settembre 1928* (Bologna: Zanichelli), I, 135–141; see also *1929a*.

1929a   Reprint, with emendations and additions, of *Hilbert 1929*, in *Mathematische Annalen 102*, 1–9.

1930   Naturerkennen und Logik, *Naturwissenschaften 18*, 959–963.

1930a   Seventh, revised edition of *Hilbert 1899*.

1930b   Reprint of *Hilbert 1929a*, in *Hilbert 1930a*, 313–323.

1931   Die Grundlegung der elementaren Zahlenlehre, *Mathematische Annalen 104*, 485–494; reprinted in part in *Hilbert 1935*, 192–195.

1931a   Beweis des tertium non datur, *Nachrichten von der Gesellschaft der Wissenschaften zu Göttingen, mathematisch-physikalische Klasse*, 120–125.

1935   *Gesammelte Abhandlungen* (Berlin: Springer), vol. 3.

Hilbert, David, and Wilhelm Ackermann

1928   *Grundzüge der theoretischen Logik* (Berlin: Springer).

1938   Second, revised edition of *Hilbert and Ackermann 1928*.

Hilbert, David, and Paul Bernays

1934   *Grundlagen der Mathematik*, vol. I (Berlin: Springer).

1939   *Grundlagen der Mathematik*, vol. II (Berlin: Springer).

1968   Second edition of *Hilbert and Bernays 1934*.

1970   Second edition of *Hilbert and Bernays 1939*.

Hinata, Shigeru

1967   Calculability of primitive recursive functionals of finite type, *Science reports of the Tokyo Kyoiku Daigaku, section A, 9*, 218–235.

Hoensbroech, Franz G.

1931   Beziehungen zwischen Inhalt und Umfang von Begriffen, *Erkenntnis 2*, 291–300.

Holton, Gerald, and Yehuda Elkana

1982   (eds.) *Albert Einstein: historical and cultural perspectives. The centennial symposium in Jerusalem* (Princeton: Princeton University Press).

Hosoi, Tsutomu, and Hiroakira Ono

1973   Intermediate propositional logics (a survey), *Journal of Tsuda College 5*, 67–82.

Howard, William A.

1968   Functional interpretation of bar induction by bar recursion, *Compositio mathematica 20*, 107–124.

1970   Assignment of ordinals to terms for primitive recursive functionals of finite type, in *Myhill et alii 1970*, 443–458.

1972   A system of abstract constructive ordinals, *The journal of symbolic logic 37*, 355–374.

1980   Ordinal analysis of terms of finite type, *ibid. 45*, 493–504.

1981   Ordinal analysis of bar recursion of type zero, *Compositio mathematica 42*, 105–119.

1981a   Ordinal analysis of simple cases of bar recursion, *The journal of symbolic logic 46*, 17–30.

Hubble, Edwin

1934   The distribution of extra-galactic nebulae, *Astrophysical journal 79*, 8–76.

Huntington, Edward V.
1932   A new set of independent postulates for the algebra of logic with special reference to Whitehead and Russell's *Principia mathematica, Proceedings of the National Academy of Sciences, U.S.A. 18*, 179–180.
1934   Independent postulates related to C. I. Lewis' theory of strict implication, *Mind* (n.s.) *43*, 181–198.

Hurewicz, Witold
1932   Une remarque sur l'hypothèse du continu, *Fundamenta mathematicae 19*, 8–9.

Jeans, James
1936   Man and the universe, *Scientific progress* (Sir Halley Stewart lecture, 1935), edited by James Jeans et alii, 11–38.

Jech, Thomas
1974   (ed.) *Axiomatic set theory*, Proceedings of symposia in pure mathematics, vol. 13, part 2 (Providence, R.I.: American Mathematical Society).

Johansson, Ingebrigt
1936   Der Minimalkalkül, ein reduzierter intuitionistischer Formalismus, *Compositio mathematica 4*, 119–136.

Kaczmarz, Stefan
1932   Axioms for arithmetic, *The journal of the London Mathematical Society 7*, 179–182.

Kahr, Andrew S., Edward F. Moore and Hao Wang
1962   Entscheidungsproblem reduced to the ∀∃∀ case, *Proceedings of the National Academy of Sciences, U.S.A. 48*, 365–377.

Kalmár, László
1929   Eine Bemerkung zur Entscheidungstheorie, *Acta litterarum ac scientiarum Regiae Universitatis Hungaricae Francisco-Josephinae, sectio scientiarum mathematicarum 4*, 248–252.
1932   Ein Beitrag zum Entscheidungsproblem, *ibid.*, *5*, 222–236.
1933   Über die Erfüllbarkeit derjenigen Zählausdrücke, welche in der Normalform zwei benachbarte Allzeichen enthalten, *Mathematische Annalen 108*, 466–484.
1934   Über einen Löwenheimschen Satz, *Acta litterarum ac scientiarum Regiae Universitatis Hungaricae Francisco-Josephinae, sectio scientiarum mathematicarum 7*, 112–121.

1955   Über ein Problem, betreffend die Definition des Begriffes der allgemein-rekursiven Funktion, *Zeitschrift für mathematische Logik und Grundlagen der Mathematik 1*, 93–96.

Kant, Immanuel
1787   *Critik der reinen Vernunft*, second revised edition (Riga: Hartknoch).

Keisler, H. Jerome, and Alfred Tarski
1964   From accessible to inaccessible cardinals: results holding for all accessible cardinal numbers and the problem of their extension to inaccessible ones, *Fundamenta mathematicae 53*, 225–308.

Ketonen, Jussi, and Robert M. Solovay
1981   Rapidly growing Ramsey functions, *Annals of mathematics 113*, 267–314.

Klanfer, Laura
1933   Über d-zyklische Quadrupel, *Ergebnisse eines mathematischen Kolloquiums 4*, 10.

Kleene, Stephen C.
1934   Proof by cases in formal logic, *Annals of mathematics (2) 35*, 529–544.

1935   A theory of positive integers in formal logic, *American journal of mathematics 57*, 153–173, 219–244.

1936   General recursive functions of natural numbers, *Mathematische Annalen 112*, 727–742; reprinted in *Davis 1965*, 236–252; for an erratum, a simplification, and an addendum, see *Davis 1965*, 253.

1936a λ-definability and recursiveness, *Duke mathematical journal 2*, 340–353.

1943   Recursive predicates and quantifiers, *Transactions of the American Mathematical Society 53*, 41–73; reprinted in *Davis 1965*, 254–287; for a correction and an addendum, see *Davis 1965*, 254 and 287.

1950   A symmetric form of Gödel's theorem, *Indagationes mathematicae 12*, 244–246.

1952   *Introduction to metamathematics* (Amsterdam: North-Holland; New York: Van Nostrand); eighth reprint, 1980.

1960   Realizability and Shanin's algorithm for the constructive deciphering of mathematical sentences, *Logique et analyse* (n.s.) *3*, 154–165.

1973   Realizability: a retrospective survey, in *Mathias and Rogers 1973*, 95–112.

| 1976 | The work of Kurt Gödel, *The journal of symbolic logic 41*, 761–778; addendum, *ibid. 43* (1978), 613. |

1976 The work of Kurt Gödel, *The journal of symbolic logic 41*, 761–778; addendum, *ibid. 43* (1978), 613.

1981 Origins of recursive function theory, *Annals of the history of computing 3*, 52–67; corrections, *Davis 1982*, footnotes 10 and 12.

1985 Kurt Gödel (1906–1978), *Proceedings of the National Academy of Sciences, U.S.A.*, to appear.

198? Gödel's impression on students of logic in the 1930s, *Gödels wissenschaftliches Weltbild* (Proceedings of a symposium on Gödel held at the Internationales Forschungszentrum, Salzburg, July 1983), to appear.

Kleene, Stephen C., and J. Barkley Rosser

1935 The inconsistency of certain formal logics, *Annals of mathematics (2) 36*, 630–636.

Klein, Fritz

1931 Zur Theorie der abstrakten Verknüpfungen, *Mathematische Annalen 105*, 308–323.

1932 Über einen Zerlegungssatz in der Theorie der abstrakten Verknüpfungen, *ibid. 106*, 114–130.

Klibansky, Raymond

1968 (ed.) *Contemporary philosophy, a survey. I, Logic and foundations of mathematics* (Florence: La Nuova Italia Editrice).

Köhler, Eckehart

198? Gödel und der Wiener Kreis: Platonismus gegen Formalismus, to appear.

Kolestos, George

198? Functional interpretation of the β-rule, *The journal of symbolic logic*, to appear.

Kolmogorov, Andrei Nikolayevich (Kolmogoroff; Колмогоров, Андрей Николаевич)

1925 On the principle of the excluded middle (Russian), *Matematicheskii sbornik 32*, 646–667; English translation by Jean van Heijenoort in *van Heijenoort 1967*, 414–437.

1932 Zur Deutung der intuitionistischen Logik, *Mathematische Zeitschrift 35*, 58–65.

König, Dénes
1926 Sur les correspondances multivoques des ensembles, *Fundamenta mathematicae 8*, 114–134.
1927 Über eine Schlussweise aus dem Endlichen ins Unendliche: Punktmengen. Kartenfärben. Verwandtschaftsbeziehungen. Schachspiel, *Acta litterarum ac scientiarum Regiae Universitatis Hungaricae Francisco-Josephinae, sectio scientiarum mathematicarum 3*, 121–130.

König, Julius
1905 Zum Kontinuum-Problem, *Mathematische Annalen 60*, 177–180, 462.

Kreisel, Georg
1951 On the interpretation of non-finitist proofs—Part I, *The journal of symbolic logic 16*, 241–267.
1952 On the interpretation of non-finitist proofs—Part II. Interpretation of number theory. Applications, *ibid. 17*, 43–58.
1953 Note on arithmetic models for consistent formulae of the predicate calculus, II, *Actes du XIème Congrès international de philosophie, Bruxelles, 20–26 août 1953* (Amsterdam: North-Holland), vol. 14, 39–49.
1958 Hilbert's programme, *Dialectica 12*, 346–372; revised version in *Benacerraf and Putnam 1964*, 157–180.
1959 Interpretation of analysis by means of constructive functionals of finite types, in *Heyting 1959*, 101–128.
1959a Inessential extensions of Heyting's arithmetic by means of functionals of finite type (abstract), *The journal of symbolic logic 24*, 284.
1959b Inessential extensions of intuitionistic analysis by functionals of finite type (abstract), *ibid.*, 284–285.
1960 Ordinal logics and the characterization of informal concepts of proof, *Proceedings of the International Congress of Mathematicians, 14–21 August 1958* (Cambridge, U.K.: Cambridge University Press), 289–299.
1960a La prédicativité, *Bulletin de la Société mathématique de France 88*, 371–391.
1962 The axiom of choice and the class of hyperarithmetic functions, *Indagationes mathematicae 24*, 307–319.
1965 Mathematical logic, *Lectures on modern mathematics*, edited by Thomas L. Saaty (New York: Wiley), vol. 3, 95–195.
1967 Mathematical logic: what has it done for the philosophy of mathematics?, in *Schoenman 1967*, 201–272.

1968 A survey of proof theory, *The journal of symbolic logic 33*, 321–388.

1968a Functions, ordinals, species, in *Staal and van Rootselaar 1968*, 143–158.

1970 Church's thesis: a kind of reducibility axiom for constructive mathematics, in *Myhill et alii 1970*, 121–150.

1976 What have we learnt from Hilbert's second problem?, in *Browder 1976*, 93–130.

1980 Kurt Gödel, 28 April 1906–14 January 1978, *Biographical memoirs of Fellows of the Royal Society 26*, 148–224; corrections, *ibid. 27*, 697, and *28*, 718.

1982 Finiteness theorems in arithmetic: an application of Herbrand's theorem for Σ_2-formulas, in *Stern 1982*, 39–55.

Kreisel, Georg, and Angus MacIntyre
1982 Constructive logic versus algebraization I, in *Troelstra and van Dalen 1982*, 217–258.

Kreisel, Georg, and Anne S. Troelstra
1970 Formal systems for some branches of intuitionistic analysis, *Annals of mathematical logic 1*, 229–387.

Kripke, Saul
1965 Semantical analysis of intuitionistic logic I, in *Crossley and Dummett 1965*, 92–130.

Krivine, Jean-Louis
1968 *Introduction à la théorie axiomatique des ensembles* (Paris: Presses Universitaires de France); translated into English as *Krivine 1971*.

1971 *Introduction to axiomatic set theory* (Dordrecht: Reidel), English translation by David Miller of *Krivine 1968*.

Kuczyński, Jerzy
1938 O twierdzeniu Gödel (On Gödel's theorem; Polish, with a French summary), *Kwartalnik filozoficzny 15*, 74–80.

Kunen, Kenneth
1970 Some applications of iterated ultrapowers in set theory, *Annals of mathematical logic 1*, 179–227.

Kuratowski, Kazimierz
1933 *Topologie I*, Monografie Matematyczne, vol. 3 (Warsaw: Garasiński).

1948   Ensembles projectifs et ensembles singuliers, *Fundamenta mathematicae 35*, 131–140.

1951   Sur une caractérisation des alephs, *ibid. 38*, 14–17.

Kuroda, Sigekatu

1951   Intuitionistische Untersuchungen der formalistischen Logik, *Nagoya mathematical journal 2*, 35–47.

Ladrière, Jean

1957   *Les limitations internes des formalismes. Etude sur la signification du théorème de Gödel et des théorèmes apparentés dans la théorie des fondements des mathématiques* (Louvain: Nauwelaerts; Paris: Gauthier-Villars).

Langford, Cooper H.

1927   On inductive relations, *Bulletin of the American Mathematical Society 33*, 599–607.

Lawvere, William

1971   Quantifiers and sheaves, *Actes du Congrès international des mathématiciens, 1–10 septembre 1970, Nice, France* (Paris: Gauthier-Villars), vol. I, 329–334.

Leibniz, Gottfried W.

1890   *Die philosophischen Schriften von Gottfried Wilhelm Leibniz*, edited by C. J. Gerhardt (Berlin: Weidmann), vol. 7.

1923   *Sämtliche Schriften und Briefe*, edited by Preussischen Akademie der Wissenschaften (Darmstadt: O. Reichl), series 1, vol. 1.

Leivant, Daniel

198?   Syntactic translations and provably recursive functions, *The journal of symbolic logic*, to appear.

Lemmon, Edward J.

1977   *An introduction to modal logic*, in collaboration with Dana Scott, edited by Krister Segerberg, American Philosophical Quarterly monograph series, no. 11 (Oxford: Blackwell).

Levy, Azriel

1960   Axiom schemata of strong infinity in axiomatic set theory, *Pacific journal of mathematics 10*, 223–238.

1960a   Principles of reflection in axiomatic set theory, *Fundamenta mathematicae 49*, 1–10.

1965   Definability in axiomatic set theory I, in *Bar-Hillel 1965*, 127–151.

Levy, Azriel, and Robert M. Solovay
1967 Measurable cardinals and the continuum hypothesis, *Israel journal of mathematics 5*, 234–248.

Lewis, Clarence I.
1918 *A survey of symbolic logic* (Berkeley: University of California Press); reprinted by Dover (New York).
1932 Alternative systems of logic, *The monist 42*, 481–507.

Löb, Martin H.
1955 Solution of a problem of Leon Henkin, *The journal of symbolic logic 20*, 115–118.

Lorenzen, Paul
1951 Algebraische und logistische Untersuchungen über freie Verbände, *The journal of symbolic logic 16*, 81–106.
1951a Die Widerspruchsfreiheit der klassischen Analysis, *Mathematische Zeitschrift 54*, 1–24.
1951b Maß und Integral in der konstruktiven Analysis, *ibid.*, 275–290.
1955 *Einführung in die operative Logik und Mathematik* (Berlin: Springer).
1969 Second edition of *Lorenzen 1955*.

Łoś, Jerzy
1955 Quelques remarques, théorèmes, et problèmes sur les classes définissables d'algèbres, *Mathematical interpretations of formal systems*, edited by Thoralf Skolem and others (Amsterdam: North-Holland), 98–113.

Löwenheim, Leopold
1915 Über Möglichkeiten im Relativkalkül, *Mathematische Annalen 76*, 447–470; English translation by Stefan Bauer-Mengelberg in *van Heijenoort 1967*, 228–251.

Luckhardt, Horst
1973 *Extensional Gödel functional interpretation. A consistency proof of classical analysis*, Springer lecture notes in mathematics, no. 306 (Berlin: Springer).

Łukasiewicz, Jan, and Alfred Tarski
1930 Untersuchungen über den Aussagenkalkül, *Sprawozdania z posiedzeń Towarzystwa Naukowego Warszawskiego*, wydział III, *23*, 30–50; English translation by J. H. Woodger in *Tarski 1956*, 38–59.

Luzin, Nikolai (Lusin, Nicolas; Лузин, Николай Николаевич)
1914   Sur un problème de M. Baire, *Comptes rendus hebdomadaires des séances de l'Académie des sciences, Paris 158*, 1258–1261.
1930   *Leçons sur les ensembles analytiques et leurs applications* (Paris: Gauthier-Villars).
1935   Sur les ensembles analytiques nuls, *Fundamenta mathematicae 25*, 109–131.

Luzin, Nikolai, and Wacław Sierpiński
1918   Sur quelques propriétés des ensembles (*A*), *Bulletin international de l'Académie des sciences de Cracovie, classe des sciences mathématiques et naturelles, série A*, 35–48; reprinted in *Sierpiński 1975*, 192–204.

Maaß, Wolfgang
1976   Eine Funktionalinterpretation der prädikativen Analysis, *Archiv für mathematische Logik und Grundlagenforschung 18*, 27–46.

Mac Lane, Saunders
1961   Locally small categories and the foundations of set theory, *Infinitistic methods. Proceedings of the symposium on foundations of mathematics, Warsaw, 2–9 September 1959* (Warsaw: PWN; Oxford: Pergamon), 25–43.

MacDowell, Robert, and Ernst Specker
1961   Modelle der Arithmetik, *Infinitistic methods, Proceedings of the symposium on foundations of mathematics, Warsaw, 2–9 September 1959* (Warsaw: PWN; Oxford: Pergamon), 257–263.

MacIntyre, Angus
See Kreisel, Georg, and Angus MacIntyre.

Maehara, Shôji
1954   Eine Darstellung der intuitionistischen Logik in der Klassischen, *Nagoya mathematical journal 7*, 45–64.

Mahlo, Paul
1911   Über lineare transfinite Mengen, *Berichte über die Verhandlungen der Königlich Sächsischen Gesellschaft der Wissenschaften zu Leipzig, Mathematisch-physische Klasse 63*, 187–225.
1912   Zur Theorie und Anwendung der ρ_0-Zahlen, *ibid. 64*, 108–112.
1913   Zur Theorie und Anwendung der ρ_0-Zahlen. II., *ibid. 65*, 268–282.

Malmnäs, P.-E.
See Prawitz, Dag, and P.-E. Malmnäs.

Maltsev, Anatolii Ivanovich (Malcev; Мальцев, Анатолий Иванович)
1936   Untersuchungen aus dem Gebiete der mathematischen Logik, *Matematicheskii sbornik 1*, 323–336; English translation by Benjamin F. Wells III in *Maltsev 1971*, 1–14.
1941   On a general method for obtaining local theorems in group theory (Russian), *Ivanovskii Gosudarstvennii Pedagogicheskii Institut im. D. A. Furmanova. Ivanovskoye matematicheskoye obshchestvo. Ucheniye zapiski 1*, 3–9; English translation by Benjamin F. Wells III in *Maltsev 1971*, 15–21.
1971   *The metamathematics of algebraic systems: collected papers, 1936–1967*, edited and translated by Benjamin F. Wells III (Amsterdam: North-Holland).

Martin-Löf, Per
1971   Hauptsatz for the theory of species, in *Fenstad 1971*, 217–233.

Mathias, Adrian R. D., and Hartley Rogers
1973   (eds.) *Cambridge summer school in mathematical logic*, Springer lecture notes in mathematics, no. 337 (Berlin: Springer).

Matiyasevich, Yuri (Matijacevič; Матиясевич, Юри)
1970   Enumerable sets are Diophantine (Russian), *Doklady Akademii Nauk S.S.S.R. 191*, 279–282; English translation, with revisions, in *Soviet mathematics doklady 11* (1970), 354–358.
See also Davis, Martin, Yuri Matiyasevich and Julia Robinson.

McAloon, Kenneth
1966   *Some applications of Cohen's method* (doctoral dissertation, University of California at Berkeley).
1971   Consistency results about ordinal definability, *Annals of mathematical logic 2*, 449–467.

McKinsey, John C. C., and Alfred Tarski
1948   Some theorems about the sentential calculi of Lewis and Heyting, *The journal of symbolic logic 13*, 1–15.

McTaggart, J. Ellis
1908   The unreality of time, *Mind* (n.s.) *17*, 457–474.

Mehrtens, Herbert
1979   *Die Entstehung der Verbandstheorie* (Hildesheim: Gerstenberg).

Menas, Telis K.
1973   *On strong compactness and supercompactness* (doctoral disser-
        tation, University of California at Berkeley).

Menger, Karl
1928   Untersuchungen über allgemeine Metrik, *Mathematische An-
        nalen 100*, 75–163.
1928a   Bemerkungen zu Grundlagenfragen IV. Axiomatik der end-
        lichen Mengen und der elementargeometrischen Verknüpfungs-
        beziehungen, *Jahresbericht der Deutschen Mathematiker-
        Vereinigung 37*, 309–325.
1930   Untersuchungen über allgemeine Metrik. Vierte Untersuchung.
        Zur Metrik der Kurven, *Mathematische Annalen 103*, 466–501.
1931   Metrische Untersuchungen. II: Die euklidische Metrik, *Ergeb-
        nisse eines mathematischen Kolloquiums 1*, 20–22.
1932   Probleme der allgemeinen metrischen Geometrie, *ibid. 2*, 20–
        22.
1932a   Bericht über die mengentheoretischen Überdeckungssätze,
        ibid., 23–27.
1936   (In collaboration with Franz Alt and Otto Schreiber) New
        foundations of projective and affine geometry. Algebra of ge-
        ometry, *Annals of mathematics (2) 37*, 456–482.
1940   On algebra of geometry and recent progress in non-Euclidean
        geometry, *The Rice Institute pamphlet 27*, 41–79.
1952   The formative years of Abraham Wald and his work in geom-
        etry, *Annals of mathematical statistics 23*, 14–20.
See also *Gödel 1933h*.

Menger, Karl, and Leonard M. Blumenthal
1970   *Studies in Geometry* (San Francisco: Freeman).

Minari, Pierluigi
1983   Intermediate logics. A historical outline and a guided bibli-
        ography, *Rapporto matematico 79*, 1–71 (Dipartimento di
        Matematica, Università degli studi di Siena).

Mints, Gregory E. (Минц, Григорий Е.)
1974   On *E*-theorems (Russian), *Zapiski nauchnyk seminarov
        Leningradskogo otdeleniya Matematicheskogo Instituta im. V.
        A. Steklova, Akademii nauk S.S.S.R. (Leningrad) 40*, 110–118,
        158–159.
1975   Finite investigations of transfinite derivations (Russian), *ibid.
        49*, 67–122; English translation in *Journal of soviet mathemat-
        ics 10* (1978), 548–596.

| 1978 | On Novikov's hypothesis (Russian), *Modal and intensional logics* (Moscow), 102–106; photocopied proceedings of a conference held by the Institute of Philosophy of the Soviet Academy of Sciences. |

1978   On Novikov's hypothesis (Russian), *Modal and intensional logics* (Moscow), 102–106; photocopied proceedings of a conference held by the Institute of Philosophy of the Soviet Academy of Sciences.

1979   Stability of E-theorems and program verification (Russian), *Semiotika i informatika 12*, 73–77.

Mirimanoff, Dmitry

1917   Les antinomies de Russell et de Burali-Forti et le problème fondamental de la théorie des ensembles, *L'enseignement mathématique 19*, 37–52.

1917a   Remarques sur la théorie des ensembles et les antinomies cantoriennes. I, *ibid.*, 209–217.

1920   Remarques sur la théorie des ensembles et les antinomies cantoriennes. II, *ibid. 21*, 29–52.

Moeschlin, Otto
See Henn, Rudolph, and Otto Moeschlin.

Mongré, Paul (pseudonym of Felix Hausdorff)

1898   *Das Chaos in kosmischer Auslese* (Leipzig: Naumann).

Montgomery, Deane

1963   Oswald Veblen, *Bulletin of the American Mathematical Society 69*, 26–36.

Moore, Edward F.
See Kahr, Andrew S., Edward F. Moore, and Hao Wang.

Moore, Gregory H.

1980   Beyond first-order logic: the historical interplay between mathematical logic and axiomatic set theory, *History and philosophy of logic 1*, 95–137.

1982   *Zermelo's axiom of choice: its origins, development, and influence*, Studies in the history of mathematics and physical sciences, vol. 8 (New York: Springer).

Morel, Anne
See Frayne, Thomas, Anne Morel and Dana S. Scott.

Morgenstern, Oskar
See von Neumann, John, and Oskar Morgenstern.

Mostowski, Andrzej
1947   On definable sets of positive integers, *Fundamenta mathematicae 34*, 81–112.
1951   Review of *Wang 1950*, *The journal of symbolic logic 16*, 142–143.
1952   *Sentences undecidable in formalized arithmetic: an exposition of the theory of Kurt Gödel* (Amsterdam: North-Holland).
1955   A formula with no recursively enumerable model, *Fundamenta mathematicae 42*, 125–140.
1959   On various degrees of constructivism, in *Heyting 1959*, 178–194.
1965   *Thirty years of foundational studies: lectures on the development of mathematical logic and the study of the foundations of mathematics in 1930–1964* (= no. 17 of *Acta philosophica fennica*); reprinted in 1966 (New York: Barnes and Noble; Oxford: Blackwell).
See also Tarski, Alfred, Andrzej Mostowski, and Raphael M. Robinson.

Müller, Gert H.
See Diller, Justus, and Gert H. Müller.

Müller, Gert H., and Dana S. Scott
1978   (eds.) *Higher set theory. Proceedings, Oberwolfach, Germany, April 13–23, 1977*, Springer lecture notes in mathematics, no. 669 (Berlin: Springer).

Mulvey, C. J.
See Fourman, Michael P., C. J. Mulvey and Dana S. Scott.

Mycielski, Jan
1964   On the axiom of determinateness, *Fundamenta mathematicae 53*, 205–224.
See also Ehrenfeucht, Andrzej, and Jan Mycielski.

Myhill, John
1970   Formal systems of intuitionistic analysis II: the theory of species, in *Myhill et alii 1970*, 151–162.
1974   The undefinability of the set of natural numbers in the ramified *Principia*, in *Nakhnikian 1974*, 19–27.
1974a   "Embedding classical type theory in 'intuitionistic' type theory": a correction, in *Jech 1974*, 185–188.

Myhill, John, Akiko Kino and Richard E. Vesley
1970   (eds.) *Intuitionism and proof theory* (Amsterdam: North-Holland).

Myhill, John, and Dana S. Scott
1971   Ordinal definability, in *Scott 1971*, 271–278.

Nahm, Werner
See Diller, Justus, and Werner Nahm.

Nakhnikian, George
1974   (ed.) *Bertrand Russell's philosophy* (London: Duckworth).

Neder, Ludwig
1931   Über den Aufbau der Arithmetik, *Jahresbericht der Deutschen Mathematiker-Vereinigung 40*, 22–37.

Notcutt, Bernard
1934   A set of axioms for the theory of deduction, *Mind* (n.s.) *43*, 63–77.

Ono, Hiroakira
See Hosoi, Tsutomu, and Hiroakira Ono.

Parikh, Rohit
1971   Existence and feasibility in arithmetic, *The journal of symbolic logic 36*, 494–508.
1973   Some results on the length of proofs, *Transactions of the American Mathematical Society 177*, 29–36.

Paris, Jeff, and Leo Harrington
1977   A mathematical incompleteness in Peano arithmetic, in *Barwise 1977*, 1133–1142.

Parry, William T.
1933   Ein Axiomensystem für eine neue Art von Implikation (analytische Implikation), *Ergebnisse eines mathematischen Kolloquiums 4*, 5–6.
1933a   Zum Lewisschen Aussagenkalkül, *ibid.*, 15–17.

Parsons, Charles D.
1970   On a number-theoretic choice schema and its relation to induction, in *Myhill et alii 1970*, 459–473.

Pasch, Moritz
1882   *Vorlesungen über neuer Geometrie* (Leipzig: Teubner).
1926   Second edition of *Pasch 1882*, with an appendix by Max Dehn (Berlin: Springer).

Peano, Giuseppe
1889   *Arithmetices principia, nova methodo exposita* (Turin: Bocca);
        partial English translation by Jean van Heijenoort in *van Hei-*
        jenoort 1967, 83–97.
1891   Sul concetto di numero, *Rivista di matematica 1*, 87–102, 256–
        267.

Pears, David F.
1972   (ed.) *Bertrand Russell: a collection of critical essays* (Garden
        City, N.Y.: Anchor).

Peirce, Charles S.
1897   The logic of relatives, *The monist 7*, 161–217; reprinted in
        1933, 288–345.
1933   *Collected papers of Charles Sanders Peirce*, edited by Charles
        Hartshorne and Paul Weiss, vol. III: *Exact Logic* (Cambridge,
        Mass.: Harvard University Press).
1976   *The new elements of mathematics*, edited by Carolyn Eisele
        (The Hague: Mouton), vols. I–V.

Perelman, Charles
1936   L'antinomie de M. Gödel, *Académie royale de Belgique, Bul-*
        letin de la classe des sciences (*5*) *22*, 730–736.

Pohlers, Wolfram
See Buchholz, Wilfried, et alii.

Post, Emil L.
1921   Introduction to a general theory of elementary propositions,
        American journal of mathematics 43, 163–185; reprinted in
        van Heijenoort 1967, 264–283.
1936   Finite combinatory processes—formulation 1, *The journal of*
        symbolic logic 1, 103–105; reprinted in *Davis 1965*, 288–291.
1941   Absolutely unsolvable problems and relatively undecidable
        propositions: account of an anticipation, in *Davis 1965*, 338–
        433.
1944   Recursively enumerable sets of positive integers and their deci-
        sion problems, *Bulletin of the American Mathematical Society*
        50, 284–316.
1953   A necessary condition for definability for transfinite von
        Neumann–Gödel set theory sets, with an application to the
        problem of the existence of a definable well-ordering of the
        continuum (preliminary report), *ibid. 59*, 246.

Powell, William C.
1975   Extending Gödel's negative interpretation to *ZF*, *The journal of symbolic logic 40*, 221–229.

Prawitz, Dag
1971   Ideas and results in proof theory, in *Fenstad 1971*, 235–307.

Prawitz, Dag, and P.-E. Malmnäs
1968   A survey of some connections between classical, intuitionistic and minimal logic, in *Schmidt et alii 1968*, 215–229.

Presburger, Mojżesz
1930   Über die Vollständigkeit eines gewissen Systems der Arithmetik ganzer Zahlen, in welchen die Addition als einzige Operation hervortritt, *Sprawozdanie z I Kongresu matematyków krajów słowiańskich, Warszawa 1929* (Warsaw, 1930), 92–101, 395.

Princeton University
1947   *Problems of mathematics*, Princeton University bicentennial conferences, series 2, conference 2.

Putnam, Hilary
1957   Arithmetic models for consistent formulae of quantification theory, *The journal of symbolic logic 22*, 110–111; abstract of a paper presented at the 27 December 1956 meeting of the Association for Symbolic Logic.
1961   *Trial and error predicates and the solution to a problem of Mostowski's* (New York: Courant Institute).
1965   Trial and error predicates and the solution to a problem of Mostowski, *The journal of symbolic logic 30*, 49–57.
See also Benacerraf, Paul, and Hilary Putnam.

Quine, Willard V.
1933   A theorem in the calculus of classes, *The journal of the London Mathematical Society 8*, 89–95.
1937   New foundations for mathematical logic, *American mathematical monthly 44*, 70–80.
1941   Whitehead and the rise of modern logic, in *Schilpp 1941*, 125–163; reprinted in *Quine 1966a*, 3–36.
1943   Notes on existence and necessity, *Journal of philosophy 40*, 113–127.
1947   The problem of interpreting modal logic, *The journal of symbolic logic 12*, 43–48.

1953 *From a logical point of view. 9 logico-philosophical essays* (Cambridge, Mass.: Harvard University Press).

1953a Three grades of modal involvement, *Actes du XIème Congrès international de philosophie, Bruxelles, 20–26 août 1953* (Amsterdam: North-Holland), vol. XIV, 65–81; reprinted in *Quine 1976*, 158–176.

1955 On Frege's way out, *Mind* (n.s.) *64*, 145–159.

1960 Carnap and logical truth, *Synthèse 12*, 350–374.

1963 Reprint of *Quine 1960* in *Schilpp 1963*, 385–406.

1966 *The ways of paradox and other essays* (New York: Random House).

1966a *Selected logic papers* (New York: Random House).

1976 Second, enlarged edition of *Quine 1966* (Cambridge, Mass: Harvard University Press).

1979 Kurt Gödel (1906–1978), *Year book 1978 of the American Philosophical Society*, 81–84.

Ramsey, Frank P.

1926 The foundations of mathematics, *Proceedings of the London Mathematical Society (2) 25*, 338–384; reprinted in *Ramsey 1931*, 1–61.

1929 On a problem of formal logic, *Proceedings of the London Mathematical Society (2) 30*, 264–286.

1931 *The foundations of mathematics and other logical essays* (London: Kegan Paul).

Rasiowa, Helena, and Roman Sikorski

1950 A proof of the completeness theorem of Gödel, *Fundamenta mathematicae 37*, 193–200.

1951 A proof of the Skolem–Löwenheim theorem, *ibid. 38*, 230–232.

1953 Algebraic treatment of the notion of satisfiability, *ibid. 40*, 62–95.

1963 *The mathematics of metamathematics* (Warsaw: PWN).

Rath, Paul

1978 *Eine verallgemeinerte Funktionalinterpretation der Heyting Arithmetik endlicher Typen* (doctoral dissertation, Münster).

Rautenberg, Wolfgang

1979 *Klassische und nichtklassische Aussagenlogik* (Braunschweig: Vieweg).

Raychaudhuri, A. K.

1979 *Theoretical cosmology* (Oxford: Clarendon Press).

Reid, Constance
1970　　　*Hilbert* (New York: Springer).

Reidemeister, Kurt
See *Hahn et alii 1931*.

Robertson, Howard P.
1933　　　Relativistic cosmology, *Reviews of modern physics 5*, 62–90.

Robinson, Abraham
1965　　　Formalism 64, in *Bar-Hillel 1965*, 228–246.
1966　　　*Non-standard analysis* (Amsterdam: North-Holland).
1974　　　Second edition of *Robinson 1966*.
1975　　　Concerning progress in the philosophy of mathematics, in *Rose and Shepherdson 1975*, 41–52.

Robinson, Julia
See Davis, Martin, Yuri Matiyasevich and Julia Robinson.

Robinson, Raphael M.
1937　　　The theory of classes. A modification of von Neumann's system, *The journal of symbolic logic 2*, 29–36.
See also Tarski, Alfred, Andrzej Mostowski, and Raphael M. Robinson.

Rose, Harvey E., and John C. Shepherdson
1975　　　(eds.) *Logic colloquium '73* (Amsterdam: North-Holland).

Rosser, J. Barkley
1935　　　A mathematical logic without variables. I, *Annals of mathematics (2) 36*, 127–150.
1935a　　A mathematical logic without variables. II, *Duke mathematics journal 1*, 328–355.
1936　　　Extensions of some theorems of Gödel and Church, *The journal of symbolic logic 1*, 87–91; reprinted in *Davis 1965*, 230–235.
1937　　　Gödel theorems for non-constructive logics, *The journal of symbolic logic 2*, 129–137.
1939　　　An informal exposition of proofs of Gödel's theorems and Church's theorem, *ibid. 4*, 53–60; reprinted in *Davis 1965*, 223–230.
See also Church, Alonzo, and J. Barkley Rosser.
See also Kleene, Stephen C., and J. Barkley Rosser.

Russell, Bertrand
1903   *The principles of mathematics* (London: Allen and Unwin).
1906   On some difficulties in the theory of transfinite numbers and order types, *Proceedings of the London Mathematical Society* *(2) 4*, 29–53; reprinted in *Russell 1973*, 135-164.
1906a   Les paradoxes de la logique, *Revue de métaphysique et de morale 14*, 627–650.
1908   Mathematical logic as based on the theory of types, *American journal of mathematics 30*, 222–262; reprinted in *van Heijenoort 1967*, 150–182.
1919   *Introduction to mathematical philosophy* (London: Allen and Unwin; New York: Macmillan).
1920   Second edition of *Russell 1919*.
1924   Reprint of *Russell 1920*.
1940   *An inquiry into meaning and truth* (London: Allen and Unwin).
1968   *The autobiography of Bertrand Russell, 1914–1944* (London: Allen and Unwin; Boston: Little, Brown and Co.)
1973   *Essays in analysis*, edited by Douglas Lackey (New York: Braziller).
See also Whitehead, Alfred North, and Bertrand Russell.

Sanchis, Luis E.
1967   Functionals defined by recursion, *Notre Dame journal of formal logic 8*, 161–174.

Scanlon, Thomas M.
1973   The consistency of number theory via Herbrand's theorem, *The journal of symbolic logic 38*, 29–58.

Schilpp, Paul A.
1941   (ed.) *The philosophy of Alfred North Whitehead*, Library of living philosophers, vol. 3 (Evanston: Northwestern University): second edition (New York: Tudor).
1944   (ed.) *The philosophy of Bertrand Russell*, Library of living philosophers, vol. 5 (Evanston: Northwestern University); third edition (New York: Tudor, 1951).
1949   (ed.) *Albert Einstein, philosopher-scientist*, Library of living philosophers, vol. 7 (Evanston: Library of living philosophers).
1955   (ed.) *Albert Einstein als Philosoph und Naturforscher*, German translation (with additions) of *Schilpp 1949* (Stuttgart: Kohlhammer).
1963   (ed.) *The philosophy of Rudolf Carnap*, Library of living philosophers, vol. 11 (La Salle, Illinois: Open Court; London: Cambridge University Press).

Schlesinger, Karl
1935   Über die Produktionsgleichungen der ökonomischen Wertlehre, *Ergebnisse eines mathematischen Kolloquiums 6*, 10–11.

Schmidt, H. Arnold, Kurt Schütte and H.-J. Thiele
1968   (eds.)   *Contributions to mathematical logic* (Amsterdam: North-Holland).

Schoenman, Ralph
1967   (ed.)   *Bertrand Russell: philosopher of the century* (London: Allen and Unwin).

Scholz, Arnold
See *Hahn et alii 1931*.

Scholz, Heinrich
See Hasse, Helmut, and Heinrich Scholz.

Schönfinkel, Moses
See Bernays, Paul, and Moses Schönfinkel.

Schreiber, Otto
See *Menger 1936*.

Schütte, Kurt
1934   Untersuchungen zum Entscheidungsproblem der mathematischen Logik, *Mathematische Annalen 109*, 572–603.
1934a Über die Erfüllbarkeit einer Klasse von logischen Formeln, *ibid. 110*, 161–194.
1954   Kennzeichnung von Ordnungszahlen durch rekursiv erklärte Funktionen, *ibid. 127*, 15–32.
1965   Predicative well-orderings, in *Crossley and Dummett 1965*, 280–303.
1965a Eine Grenze für die Beweisbarkeit der transfiniten Induktion in der verzweigten Typenlogik, *Archiv für mathematische Logik und Grundlagenforschung 7*, 45–60.
1977   *Proof theory* (Berlin: Springer).
See also Diller, Justus, and Kurt Schütte.
See also Schmidt, H. Arnold, Kurt Schütte and H. -J. Thiele.

Schützenberger, Marcel P. (Marco; Maurice)
1945   Sur certains axiomes de la théorie des structures, *Comptes rendus hebdomadaires des séances de l'Académie des Sciences, Paris 221*, 218–220.

Schwichtenberg, Helmut

1973   *Einige Anwendungen von unendlichen Termen und Wert-funktionalen* (Habilitationsschrift, Münster).

1975   Elimination of higher type levels in definitions of primitive recursive functionals by means of transfinite recursion, in *Rose and Shepherdson 1975*, 279–303.

1977   Proof theory: some applications of cut-elimination, in *Barwise 1977*, 867–895.

1979   On bar recursion of types 0 and 1, *The journal of symbolic logic 44*, 325–329.

Scott, Dana S.

1961   Measurable cardinals and constructible sets, *Bulletin de l'Académie polonaise des sciences, série des sciences mathématiques, astronomiques, et physiques 9*, 521–524.

1971   (ed.) *Axiomatic set theory*, Proceedings of symposia in pure mathematics, vol. 13, part 1 (Providence, R.I.: American Mathematical Society).

See also Fourman, Michael P., C. J. Mulvey and Dana S. Scott.

See also Fourman, Michael P., and Dana S. Scott.

See also Frayne, Thomas, Anne Morel and Dana S. Scott.

See also Hanf, William P., and Dana S. Scott.

See also *Henkin et alii 1974*.

See also *Lemmon 1977*.

See also Müller, Gert H., and Dana S. Scott.

See also Myhill, John, and Dana S. Scott.

Scott, Philip J.

1978   The "Dialectica" interpretation and categories, *Zeitschrift für mathematische Logik und Grundlagen der Mathematik 24*, 553–575.

Sheffer, Henry M.

1926   Review of *Whitehead and Russell 1925*, *Isis 8*, 226–231.

Shelah, Saharon

See Gurevich, Yuri, and Saharon Shelah.

Shepherdson, John C.

See Rose, Harvey E., and John C. Shepherdson.

Shoenfield, Joseph R.

1967   *Mathematical logic* (Reading, Mass.: Addison-Wesley).

Sieg, Wilfried
See Buchholz, Wilfried, et alii.

Sierpiński, Wacław
1919   Sur un théorème équivalent à l'hypothèse du continu ($2^{\aleph_0}$ = $\aleph_1$), *Biuletyn Polskiej Akademii Umiejętności, Kraków* (= *Bulletin international de l'Académie des sciences et des lettres, Cracovie*), 1–3; reprinted in *Sierpiński 1975*, 272–274.
1924   Sur l'hypothèse du continu ($2^{\aleph_0}$ = $\aleph_1$), *Fundamenta mathematicae 5*, 177–187; reprinted in *Sierpiński 1975*, 527–536.
1934   *Hypothèse du continu*, Monografie Matematyczne, vol. 4 (Warsaw: Garasiński).
1934a   Sur une extension de la notion de l'homéomorphie, *Fundamenta mathematicae 22*, 270–275; reprinted in *Sierpiński 1976*, 201–206.
1935   Sur une hypothèse de M. Lusin, *Fundamenta mathematicae 25*, 132–135; reprinted in *Sierpiński 1976*, 269–272.
1935a   Sur deux ensembles linéaires singuliers, *Annali della Scuola Normale Superiore di Pisa (2) 4*, 43–46.
1951   Sur quelques propositions concernant la puissance du continu, *Fundamenta mathematicae 38*, 1–13; reprinted in *Sierpiński 1976*, 654–664.
1956   Second, expanded edition of *Sierpiński 1934* (New York: Chelsea).
1975   *Oeuvres choisies. Tome II: Théorie des ensembles et ses applications. Travaux des années 1908–1929* (Warsaw: PWN).
1976   *Oeuvres choisies. Tome III: Théorie des ensembles et ses applications. Travaux des années 1930–1966* (Warsaw: PWN).
See also Braun, Stefania, and Wacław Sierpiński.
See also Luzin, Nikolai, and Wacław Sierpiński.

Sierpiński, Wacław, and Alfred Tarski
1930   Sur une propriété caractéristique des nombres inaccessibles, *Fundamenta mathematicae 15*, 292–300; reprinted in *Sierpiński 1976*, 29–35.

Sikorski, Roman
1951   A characterization of alephs, *Fundamenta mathematicae 38*, 18–22.
See also Rasiowa, Helena, and Roman Sikorski.

Skolem, Thoralf

1920 Logisch-kombinatorische Untersuchungen über die Erfüllbarkeit oder Beweisbarkeit mathematischer Sätze nebst einem Theoreme über dichte Mengen, *Skrifter utgit av Videnskapsselskapet i Kristiania*, I. *Matematisk-naturvidenskabelig klasse*, no. 4, 1–36; reprinted in *Skolem 1970*, 103–136; partial English translation by Stefan Bauer-Mengelberg in *van Heijenoort 1967*, 252–263.

1923 Begründung der elementaren Arithmetik durch die rekurrierende Denkweise ohne Anwendung scheinbarer Veränderlichen mit unendlichem Ausdehnungsbereich, *Skrifter utgit av Videnskapsselskapet i Kristiania*, I. *Matematisk-naturvidenskabelig klasse*, no. 6, 1–38; reprinted in *Skolem 1970*, 153–188; English translation by Stefan Bauer-Mengelberg in *van Heijenoort 1967*, 302–333.

1923a Einige Bemerkungen zur axiomatischen Begründung der Mengenlehre, *Matematikerkongressen i Helsingfors 4–7 Juli 1922, Den femte skandinaviska matematikerkongressen, Redogörelse* (Helsinki: Akademiska Bokhandeln), 217–232; reprinted in *Skolem 1970*, 137–152; English translation by Stefan Bauer-Mengelberg in *van Heijenoort 1967*, 290–301.

1928 Über die mathematische Logik, *Norsk matematisk tidsskrift 10*, 125–142; reprinted in *Skolem 1970*, 189–206; English translation by Stefan Bauer-Mengelberg and Dagfinn Føllesdal in *van Heijenoort 1967*, 508–524.

1929 Über einige Grundlagenfragen der Mathematik, *Skrifter utgitt av Det Norske Videnskaps-Akademi i Oslo*, I. *Matematisk-naturvidenskapelig klasse*, no. 4, 1–49; reprinted in *Skolem 1970*, 227–273.

1931 Über einige Satzfunktionen in der Arithmetik, *Skrifter utgitt av Det Norske Videnskaps-Akademi i Oslo*, I. *Matematisk-naturvidenskapelig klasse*, no. 7, 1–28: reprinted in *Skolem 1970*, 281–306.

1932 Über die symmetrisch allgemeinen Lösungen im identischen Kalkul, *ibid. 1932*, no. 6, 1–32; also appeared in *Fundamenta mathematicae 18*, 61–76; reprinted in *Skolem 1970*, 307–336.

1933 Ein kombinatorischer Satz mit Anwendung auf ein logisches Entscheidungsproblem, *Fundamenta mathematicae 20*, 254–261; reprinted in *Skolem 1970*, 337–344.

1933a Über die Unmöglichkeit einer vollständigen Charakterisierung der Zahlenreihe mittels eines endlichen Axiomensystems, *Norsk matematisk forenings skrifter*, series 2, no. 10, 73–82; reprinted in *Skolem 1970*, 345–354.

1934   Über die Nicht-charakterisierbarkeit der Zahlenreihe mittels endlich oder abzählbar unendlich vieler Aussagen mit ausschließlich Zahlenvariablen, *Fundamenta mathematicae 23*, 150–161; reprinted in *Skolem 1970*, 355–366.
1938   Review of *Hilbert and Ackermann 1938*, *Norsk matematisk tidsskrift 20*, 67–69.
1970   *Selected works in logic*, edited by Jens E. Fenstad (Oslo: Universitetsforlaget).

Slisenko, Anatol O. (Слисенко, Анатоль О.)
1970   (ed.) *Studies in constructive mathematics and mathematical logic, part II*, Seminars in mathematics, V. A. Steklov Mathematical Institute, vol. 8 (New York: Consultants Bureau).

Smoryński, Craig A.
1977   The incompleteness theorems, in *Barwise 1977*, 821–865.

Smullyan, Arthur F.
1948   Modality and description, *The journal of symbolic logic 13*, 31–37.

Smullyan, Raymond M.
1958   Undecidability and recursive inseparability, *Zeitschrift für mathematische Logik und Grundlagen der Mathematik 4*, 143–147.

Solomon, M. K.
1981   A connection between Blum speedable sets and Gödel's speedup theorem (unpublished typescript).

Solovay, Robert M.
1963   Independence results in the theory of cardinals. I, II, *Notices of the American Mathematical Society 10*, 595.
1967   A nonconstructible Δ_3^1 set of integers, *Transactions of the American Mathematical Society 127*, 50–75.
1970   A model of set theory in which every set of reals is Lebesgue measurable, *Annals of Mathematics (2) 92*, 1–56.
1976   Provability interpretations of modal logic, *Israel journal of mathematics 25*, 287–304.
See also Ketonen, Jussi, and Robert M. Solovay.
See also Levy, Azriel, and Robert M. Solovay.

Specker, Ernst
See MacDowell, Robert, and Ernst Specker.

Spector, Clifford
    1957   Recursive ordinals and predicative set theory, in *Summaries of talks presented at the Summer Institute for Symbolic Logic. Cornell University* (Institute for Defense Analysis), *1957*, 377–382.
    1962   Provably recursive functionals of analysis: A consistency proof of analysis by an extension of principles formulated in current intuitionistic mathematics, in *Dekker 1962*, 1–27.
  See also Feferman, Solomon, and Clifford Spector.

Staal, J. F., and B. van Rootselaar
    1968   (eds.) *Logic, methodology and philosophy of science III. Proceedings of the third international congress for logic, methodology and philosophy of science, Amsterdam 1967* (Amsterdam: North-Holland).

Statman, Richard
    1978   Bounds for proof-search and speed-up in the predicate calculus, *Annals of mathematical logic 15*, 225–287.
    1981   Speed-up by theories with infinite models, *Proceedings of the American Mathematical Society 81*, 465–469.

Stein, Martin
    1976   *Interpretationen der Heyting-Arithmetik endlicher Typen* (doctoral dissertation, Münster).
    1978   Interpretationen der Heyting-Arithmetik endlicher Typen, *Archiv für mathematische Logik und Grundlagenforschung 19*, 175–189.
    1980   Interpretations of Heyting's arithmetic—an analysis by means of a language with set symbols, *Annals of mathematical logic 19*, 1–31.
    1981   A general theorem on existence theorems, *Zeitschrift für mathematische Logik und Grundlagen der Mathematik 27*, 435–452.

Stern, Jacques
    1982   (ed.) *Proceedings of the Herbrand Symposium. Logic Colloquium '81* (Amsterdam: North-Holland).

Straus, Ernst G.
    1982   Reminiscences, in *Holton and Elkana 1982*, 417–423.

Surányi, János
    1950   Contributions to the reduction theory of the decision problem. Second paper. Three universal, one existential quantifiers, *Acta Mathematica Academiae Scientiarum Hungaricae 1*, 261–271.

Tait, William W.

1965   Infinitely long terms of transfinite type, in *Crossley and Dummett 1965*, 176–185.

1965a   Functionals defined by transfinite recursion, *The journal of symbolic logic 30*, 155–174.

1967   Intensional interpretations of functionals of finite type. I, *ibid. 32*, 198–212.

1971   Normal form theorem for bar recursive functions of finite type, in *Fenstad 1971*, 353–367.

Takeuti, Gaisi

1957   Ordinal diagrams, *Journal of the Mathematical Society of Japan 9*, 386–394.

1960   Ordinal diagrams II, *ibid. 12*, 385–391.

1961   Remarks on Cantor's absolute, *Journal of the Mathematical Society of Japan 13*, 197–206.

1967   Consistency proofs of subsystems of classical analysis, *Annals of mathematics (2) 86*, 299–348.

1975   *Proof theory* (Amsterdam: North-Holland).

Tarski, Alfred

1924   Sur les principes de l'arithmétique des nombres ordinaux (transfinis), *Polskie Towarzystwo Matematyczne (Cracow), Rocznik (=Annales de la Société Polonaise de Mathématique) 3*, 148–149.

1925   Quelques théorèmes sur les alephs, *Fundamenta mathematicae 7*, 1–14.

1930   Über einige fundamentale Begriffe der Metamathematik, *Sprawozdania z posiedzeń Towarzystwa Naukowego Warszawskiego, wydział III, 23*, 22–29; English translation by J. H. Woodger, with revisions, in *Tarski 1956*, 30–37.

1932   Der Wahrheitsbegriff in den Sprachen der deduktiven Disziplinen, *Anzeiger der Akademie der Wissenschaften in Wien 69*, 23–25.

1933   Einige Betrachtungen über die Begriffe der ω-Widerspruchsfreiheit und der ω-Vollständigkeit, *Monatshefte für Mathematik und Physik 40*, 97–112; English translation by J. H. Woodger in *Tarski 1956*, 279–295.

1933a   Pojecie prawdy w jezykach nauk dedukcyjnych (The concept of truth in the languages of deductive sciences), *Prace Towarzystwa Naukowego Warszawskiego, wydział III*, no. 34; English translation by J. H. Woodger in *Tarski 1956*, 152–278.

1935   Der Wahrheitsbegriff in den formalisierten Sprachen, *Studia philosophica* (Lemberg), *1*, 261–405; German translation by L. Blaustein of *Tarski 1933a.*

1935a   Grundzüge des Systemenkalküls, Erster Teil, *Fundamenta mathematicae 25*, 503–526.

1936   Grundzüge des Systemenkalküls, Zweiter Teil, *ibid. 26*, 283–301.

1938   Über unerreichbare Kardinalzahlen, *ibid. 30*, 68–89.

1944   The semantic conception of truth and the foundations of semantics, *Philosophy and phenomenological research 4*, 341–376.

1949   On essential undecidability, *The journal of symbolic logic 14*, 75–76.

1952   Some notions and methods on the borderline of algebra and metamathematics, *Proceedings of the International Congress of Mathematicians, Cambridge, Massachusetts, August 30–September 6, 1950* (Providence, R.I.: American Mathematical Society), vol. 1, 705–720.

1956   *Logic, semantics, metamathematics: papers from 1923 to 1938*, translated into English and edited by J. H. Woodger (Oxford: Clarendon Press).

1962   Some problems and results relevant to the foundations of set theory, *Logic, methodology, and philosophy of science. Proceedings of the 1960 International Congress*, edited by Ernst Nagel, Patrick Suppes, and Alfred Tarski (Stanford: Stanford University Press), 125–135.

1983   Second edition of *Tarski 1956*, edited by John Corcoran.
See also Feferman, Solomon, and Alfred Tarski.
See also Keisler, H. Jerome, and Alfred Tarski.
See also Lukasiewicz, Jan, and Alfred Tarski.
See also McKinsey, John C. C., and Alfred Tarski.
See also Sierpiński, Wacław, and Alfred Tarski.

Tarski, Alfred, Andrzej Mostowski and Raphael M. Robinson
1953   *Undecidable theories* (Amsterdam: North-Holland).

Taussky-Todd, Olga
198?   Remembrances of Kurt Gödel, *Gödels wissenschaftliches Weltbild* (Proceedings of a symposium on Gödel held at the Internationales Forschungszentrum, Salzburg, July 1983), to appear.

Thomas, Ivo
1962   Finite limitations on Dummett's LC, *Notre Dame journal of formal logic 3*, 170–174.

Troelstra, Anne S.
1973   *Metamathematical investigation of intuitionistic arithmetic and analysis*, Springer lecture notes in mathematics, no. 344 (Berlin: Springer).
1977   *Choice sequences* (Oxford: Clarendon Press).
See also Kreisel, Georg, and Anne S. Troelstra.

Troelstra, Anne S., and Dirk van Dalen
1982   (eds.) *The L. E. J. Brouwer centenary symposium* (Amsterdam: North-Holland).

Turing, Alan M.
1937   On computable numbers, with an application to the Entscheidungsproblem, *Proceedings of the London Mathematical Society* (*2*) *42*, 230–265; correction, *ibid. 43*, 544–546, reprinted in *Davis 1965*, 116–154.
1939   Systems of logic based on ordinals, *Proceedings of the London Mathematical Society* (*2*) *45*, 161–228; reprinted in *Davis 1965*, 155–222.

Ulam, Stanislaw
1958   John von Neumann, 1903–1957, *Bulletin of the American Mathematical Society 64*, no. 3, part 2 (May supplement), 1–49.
1976   *Adventures of a mathematician* (New York: Scribner's).

Vacca, Giovanni
1903   La logica di Leibniz, *Rivista di matematica 8*, 64–74.

van Heijenoort, Jean
1967   (ed.) *From Frege to Gödel: a source book in mathematical logic, 1879–1931* (Cambridge, Mass.: Harvard University Press).
1967a   Logic as calculus and logic as language, *Boston studies in the philosophy of science 3*, 440–446.
1982   L'oeuvre logique de Jacques Herbrand et son contexte historique, in *Stern 1982*, 57–85.

Vaught, Robert L.
1974   Model theory before 1945, in *Henkin et alii 1974*, 153–172.
See also *Henkin et alii 1974*.

Vesley, Richard E.
1972   Choice sequences and Markov's principle, *Compositio mathematica 24*, 33–53.

Vogel, Helmut
1976   Ein starker Normalisationssatz für die bar-rekursiven Funktionale, *Archiv für mathematische Logik und Grundlagenforschung 18*, 81–84.
See also Diller, Justus, and Helmut Vogel.

von Juhos, Béla
1930   *Das Problem der mathematischen Wahrscheinlichkeit* (Munich: Reinhardt).

von Neumann, John
1925   Eine Axiomatisierung der Mengenlehre, *Journal für die reine und angewandte Mathematik 154*, 219–240; correction, *ibid. 155*, 128; reprinted in *von Neumann 1961*, 34–56; English translation by Stefan Bauer-Mengelberg and Dagfinn Føllesdal in *van Heijenoort 1967*, 393–413.
1927   Zur Hilbertschen Beweistheorie, *Mathematische Zeitschrift 26*, 1–46; reprinted in *von Neumann 1961*, 256–300.
1928   Über die Definition durch transfinite Induktion und verwandte Fragen der allgemeinen Mengenlehre, *Mathematische Annalen 99*, 373–391; reprinted in *von Neumann 1961*, 320–338.
1928a   Die Axiomatisierung der Mengenlehre, *Mathematische Zeitschrift 27*, 669–752; reprinted in *von Neumann 1961*, 339–422.
1929   Über eine Widerspruchsfreiheitsfrage in der axiomatischen Mengenlehre, *Journal für die reine und angewandte Mathematik 160*, 227–241; reprinted in *von Neumann 1961*, 494–508.
1931   Die formalistische Grundlegung der Mathematik, *Erkenntnis 2*, 116–121; English translation by Erna Putnam and Gerald J. Massey in *Benacerraf and Putnam 1964*, 50–54.
1961   *Collected works*, vol. I: *Logic, theory of sets, and quantum mechanics*, edited by A. H. Taub (Oxford: Pergamon).
See also *Hahn et alii 1931*.

von Neumann, John, and Oskar Morgenstern
1944   *Theory of games and economic behavior* (Princeton: Princeton University Press).
1947   Second edition of *von Neumann and Morgenstern 1944*.
1953   Third edition of *von Neumann and Morgenstern 1944*.

Waismann, Friedrich
1967   *Wittgenstein und der Wiener Kreis*, edited by Brian McGuinness (Oxford: Blackwell).

Wajsberg, Mordechaj
1933   Ein erweiterter Klassenkalkül, *Monatshefte für Mathematik und Physik 40*, 113–126.

Wald, Abraham
1931   Axiomatik des Zwischenbegriffes in metrischen Räumen, *Mathematische Annalen 104*, 476–484.
1932   Axiomatik des metrischen Zwischenbegriffes, *Ergebnisse eines mathematischen Kolloquiums 2*, 17–18.
1935   Über die eindeutige positive Lösbarkeit der neuen Produktionsgleichungen, *ibid. 6*, 12–18.
1936   Über die Produktionsgleichungen der ökonomischen Wertlehre (II. Mitteilung), *ibid. 7*, 1–6.
See also *Gödel 1933h*.

Wang, Hao
1950   Remarks on the comparison of axiom systems, *Proceedings of the National Academy of Sciences, U.S.A. 36*, 448–453.
1951   Arithmetic models for formal systems, *Methodos 3*, 217–232.
1954   The formalization of mathematics, *The journal of symbolic logic 19*, 241–266: reprinted in *Wang 1963*, 559–584.
1959   Ordinal numbers and predicative set theory, *Zeitschrift für mathematische Logik und Grundlagen der Mathematik 5*, 216–239; reprinted in *Wang 1963*, 624–651.
1962   *A survey of mathematical logic* (Peking: Science Press; also Amsterdam: North-Holland, 1963); reprinted as *Logic, computers and sets* (New York: Chelsea, 1970).
1970   A survey of Skolem's work in logic, in *Skolem 1970*, 17–52.
1974   *From mathematics to philosophy* (New York: Humanities Press).
1978   Kurt Gödel's intellectual development, *The mathematical intelligencer 1*, 182–184.
1981   Some facts about Kurt Gödel, *The journal of symbolic logic 46*, 653–659.
See also Kahr, Andrew S., Edward F. Moore and Hao Wang.

Weintraub, E. Roy
1983   On the existence of a competitive equilibrium: 1930–1954, *The journal of economic literature 21*, 1–39.

Wernick, Georg
1929   Die Unabhängigkeit des zweiten distributiven Gesetzes von den übrigen Axiomen der Logistik, *Journal für die reine und angewandte Mathematik 161*, 123–134.

Weyl, Hermann
1918   *Das Kontinuum. Kritische Untersuchungen über die Grund-lagen der Analysis* (Leipzig: Veit).
1932   Second edition of *Weyl 1918.*
1946   Review of *Schilpp 1944, American mathematical monthly 53,* 208–214; reprinted in *Weyl 1968,* 599–605.
1968   *Gesammelte Abhandlungen,* edited by K. Chandrasekharan (Berlin: Springer), vol. 4.

Whitehead, Alfred North, and Bertrand Russell
1910   *Principia mathematica* (Cambridge, U.K.: Cambridge University Press), vol. 1.
1912   *Principia mathematica,* vol. 2.
1913   *Principia mathematica,* vol. 3.
1925   Second edition of *Whitehead and Russell 1910.*

Wittgenstein, Ludwig
1921   Logisch-philosophische Abhandlung, *Annalen der Naturphilosophie 14,* 185–262.
1922   *Tractatus logico-philosophicus,* English translation of *Wittgenstein 1921* (London: Routledge and Kegan Paul).

Yasugi, Mariko
1963   Intuitionistic analysis and Gödel's interpretation, *Journal of the Mathematical Society of Japan 15,* 101–112.

Zemanek, Heinz
1978   Oskar Morgenstern (1902–1977)—Kurt Gödel (1906–1978), *Elektronische Rechenanlagen 20,* 209–211.

Zermelo, Ernst
1904   Beweis, daß jede Menge wohlgeordnet werden kann (Aus einem an Herrn Hilbert gerichteten Briefe), *Mathematische Annalen 59,* 514–516; English translation by Stefan Bauer-Mengelberg in *van Heijenoort 1967,* 139–141.
1908   Untersuchungen über die Grundlagen der Mengenlehre. I, *Mathematische Annalen 65,* 261–281; English translation by Stefan Bauer-Mengelberg in *van Heijenoort 1967,* 199–215.
1908a   Neuer Beweis für die Möglichkeit einer Wohlordnung, *Mathematische Annalen 65,* 107–128; English translation by Stefan Bauer-Mengelberg in *van Heijenoort 1967,* 183–198.
1930   Über Grenzzahlen und Mengenbereiche: Neue Untersuchungen über die Grundlagen der Mengenlehre, *Fundamenta mathematicae 16,* 29–47.

Index